Cell Biology Monographs

Volume 10

Springer-Verlag
Wien New York

Sialic Acids

Chemistry, Metabolism and Function

Edited by R. Schauer

Springer-Verlag
Wien New York

Prof. Dr. Roland Schauer
Biochemisches Institut, Christian-Albrechts-Universität, Kiel
Federal Republic of Germany

Softcover reprint of the hardcover 1st edition 1982

With 66 Figures

Library of Congress Cataloging in Publication Data. Main entry under title: Sialic acids – chemistry, metabolism and function. (Cell biology monographs; v. 10.) Includes bibliographical references and index. 1. Sialic acids. I. Schauer, R. (Roland), 1936–. II. Series. [DNLM: 1. Sialic acids. W1 CE128H v. 10/QU 84 S562.] QP801.S47S55. 1982. 574.19'2482. 82-19326

ISSN 0172-4665

ISBN-13:978-3-7091-8682-4 e-ISBN-13:978-3-7091-8680-0
DOI: 10.1007/978-3-7091-8680-0

To Elfriede

Preface

Rapid progress in the field of sialic acids has made it desirable to collect the new data about these unique sugars and to continue the series of books on this topic. In 1960, A. GOTTSCHALK wrote "The Chemistry and Biology of Sialic Acids and Related Substances" (Cambridge University Press) and in 1976, A. ROSENBERG and C.-L. SCHENGRUND published "Biological Roles of Sialic Acids" (Plenum Press). In this book emphasis is given to various modern methods used in the isolation and analysis of sialic acids. New approaches to the synthesis of free and bound sialic acids are described and the vast field of occurrence and metabolism of these substances is reviewed. Sialidoses are dealt with in one of the chapters, because sialidases have been recognized as factors of pathophysiological importance. As knowledge is increasing about the involvement of sialic acids in many aspects of cell biology, another chapter is devoted to these phenomena. With this book I intend to demonstrate modern trends in sialic acid chemistry and biochemistry, and I hope that it will be of practical use and find its place in laboratories rather than in libraries.

This publication offers an opportunity to thank all colleagues in many countries, including my coworkers at the universities of Bochum and Kiel, for their cooperation, stimulating discussions and, very important, useful criticism. The continuous cooperation with J. F. G. VLIEGENTHART and his coworkers, Utrecht, has been rewarding in many respects.

I am sorry that my thanks to the authors of this book cannot be extended to C. F. A. CULLING, who died in July 1982.

The help of WIGBERT BERG, GABRIELE GUTSCHKER-GDANIEC, ERNST MÜLLER, ULRICH NÖHLE, GERD REUTER, CORNELIA SCHRÖDER, and ASHOK K. SHUKLA, who assisted in proof-reading of the manuscripts and in the preparation of the subject index, is very much appreciated.

Kiel, September 1982 ROLAND SCHAUER

Contents

Abbreviations

AB	Alcian blue
AcCoA	Acetyl-Coenzyme A
ACTH	Adrenocorticotropic hormone
ADP	Adenosine diphosphate
AMP	Adenosine monophosphate
Ara	D-Arabinose
Asn	Asparagine
ATP	Adenosine triphosphate
ax	axial
BCG	Bacillus Calmette-Guérin
BHK	Baby hamster kidney
BSM	Bovine submandibular gland mucin
Cer	Ceramide
CHO	Chinese hamster ovary
c.i.-m.s.	Chemical ionization – mass spectrometry
CMP	Cytidine monophosphate
CoA	Coenzyme A
ConA	Concanavalin A
CTP	Cytidine triphosphate
DMSO	Dimethylsulfoxide
DNA	Deoxyribonucleic acid
DON	6-Diazo-5-oxo-L-norleucine
DSS	4-Dimethyl-4-silapentane-1-sulfonate
EDTA	Ethylenediamine-tetraacetic acid
e.i.-m.s.	Electron impact – mass spectrometry
e.p.r.	Electron paramagnetic resonance
eq	equatorial
ESM	Equine submandibular gland mucin
Fru-6-P	D-Fructose-6-phosphate
FPV	Fowl plague virus
Fuc	L-Fucose
Gal	D-Galactose
GalNAc	N-Acetyl-D-galactosamine
GalNAc-ol	N-Acetyl-D-galactosaminitol
GgOse	Oligosaccharides of the ganglio-series (see chapter B)
g.l.c.	Gas-liquid chromatography
Glc	D-Glucose
GlcN	D-Glucosamine
GlcNAc	N-Acetyl-D-glucosamine
GlcNAc-1-P	N-Acetyl-D-glucosamine-1-phosphate
Glc-6-P	D-Glucose-6-phosphate
GlcN-6-P	D-Glucosamine-6-phosphate

GM_1	Monosialoganglioside (for other ganglioside abbreviations see LEDEEN, R. W., and YU, R. K., 1982; Methods Enzymol. **83**, 139-191, and Chapter B)
Hex	Hexose
HPLC	High-Pressure (Performance) Liquid Chromatography
IA_2V	Influenza A_2 virus
J	NMR coupling constant
Lac	Lactose
LacCer	Lactosyl ceramide
LacNAc	N-Acetyllactosamine
LcnOse	Oligosaccharides of the neolacto-series (s. Chapter B)
LcOse	Oligosaccharides of the lacto-series (s. Chapter B)
Man	D-Mannose
ManNAc	N-Acetyl-D-mannosamine
ManNAc-6-P	N-Acetyl-D-mannosamine-6-phosphate
MBTH	Methyl-3-benzothiazolinone-2-hydrazone
Me	Methyl
m/e	Ratio of mass and charge
m.s.	Mass spectrometry
$NAD(P)^+$/NAD(P)H	Nicotinamide adenine dinucleotide (phosphate), oxidized or reduced form
NDV	Newcastle disease virus
Neu	D-Neuraminic Acid
Neu-β-Me	D-Neuraminic acid-β-methyl glycoside
Neu4,5Ac_2	N-Acetyl-4-0-acetyl-D-neuraminic acid
Neu4,5,9Ac_3	N-Acetyl-4,9-di-0-acetyl-D-neuraminic acid
Neu4,5$Ac_2$9Lt	N-Acetyl-4-0-acetyl-9-0-lactyl-D-neuraminic acid
Neu4Ac5Gc	N-Glycolyl-4-0-acetyl-D-neuraminic acid
Neu5Ac	N-Acetyl-D-neuraminic acid
Neu5Ac1Me	N-Acetyl-D-neuraminic acid methyl ester
Neu5Ac2en	2-Deoxy-2,3-didehydro-N-acetyl-D-neuraminic acid
Neu5Ac2Me	N-Acetyl-D-neuraminic acid β-methyl glycoside
Neu5Ac2P	N-Acetyl-D-neuraminic acid-2-phosphate
Neu5Ac4Me	N-Acetyl-4-0-methyl-D-neuraminic acid
Neu5,7Ac_2	N-Acetyl-7-0-acetyl-D-neuraminic acid
Neu5,7,8,9Ac_4	N-Acetyl-7,8,9-tri-0-acetyl-D-neuraminic acid
Neu5,7,9Ac_3	N-Acetyl-7,9-di-0-acetyl-D-neuraminic acid
Neu5,8Ac_2	N-Acetyl-8-0-acetyl-D-neuraminic acid
Neu5,8,9Ac_3	N-Acetyl-8,9-di-0-acetyl-D-neuraminic acid
Neu5Ac8Me	N-Acetyl-8-0-methyl-D-neuraminic acid
Neu5Ac8S	N-Acetyl-D-neuraminic acid-8-sulfate
Neu5,9Ac_2	N-Acetyl-9-0-acetyl-D-neuraminic acid
Neu5Ac9Lt	N-Acetyl-9-0-L-lactyl-D-neuraminic acid
Neu5Ac9N_3	9-Azido-9-deoxy-N-acetyl-D-neuraminic acid
Neu5Ac9P	N-Acetyl-D-neuraminic acid-9-phosphate
Neu7Ac5Gc	N-Glycolyl-7-0-acetyl-D-neuraminic acid
Neu7,8,9$Ac_3$5Gc	N-Glycolyl-7,8,9-tri-0-acetyl-D-neuraminic acid
Neu7,9$Ac_2$5Gc	N-Glycolyl-7,9-di-0-acetyl-D-neuraminic acid
Neu8,9$Ac_2$5Gc	N-Glycolyl-8,9-di-0-acetyl-D-neuraminic acid
Neu9Ac5Gc	N-Glycolyl-9-0-acetyl-D-neuraminic acid
Neu5Gc	N-Glycolyl-D-neuraminic acid
Neu5Gc8Me	N-Glycolyl-8-0-methyl-D-neuraminic acid
Neu5Gc8S	N-Glycolyl-D-neuraminic acid-8-sulfate

NMR	Nuclear magnetic resonance
OSM	Ovine submandibular gland mucin
PAPS	Periodic acid-phenylhydrazine-Schiff
PAS	Periodic acid-Schiff
PAT	Periodic acid-thionine
PBT	Periodic acid-borohydride technique
PEP	Phosphoenolpyruvate
PHA	Phytohemagglutinin
PPi	Pyrophosphate
ppm	parts per million
PSM	Porcine submandibular gland mucin
RNA	Ribonucleic acid
Ser	Serine
Thr	Threonine
TMS	Trimethylsilyl
UDP	Uridine diphosphate
UTP	Uridine triphosphate
VCN	*Vibrio cholerae* neuraminidase
VSV	Vesicular stomatitis virus
WGA	Wheat germ agglutinin

List of Contributors

Bauer, C., Prof. Dr., Institut für Molekularbiologie und Biochemie, Freie Universität Berlin, Arnimallee 22, D-1000 Berlin 33 (Dahlem).

Cantz, M., Prof. Dr., Institut für Pathochemie und allgemeine Neurochemie, Universität Heidelberg, Im Neuenheimer Feld 220–221, D-6900 Heidelberg 1, Federal Republic of Germany.

Corfield, A. P., Dr., Clinical Research Laboratories, University of Bristol, Department of Medicine, Bristol Royal Infirmary, Bristol BS2 8HW, U.K.

Culling†, C. F. A., Prof. Dr., Department of Pathology, Faculty of Medicine, University of British Columbia, 2211 Wesbrook Mall, Vancouver, British Columbia, Canada V6T 1W5.

Dorland, L., Dr., University Children's Hospital, "Het Wilhelmina Kinderziekenhuis", Nieuwe Gracht 137, NL-3512 LK Utrecht, The Netherlands.

Gerok, W., Prof. Dr., Medizinische Klinik, Universität Freiburg i. Br., Hugstetterstrasse 55, D-7800 Freiburg i. Br., Federal Republic of Germany.

Van Halbeek, H., Dr., Department of Bio-Organic Chemistry, State University of Utrecht, Croesestraat 79, NL-3522 AD Utrecht, The Netherlands.

Haverkamp, J., Dr., Department of Biochemistry and Immunochemistry, Unilever Research Laboratory, Olivier van Noortlaan 120, NL-3133 AT Vlaardingen, The Netherlands.

Kamerling, J. P., Dr., Department of Bio-Organic Chemistry, State University of Utrecht, Croesestraat 79, NL-3522 AD Utrecht, The Netherlands.

Köttgen, E., Prof. Dr., Universitätsklinikum Charlottenburg, Freie Universität Berlin, Spandauer Damm 130, D-1000 Berlin 19.

Reid, P. E., Prof. Dr., Department of Pathology, Faculty of Medicine, University of British Columbia, 2211 Wesbrook Mall, Vancouver, British Columbia, Canada V6T 1W5.

Reutter, W., Prof. Dr., Institut für Molekularbiologie und Biochemie, Freie Universität Berlin, Arnimallee 22, D-1000 Berlin 33 (Dahlem).

Schauer, R., Prof. Dr., Biochemisches Institut, Christian-Albrechts-Universität, Olshausenstrasse 40–60, D-2300 Kiel, Federal Republic of Germany.

Vliegenthart, J. F. G., Prof. Dr., Department of Bio-Organic Chemistry, State University of Utrecht, Croesestraat 79, NL-3522 AD Utrecht, The Netherlands.

A. Introduction

ROLAND SCHAUER and JOHANNES F. G. VLIEGENTHART

Biochemisches Institut, Christian-Albrechts-Universität, Kiel, Federal Republic of Germany, and Bio-Organic Chemistry Department, University of Utrecht, Utrecht, The Netherlands

Sialic acids comprise a family of derivatives of neuraminic acid (5-amino-3,5-dideoxy-**D**-*glycero*-**D**-*galacto*-nonulosonic acid) occurring widespread in nature, especially in the animal kingdom. From an evolutionary point of view, they probably have a long history, dating back to precambrian times, as they have already been found in *Echinodermata* (chapter B). It took a number of years (1936–1962) to establish the structure of N-acetylneuraminic acid by chemical and enzymic means, as has been reviewed by GOTTSCHALK (1972), LEDEEN and YU (1976) and TUPPY and GOTTSCHALK (1972).

In the past 20 years the number of well-characterized sialic acids has increased to 23. Largely due to the enormous progress in analytical instrumentation such as gas-liquid chromatography in combination with mass-spectrometry (chapter F) and nuclear magnetic resonance spectroscopy (chapter G), many sialic acids could be identified, even when present in small amounts in more or less complex mixtures. As substituents at the amino group acetyl and glycolyl residues have been found. The hydroxyl functions show a greater variety in the type of substituents, since acetyl, lactyl, phosphate, sulfate and methyl groups have been shown to occur. Furthermore, unsaturated N-acetylneuraminic acid has been detected as a natural compound.

The different sialic acids exhibit an interesting species- and tissue-specific distribution. However, the biological implications of the various sialic acids and of the different substitution patterns are not fully understood. There is some evidence that different N and O substituents influence enzymic reactions, particularly in the catabolic pathways of sialoglycoconjugates (chapter I). Moreover, effects of substitution of sialic acids on immunological properties of sialoglycoconjugates have been observed (chapter J). Further elucidation of the role of sialic acid modifications remains a challenge for future research.

Sialic acids rarely occur as free molecules, they are usually α-glycosidically linked to carbohydrate chains. They can be attached to different positions of various monosaccharides including sialic acids. By consequence, they can occupy terminal as well as internal positions in the carbohydrate chain. The variability in binding type, too, gives rise to physico-chemically distinguishable sialic acids and may thus contribute to the great diversity of biological phenomena sialic acids are involved in. A further structural and functional modulation may be obtained by introducing substituents on sialic acid residues. In higher organisms, sialic acids frequently occur in terminal positions in carbohydrate chains of glycoconjugates. This feature may be relevant for the architecture and function of the cell membrane. The spatial arrangement in conjunction with the different chemical aspects of glycosidically-bound sialic acids can be held responsible for phenomena such as the masking of antigenic sites or of recognition markers on cell surfaces. In addition, such sialic acids affect the dynamic properties of cell membranes, including their communication with the environment. However, these and possible other functions are only partially understood at the molecular level.

In evolutionary low organisms, which have no sialic acid, it would be interesting to study which compounds substitute for sialic acids. The solution of this question might provide further insight into the essential roles of sialic acids in cell function and may clarify the importance of the negative charge of sialic acids at physiological pH. The availability of mutants of cells from higher organisms lacking sialic acid would also be helpful in the investigation of the precise biological role of this sugar. The fact that such mutants have never been observed suggests that such a mutation is lethal.

The regulation of biological processes requires, as far as sialic acids are involved, effective control mechanisms of the metabolism of these substances. This can take place at genetic, enzymic and/or hormonal levels. Little is known of this fascinating area, although some feedback mechanisms and hormonal influences on the metabolism of sialic acids have been demonstrated (chapter I). Elucidation of metabolic regulations is relevant for an understanding of cell and tissue growth and differentiation, as well as of malignant transformation. There are indications that changes occur in sialic acid metabolism during the aging of mature cells. Several diseases are known, which involve abnormalities in sialic acid metabolism. Remarkable examples are sialuria and Salla's disease, which are accompanied by an urinary excretion of unusually large amounts of free sialic acids. The causes of these diseases have only been speculated upon. More information is available on sialidoses (chapter K), where genetic defects in sialidases are involved.

The introduction of exogeneous sialidase e.g. due to bacterial or viral infections can lead to a decrease in the amount of glycosidically bound sialic acids. Consequently, an exposure of cellular antigens may occur as well as an enhanced clearance of serum glycoproteins and cells (chapter J). In general, alterations in the activity of anabolic (sialyltransferases) or catabolic (exogeneous and endogeneous sialidases) enzymes of sialic acid metabolism may play a part in chronic, (auto)immunological diseases.

It is noteworthy that some low organisms like a few bacterial and protozoan species possess sialic acids. This may help these parasites to survive in the host. For a similar reason it may be advantageous for some microorganisms to have at their disposal catabolic enzymes of sialic acid metabolism like sialidase and N-acetylneuraminate lyase. Strikingly, many of these organisms are pathogenic for mammals. It could well be that sialidase is required for a faster spreading of bacteria in tissues by an attack on sialic acid residues of cell surfaces. The role of sialidase for viral attachment to the host cell has long been known.

In connexion with these considerations, it is intriguing to ask whether the bacteria developed biochemical pathways for the synthesis of sialic acids independently, or whether they acquired the genome for the enzymes involved from host organisms. The similarity in the biosynthetic pathways leading to sialic acid in bacteria and in higher animals lends support to the latter possibility. It is tempting to presume that such a genome transfer is also responsible for the acquisition of sialidase and N-acetylneuraminate lyase by some bacteria. Comparison of the properties of these enzymes obtained from mammalian and bacterial sources does not contradict such an event.

Further research on the manifold chemical and biological aspects of sialic acids will contribute to the understanding of the molecular organization of living systems.

Bibliography

GOTTSCHALK, A. (ed.), 1972: Glycoproteins, Their Composition, Structure and Function, 2nd ed., parts A and B. Amsterdam: Elsevier.

LEDEEN, R. W., YU, R. K., 1976: In: Biological Roles of Sialic Acid (ROSENBERG, A., SCHENGRUND, C.-L., eds.), pp. 1—58. New York-London: Plenum Press.

TUPPY, H., GOTTSCHALK, A., 1972: In: Glycoproteins, Their Composition, Structure and Function (GOTTSCHALK, A., ed.), 2nd ed., part A, pp. 403—449. Amsterdam: Elsevier.

B. Occurrence of Sialic Acids

ANTHONY P. CORFIELD and ROLAND SCHAUER

Biochemisches Institut, Christian-Albrechts-Universität, Kiel, Federal Republic of Germany

With 2 Figures

Contents

I. Introduction

The discovery and the widespread occurrence of the sialic acids in mammalian tissues (BLIX 1936, KLENK 1941) has been correlated with a range of different biological functions (see chapter J) which continues to expand. Although the literature on sialic acid occurrence is large (GOTTSCHALK 1960, BLIX and JEANLOZ 1969, TUPPY and GOTTSCHALK 1972, SCHAUER 1973, NG and DAIN 1976), it has been essentially limited to the vertebrates and some examples from the *Echinodermata* and *Bacteria*. Only one survey of sialic acid distribution in different phyla has been published (WARREN 1963). Current data on sialic acid structure is compiled in Table 1, introducing the known spectrum of derivatives covered in this chapter.

The study carried out by WARREN (1963) illustrated that the use of one method for identification of sialic acids is insufficient and can yield false positive results, as discussed in chapter E. Conclusive identification of sialic acids from natural

Table 1. *Established structures of naturally occurring sialic acids*

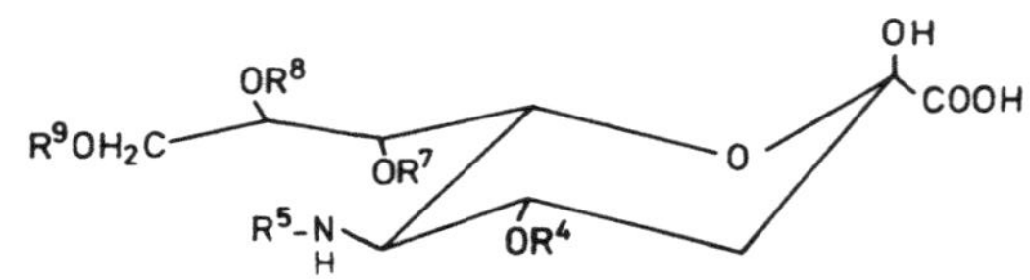

Name	Abbreviation	Substituent				
		R^4	R^5	R^7	R^8	R^9
N-Acetylneuraminic acid	Neu5Ac	H	acetyl	H	H	H
N-Acetyl-4-O-acetylneuraminic acid	$Neu4,5Ac_2$	acetyl	acetyl	H	H	H
N-Acetyl-7-O-acetylneuraminic acid	$Neu5,7Ac_2$	H	acetyl	acetyl	H	H
N-Acetyl-8-O-acetylneuraminic acid	$Neu5,8Ac_2$	H	acetyl	H	acetyl	H
N-Acetyl-9-O-acetylneuraminic acid	$Neu5,9Ac_2$	H	acetyl	H	H	acetyl
N-Acetyl-4,9-di-O-acetylneuraminic acid	$Neu4,5,9Ac_3$	acetyl	acetyl	H	H	acetyl
N-Acetyl-7,9-di-O-acetylneuraminic acid	$Neu5,7,9Ac_3$	H	acetyl	acetyl	H	acetyl
N-Acetyl-8,9-di-O-acetylneuraminic acid	$Neu5,8,9Ac_3$	H	acetyl	H	acetyl	acetyl
N-Acetyl-7,8,9-tri-O-acetylneuraminic acid	$Neu5,7,8,9Ac_4$	H	acetyl	acetyl	acetyl	acetyl
N-Acetyl-9-O-L-lactylneuraminic acid	Neu5Ac9Lt	H	acetyl	H	H	L-lactyl
N-Acetyl-4-O-acetyl-9-O-lactylneuraminic acid	$Neu4,5Ac_29Lt$	acetyl	acetyl	H	H	lactyl
N-Acetyl-8-O-methylneuraminic acid	Neu5Ac8Me	H	acetyl	H	methyl	H
N-Acetyl-8-O-sulphoneuraminic acid	Neu5Ac8S	H	acetyl	H	sulphate	H
N-Acetyl-9-O-phosphoroneuraminic acid	Neu5Ac9P	H	acetyl	H	H	phosphate
N-Acetyl-2-deoxy-2,3-dehydroneuraminic acid	Neu5Ac2en	H	acetyl	H	H	H
N-Glycolylneuraminic acid	Neu5Gc	H	glycolyl	H	H	H
N-Glycolyl-4-O-acetylneuraminic acid	Neu4Ac5Gc	acetyl	glycolyl	H	H	H
N-Glycolyl-7-O-acetylneuraminic acid	Neu7Ac5Gc	H	glycolyl	acetyl	H	H
N-Glycolyl-9-O-acetylneuraminic acid	Neu9Ac5Gc	H	glycolyl	H	H	acetyl
N-Glycolyl-7,9-di-O-acetylneuraminic acid	$Neu7,9Ac_25Gc$	H	glycolyl	acetyl	H	acetyl
N-Glycolyl-8,9-di-O-acetylneuraminic acid	$Neu8,9Ac_25Gc$	H	glycolyl	H	acetyl	acetyl
N-Glycolyl-7,8,9-tri-O-acetylneuraminic acid	$Neu7,8,9Ac_35Gc$	H	glycolyl	acetyl	acetyl	acetyl
N-Glycolyl-8-O-methylneuraminic acid	Neu5Gc8Me	H	glycolyl	H	methyl	H
N-Glycolyl-8-O-sulphoneuraminic acid	Neu5Gc8S	H	glycolyl	H	sulphate	H

material requires gas-liquid chromatography and mass spectrometry (g.l.c./m.s.) analysis (chapter F) or n.m.r. spectroscopy (chapter G).

A biosynthetic approach for sialic acid identification, described in chapter I, is of value where the size of the organism imposes a limit on the amount of material

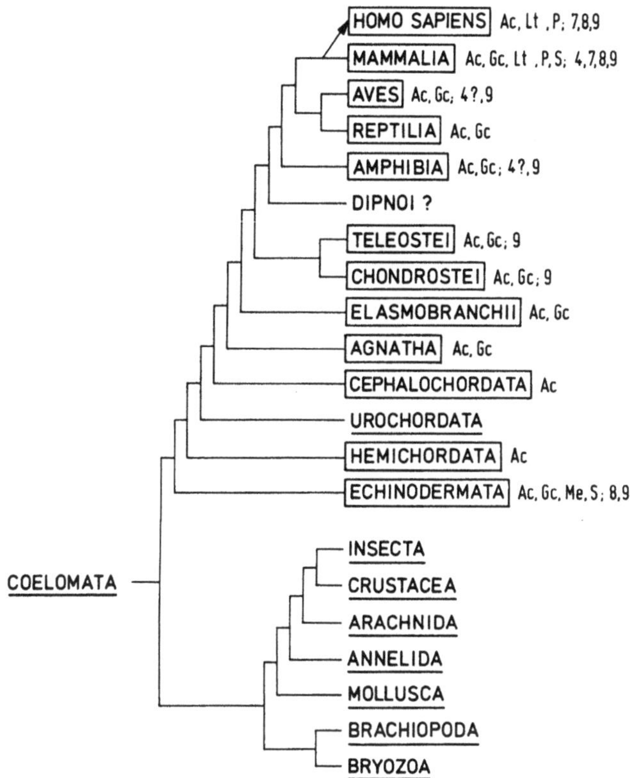

Fig. 1. Evolution of the sialic acids.

General relationships between the phyla are diagramatically represented. The presence of sialic acid in all examples within one group is indicated by a solid box, *e.g.* AGNATHA. Groups found to contain no sialic acid are underlined, *e.g.* BRYOZOA. The occurrence of Neu5Ac and Neu5Gc is indicated by Ac and Gc, respectively. Other substituents are lactyl, Lt; sulphate, S; and phosphate, P. O-Acyl substitution at positions 4, 7, 8, and 9 is shown by numbers, *e.g.* MAMMALIA Ac, Gc, Lt, P, S; 4, 7, 8, 9. A question mark shows where identification is not conclusive or missing.

Helpful advice for the construction of this figure was kindly given by Prof. Dr. U. Welsch, Dipl.-Biol. G. Gutschker-Gdaniec (Kiel), and Dr. J. Dodson (Bristol).

available for study and where the presence of sialic acid as part of the normal diet cannot be ruled out. In these cases the use of radioactive tracers such as N-acetylmannosamine (ManNAc) may be used to illustrate a metabolism of sialic acids in tissue extracts.

The occurrence of sialic acids in nature is presented in section XII in the form of a phylogenetic chart (Fig. 1). This serves as a basis for comparison in the

Table 2. *Linkages of sialic acids in complex carbohydrates*

The frequency of occurrence is indicated by the number of + symbols. + + +, common; + +, several examples identified; +, few examples known; —, not identified; ?, structure not confirmed; *, see also bacterial polysaccharide structures in Table 3. The data are taken from MONTREUIL 1975, 1980, SLOMIANY and SLOMIANY 1978, KOBATA 1979, SLOMIANY *et al.* 1980

Linkage monosaccharide	Partial structure	Glycoprotein		Ganglioside	Milk oligo-saccharide
		N-glycosidic	O-glycosidic		
Galactose	Neu5Acα(2-3)Galβ(1-4)GlcNAc-	+ + +	—	+ +	+ +
	Neu5Acα(2-4)Galβ(1-3/4)GlcNAc-	+	—	—	?
	Neu5Acα(2-6)Galβ(1-4)GlcNAc-	+ + +	+	+	+ +
	Neu5Acα(2-6)Galβ(1-3)GlcNAc-	—	—	—	+ +
	Neu5Acα(2-3)Galβ(1-3)GalNAc-	—	+ + +	+ + +	—
N-Acetylgalactosamine	Neu5Acα(2-6)GalNAc- |β(1-3) Gal	—	+ + +	—	—
N-Acetylglucosamine	Neu5Acα(2-4)GlcNAcβ(1-3)Gal-	—	+	—	—
	Neu5Acα(2-6)GlcNAcβ(1-3)Gal- |β(1-3/4) Gal	—	—	—	+ +
	Neu5Acα(2-4)Galβ(1-3)GlcNAcβ(1-2)Man- |α(2-6) Neu5Ac	+	—	—	—
Sialic acid*	Neu5Acα(2-8)Neu5Acα(2-3)Gal-	+ ?	+	+ + +	+ +
	[Neu5Acα(2-8)]$_n$Neu5Acα(2-3)Gal-	?	+	+ + +	—

discussion below, and a final consideration of the significance of sialic acid occurrence in this scheme. The nature of the linkage of sialic acids in glycoconjugates is presented in Table 2 as a guide in the sections following this introduction.

II. Viruses

The presence of sialic acid in the viruses is not widespread. The few well characterized examples are the Rhabdoviruses, vesicular stomatitis virus (VSV) (HUNT and SUMMERS 1976a, b) and rabies virus (DIETZSCHOLD 1977); the sarcoma virus, Rous sarcoma virus (KRANTZ *et al.* 1974); the Toga virus, Sindbis virus (KEEGSTRA *et al.* 1975); hepatitis B surface antigen (SHIRAISHI *et al.* 1977) and avian myeloblastosis virus (PORTER and WINZLER 1975). It is interesting to note that those viruses known to possess a sialidase as part of the membrane, *e.g.* some members of the myxoviruses, paramyxoviruses and influenza viruses, contain no sialic acid (ROTHMAN and LODISH 1977). The sialic acids in viruses have been found in glycoproteins containing N-glycosidically bound oligosaccharide chains linked via asparagine to the polypeptide core. No report of sialic acids other than Neu5Ac exists.

The VSV and Sindbis virus have been more extensively studied with regard to the glycosylation of the membrane glycoproteins and their insertion into cell membranes. Vesicular stomatitis virus contains only one membrane glycoprotein, the G protein, and this has served as a model for study of glycosylation by host cells and insertion into membranes with sialic acid addition being a final step (ROTHMAN and LODISH 1977). The processing of the oligosaccharide chains before conversion into the sialylated complex type has also been studied using VSV (KORNFELD *et al.* 1978). In Sindbis virus two glycoproteins E_1 and E_2 have been identified and the presence of both high-mannose and sialylated complex oligosaccharide chains demonstrated (KEEGSTRA *et al.* 1975).

The mechanism of virus glycosylation is a viral genome-directed synthesis of virus-specific glycoproteins using glycosyltransferases present in the host cell (COMPANS and KEMP 1978). This has been demonstrated in cases where incomplete glycosylation of virus glycoproteins, due to the absence of glycosyltransferases, has been found. For example, chicken embryo fibroblasts were unable to add galactose and sialic acid to Sindbis virus glycoproteins (KEEGSTRA *et al.* 1975), and VSV grown in mosquito cells, which do not synthesize sialic acids, were deficient in sialylated glycoproteins (SCHLOEMER and WAGNER 1975).

Sialic acid has also been found in glycolipids associated with Sindbis virus membranes (HIRSCHBERG and ROBBINS 1974) and in VSV membranes (SCHLOEMER and WAGNER 1974). However, this may be due to a non-specific incorporation of these components from the host cell and may vary with the cell used to host the virus during culture (HIRSCHBERG and ROBBINS 1974).

III. Bacteria

The detection of sialic acids in bacteria is limited to a few examples, strikingly all of them being pathogenic. These include some *Escherichia coli* strains, some *Neisseria meningitidis* strains and bacteria of the O serotype including *Salmonella* (BARRY *et al.* 1962, BARRY 1965).

Colominic acid, a homopolymer of Neu5Ac (Table 3), was found in *E. coli* K 235 of K 1 serotype (BARRY and GOEBEL 1957, BARRY 1958). This polysaccharide was found to contain ester or lactone linkages between the carboxyl group of one Neu5Ac and the C-7 or C-9 hydroxyl of the adjacent residue, while the glycosidic connection was through α(2-8) bonds. Other studies with *E. coli* K 1 serotypes led to the identification of O-acetylated sialic acids (DEWITT and ROWE 1959, 1961, DEWITT and ZELL 1961). An n.m.r. study of the Neu5Ac homopolymer isolated

Table 3. *Bacterial polymers containing sialic acid*
Data compiled from JENNINGS *et al.* 1977, LIU *et al.* 1977, GLODE *et al.* 1979, ØRSKOV *et al.* 1979

Bacteria	Repeating unit	O-acetyl-sialic acids/unit
Neisseria meningitidis		
Serotype B	-8)Neu5Acα(2-8)Neu5Acα(2-	none
Serotype C	-9)Neu5Acα(2-9)Neu5Acα(2-	~ 2
Serotype W 135	-4)Neu5Acα(2-6)Galβ(1-	none
Serotype Y	-4)Neu5Acα(2-6)Glcβ(1-	~ 1
Escherichia coli Bos 12	-9)Neu5Acα(2-9)Neu5Acα(2- -8)Neu5Acα(2-8)Neu5Acα(2-	none
Escherichia coli K 1	-8)Neu5Acα(2-8)Neu5Acα(2-	~ 2 or none

from *E. coli* K 92 (Bos 12) revealed a mixture of α(2-8) and (2-9) linkages (LIU *et al.* 1977, EGAN *et al.* 1977, Table 3). Other strains of *E. coli* have been found to contain no sialic acid (BARRY *et al.* 1963). Several serotypes of *N. meningitidis* have been extensively studied in relation to their antigenic properties and to vaccines (JENNINGS *et al.* 1977, LIU *et al.* 1977, GLODE *et al.* 1979). In these cases extensive chemical studies have elucidated the structure of the polymers and the sialic acids. Neu5Ac has been found in serogroups B, C, W 135 and Y. The nature of the sialic acids was determined in several studies as Neu5Ac, Neu5,7Ac_2 in serogroups B and W 135, Neu5,8Ac_2 in serogroup C and another, unidentified mono-O-acetyl-Neu5Ac in serogroup Y (LIU *et al.* 1971, BHATTACHARJEE *et al.* 1975, 1976, JENNINGS *et al.* 1977). The occurrence of sialic acid in bacteria of O serotype has been demonstrated for *Salmonella toucra* O 48 (KĘDZIERSKA 1978), *S. Arizona* O 5 and O 29 and *Citrobacter freundii* O 5 (BARRY *et al.* 1963). In the case of *S. toucra* O 48 Neu5Ac was found associated with the lipopolysaccharide O-antigen (KĘDZIERSKA 1978) but not with the K type antigens.

IV. Fungi, Algae, and Plants

WARREN (1963) could not find sialic acids in yeast, slime mould and the edible mushroom *Agaricus bisporus*. Higher plants of the phyla *Spermatophytes* (see also ROST and SCHAUER 1977), *Pterophytes* and *Bryophytes* were also negative in these studies, as were a lichen and individuals from all algal classes (WARREN 1963).

Fungi and yeast were also studied in this respect with negative results (AARONSON and LESSIE 1960).

Reports of sialic acids in plants exist in the literature (MAYER *et al.* 1964, ONODERA *et al.* 1966), but the analytical methods employed were insufficient to exclude other compounds such as 2-keto-3-deoxy acids noted earlier in this chapter. Several investigations produced negative results, although a reaction in the periodic acid/thiobarbituric acid assay was obtained. The compounds in question gave no colour in the direct Ehrlich, orcinol/Fe^{3+} and resorcinol assays for sialic acids (GIELEN 1968, CABEZAS 1968, 1973, CABEZAS and FEO 1969, UNGER 1981).

The conclusive demonstration of sialic acid in these materials must be by g.l.c./m.s. studies. In addition, no evidence for sialic acid biosynthesis in these phyla exists. The presence of sialidase in some fungi (UCHIDA *et al.* 1974, MÜLLER 1975) has been related to their nutritional requirements, as no bound sialic acid could be detected in the organisms.

V. Protozoa

A marine diatom (SULLIVAN and VOLCANI 1974) and the malarial parasite *Plasmodium berghei* (SEED *et al.* 1974) were found to contain membrane-bound sialic acid. Studies in the authors' laboratory have confirmed the presence of Neu5Ac and Neu5Gc in a ratio of 9 : 1 in *Trypanosoma cruzi*.

The presence of sialic acids in *Amoeba* remains uncertain. Treatment of *Amoeba proteus* with bacterial sialidase led to a reduction in pinocytotic activity compared to non-treated organisms (CHATTERJEE and RAY 1975). The same authors could also enhance lysolecithin-induced intercellular adhesion in *A. proteus* by sialidase treatment (RAY and CHATTERJEE 1975), although the release of Neu5Ac was not measured. These studies stand in opposition to those conducted with *A. discoides* (ALLEN *et al.* 1974, 1976), where no sialic acid was detected in biosynthetic and structural analytical experiments.

VI. Porifera, Platyhelminthes, Nemertinea, Ctenophora, and Coelenterata

The sponges (*Porifera*) belong to the mesozoans, and three species investigated by WARREN (1963) contained no sialic acid. A primitive platyhelminth, *Polychoerus carmelensis*, an acoel turbellarian, contains Neu5Ac (WARREN 1963). The sialic acid was endogeneous, contained no glycolyl derivative and was cleaved by acylneuraminate pyruvate-lyase. Other species of *Turbellaria* contained no sialic acids, but a trematode, *Fascioloides magna*, was found to contain small amounts. This example may be a false positive result, due to the presence of dietary sialic acid taken up from the blood of the host animal, and as a closely related species was found not to contain sialic acid (WARREN 1963). A recent survey of sialic acid occurrence in *Protostomia* (*Bryozoa* and *Brachiopoda*) and *Deuterostomia* showed no sialic acid in the triclad turbellarian *Euplanaria gonocephala*, and raised the question whether the positive result obtained by WARREN (1963) was representative of genuine endogeneous sialic acid (SEGLER *et al.* 1978). Investigation of organisms from the *Nemertinea* (four species), *Ctenophora* (one example) and *Coelenterata* (ten species) all yielded negative results. Several species of *Coelenterata* examined in the authors' laboratory were also devoid of sialic acid.

VII. Sipuncula, Annelida, Arthropoda, and Mollusca

The branch of the evolutionary tree leading to the *Arthropoda* and *Mollusca* is essentially devoid of sialic acid (Fig. 1). Reinvestigation of the reports of free sialic acid in the digestive glands of some crustaceans (WARREN 1963) confirmed this work and demonstrated that this free sialic acid was of dietary nature (SEGLER *et al.* 1978).

Two lower phyla, *Sipuncula* and *Annelida* do not possess sialic acid (WARREN 1963). The *Arthropoda* have not yielded conclusive evidence for the presence of endogeneous sialic acid. All species of insects, arachnids and crustaceans studied by WARREN (1963) were negative with the exception of the digestive glands of the lobster *Homarus americanus*. This result was confirmed with the decapod crustaceans *Astacus leptodactylus* and *Uca tangeri*, no sialic acids being found in the tissues, but significant amounts in the digestive gland (SEGLER *et al.* 1978). These workers could not demonstrate the biosynthesis of sialic acid in *A. leptodactylus* and concluded that digestive gland sialic acid was due to dietary sialic acid. The presence of sialidase activity in the digestive gland of the lobster (WARREN 1963) is thus of interest from the viewpoint of evolution and environmental adaptation.

The presence of gangliosides in crab eye stalks (ISHIZUKA *et al.* 1970, SVENNERHOLM 1970) appears uncertain (WIEGANDT 1971), while the isolation of a haemagglutinin from a lobster yielded a sialic acid component (ACTON *et al.* 1973). Analyses of several insects including *Musca domestica*, *Calliphora erythrocephala* (WIEGANDT 1971), *Drosophila melanogaster* larval and adult forms and the eyes and ganglia of several other flies (unpublished) showed no sialic acids.

The occasional report of very low levels of sialic acid could not be reproduced in the molluscs. Furthermore, the demonstration of free sialic acid in the digestive gland of the squid *Loligo pealii* parallels the situation in the lobster and is almost certainly of dietary origin (WARREN 1963).

The occurrence of gangliosides in the squid and the octopus has been cited (ISHIZUKA *et al.* 1970, SVENNERHOLM 1970), but levels were very low and confirmation of these results is necessary. Sialic acid identified in a snail (ELDREDGE *et al.* 1963) has not been confirmed (SEGLER *et al.* 1978). Finally, the haemagglutinin isolated from the oyster *Crassostrea virginica* was found to contain one mole of sialic acid per mole protein (ACTON *et al.* 1973).

Conclusive evidence is still missing and until these reports are corroborated, the main body of evidence is in favour of the absence of sialic acids in these organisms.

VIII. Bryozoa, Brachiopoda, Chaetognatha, and Echinodermata

Organisms from the *Bryozoa* and *Brachiopoda* (five species) and *Chaetognatha* (one species) contained no sialic acid (WARREN 1963). The *Echinodermata* are the first group of animals in this branch of the evolutionary tree with certainty found to contain sialic acids common to all species investigated and were found in most tissues within one animal (ELDREDGE *et al.* 1963, WARREN 1963, VASKOVSKY *et al.* 1970, HOTTA 1977). A diverse group of gangliosides (Table 4) occurs in these animals (VASKOVSKY *et al.* 1970, NAGAI and HOSHI 1975, KOCHETKOV *et al.* 1976). Most of the sialic acids found in echinoderm species were Neu5Gc and its derivatives indicating the presence of Neu5Ac monooxygenase in these organisms. The *Echinodermata* also possess O-acetylated sialic acids and other derivatives summarized in Table 4. The occurrence of these sialic acids in high amounts in this phylum points to an important evolutionary development which took place in the precambrian era (see section XII).

IX. Hemichordata, Urochordata, and Cephalochordata

Single species from the *Hemichordata* (*Dolichoglossus kowalevskii*) and the *Cephalochordata* (*Amphioxus*) were found by WARREN (1963) to contain sialic acid. In eight urochordate species no sialic acid was found; this is surprising, as the urochordates (= tunicates) developed after the *Echinodermata* and the *Hemichordata*. Additional support for this finding has been put forward by RÖSNER and RAHMANN (1981), who found no sialic acid in the tunicates *Phallusia mammillata*, *Distomum adriatica* and *Microcosmos sulcatus*. These authors argue that the tunicates may have selected against the sialic acid coding genome and adopted an alternative life style.

X. Vertebrata

The sialic acids occur throughout the vertebrates. The majority of structural studies on isolated complex carbohydrates have been carried out with this class of animals, especially the *Mammalia*.

1. Pisces

Two primitive fishes, the hagfish (*Myxine glutinosa*) and the lamprey (*Petromyzon marinus*), both from the *Cyclostomata*, contain sialic acid. Although early reports were negative for hagfish mucus (WESSLER and WERNER 1957), a reinvestigation showed sialic acid constituting 0.5% of the mucus dry weight (unpublished). The lamprey has been more extensively investigated, and Neu5Ac and an O-acetylated derivative have been detected in eggs and liver (CABEZAS and FROIS 1966). Neu5Ac and Neu5Gc were found in lamprey blood cells (EYLAR *et al.* 1962) and Neu5Ac in lamprey fibrinopeptide B (DOOLITTLE and COTTRELL 1974).

Both the elasmobranches and teleosts contain sialic acids (WARREN 1963). Studies have concentrated mainly on skin mucus glycoproteins, spermatozoa, eggs, blood and brain tissue, although evidence exists for the presence of sialic acids in most organs (WARREN 1963).

Extensive studies have been carried out with gangliosides in the fishes (ELDREDGE *et al.* 1963, ISHIZUKA *et al.* 1970, AVROVA 1971, MCCLUER and AGRANOFF 1972, TETTAMANTI *et al.* 1972, LEDEEN 1978, HILBIG and RAHMANN 1980, RAHMANN 1980). The occurrence in brain tissue of polysialylated, highly polar gangliosides is common (YU and ANDO 1980), in contrast to the mammals,

Table 4. *Occurrence of sialic acids in the Echinodermata*

Class and species	Source	Sialic acids	Structure	Literature
Asteroidea (starfish)				
Asterias forbesi	many tissues	Neu5Gc, Neu5Gc8Me	—	Warren 1963
Asterias pectinifera	whole animal	Neu5Gc, Neu5Gc8Me	Ara(1-6)Galβ(1-4)[Galβ(1-8)]Neu5Gcα(2-3)Galβ(1-4)GlcCer	Sugita and Hori 1976
			Ara(1-6)Galβ(1-4)Neu5Gc8Meα(2-3)Galβ(1-4)GlcCer	Sugita 1979, a, b
			Ara(1-6)Galβ(1-4)Neu5Gcα(2-3)Galβ(1-4)GlcCer	
Distolasterias nipon	whole animal	Neu5Gc	—	Kochetkov *et al.* 1970, 1973
Pativia pectinifera	whole animal	Neu5Gc8Me	—	Kochetkov *et al.* 1970, 1973
Echinoidea (sea urchin)				
Pseudocentrotus depressus	eggs	Neu5Gc > Neu5Ac Neu9Ac5Gc	Fucα(1-4)Neu5Gc— (glycoprotein)	Hotta 1977 Kamerling *et al.* 1980
		Neu5Gc	sialosphingolipid	Nagai and Hoshi 1975
Anthocidaris crassispina	eggs, spermatozoa	Neu5Gc > Neu5Ac	glycoprotein Neu5Acα(2-8)Neu5Acα(2-6)GlcCer (spermatozoa)*	Hotta 1977 Nagai and Hoshi 1975
			Neu5Acα(2-6)GlcCer (eggs)	Hoshi and Nagai 1975

Hemicentrotus pulcherrimus	eggs, spermatozoa	Neu5Gc > Neu5Ac	glycoprotein	Hotta 1977
			sialosphingolipid	Nagai and Hoshi 1975
Strongylocentrotus intermedius	gonads	Neu5Ac > Neu5Gc	Neu5Acα(2-6)Glcβ(1-8)Neu5Acα(2-6)GlcCer	Kochetkov *et al.* 1973
			Neu5Acα(2-8)Neu5Acα(2-6)Glcβ(1-6)GlcCer	Kochetkov *et al.* 1973
Echinocardium cordatum	gonads	Neu5Gc > Neu5Ac Neu5Gc8S	Neu5Gc8Sα(2-6)GlcCer	Kochetkov *et al.* 1976
Clypeastroidea (sea urchin)				
Clypeaster japonicus	gonads	Neu5Gc, Neu5Ac	sialosphingolipid	Nagai and Hoshi 1975
Holothuria (sea cucumber)				
Thyrone briareus	whole animal	Neu5Gc	—	Warren 1963
	cuverian tubules	Neu5Gc	Fucα(1-4)Neu5Gc— (glycoprotein)	Isemura *et al.* 1973 (unpublished)
Ophiuroidea (brittle star)				
Ophioderma brevispinum	whole animal	Neu5Gc	sialosphingolipids and glycoproteins	Warren 1963, Vaskovsky *et al.* 1970
Crinoidea (sea lily)				
Nemaster species	whole animal	Neu5Gc	—	Warren 1963

* Determined by g.l.c. and reacetylated. Assumed to be Neu5Gc on the basis of other experiments; —, structure not determined

and the occurrence of mono, di, tri, and higher sialylated sphingolipids has been associated with phylogenetic (AVROVA 1980, HILBIG *et al.* 1981) and ontogenetic development (HILBIG *et al.* 1981) and thermal adaptation (AVROVA 1980, RAHMANN 1980, RAHMANN and HILBIG 1981). A similar trend in gangliosides isolated from central nervous system myelin has also been reported (LEDEEN *et al.* 1980).

Neu5Ac and Neu5,9Ac$_2$ have been identified in fish brain gangliosides. The O-acetylated sialic acid was originally designated 8-O-acetyl (ISHIZUKA *et al.* 1970), but studies on codfish brain gangliosides have identified Neu5,9Ac$_2$ (HAVERKAMP *et al.* 1977). No Neu5Gc has been described in fish gangliosides. High amounts (50%) of free sialic acids, as Neu5Ac and Neu5Gc exist in the eggs of two teleosts (WARREN 1960). RAHMANN and BREER (1976) found the level of free sialic acid to vary with the developmental stage of the fish. Bound sialic acid was found almost entirely associated with glycoproteins. Reports on the structure of O-glycosidically linked oligosaccharide chains (Thr/Ser-GalNAc type) from a glycoprotein associated with ovulated eggs in *Salmo irideus* showed evidence for polysialyl chains containing up to 15 molecules of Neu5Gc in α(2-8) linkage (INOUE and MATSUMURA 1979, INOUE and IWASAKI 1980).

The use of metabolic labelling techniques with radioactive ManNAc has proved particularly valuable in the study of brain ganglioside and glycoproteins in fishes (RÖSNER *et al.* 1973). The epithelial mucus glycoproteins of six teleost species contain sialic acid (TURUMI and SAITO 1953, WESSLER and WERNER 1957), although an elasmobranch (ray) was negative. This was confirmed by structural and biosynthetic studies (PICKERING 1976, WOLD and SELSET 1977, 1978). Fresh- and saltwater teleosts were found to contain sialic acid in their epithelial tissues, and the content was correlated with their environmental situation (HENTSCHEL and MÜLLER 1979). Further evidence for the involvement of sialic acid containing glycoproteins in adaptation to fresh- and saltwater has been furnished with experiments on eel epidermis (LEMOINE and OLIVEREAU 1973, 1974, LEMOINE 1974). Sulphated and sialylated glycoproteins which are transported down the nerve were isolated from the garfish and goldfish optic and olfactory nerve (see ELAM 1979 for a review). The Na^+/K^+-ATPase from dogfish salt gland has been found to contain sialic acid (MARSHALL 1976).

2. *Amphibia*

The presence of sialic acids in frogs and toads has been described by BÖHM and BAUMEISTER 1956, WARREN 1963, ELDREDGE *et al.* 1963, BOLOGNANI *et al.* 1966, AVROVA 1971, TETTAMANTI *et al.* 1972 and MARTINEZ and OLAVARRIA 1973, 1977. Sialic acid was found in several tissues of the frog *Rana catasbeiana* including skin, muscle, eggs and blood (WARREN 1963). ELDREDGE *et al.* (1963) also found Neu5Ac in the brain of frogs, gangliosides being the main source (TETTAMANTI *et al.* 1965, AVROVA 1971, RAHMANN 1978, 1980). Three new monosialogangliosides and two disialogangliosides were found in the fat body of the frog *Rana catasbeiana* (OHASHI 1980). The pattern of gangliosides was similar to that detected in fishes with relatively large amounts of polysialogangliosides (AVROVA 1971, NG and DAIN 1976, RAHMANN 1980). This was also the case for gangliosides isolated from frog retina (URBAN *et al.* 1980) and myelin (LEDEEN *et al.* 1980). These molecules from the brain contain little or no Neu5Gc, again similar to those found in fishes.

Sialic acid was found in the egg jelly coat of the frog species *Latastei* and the toad *Bufo vulgaris* (BOLOGNANI *et al.* 1966). Investigations on the oviduct glycoprotein from the toad *Bufo avenarum* revealed the presence of several sialic acids tentatively identified as Neu5Ac, Neu4,5Ac$_2$ and Neu5Gc (MARTINEZ and OLAVARRIA 1973). A dialyzable fraction of sialic acid was found in oviduct, which fluctuated with the ovulatory cycle, analogous to the situation reported in fish eggs (MARTINEZ *et al.* 1975, MARTINEZ and OLAVARRIA 1977). Palatal epithelium mucus glycoprotein from frog (MEYER 1977) and the oxyntic cell microsomes from frog and bullfrog contain glycoprotein-bound sialic acid (BEESLEY and FORTE 1973, 1974). The nucleated red blood cells from the salamander *Amphiuma means* contain Neu5Ac (PAPE *et al.* 1975).

3. *Reptilia*

The reptiles have not yet systematically been screened for sialic acid occurrence. The few species which have been studied all contain sialic acid. Examples are turtle skin (WARREN 1963) and brain (ELDREDGE *et al.* 1963). The reptile species show a high percentage of mono-, di-, and trisialogangliosides, differing from the fishes and amphibians as was studied in the slow worm *Anguis fragilis*, the tortoise *Testuclo horsfieldi* and the turtle *Pseudemys scripta elegans* (AVROVA 1968, 1971, RAHMANN 1980, RAHMANN and HILBIG 1981). Sialic acid was detected in reptilian gonadotropin and found to be necessary for the full biological action of this hormone (LICHT and PAPKOFF 1974).

Some histochemical evidence for sialic acids in lizards, teiids and dibamids (*Lacertilia*, *Teiidae*, and *Ophiadae*), cobras and vipers (*Elipidae* and *Viperidae*) has been reported, but isolation of sialic acids is still outstanding (LOPES *et al.* 1973, 1974).

4. *Aves*

Initial reports of sialic acids in birds were made by BÖHM and BAUMEISTER (1956) in hen serum, and in goose and pheasant mucus by WESSLER and WERNER (1957). Ovomucoids from 26 different species were investigated by FEENEY *et al.* (1960). The order *Casuariformes* (*e.g.* cassowary and emu) not only contained the highest amounts of ovomucoid in egg white, but also had the highest percentage of sialic acid in this molecule (emu 10.5‰, cassowary 5.4‰). Ovomucoid from anseriforms (duck and goose) was free from sialic acid, although total egg white contained about 0.1‰. Only Neu5Ac was found in ovomucoid.

Several mucins have been identified as sialoglycoproteins in avian species. The glycoprotein from the salivary gland of the chinese swift (*Collocalia*) was reported to contain Neu5Ac and Neu4,5Ac$_2$ (KATHAN and WEEKS 1969). Subsequent studies, however, presented no evidence for the presence of Neu4,5Ac$_2$ (HOUDRET *et al.* 1975, unpublished). The mucin from the European swiftlet species *Chelidon urbica* contains exclusively Neu5Ac (unpublished). Mucus glycoproteins derived from organ culture of goose trachea possess sialic acid (PHIPPS *et al.* 1977) and the lung lavage and lamellar organs of the chicken includes a well characterized sialoglycoprotein (BHATTACHARYYA *et al.* 1976).

Sialoglycoproteins occur in avian blood serum and in erythrocyte membranes. The sera of chicken, turkey, goose, and duck contain sialic acids as Neu5Ac with less than 10% as Neu5Gc. No O-acetylated derivatives were reported (FAILLARD and CABEZAS 1963, DŻUŁYŃSKA *et al.* 1969). A hepatic receptor, responsible for the removal of avian serum glycoproteins contains sialic acid, which in contrast to the analogous receptor in mammalian liver tissues remains active after removal of the sialic acid by sialidase (LUNNEY and ASHWELL 1976, KAWASAKI and ASHWELL 1977).

Incorporation studies with radioactive glucosamine, galactosamine and mannosamine confirmed the biosynthesis of sialoglycoproteins in avian tissues (CHANNON *et al.* 1972, SCOTT and ANASTASSIADIS 1978).

Chicken tissues, mainly brain and retina, have been widely used for analysis of gangliosides and assay of ganglioside metabolic enzymes (ELDREDGE *et al.* 1963, TETTAMANTI *et al.* 1965, AVROVA 1971, ANDO *et al.* 1978, DREYFUS *et al.* 1975, 1980). These gangliosides show similar patterns to those in reptiles and mammals, but total amounts were higher than in the mammals (AVROVA 1968, RAHMANN 1978, 1980, HILBIG *et al.* 1981). The developmental patterns observed support a change from polysialogangliosides in the fishes to mono- and trisialogangliosides in birds and mammals. Ganglioside patterns in isolated chicken myelin show characteristic phylogenetic patterns (HILBIG *et al.* 1981, RÖSNER 1981). Furthermore, gangliosides in the optic tecta of chicken embryos vary with development, showing an increase in polysialoganglioside concentration between day 6 and day 10 (IRWIN *et al.* 1976).

These gangliosides contain mainly Neu5Ac. Neu5,9Ac_2 (< 20% of total sialic acid) was found in total chicken brain ganglioside (HAVERKAMP *et al.* 1977).

The use of metabolic labelling with radioactive ManNAc has corroborated many of the structural studies and provided an in vivo method of identifying the biosynthetic pathways and subcellular localization of the gangliosides (*e.g.* CAPUTTO *et al.* 1974, DREYFUS *et al.* 1975, 1976, RÖSNER 1975, LANDA *et al.* 1979).

5. *Mammalia*

a) Introduction

No mammalian species has yet been found without sialic acids. The volume of literature supporting this conclusion is vast, and the reader is referred to more detailed reviews on specialized topics, *e.g.* glycoproteins in general (GOTTSCHALK 1966, 1972, MONTREUIL 1975, 1980, KORNFELD and KORNFELD 1976, LENNARZ 1980), serum glycoproteins (ALLISON 1974, 1976, PUTNAM 1975, 1977), mucus glycoproteins (ELSTEIN and PARKE 1977, BRITISH MEDICAL BULLETIN 1978, *CIBA symposium* 1978), membrane and cell surface structures (COOK and STODDART 1973, HUGHES 1976, HARMON 1978), complex carbohydrates in general (SHARON 1975, HOROWITZ and PIGMAN 1977, 1978) and sialic acids in general (GOTTSCHALK 1960, CABEZAS 1973, ROSENBERG and SCHENGRUND 1976, SCHAUER 1982).

b) Oligosaccharides

α) *Milk and Colostrum*

Acidic compounds in the milk and colostrum of mammals are found in high concentration at lactation (MONTREUIL and MULLET 1960). Isolation and characterization of these oligosaccharides have been largely achieved in the laboratories of Kuhn, Montreuil, and Kobata (Kuhn 1959, Montreuil 1975, Kobata 1977).

The acidic oligosaccharides from human, bovine, ovine, canine and porcine colostrum consist mainly of II^{3}Neu5AcLac and II^{6}Neu5AcLac (Kuhn and Brossmer 1956, Carubelli *et al.* 1961). Abbreviations for commonly occurring oligosaccharides have been adopted as suggested for glycolipids (IUPAC-IUB 1977). Bovine colostrum was found to contain large amounts of sialyllactose with a mono-O-acetylated sialic acid, Neu5,9Ac$_2$ (Kuhn and Brossmer 1956). II^3(Neu5Ac)$_2$Lac and sialyl-N-acetyllactosamine, II^{6}Neu5AcLacNAc, also occur as major components in this fluid (Kuhn and Gauhe 1965, Veh *et al.* 1981). The detection of Neu5Gc in colostral sialyloligosaccharides was traced to the presence of II^{3}Neu5GcLac (Kuhn and Gauhe 1965, Veh *et al.* 1981). Other minor components were tentatively identified as II^{6}Neu5GcLacNAc and II^{6}Neu5GcLac (Veh *et al.* 1981). Sialyloligosaccharides from human, bovine and ovine milk included α(2-3) or (2-6)-linked sialyllactose (Kuhn and Brossmer 1956).

Human milk has been more extensively studied and larger oligosaccharides detected. The main sialyloligosaccharides are sialyllacto-N-tetraose a (IV^{3}Neu5AcLcOse$_4$) and b (III^{6}Neu5AcLcOse$_4$) (Kuhn and Gauhe 1965), disialyllacto-N-tetraose (IV^{3}Neu5AcIII^{6}Neu5AcLcOse$_4$) (Grimmonprez and Montreuil 1968) and sialyllacto-N-neotetraose (IV^{6}Neu5AcLcnOse$_4$) (Kuhn and Gauhe 1962). Larger, branched oligosaccharides of the lacto-N-hexaose and lacto-N-neohexaose series have also been detected, with mono- and disialyl substitution, in some cases with fucose (Kobata 1977). Sialyllactoses have also been found in rat milk and mammary gland (Carubelli *et al.* 1961, Kuhn 1972), one of these isomers containing galactose-6-sulphate (Ryan *et al.* 1965). Of special interest is the sialyllactose isolated from the milk of the monotreme *Echidna*, since it contains Neu4,5Ac$_2$ (Kamerling *et al.* 1982a). This animal has primitive mammary glands and sialyllactose (II^{3}Neu4,5Ac$_2$Lac) constitutes 50% of the total free milk carbohydrates (Messer and Kerry 1973, Messer 1974).

β) *Urine*

Normal urine from rat and man contains sialyllactose and sialyl-N-acetyllactosamine isomers. In human urine Neu5Ac is found in α(2-3) and (2-6) glycosidic linkage to lactose and in α(2-6) linkage to N-acetyllactosamine (Huttunen 1966, Strecker and Montreuil 1979). The presence of mono- and disialyloligosaccharide derivatives of Galβ(1-3)GalNAc were also reported as Neu5Acα(2-8)Neu5Acα(2-3)Galβ(1-3)GalNAc and the corresponding α(2-3)-monosialyltrisaccharide (Huttunen and Miettinen 1969, Huttunen 1966, Bourrillon 1972). In addition, the tetrasaccharide with the structure

```
Neu5Acα(2-3)Galβ(1-3)GalNAc
                       |
                       |α(2-6)
                       |
                    Neu5Ac
```

(sialylated T-antigen) has been found and also isolated from a pathological urine (Michalski, personal communication).

In rat urine at least 10 different sialyloligosaccharides could be identified (MAURY 1971a, b). These include sialyllactose and sialyl-N-acetyllactosamine isomers containing Neu5Ac, Neu5Gc and O-acetylated sialyllactose. Sialyloligosaccharides containing GalNAc were also reported (MAURY 1971a).

Human pregnancy urine was found to contain increased levels of glycosidically bound Neu5Ac in the form of sialyl-N-acetyllactosamine, II^3Neu5AcLac (MAURY 1976) and IV^3Neu5AcIII^6Neu5AcLcOse$_4$ (LEMONNIER *et al.* 1975, 1977, 1978, LEMONNIER and BOURRILLON 1976). Only II^3Neu5AcLac was increased during pregnancy in the rat (MAURY 1972).

The sialidoses, a group of diseases characterized by the absence or partial deletion of sialidase activity, produce a spectrum of sialyloligosaccharides related to serum glycoprotein oligosaccharides, as described in chapter K. More than 20 of these oligosaccharides have been isolated and characterized (STRECKER *et al.* 1976, MICHALSKI *et al.* 1977, DORLAND *et al.* 1979, STRECKER and MONTREUIL 1979). Several novel sialyloligosaccharides have been found in the urine of asparaginyl-glucosaminuria patients (POLLIT and PRETTY 1974, LUNDBLAD *et al.* 1976, SUGAHARA *et al.* 1976).

γ) *Nucleotide Sugars*

The activated forms of sialic acids, prerequisites to sialyltransfer, are the CMP-β-glycosides of Neu5Ac, Neu5Gc, their 4- and 9-mono-O-acetyl derivatives and Neu5Ac4Me (SCHAUER *et al.* 1972, SCHAUER and WEMBER 1973, CORFIELD *et al.* 1976, BUSCHER *et al.* 1977, CAREY and HIRSCHBERG 1979, BEAU and SCHAUER 1980). N.m.r. studies have confirmed that the glycosidic bond is β in CMP-Neu5Ac (HAVERKAMP *et al.* 1979).

c) Glycoproteins

α) *Serum*

Sialic acids from the blood, sera or plasma of mammals have been reported (GOTTSCHALK 1960, CABEZAS 1973). Human sera contain mainly Neu5Ac (80%), Neu5Ac9Lt (20%) and traces of Neu5,9Ac$_2$. Neu5Gc is completely absent (HAVERKAMP *et al.* 1976). Primate sera were found to contain mainly Neu5Ac and only small amounts of Neu5Gc (UHLENBRUCK and SCHMITT 1965), although a ratio of Neu5Ac : Neu5Gc of 61 : 39 has also been reported (DŻUŁYŃSKA *et al.* 1966). Calf serum contains between 58–90% Neu5Ac with the remainder as Neu5Gc (FAILLARD and CABEZAS 1963, DŻUŁYŃSKA *et al.* 1968), whereas in serum from adult animals a higher ratio of Neu5Gc to Neu5Ac (7 : 3) occurs (MÅRTENSSON *et al.* 1958) suggesting a developmental change in favour of Neu5Gc (DŻUŁYŃSKA *et al.* 1968). Horse serum contains an overall ratio of Neu5Ac : Neu5Gc of 85 : 15 (DŻUŁYŃSKA *et al.* 1966, MÅRTENSSON *et al.* 1958, PEPPER 1968); at least 60% of these sialic acids are 4-O-acetylated (PEPPER 1968, BUSCHER *et al.* 1974). Neu5Ac, Neu5Gc, and Neu4,5Ac$_2$ have been detected by g.l.c./m.s. in donkey serum, and Neu5Ac, Neu5Gc, and Neu5,9Ac$_2$ in rabbit, rat and mouse serum. These sialic acids are also produced in rat hepatocytes (unpublished).

The number of known serum glycoproteins is large and has been reviewed (ALLISON 1974, 1976, PUTNAM 1975, 1977, SCHWICK *et al.* 1977). Their sialic acid content varies between 14.3% (C1 inactivator) and 0% (Gc globulin), while most have between 3 and 7% as sialic acid in man (SCHWICK *et al.* 1977). The structures of serum glycoproteins are varied as are the functions (considered in chapter J). Some structural aspects are presented in section XI.3.b).

Fetuin obtained from fetal calf serum contains about 7% of the total sialic acid (9% of total carbohydrate) as Neu5Gc, the remainder is Neu5Ac (GRAHAM 1972). In human α_1-acid glycoprotein only Neu5Ac was reported (PUTNAM 1975, 1977). Bovine α_1-acid glycoprotein contains Neu5Ac, Neu5Gc and possibly an O-acyl derivative (BEZKOROVAINY 1963), while in horse α_1-acid glycoprotein Neu5Ac, Neu5Gc, and Neu4,5Ac$_2$ occur (unpublished). A high molecular weight kininogen from bovine serum has carbohydrate structures related to T-antigen, with one or two sialic acid molecules and being linked to serine or threonine (ENDO *et al.* 1977). A similar kininogen has been reported in human serum (LONDESBOROUGH and HAMBERG 1975).

The serum glycoproteins all contain sialic acid with the exception noted above, and detailed structural analyses of the oligosaccharide chains have been reported in many cases (GOTTSCHALK 1972, PUTNAM 1975, 1977, SCHWICK *et al.* 1977).

Sialic acids are responsible for the acidic nature of these glycoproteins and may give rise to electrophoretic microheterogeneity (GOTTSCHALK 1969, MONTGOMERY 1972).

β) *Epithelial or Mucus Glycoproteins*

The mucus glycoproteins have played an important part in the history of sialic acids. Neu5Ac was first isolated in crystalline form from bovine submandibular gland glycoprotein (KLENK and FAILLARD 1954). Studies with bovine, porcine, ovine and equine submandibular glands established the presence of a family of O-acetylated sialic acids based on both Neu5Ac and Neu5Gc (BLIX *et al.* 1955, 1956, BLIX and LINDBERG 1960, GOTTSCHALK 1960, TUPPY and GOTTSCHALK 1972, BUSCHER *et al.* 1974, SCHAUER 1978a, 1982; see also section XI.3. and Table 1). Therefore, the majority of later structural analyses were concentrated on bovine and equine submandibular glands and the spectrum of sialic acids appreciably extended (BUSCHER *et al.* 1974, CORFIELD *et al.* 1976, SCHAUER *et al.* 1976, SCHAUER 1978a, b, 1982).

In spite of the abundance of sialic acids in these mucus glycoproteins, up to 35% in purified bovine submandibular gland major mucus glycoprotein (TETTAMANTI and PIGMAN 1968), the structure of these glycoproteins is poorly understood. They are characterized by O-glycosidic linkages of GalNAc to serine and threonine in the polypeptide chain, a high percentage of α(2-6) linkages of sialic acid to GalNAc and by blood group reactivity (PIGMAN 1977a). The structure of some of the oligosaccharide chains has been reported for cow, sheep, pig, rat, armadillo, hamster, and dog salivary gland mucus glycoproteins (WINZLER 1973, PIGMAN 1977b, HERP *et al.* 1979). The armadillo glycoprotein is believed to represent the most primitive of the mammalian structures (HERP *et al.* 1979); it contains only Neu5Ac in the disaccharide units (WU and PIGMAN 1977, WU *et al.* 1978, 1979). Both rat and hamster sublingual gland mucus glycoproteins have O-acetylated sialic acids; in the rat these were tentatively identified as Neu4,5Ac$_2$ and Neu4,5(7 or 8)Ac_3 (MOSCHERA and PIGMAN 1975). Most other salivary gland mucus glycoproteins, *e.g.* canine (LOMBART and WINZLER 1974), murine (ROUKEMA *et al.* 1976), and monkey (HERZBERG *et al.* 1979) contain only Neu5Ac. Both Neu5Ac and Neu5Gc occur in feline salivary gland glycoproteins (PHELPS and STEVENS 1978, unpublished). Several new features involving sialic acids of mucus glycoprotein structures have come to light recently (see section XI.3.b).

Mucus glycoproteins from the trachea and lung of many mammals have been studied (HOLDEN and GRIGGS 1977). Experiments with glycoprotein fractions obtained from normal and diseased animals and patients have emphasized the problems of isolating pure glycoproteins (KENT 1978, LAMBLIN *et al.* 1980). Neu5Ac has been found in human (ROBERTS 1974, BOAT *et al.* 1976, FELDHOFF *et al.* 1979, LAMBLIN *et al.* 1979, ROSE *et al.* 1979), feline (PHELPS and STEVENS 1978) and canine (LIAO *et al.* 1979) tracheal glycoproteins. Experiments with explants and radiolabelled tracers have yielded similar results in canine (CHAKRIN *et al.* 1972, ELLIS and STAHL 1973), rat (YEAGER *et al.* 1971, BONANNI and DE LUCA 1974), feline (GALLAGHER *et al.* 1975) and rabbit (GALLAGHER and KENT 1975) systems.

Gastric mucus glycoproteins have generally been found to contain low levels of sialic acids (STARKEY *et al.* 1974, OATES *et al.* 1974, HOROWITZ 1977, CLAMP *et al.* 1978, LABAT-ROBERT and DECAENS 1979). Neu5Ac occurs in human gastric glycoproteins (CLAMP *et al.* 1978), while in porcine stomach Neu5Gc is probably present, although measured as Neu5Ac by g.l.c. (SLOMIANY and MEYER 1972, STARKEY *et al.* 1974, CLAMP *et al.* 1978). Rat (*e.g.* WALDRON-EDWARD *et al.* 1976, SPEE-BRAND *et al.* 1980) and canine gastric mucus glycoproteins (*e.g.* WOUSSEN-COLLE *et al.* 1975a, b) contain Neu5Ac. Histochemical data support the presence of sialic acids in gastric tissues (see chapter H), and in some human diseased states O-acylated sialic acids may occur (*e.g.* JASS 1980, JASS and FILIPE 1980). Evidence for elevated amounts of Neu5Ac9Lt in ulcerous human gastric tissue has been obtained (unpublished).

Chemical study of human colonic mucus glycoproteins has revealed the presence of Neu5Ac and O-acetylated derivatives. The latter have been confirmed using histochemical methods (*e.g.* FILIPE and BRANFOOT 1976, and the work of CULLING, see chapter H). They have been isolated and partially characterized to show Neu5Ac, Neu5,9Ac_2, Neu5,7,9Ac_3 and at least three other components probably including Neu5,8,9Ac_3 (ROGERS *et al.* 1978, 1979). Rat colonic mucus glycoproteins contain both Neu5Ac and Neu5Gc (ratio 4 : 1) as parent sialic acids (WOLD *et al.* 1974, CARLSSON *et al.* 1978, MURTY *et al.* 1978). O-Acetylated sialic acids were detected in the isolated glycoprotein at a ratio of 1 mole O-acetyl group per mole of sialic acid. The structure of the oligosaccharide chains has also been determined (SLOMIANY *et al.* 1980 and section XI.3b). Ovine colonic mucus glycoproteins also contain Neu5Ac, Neu5Gc and O-acetyl derivatives, as determined in radiolabelled tracer experiments (KENT and DRAPER 1968, KENT 1973). Porcine colonic mucus sialic acids have similarly been found to include Neu5Ac, Neu5Gc and O-acyl derivatives (unpublished). The isolated glycoprotein contains 10% sialic acids (MARSHALL and ALLEN 1978).

A glycoprotein isolated from the cervical mucus of the bonnet monkey has Neu5Ac and Neu5Gc in a ratio of 3 : 1, but no O-acetylated sialic acids (HATCHER *et al.* 1977). Neu5Ac and Neu5Gc also occur in bovine cervical mucus (see CLAMP *et al.* 1978), and Neu5Ac only in the human glycoprotein (CLAMP *et al.* 1978, WOLF *et al.* 1980).

Sialic acids have been isolated from human, porcine, ovine and bovine bile secretions (CABEZAS and RAMOS 1972, CABEZAS 1973). O-Acetylated derivatives occur in all species investigated, and with the exception of human bile, both Neu5Ac and Neu5Gc were present (CABEZAS and RAMOS 1972, CABEZAS 1973). Isolated bile mucus glycoproteins contain sialic acids, but no sialic acid identification has been reported (see HOROWITZ 1977).

Sialic acids are also present in human meconium as Neu5Ac (ODIN 1955, FRASER and CLAMP 1975).

γ) *Membrane Glycoproteins*

Sialic acid is an important component of membrane glycoproteins, which has been described in several reviews (*e.g.* COOK and STODDART 1973, HUGHES 1976, JEANLOZ and CODINGTON 1976, WARREN 1976, GLICK and FLOWERS 1978, and many others). Analysis of subcellular membrane fractions has demonstrated the presence of Neu5Ac in all fractions of cultured cells and tissue extracts.

Recent interest in the relationship of sialic acid to the half-life of circulating erythrocytes in various mammals has led to a more detailed analysis of erythrocyte membrane glycoproteins and their sialic acids. Human erythrocytes contain only Neu5Ac linked to glycoproteins, no evidence for other types has been obtained (REUTER *et al.* 1980). In contrast, three different murine strains were found to contain Neu5Ac, Neu5Gc, Neu5,9Ac$_2$, Neu9Ac5Gc and other derivatives in varying proportions related to the strain (SARRIS and PALADE 1979). An inverse relationship between total sialic acid content and O-acetylated sialic acids in murine erythrocyte membranes from different strains was observed, and a relationship between O-acetylation of sialic acids and complement fixation revealed (VARKI and KORNFELD 1980). The erythrocytes of the mouse BALB/c strain contain almost exclusively Neu5,9Ac$_2$ associated with the membranes (REUTER *et al.* 1980). High proportions of Neu5Gc as parent sialic acid were found in rat, cat, pig, sheep, cow, horse, mule, and donkey erythrocytes (CABEZAS 1973, CABEZAS and SECO 1975, unpublished). Mouse, rat, rabbit, and rhesus monkey erythrocytes were found to contain higher relative concentrations of Neu5Ac and in addition Neu5,9Ac$_2$. Rabbit erythrocytes also contain Neu9Ac5Gc (PFEIL *et al.* 1980), and equine erythrocytes Neu4Ac5Gc (BUSCHER *et al.* 1974). A part of the Neu5,9Ac$_2$ in rat erythrocyte membranes has been localized in gangliosides (unpublished).

Human erythrocyte glycoprotein oligosaccharides have been found to include polyglycosyl components with sialic acid in various [α(2-3), (2-6), (2-8)] linkages (JÄRNEFELT *et al.* 1978, KRUSIUS *et al.* 1978b). Isolated glycophorin, the major sialoglycoprotein of the red cell membrane, contains 15 O-glycosidically linked chains related to the sialylated "T-antigen" structure (MARCHESI *et al.* 1976) and one biantennary complex carbohydrate chain, N-glycosidcally linked to asparagine (YOSHIMA *et al.* 1980).

In blood platelet membranes sialoglycoproteins occur, and the sialic acids isolated from pig, horse, donkey, and mule platelets consist of Neu5Gc with traces of Neu5Ac. Calf platelets contain mainly Neu5Ac and lamb platelets Neu5Ac : Neu5Gc in a ratio of approximately 1 : 1 (CABEZAS and CABEZAS 1973).

Leukocytes from pig, horse, donkey, and mule all contain Neu5Gc and smaller amounts of Neu5Ac in their surface membranes. In addition, about 10% of the sialic acids were O-acetylated, and 20% occurred as a "non-identified" form (ROCHA *et al.* 1975).

Mouse lymphocytes exhibit 75% of their total sialic acid as Neu5Gc (KAUFMANN *et al.* 1981), while human tonsil B lymphocytes contain 50% in O-acetylated form as Neu5,9Ac$_2$ (KAMERLING *et al.* 1982b). In contrast, T lymphocytes show only traces of O-acetylated sialic acids.

Well characterized membrane glycoproteins from cell surfaces have complex oligosaccharide structures linked via asparagine, *e.g.* hamster embryo fibroblast galactoprotein (CARTER and HAKOMORI 1979), fibronectin with Neu5Acα(2-4)Gal linkages (KOBATA 1979), platelet glycocalicin (OKUMURA *et al.* 1976) and epiglycanin from TA 3 ascites cells containing sialyloligosaccharides related to the T-antigen structure (VAN DEN EIJNDEN *et al.* 1979).

δ) *Brain Glycoproteins*

Analysis of sialic acids in brain tissues has revealed about 30% associated with glycoprotein fractions (DI BENEDETTA *et al.* 1969, MORGAN *et al.* 1977, MARGOLIS and MARGOLIS 1979). In human, rat and rabbit brain glycoproteins only Neu5Ac was found (MARGOLIS and MARGOLIS 1979). Rat and human glycoproteins include O-glycosidically linked oligosaccharides of the sialylated T-antigen type and smaller oligosaccharides (FINNE 1975, FINNE and RAUVALA 1977). N-glycosidically linked chains with sialic acids linked via α(2-3) and (2-6) bonds to galactose occur and a characteristic structural component could be identified as

```
Neu5Acα(2-3)Galβ(1-4)GlcNAcβ(1-
                        |
                        |α(1-3)
                        |
                       Fuc
```

(KRUSIUS and FINNE 1978).

An important discovery was made by FINNE *et al.* (1977) demonstrating the presence of Neu5Acα(2-8)Neu5Ac sequences in rat and rabbit brain glycoproteins. Soluble glycoproteins from adult rat brain exhibit 1.6% of the total sialic acid in this form, while microsomal, mitochondrial and nuclear fractions contain about 9%. In 8 day old rats these percentages were between 14 and 19% (FINNE *et al.* 1977, KRUSIUS *et al.* 1978a). The presence of sialyloligosaccharides, linked through mannose to chondroitin sulphate proteoglycan, has been reported (FINNE *et al.* 1979).

ε) *Milk, Colostrum and Urine Glycoproteins and Glycopeptides*

Milk and colostrum glycoproteins contain sialic acids. The caseins are the major milk proteins, but only κ-casein is a glycoprotein, with 5% carbohydrate. It has been isolated from human, bovine, ovine and goat sources (see KOBATA 1977).

Sheep and goat glycoproteins contain Neu5Ac and Neu5Gc (ALAIS and JOLLÈS 1961). The structure of the carbohydrate moiety has been identified as the sialylated T-antigen (FOURNET *et al.* 1979, VAN HALBEEK *et al.* 1980). In lactotransferrin Neu5Ac α(2-6)-linked to galactose occurs as a component of biantennary, N-glycosidically linked oligosaccharides. Bovine lactotransferrin has the same structure plus one fucose moiety (MONTREUIL 1980). The so-called M_1-glycoproteins occur in human and bovine milk and show varying molecular weights from ~ 4,800 to 31,000. Sialic acid comprises between 10 and 25% of the dry weight of these glycoprotein-glycopeptide fragments (KUHN and EKONG 1963, BEZKOROVAINY and GROHLICH 1969, NICHOLS and BEZKOROVAINY 1973, NICHOLS *et al.* 1975). The carbohydrate chains are linked to serine and threonine residues via O-glycosidic linkage to GalNAc, and are more complex than those found for κ-casein (NICHOLS and BEZKOROVAINY 1973, NICHOLS *et al.* 1975). The milk fat globule membrane contains a number of glycoproteins (PATTON and KEENAN 1975, KEENAN *et al.* 1977) having N- and O-glycosidically linked oligosaccharides (SNOW *et al.* 1977, FARRAR and HARRISON 1978). The O-glycosidic oligosaccharides are sialylated T-antigen structures.

Several reports of sialic acid-rich glycoproteins and glycopeptides in urine have appeared (see LUNDBLAD 1977 for a review). These comprise a mixture of components of unknown origin. Some glycoproteins and glycopeptides are probably derived from blood plasma (LUNDBLAD 1977, STRECKER and MONTREUIL 1979). One typical urinary glycoprotein is the Tamm-Horsfall glycoprotein, synthesized in the kidney tubular system (SCHENK *et al.* 1971). The molecular weight is high (~ 80,000 in human and rabbit), and it contains about 5% sialic acid as Neu5Ac in the case of man (FLETCHER 1972), and as Neu5Ac and two O-acetylated sialic acids in the case of rabbit (NEUBERGER and RATCLIFFE 1972).

A glycopeptide with a carbohydrate moiety consisting of Neu5Ac : Gal : GalNAc in the relationship 2 : 1 : 1 and possibly linked to serine, has been isolated from the urine of pregnant women (LEMONNIER and BOURRILLON 1975). This type of structure, related to the red cell membrane glycoproteins, has also been isolated as larger glycopeptides (LUNDBLAD 1977).

ζ) *Hormone and Enzyme Glycoproteins*

A large number of enzymes are glycoproteins. These belong mainly to the oxidoreductases, transferases and hydrolases. About 50% of the glycoprotein enzymes have sialic acid as Neu5Ac and Neu5Gc (JUTISZ and DE LA LLOSA 1972, BAHL and SHAH 1977).

Most glycoprotein enzymes characterized so far contain N-glycosidic linkages between a protein asparagine residue and GlcNAc (JUTISZ and DE LA LLOSA 1972, BAHL and SHAH 1977). The role of the sialic acid moieties in enzymes corresponds to those found for other glycoproteins (see chapter J) and may be involved in electrophoretic heterogeneity of *e.g.* phosphatase (STEPAN and FERWERDA 1973), arylsulphatase B (FAROOQUI 1976), γ-glutamyltransferases (KÖTTGEN *et al.* 1976) and renal angiotensin I converting enzyme (OSHIMA *et al.* 1976), which in turn may be related to the survival of such enzymes in the circulation, studied *e.g.* with ribonuclease (BAYNES and WOLD 1976), plasminogen (SIEFRING and CASTELLINO

1974) and ceruloplasmin (ASHWELL and MORELL 1974). Human plasminogen contains N- and O-glycosidically linked oligosaccharides, the typical complex type chains with Neu5Acα(2-6)Gal linkages and the sialylated T-antigen tetrasaccharide structure with α(2-3)- and (2-6)-linked sialic acids (HAYES and CASTELLINO 1979a, b, c).

Although hormones have been isolated from mammals other than man, including mouse, pig, sheep, cow, and horse, only Neu5Ac has so far been identified. Human chorionic gonadotropin contains both N- and O-glycosidically linked oligosaccharide chains. Four "T-antigen" type residues are linked to serine and are sialylated on the Gal(α(2-3)) and GalNAc(α(2-6)) groups. Sialylated "complex" type chains are found linked to asparagine (ENDO *et al.* 1979, KESSLER *et al.* 1979a, b).

Chorionic gonadotropin, luteinizing hormone, follicle-stimulating hormone and thyroid-stimulating hormone are comprised of α- and β-subunits, both of which are glycosylated. The significance of sialic acid-containing oligosaccharides to subunit structure is not known, although structural differences have been reported (MAGHUIN-ROGISTER *et al.* 1975, BAHL and SHAH 1977).

η) *Lipoproteins*

The lipoprotein complexes which are involved in lipid transport (MORRISETT *et al.* 1975, SCANU *et al.* 1975) consist of well-defined apoproteins which in part contain sialic acids in addition to gangliosides in smaller amounts. These functionally important sialic acids appear to be associated with complex type oligosaccharide chains linked to asparagine (SWANINATHAN and ALADJEM 1976), and study of the different apoproteins has revealed a specific distribution with 0, 1 or 2 sialic acid molecules per mole of apoprotein (SCANU *et al.* 1975).

ϑ) *Miscellaneous*

Sialoglycoproteins have been isolated from a number of connective tissues (see PIGMAN 1977c) including cartilage (SHIPP and BOWNESS 1975, WHITE *et al.* 1975), bovine pulmonary tissue (FRANCIS and THOMAS 1975), rabbit palatal mucosa (GOLDITCH *et al.* 1974), arterial wall (SAITO and YOSIZAWA 1975, BAIG and AYOUB 1976), human teeth (HOLBROOK and LEAVER 1976), bovine cortical bone (ANDREWS *et al.* 1967, ASHTON *et al.* 1974) and bovine bone matrix and tendon (HERRING 1976). Basement membranes from several mammalian species also contain sialoglycoproteins (*e.g.* LIU and KALANT 1974, BARDOS *et al.* 1976, GIBBONS *et al.* 1976). Most of these studies analyzed sialic acid as Neu5Ac by g.l.c. after methanolysis, or by non-discriminating colorimetric methods. Neu5Gc was detected in rabbit palatal mucosa (GOLDITCH *et al.* 1974) and bovine liver basement membrane (GIBBONS *et al.* 1976) but was absent in rat basement membrane (LIU and KALANT 1974).

The presence of both O- and N-glycosidically linked oligosaccharides containing sialic acid was shown in proteoglycans from swarm rat chondrosarcoma and chick limb bud chondrocyte cultures by LOHMANDER *et al.* (1980) and DE LUCA *et al.* (1980). The O-glycosidically linked chains were all related to the sialylated "T-antigen" structure (LOHMANDER *et al.* 1980).

Human saliva contains glycoproteins in addition to the saliva-specific sialoglycoproteins (see MANDEL 1977, KLEINBERG *et al.* 1979). Both gastroferrin and intrinsic factor, from human (Neu5Ac) and porcine (Neu5Gc : Neu5Ac 6 : 4) gastric tissues and juice are sialoglycoproteins (FAILLARD *et al.* 1962, BUDDECKE 1972, RUDZKI and DELLER 1973). A glycopeptide isolated from human gastric juice contained both O- and N-glycosidically linked chains (GOSO and HOTTA 1977). Sialoglycoproteins including mucus glycoproteins have been identified in human tears and lacrimal gland explant secretions (WRIGHT and MACKIE 1977, CHAO *et al.* 1980). Neu5Ac was identified in human (CABEZAS *et al.* 1964) and rabbit secretions (KRENGER *et al.* 1976). Sialoglycoproteins have been isolated from the seminal plasma and spermatozoal membranes (HARTREE 1962, HUDSON *et al.* 1965, HERRMAN and UHLENBRUCK 1972, SRIVASTAVA *et al.* 1974) and from the ova (SOUPART and NOYES 1964) of many mammalian species. Other sources of sialic acid-containing glycoproteins are sweat (PALLAVICINI *et al.* 1963), amniotic fluid (LAMBOTTE and UHLENBRUCK 1966), synovial fluid (SWANN *et al.* 1977) and cerebrospinal fluid (BOGOCH 1958). In general, human tissues have yielded only Neu5Ac, while other mammals, especially pig, cow, sheep, rabbit, rat, and goat have shown significant amounts of Neu5Gc (FAILLARD and SCHAUER 1972).

d) Gangliosides

The gangliosides are a large group of sialic acid-containing glycolipids, composed of sialyloligosaccharide and lipid (ceramide) moieties. They are found mainly in nervous tissue, and early isolation techniques for sialic acid itself were centred on brain glycolipids (KLENK 1941). Due to their relatively small molecular weight, structural analysis has been possible and a number of reviews should be consulted (SWEELEY and SIDDIQUI 1977, LEDEEN 1978, 1979, YAMAKAWA and NAGAI 1978, NAGAI and IWAMORI 1980, WIEGANDT 1980).

Apart from the gangliosides I^6Neu5AcGlcCer and I^3Neu5AcGalCer (WIEGANDT 1980) all gangliosides so far detected belong to two series, having a typical basic oligosaccharide unit. These are the ganglio-series (GgOseCer; GalNAcβ(1-4)Galβ(1-4)Glcβ(1-1)Cer) and the lacto-series (LcOseCer; GlcNAcβ(1-3)Galβ(1-4)Glcβ(1-1)Cer), the former also found with terminal β(1-3)-linked galactose ($GgOse_4$) and the latter occurring in two tetraglycosyl structures, with terminal β(1-3)-linked galactose ($LcOse_4$) or with terminal β(1-4)-linked galactose, the neo-tetraglycosyl form ($LcnOse_4$). A further subdivision may be made for gangliosides containing only the Galβ(1-4)GlcCer (LacCer) unit present in both ganglio and lacto structures. In most series, especially $GgOse_4$Cer and LacCer, mono-, di- and trisialyl linkages to individual galactose residues have been desribed.

Variation of sialic acids in gangliosides is well documented. All human gangliosides contain Neu5Ac and in addition Neu5,9Ac_2 in brain (HAVERKAMP *et al.* 1977), where it constitutes between 4-15% of the total ganglioside sialic acid. The presence of O-acetylated sialic acids in gangliosides other than the human brain has been confirmed in horse erythrocyte stroma, II^3Neu4Ac5GcLacCer being found (HAKOMORI and SAITO 1969, VEH *et al.* 1979, SCHAUER *et al.* 1980), in the mouse brain ganglioside IV^3Neu5AcII^3(Neu5,9Ac_2-Neu5Ac)$GgOse_4$Cer

Table 5. *Mammalian gangliosides containing N-glycolylneuraminic acid*

Structure	Abbreviation	Source	Literature
GalNAcβ(1-4)Galβ(1-4)GlcCer α(2-3) (on Gal) Neu5Gc	II^{3}Neu5GcGgOse$_3$Cer	bovine spleen, kidney	WIEGANDT 1973
Galβ(1-3)GalNAcβ(1-4)Galβ(1-4)GlcCer α(2-3) (on Gal) Neu5Gc	II^{3}Neu5GcGgOse$_4$Cer	bovine spleen, kidney	WIEGANDT 1973
Galβ(1-3)GalNAcβ(1-4)Galβ(1-4)GlcCer α(1-2) (on terminal Gal) α(2-3) (on inner Gal) Fuc Neu5Gc	IV^{2}FucII^{3}Neu5GcGgOse$_4$Cer	bovine liver	WIEGANDT 1973
Galβ(1-3)GalNAcβ(1-4)Galβ(1-4)GlcCer α(2-3) (on terminal Gal) α(2-3) (on inner Gal) Neu5Ac Neu5Gc	IV^{3}Neu5AcII^{3}Neu5GcGgOse$_4$Cer	bovine brain	GHIDONI *et al.* 1976
Galβ(1-3)GalNAcβ(1-4)Gal(1-4)GlcCer α(2-3) (on terminal Gal) α(2-3) (on inner Gal) Neu5Gc Neu5Ac	IV^{3}Neu5GcII^{3}Neu5AcGgOse$_4$Cer	bovine brain	GHIDONI *et al.* 1976
Galβ(1-3)GalNAcβ(1-4)Galβ(1-4)GlcCer α(2-3) (on terminal Gal) α(2-3) (on inner Gal) Neu5Gc Neu5Gc	IV^{3}Neu5GcII^{3}Neu5GcGgOse$_4$Cer	bovine liver, kidney, spleen	WIEGANDT 1973

Galβ(1-4)GlcNAcβ(1-3)Galβ(1-4)GlcCer │α(2-3) Neu5Gc	IV^{3}Neu5GcLcnOse$_4$Cer	bovine spleen, kidney rabbit thyroid	Wiegandt 1973 Nagai and Iwamori 1980
Galβ(1-4)GlcNAcβ(1-3)Galβ(1-4)GlcCer │β(1-3) GlcNAcβ(1-4)Galα(2-3)Neu5Gc	VI^{3}Neu5GcLcnOse$_6$Cer	rabbit thyroid	Nagai and Iwamori 1980
Neu5Gcα(2-3)Galβ(1-4)GlcCer	II^{3}Neu5GcLacCer	bovine spleen	Yamakawa and Suzuki 1951
Neu4Ac5Gcα(2-3)Galβ(1-4)GlcCer	II^{3}Neu4Ac5GcLacCer	equine erythrocytes	Hakomori and Saito 1969, Veh *et al.* 1979
Neu5Gcα(2-8)Neu5Acα(2-3)Galβ(1-4)GlcCer	II^3(Neu5Gc—Neu5Ac)LacCer	bovine spleen, liver, kidney	Wiegandt 1973
Neu5Acα(2-8)Neu5Gcα(2-3)Galβ(1-4)GlcCer	II^3(Neu5Ac—Neu5Gc)LacCer	bovine spleen, liver, kidney	Wiegandt 1973
Neu5Gcα(2-8)Neu5Gcα(2-3)Galβ(1-4)GlcCer	II^3(Neu5Gc)$_2$LacCer	feline erythrocytes	Handa and Yamakawa 1964

Table 6. *Some novel*

Structure
Neu5Acα(2-6)GlcCer
Neu5Ac8Sα(2-8)Neu5Acα(2-3)Galβ(1-4)GlcCer
GalNAcβ(1-4)Galβ(1-4)GlcCer │α(2-3) Neu5Ac8S
Neu5Acα(2-6)Galβ(1-4)GlcNAcβ(1-3)Galβ(1-4)GlcCer
Neu5Acα(2-3)GalNAcβ(1-3)Galβ(1-4)GlcNAcβ(1-3)Galβ(1-4)GlcCer
Neu5Acα(2-6)Galβ(1-4)GlcNAcβ(1-3)Galβ(1-4)GlcNAcβ(1-3)Galβ(1-4)GlcCer
Fucα(1-2)Galβ(1-4)GlcNAcβ(1-6)╲ Galβ(1-4)GlcNAcβ(1-3)Galβ(1-4)GlcCer Neu5Acα(2-3)Galβ(1-4)GlcNAcβ(1-3)╱
Neu5Acα(2-8)Neu5Acα(2-3)Galβ(1-4)GlcNAcβ(1-3)Galβ(1-4)GlcCer
Neu5Acα(2-8)Neu5Acα(2-8)Neu5Acα(2-3)Galβ(1-4)GlcNAcβ(1-3)Galβ(1-4)GlcCer
Neu5Acα(2-3)Galβ(1-4)GlcNAcβ(1-3)Galβ(1-4)GlcCer │α(1-3) (on GlcNAc) Fuc
Neu5Acα(2-3)Galβ(1-3)GalNAcβ(1-4)Galβ(1-4)GlcCer
Fucα(1-2)Galβ(1-3)GalNAcβ(1-4)Galβ(1-4)GlcCer │α(2-3) (on inner Gal) Neu5Ac
GalNAcβ(1-4)Galβ(1-3)GalNAcβ(1-4)Galβ(1-4)GlcCer │α(2-3) (on outer Gal) │α(2-3) (on inner Gal) Neu5Ac Neu5Ac
Neu5Acα(2-3)Galβ(1-3)GalNAcβ(1-4)Galβ(1-4)GlcCer │α(2-3) (on inner Gal) Neu5,9Ac₂α(2-8)Neu5Ac

(GHIDONI *et al.* 1980) and in a sample of IV³Neu5AcII³Neu5AcGgOse$_4$Cer isolated from bovine brain (VEH *et al.* 1977). Furthermore, Neu5,9Ac$_2$ was detected in the ganglioside mixtures of cow, horse, pig, sheep, cat and rabbit brain (HAVERKAMP *et al.* 1977) and rat erythrocytes (unpublished). The discovery of bovine gastric mucosal gangliosides containing Neu5Ac8S has been reported (SLOMIANY *et al.* 1981, a, b). Two gangliosides were found, constituting about 11% of the total sulphatide fraction in this tissue, and had the structures II³(Neu5Ac8S-

ganglioside structures

Abbreviation	Source	Literature
I^6Neu5AcGlcCer	—	WIEGANDT 1980
II^3(Neu5Ac8S—Neu5Ac)LacCer	bovine gastric mucosa	SLOMIANY *et al.* 1981 a
II^3Neu5Ac8SGgOse$_3$Cer	bovine gastric mucosa	SLOMIANY *et al.* 1981 b
IV^6Neu5AcLcnOse$_4$Cer	bovine spleen, kidney	WIEGANDT 1973
V^3Neu5AcIV^3βGalNAcLcnOse$_4$Cer	human erythrocyte	WATANABE and HAKOMORI 1979
VI^6Neu5AcLcnOse$_6$Cer	human erythrocyte	WATANABE *et al.* 1979
based on LcnOse$_4$	human erythrocyte	WATANABE *et al.* 1978
IV^3(Neu5Ac)$_2$LcnOse$_4$Cer	human kidney	RAUVALA *et al.* 1978
IV^3(Neu5Ac)$_3$LcnOse$_4$Cer	human kidney	RAUVALA *et al.* 1978
IV^3Neu5AcIII^3FucLcnOse$_4$Cer	human kidney	RAUVALA 1976
IV^3Neu5AcGgOse$_4$Cer	rat ascites hepatoma cells	HIRABAYASHI *et al.* 1979
IV^2FucII^3Neu5AcGgOse$_4$Cer	boar testis	SUZUKI *et al.* 1975
	bovine liver	WIEGANDT 1973
IV^4βGalNAcIV^3Neu5Ac—II^3Neu5AcGgOse$_4$Cer	brain, Tay-Sachs disease	IWAMORI and NAGAI 1979
IV^3Neu5AcII^3(Neu5,9Ac$_2$—Neu5Ac)—GgOse$_4$Cer	mouse brain	GHIDONI *et al.* 1980

Neu5Ac)LacCer and II^3Neu5Ac8SGgOse$_3$Cer. Most of the parent sialic acid (~ 70%) was present as Neu5Ac. Gangliosides from several mammalian species contain Neu5Gc. In horse, cow, pig, and sheep brain this amounts to less than 2% of total sialic acids (YU and LEDEEN 1970). However, in some cases, e.g. in the erythrocytes of horse and pig, the relative content of Neu5Gc approaches 95-100%. The structures of some of the Neu5Gc-containing gangliosides are analogues of the Neu5Ac-containing species while others have both Neu5Gc and Neu5Ac in the same molecule, even linked to another through α(2-8) bonds. A survey of known ganglioside structures with Neu5Gc is given in Table 5.

In the vast majority of mammalian gangliosides so far discovered the sialic acids are α(2-3)-linked to galactose, or α(2-8)-linked to other sialic acid moieties, and were originally thought to be the only types of linkage in ganglioside structures. A more detailed analysis of minor ganglioside fractions, especially after the introduction of the ganglioside mapping technique (ANDO *et al.* 1976, MOMOI *et al.* 1976, NAGAI and IWAMORI 1980), revealed the presence not only of α(2-6) linkages, but confirmed a general occurrence of "unusual" structures with fucose, additional GalNAc and new "core sequences". A selection of such structures is given in Table 6.

Additional gangliosides having large, branched molecules have been isolated from bovine erythrocytes. These contain 90% of total sialic acid as Neu5Gc (MILLER-PODRAZA 1979). Human erythrocyte glycolipids containing sialic acid with I and i blood group activity have been found. The I and i activity could be increased by sialidase digestion (FEIZI *et al.* 1978).

The distribution of gangliosides in mammalian tissues has received a great deal of attention and has been extensively reviewed, including the tissue-specific occurrence of ganglioside patterns within one animal and between different species (SWEELEY and SIDDIQUI 1977, IWAMORI and NAGAI 1978a, b, LEDEEN 1978, 1979, YAMAKAWA and NAGAI 1978, NAGAI and IWAMORI 1980, SVENNERHOLM *et al.* 1980). A general division may be made as follows: the majority of mammalian brain gangliosides are derived from the $GgOse_4Cer$ structure, while the extraneural tissues contain gangliosides being related to LacCer and including the $LcOse_4Cer$, $LcnOse_4Cer$ and $LcnOse_6Cer$ structures (LEDEEN 1978, 1979, WIEGANDT 1980).

XI. The Nature of Sialic Acids

1. Introduction

The preceding parts of this chapter have presented the different forms of sialic acid under the aspects of phylogenetic and ontogenetic occurrence and as components of complex carbohydrate families. This has resulted in a somewhat fragmentary catalogue of sialic acids. The following section is, therefore, intended to present a survey of the occurrence with comments and trends relating to the type of sialic acid and the structure of sialyloligosaccharides in complex carbohydrates. A final correlation of these observations with evolution follows in section XII.

2. Free Sialic Acids

The majority of sialic acids have been detected in glycosidically linked form, but there are tissues and secretions containing relatively constant, although rather variable levels of free sialic acids. Free sialic acids are generally present in low concentrations in tissues. Significant amounts of free sialic acid were found in trout eggs (WARREN 1960), and this varied with the stage of development of these fishes (RAHMANN and BREER 1976). The seminal vesicle secretion of the Chinese hamster contains small amounts of free sialic acid (FOUQUET 1971), and the

Golden hamster has very high levels (FOUQUET 1972). Neu5Ac and Neu5Gc were found in porcine gastric extracts (ATTERFELT *et al.* 1958). Free sialic acid in mammalian brain tissue amounted to less than 3% of the total (DI BENEDETTA *et al.* 1969, MARGOLIS and MARGOLIS 1979). Human cerebrospinal fluid also contains free sialic acid (JAKOBY and WARREN 1961). Concentrations in normal human urine and serum are in the range of 1-3 μM and in saliva about 25 μM (HAVERKAMP *et al.* 1977). Tissues containing acylneuraminate pyruvate-lyase activity should contain low levels of free sialic acids (*e.g.* liver, kidney and intestinal tract), while in submandibular glands from cow, pig, and horse, where no lyase activity occurs, significant levels of free sialic acids are found (*e.g.* 50 μM in bovine submandibular glands).

Free sialic acids found in human serum, urine and saliva include O-acyl derivatives ($Neu5,9Ac_2$ and Neu5Ac9Lt) in addition to Neu5Ac (HAVERKAMP *et al.* 1976). Free submandibular gland sialic acids of bovine origin contain a range of mono- and di-O-acetylated derivatives of both Neu5Ac and Neu5Gc (BUSCHER *et al.* 1974, SCHAUER *et al.* 1974, CORFIELD *et al.* 1976).

The sialic acid concentrations in the urine, serum and saliva of a sialuria patient are elevated up to 10,000 times of those in normal individuals corresponding to a daily excretion rate of several grammes. The concentration of Neu5Ac in urine was found to be 45 mM (KAMERLING *et al.* 1979). These studies also identified Neu5Ac2en as a natural product for the first time, detected as a component of serum, urine (KAMERLING *et al.* 1975) and saliva (HAVERKAMP *et al.* 1976). The latter study confirmed that Neu5Ac2en is also a component of normal urine and serum, and in saliva may reach levels high enough to cause inhibition of bacterial sialidase activity (HAVERKAMP *et al.* 1976). Experiments under physiological conditions of pH and temperature showed non-enzymic formation of Neu5Ac2en from CMP-Neu5Ac (unpublished).

Increased levels of sialic acid in human urine have been reported in Salla disease, 165-350 μmole Neu5Ac being excreted per day, thus corresponding to a 10-fold increase over normal levels (RENLUND *et al.* 1979).

3. Glycosidically Linked Sialic Acids

a) The Type of Sialic Acid

The current knowledge on sialic acid derivatives in nature is presented in Table 1. Two parent molecules, differing in their N-acylation, Neu5Ac and Neu5Gc, are found. In addition, substitution of the hydroxyl groups at C-4, C-7, C-8 and C-9 by acetyl, glycolyl, lactyl, methyl, sulphate and phosphate moieties occurs and the number of possible derivatives becomes very large. This is important because of its potential for informational representation, as is discussed in section XI.3.b). Evolutionary aspects are discussed in section XII.

b) Sialyloligosaccharide Structural Considerations

α) *Terminal Linear Position*

The linkage of single sialyl units to oligosaccharide chains (see Table 2) involves α-glycosidic bonds between the C-2 anomeric hydroxyl group of sialic acid and the C-3, C-4 or C-6 hydroxyl groups of the penultimate non-sialic acid monosaccharide moiety. These linkages may involve Gal, GlcNAc, GalNAc and in some unique gangliosides Glc. The most common linkages found are α(2-3) to Gal and α(2-6) to Gal or GalNAc. The presence of α(2-4) linkages to Gal and GlcNAc has now been established (see SLOMIANY and SLOMIANY 1978, KOBATA 1979), but these linkages are not widely distributed as far as is known at present.

A further subdivision of structures can be made according to the type of oligosaccharide chain involved. Thus, sialic acids may be found in the N-glycosidically linked oligosaccharides of numerous glycoproteins of the "complex" type. This group includes α(2-3) and (2-6) bonds to Gal in the Galβ(1-4)GlcNAc- unit, and single oligosaccharides with up to four "antennae" have been described (MONTREUIL 1975, 1980). The linkages in these units may involve only α(2-3) or (2-6) bonds, or a combination of both (MONTREUIL 1975, 1980). Linkage to O-glycosidically bound oligosaccharides has been found largely based on the Galβ(1-3)GalNAc-protein unit with α(2-3) linkage to Gal and α(2-6) to GalNAc residues. The α(2-6) linkages is an example of a "side chain" terminal sialic acid (see below). A comparison of the large number of data reported for both N- and O-glycosidically linked units has been presented by STRECKER and MONTREUIL (1979) and MONTREUIL (1980).

The ganglioside structures known (see section X.5.d) derive from the ganglio-, lacto- or lactose series and contain almost exclusively α(2-3) sialic acid linkages to galactose, in addition to oligosialyl side chains (as described below). The presence of α(2-6) linkages is less common. Comparative structural studies compiled by RAUVALA and FINNE (1979) show a widespread similarity between glycoproteins and glycolipids in oligosaccharide structure apart from the "core" or "carrier" unit. The significance of this observation for biosynthesis is discussed in chapter I.

β) *Terminal "Side-Chain Position"*

In some sialyloligosaccharide chains sialic acids are found in terminal, but "side chain" or branch positions. The two most common examples are the α(2-6)-linked sialic acid bound to GalNAc in the disaccharide Galβ(1-3)GalNAc-protein, and the sialic acid unit in α(2-3) linkages to the internal Gal residue of II^{3}Neu5AcGgOse$_4$Cer,

Galβ(1-3)GalNAcβ(1-4)Galβ(1-4)GlcCer,
| α(2-3)
Neu5Ac

and related gangliosides.

Linkage of sialic acid to GlcNAc via α(2-6) bonds has been found in milk oligosaccharides of the lacto-N-tetraose (LcOse$_4$) type,

Galβ(1-3)GlcNAcβ(1-3)Galβ(1-4)Glc
| α(2-6)
Neu5Ac

A similar structure has been reported for the N-glycosidic oligosaccharide units in bovine cold-insoluble globulin (KOBATA 1979)

Neu5Ac
|α(2-6)
Neu5Acα(2-4)Galβ(1-3)GlcNAcβ(1-2)Manα(1-6)
\
Manβ(1-4)GlcNAc—Asn.
/
Neu5Acα(2-4)Galβ(1-3)GlcNAcβ(1-2)Manα(1-3)
|α(2-6)
Neu5Ac

Other examples of "side chain" sialic acids have been described in oligosaccharides isolated from mucus glycoproteins. Investigation of rat sublingual gland and colon mucus glycoproteins revealed a range of oligosaccharides in each case, showing "side chain" sialic acid attached to Gal and GalNAc (SLOMIANY and SLOMIANY 1978, SLOMIANY *et al.* 1980) as follows: Rat sublingual gland glycoprotein,

protein
|
Neu5Acα(2-4)GlcNAcβ(1-3)Galβ(1-4)GlcNAcβ(1-3)Galβ(1-4)GlcNAcβ(1-3)GalNAc
|α(2-6) |α(2-6) |α(2-6)
Neu5Ac Neu5Ac Neu5Ac

Rat colon glycoprotein,

GalNAcα(1-3)Galβ(1-3/4)GlcNAcβ(1
\
3/6)Galβ(1-4)GlcNAcβ(1-3)GalNAc-protein
/ |α(2-6)
GalNAcα(1-3)Galβ(1-4)GlcNAcβ(1 Neu5Ac
|α(2-6)
Neu5Ac

γ) *Internal Sialic Acids in Oligosialylglycoconjugates and Sialylpolysaccharides*

Polysialyl and oligosialyl structures occur in bacterial polysaccharides, glycoproteins and gangliosides. In these oligosaccharides sialic acid represents an internal monosaccharide, linked either to other sialyl residues or to a core oligosaccharide. The best known examples are gangliosides, where Neu5Acα(2-8)Neu5Ac-units are common in many phyla. Even larger sequences appear to be associated with the "ganglio" series, *e.g.* $IV^3Neu5AcII^3(Neu5Ac)_3GgOse_4Cer$ from fish brain, and evidence for the occurrence of hexa- and septasialogangliosides was obtained by RÖSNER (1981).

Sialic Acids

N- and O-acyl derivatives

↓

Type of Linkage

α(2-3) α(2-4) α(2-6) α(2-8) α(2-9)

terminal, linear or side chain sialic acids — internal sialic acids — oligo- and polysialyl groups

↓

Oligosaccharide Chain

glycoprotein — ganglioside — polysaccharide

N-glycosidic: number of antennae

O-glycosidic: Galβ(1-3)GalNAc— ; GlcNAcβ(1-6)GalNAc—

polysaccharide: homopolymer, heteropolymer

↓

Macromolecular Structure

number of chains — density of chains — three-dimensional arrangement

↓

Cellular Topography

intra-membrane — inter-membrane — cytosol-membrane — cytosol — extracellular

Fig. 2. Levels of complexity in the structure of complex carbohydrates containing sialic acids.

The α(2-8) linkage has also been identified in glycoproteins from brain and submandibular glands (FINNE *et al.* 1977, SLOMIANY *et al.* 1978) and polysialyloligosaccharide chains with (2-8) sialyl linkages found in fish eggs (INOUE and IWASAKA 1980).

Sialic acid-containing polymers found in bacterial surface antigens show characteristic structures not found in other phyla. The best known example is the homopolymer of α(2-8)-linked sialic acid, colominic acid, occurring in some strains of *E. coli* (section III). Structural studies with meningococcal serotypes introduced new units containing α(2-9) linkages (Table 3).

δ) *Internal Sialic Acids in the Echinodermata*

A number of unique carbohydrates containing sialic acids have so far been found only in the *Echinodermata*. These oligosaccharides include non-terminal sialic acids, which are not in polymer form as found in bacteria and vertebrates, and are constituents of gangliosides (see Table 4) typified by the sequence Neu5Acα(2-6)Glcβ(1-8)Neu5Acα(2-6)GlcCer. Such gangliosides occur in the gonads of the sea urchin *Strongylocentrotus intermedius* (KOCHETKOV *et al.* 1973). Not only is the sialic acid internal, it is also found in α(2-6) linkage, uncommon for gangliosides. Other examples are noted in Table 4.

Another structure so far found only in the *Echinodermata* is the disaccharide Fucα(1-4)Neu5Gcα(2-) from sea urchin eggs (HOTTA 1977) and the cuverian tubules of the sea cucumber *Holothuria forskali* (unpublished); again the sialic acid residue is internal.

It is clear that an enormous variability exists in the nature of sialic acid in glycosidic linkage. This includes the nature of the linkage, the formation of oligosialyl units, the type of oligosaccharide chain and the "carrier" (protein, lipid) molecule. A further flexibility is imposed on this by variation of the type of sialic acid. Thus, the number of possible combinations is vast and is further potentiated on a molecular level in three dimensions (MONTREUIL 1975, 1980) in addition to considerations of the number of sialyloligosaccharide chains per molecule and the density of these chains on the molecule and in *e.g.* membrane structures. A flow diagram of this complexity is given in Fig. 2 and includes a final level of complexity summarizing membrane and cytosol interactions. The number of possible structures represents a powerful potential for expression of information, and this is reflected in the manifold functional roles ascribed to sialic acids in complex carbohydrates (see chapter J).

XII. Evolution of the Sialic Acids

The sialic acids are a relatively young innovation in evolution, appearing in the *Echinodermata* about 650 million years ago, during the precambrian era (Fig. 1). A number of organisms, such as bacteria, protozoa and viruses, have been found to contain sialic acids which belong to phyla considerably older than this date, but it is likely that evolution has functioned to adapt these organisms in the face of "new" life forms. In each of these cases parasitic or symbiotic relationships exist with hosts which possess sialic acids. A possible explanation for the occurrence of sialic acid in a primitive acoel turbellarian living on plants, has been put forward by SEGLER *et al.* (1978). These authors suggested that bacteria in the organism may contain sialic acid. This possibility awaits practical verification; and, if correct, is a situation which may well occur in other organisms. Adaptation of the microorganisms to the new environments may have included the development of the ability to synthesize sialic acids and incorporate them into macromolecules. This is most vividly demonstrated in viruses where their host's biosynthetic machinery is utilized to produce sialylglycoproteins in the viral membrane.

The first phylum possessing sialic acid in all classes is the *Echinodermata*, and it appears that the sialic acids were first introduced at this stage in evolution. These animals present a great array of neuraminic acid derivatives in glycoproteins and gangliosides. A number of "experiments" with sialic acid structures were made at this time. Molecules which have not been found later in evolution exist here, these include 8-O-methylated sialic acids and internal sialic acids in glycoproteins and gangliosides (see Table 4). In many cases the predominant sialic acid is Neu5Gc. This is important, as studies with higher animals have shown that Neu5Gc is formed enzymically from Neu5Ac (chapter I), and thus the gene for the monooxygenase is implicated at this stage in development and persists throughout evolution before being abandoned in the human species. Recent work also elucidated the presence of Neu9Ac5Gc (Kamerling *et al.* 1980), and this is again important from a biosynthetic point of view, as enzymic acetylation in the side chain also may have been introduced at this stage in evolution.

The reasons for the experimentation with different sialic acid structures in *Echinodermata* are not apparent, however, some speculation can be made. The introduction of a new monosaccharide with a negative charge may increase the versatility of complex carbohydrates, as only one step is required to remove it (*i.e.* sialidase action, chapter I). However, the biological function may demand a longer half-life of the sialic acid moiety in the molecule, and thus a resistance to sialidase action would be required. This may be achieved by substituting the C-4 hydroxyl group of sialic acid. In the *Echinodermata*, there is no evidence for a 4-O-acetyl-sialic acid species found in higher organisms, instead sialic acids occur in "internal" positions. Some of these links are to the sialic acid side chain, *e.g.* Neu5Acα(2-6)Glcβ(1-8)Neu5Acα(2-6)GlcCer, a sea urchin ganglioside (Kochetkov *et al.* 1973), while others are linked via the C-4 hydroxyl, *e.g.* Ara(1-6)Galβ(1-4)Neu5Gcα(2-3)Galβ(1-4)GlcCer, a starfish ganglioside (Sugita 1979 a, b) and Fucα(1-4)Neu5Gcα(2-) in starfish (Hotta 1977) and sea cucumber (unpublished) glycoproteins. Such residues at C-4 of sialic acid are resistant to sialidase action, while O-acylation or glycosidic linkage to the side chain C-7-C-9 or substitution of N-acetyl by N-glycolyl may result in reduced cleavage rates. Also of interest in this respect is the lack of cleavage of Neu5Gc8S and Neu5Ac8S from sea urchin gonad and bovine gastric ganglioside, respectively, by bacterial sialidase (Kochetkov *et al.* 1976, Slomiany *et al.* 1981 a, b).

Evolution from the echinoderms towards the vertebrates and man shows a diminution in the types of O-substituted sialic acids, *e.g.* disappearance of Neu5Gc8Me, although some derivatives, *e.g.* Neu5Gc8S, are found both in the echinoderms and mammals. Neu5Gc constitutes a high proportion of total echinoderm sialic acids. This trend persists into the *Mammalia* although some species exhibit low levels or the absence of this sialic acid and in man Neu5Gc has never been conclusively identified. Parallel to this trend is the evolution of positional isomers of O-acetyl-sialic acids (*e.g.* Neu5,9Ac$_2$, Neu4,5Ac$_2$, Neu5,8,9Ac$_3$, Neu5,7,8,9Ac$_4$, etc.) reaching a peak in the bovine submandibular gland. The O-acetylation of the C-7-C-9 side chain can be traced back to the *Echinodermata* and appears continually in fishes, birds and throughout the *Mammalia* including man. The presence of Neu4,5Ac$_2$ has been confirmed by g.l.c./m.s. in the primitive marsupial *Echidna* and both Neu4,5Ac$_2$ and Neu4Ac5Gc in horse and donkey. Reports of 4-O-acetyl-sialic acids in avian and amphibian species await corroboration. Sialic acids with 9-O-lactyl groups have only been detected in cow, horse and man.

Information on the acranians (hemichordates and cephalochordates) is very limited and, although Neu5Ac was demonstrated, different sialic acid types have not been identified. The tunicates (urochordates) have been more widely studied and no sialic acid was found. This is unexpected as the tunicates are considered to have developed later in evolution. RÖSNER and RAHMANN (1981) have suggested that a selection against the genome coding for sialic acid may have occurred.

Investigation of protostomates, believed to be older than the echinoderms, *e.g.* *Brachiopoda* and *Bryozoa* also gave negative results for sialic acid. In order to confirm the *Echinodermata* as the initial sialic acid-containing phylum, further comparative studies of phyla believed to have developed at this stage in evolution must be carried out.

Bibliography

AARONSON, S., LESSIE, T., 1960: Nature **186**, 719.

ACTON, R. T., WEINHEIMER, P. F., NIEDERMEIER, W., 1973: Comp. Biochem. Physiol. **44 B**, 185—189.

ALAIS, C., JOLLÈS, P., 1961: Biochim. Biophys. Acta **51**, 315—322.

ALLEN, H. J., AULT, C., WINZLER, R. J., DANIELLI, J. F., 1974: J. Cell Biol. **60**, 26—38.

— WINZLER, R. J., DANIELLI, J. F., 1976: Exp. Cell Res. **100**, 408—411.

ALLISON, A. C. (ed.), 1974: Structure and Function of Plasma Proteins, Vol. 1. New York: Plenum Press.

— (ed.), 1976: Structure and Function of Plasma Proteins, Vol. 2. New York: Plenum Press.

ANDO, S., CHANG, N.-C., YU, R. K., 1978: Anal. Biochem. **89**, 437—450.

— ISOBE, M., NAGAI, Y., 1976: Biochim. Biophys. Acta **424**, 98—105.

ANDREWS, A. T. de B., HERRING, G. M., KENT, P. W., 1967: Biochem. J. **104**, 705—715.

ASHTON, B. A., TRIFFITT, J. T., HERRING, G. M., 1974: Eur. J. Biochem. **45**, 525—533.

ASHWELL, G., MORELL, A., 1974: Adv. Enzymol. **41**, 99—128.

ATTERFELT, P., BLOHME, I., NORBY, A., SVENNERHOLM, L., 1958: Acta Chem. Scand. **12**, 359—360.

AVROVA, N. F., 1968: Zh. Evol. Biochim. Fiziol. **4**, 128—135.

— 1971: J. Neurochem. **18**, 667—674.

— 1980: In: Structure and Function of Gangliosides (SVENNERHOLM, L., MANDEL, P., DREYFUS, H., URBAN, P. F., eds.), pp. 177—183. New York: Plenum Press.

BAHL, O. P., SHAH, R. H., 1977: In: The Glycoconjugates (HOROWITZ, M. I., PIGMAN, W., eds.), Vol. 1, pp. 385—422. New York: Academic Press.

BAIG, M. M., AYOUB, E. M., 1976: Biochemistry **15**, 2585—2590.

BARDOS, P., MUH, J. P., LUTHIER, B., DEVULDER, B., TACQUET, A., 1976: Comp. Biochem. Physiol. **53 B**, 49—56.

BARRY, G. T., 1958: J. Exptl. Med. **107**, 507—521.

— 1965: Bull. Soc. Chim. Biol. **47**, 529—533.

— GOEBEL, W. F., 1957: Nature **179**, 206.

— ABBOT, V., TSAI, T., 1962: J. Gen. Microbiol. **29**, 335—352.

— HAMM, J. D., GRAHAM, M. G., 1963: Nature **200**, 806—807.

BAYNES, J. W., WOLD, F., 1976: J. Biol. Chem. **251**, 6016—6024.

BEAU, J.-M., SCHAUER, R., 1980: Eur. J. Biochem. **106**, 531—540.

BEESLEY, R. C., FORTE, J. G., 1973: Biochim. Biophys. Acta **307**, 372—385.

— — 1974: Biochim. Biophys. Acta **356**, 144—155.

BEZKOROVAINY, A., 1963: Biochemistry **2**, 10—16.
— GROHLICH, D., 1969: Biochem. J. **115**, 817—822.
BHATTACHARJEE, A. K., JENNINGS, H. J., KENNEY, C. P., MARTIN, A., SMITH, I. C. P., 1975: J. Biol. Chem. **250**, 1926—1932.
— — — — — 1976: Can. J. Biochem. **54**, 1—8.
BHATTACHARYYA, S. N., ROSE, M. C., LYNN, M. G., MACLEOD, C., ALBERTS, M., LYNN, W. S., 1976: Am. Rev. Resp. Dis. **114**, 843—850.
BLIX, G., 1936: Hoppe-Seyler's Z. Physiol. Chem. **240**, 43—54.
— JEANLOZ, R. W., 1969: In: The Amino Sugars (JEANLOZ, R. W., BALAZS, E. A., eds.), pp. 213—265. New York: Academic Press.
— LINDBERG, E., 1960: Acta Chem. Scand. **14**, 1809—1814.
— — ODIN, L., WERNER, I., 1955: Nature **175**, 340—341.
— — — — 1956: Acta Soc. Med. Upsalien. **61**, 1—25.
BOAT, T. F., CHENG, P. W., IYER, R. N., CARLSON, D. M., POLONY, I., 1976: Arch. Biochem. Biophys. **177**, 95—104.
BOGOCH, S., 1958: Arch. Neurol. Psychiat. **80**, 221—227.
BOLOGNANI, L., BOLOGNANI, A. M., LUSIGNANI, R., ZONTA, L., 1966: Experientia **22**, 601—603.
BONANNI, F., DE LUCA, L., 1974: Biochim. Biophys. Acta **343**, 632—637.
BOURRILLON, R., 1972: In: Glycoproteins, Their Composition, Structure and Function (GOTTSCHALK, A., ed.), 2nd ed., part B, pp. 909—925. Amsterdam: Elsevier.
BÖHM, P., BAUMEISTER, L., 1956: Hoppe-Seyler's Z. Physiol. Chem. **305**, 42—50.
BRITISH MEDICAL BULLETIN, 1978: Vol. 34.
BUDDECKE, E., 1972: In: Glycoproteins: Their Composition, Structure and Function (GOTTSCHALK, A., ed.), 2nd ed., part A, pp. 535—564. Amsterdam: Elsevier.
BUSCHER, H.-P., CASALS-STENZEL, J., SCHAUER, R., 1974: Eur. J. Biochem. **50**, 71—82.
— — — MESTRES-VENTURA, P., 1977: Eur. J. Biochem. **77**, 297—310.
CABEZAS, J. A., 1968: An Real. Acad. Farm. **34**, 155—172.
— 1973: Rev. Esp. Fisiol. **29**, 307—322.
— FEO, F., 1969: Rev. Esp. Fisiol. **25**, 153—156.
— FROIS, M. D., 1966: Rev. Esp. Fisiol. **22**, 147—152.
— RAMOS, M., 1972: Carbohyd. Res. **24**, 486—488.
— SECO, A., 1975: Comp. Biochem. Physiol. **51 B**, 243—245.
— PORTO, J. V., FROIS, M. D., MARINO, C., ARZUA, J., 1964: Biochim. Biophys. Acta **83**, 318—325.
CABEZAS, M., CABEZAS, J. A., 1973: Rev. Esp. Fisiol. **29**, 323—328.
CAPUTTO, R., MACCIONI, H. J. F., ARCE, A., 1974: Mol. Cell Biochem. **4**, 97—106.
CAREY, D. J., HIRSCHBERG, C. B., 1979: Biochemistry **18**, 2086—2092.
CARLSSON, H. E., SUNBLAD, G., HAMMARSTRÖM, S., PERLMANN, P., GUSTAFSSON, B. E.: 1978: Arch. Biochem. Biophys. **187**, 366—375.
CARTER, W. G., HAKOMORI, S. I., 1979: Biochemistry **18**, 730—738.
CARUBELLI, R., RYAN, L. C., TRUCCO, R. E., CAPUTTO, R., 1961: J. Biol. Chem. **236**, 2381—2388.
CHAO, C.-C. W., VERGNES, J.-P., FREEMAN, I. L., BROWN, S. I., 1980: Exp. Eye Res. **30**, 411—425.
CHAKRIN, L. W., BAKER, A. P., SPICER, S. S., WARDELL, J. R., jr., DESANCTIS, N., DRIES, C., 1972: Am. Rev. Resp. Dis. **105**, 368—381.
CHANNON, M., HENNEBERRY, G. O., ANASTASSIADIS, P. A., 1972: Poult. Sci. **51**, 1740—1743.
CHATTERJEE, S., RAY, P. K., 1975: Ind. J. Exp. Biol. **13**, 394—395.
CIBA Foundation Symposium 54, 1978. Respiratory Tract Mucus. Amsterdam: Elsevier.

CLAMP, J., ALLEN, A., GIBBONS, R. A., ROBERTS, G. P., 1978: Br. Med. Bull. **34**, 25—41.

COMPANS, R. W., KEMP, M. C., 1978: In: Current Topics in Membranes and Transport (BRONNER, F., KLEINZELLER, A., JULIANO, R. L., ROTHSTEIN, A., eds.), Vol. 11, pp. 233—277. New York: Academic Press.

COOK, G. M. W., STODDART, R. W. (eds.), 1973: Surface Carbohydrates of the Eukaryotic Cell. London: Academic Press.

CORFIELD, A. P., FERREIRA DO AMARAL, C., WEMBER, M., SCHAUER, R., 1976: Eur. J. Biochem. **68**, 597—610.

DE LUCA, S., LOHMANDER, S., NILSSON, B., HASCALL, V. C., CAPLAN, A. I., 1980: J. Biol. Chem. **255**, 6077—6083.

DEWITT, C. W., ROWE, J. A., 1959: Nature **184**, 381—382.

— — 1961: J. Bacteriol. **82**, 838—848.

— ZELL, E. A., 1961: J. Bacteriol. **82**, 849—856.

DI BENEDETTA, C., BRUNNGRABER, E. G., WHITNEY, G., BROWN, B. D., ARO, A., 1969: Arch. Biochem. Biophys. **131**, 404—413.

DIETZSCHOLD, B., 1977: J. Virol. **23**, 286—293.

DOOLITTLE, R. F., COTTRELL, B. A., 1974: Biochem. Biophys. Res. Commun. **60**, 1090—1096.

DORLAND, L., VAN HALBEEK, H., HAVERKAMP, J., VELDINK, G. A., VLIEGENTHART, J. F. G., FOURNET, B., MONTREUIL, J., AMINOFF, D., 1979: In: Glycoconjugates (SCHAUER, R., BOER, P., BUDDECKE, E., KRAMER, M. F., VLIEGENTHART, J. F. G., WIEGANDT, H., eds.), pp. 29—30. Stuttgart: G. Thieme.

DREYFUS, H., URBAN, P. F., EDEL-HARTH, S., MANDEL, P., 1975: J. Neurochem. **25**, 245—250.

— — HARTH, S., PRETI, A., MANDEL, P., 1976: In: Ganglioside Function (PORCELLATI, G., CECCARELLI, B., TETTAMANTI, G., eds.), pp. 163—188. New York: Plenum Press.

— HARTH, S., YUSUFI, A. N. K., URBAN, P. F., MANDEL, P., 1980: In: Structure and Function of Gangliosides (SVENNERHOLM, L., MANDEL, P., DREYFUS, H., URBAN, P. F., eds.), pp. 227—237. New York: Plenum Press.

DŻUŁYŃSKA, J., KRAJEWSKA, J., STARZYŃSKY, W., 1966: Bull. Acad. Pol. Sci. **14**, 527—532.

— WALKOWIAK, A., SZUBISZEWSKI, B., 1968: Bull. Acad. Pol. Sci. **16**, 501—506.

— POTEMKOWSKA, E. A., WALKOWIAK, H., FABIJAŃSKA, I., 1969: Bull. Acad. Pol. Sci. **17**, 523—529.

EGAN, W., LIU, T.-Y., DOROW, D., COHEN, J. S., ROBBINS, J. D., GOTSCHLICH, E. C., ROBBINS, J. B., 1977: Biochemistry **16**, 3687—3692.

ELAM, J. S., 1979: In: Complex Carbohydrates of Nervous Tissue (MARGOLIS, R. U., MARGOLIS, R. K., eds.), pp. 235—267. New York: Plenum Press.

ELDREDGE, N. F., READ, G., CUTTING, W., 1963: Med. Exp. **8**, 265—277.

ELLIS, D. B., STAHL, G. H., 1973: Biochem. J. **136**, 837—844.

ELSTEIN, M. I., PARKE, D. V. (eds.), 1977: Mucus in Health and Disease. London: Plenum Press.

ENDO, Y., YAMASHITA, K., HAN, Y. N., IWANAGA, S., KOBATA, A., 1977: J. Biochem. **82**, 545—550.

— — TACHIBANA, Y., TOJO, S., KOBATA, A., 1979: J. Biochem. **85**, 669—679.

EYLAR, E. H., DOOLITTLE, R. F., MADOFF, M. A., 1962: Nature **193**, 1183—1184.

FAILLARD, H., CABEZAS, J. A., 1963: Hoppe-Seyler's Z. Physiol. Chem. **333**, 266—271.

— SCHAUER, R., 1972: In: Glycoproteins, Their Composition, Structure and Function (GOTTSCHALK, A., ed.), 2nd ed., part B, pp. 1246—1267. Amsterdam: Elsevier.

— PRIBILLA, W., POSTH, H. E., 1962: Hoppe-Seyler's Z. Physiol. Chem. **327**, 100—108.

FARRAR, G. H., HARRISON, R., 1978: Biochem. J. **171**, 549—557.

FAROOQUI, A. A., 1976: Biochimie **58**, 759—761.

FEENEY, R. E., ANDERSON, J. S., AZARI, P. R., BENNETT, N., RHODES, M. B., 1960: J. Biol. Chem. **235**, 2307—2311.
FEIZI, T., CHILDS, R. A., HAKOMORI, S.-I., POWELL, M. E., 1978: Biochem. J. **173**, 245—254.
FELDHOFF, P. A., BHAVANANDAN, V. P., DAVIDSON, E. A., 1979: Biochemistry **18**, 2430—2436.
FILIPE, M. I., BRANFOOT, A. C., 1976: Curr. Top. Path. **63**, 143—178.
FINNE, J., 1975: Biochim. Biophys. Acta **412**, 317—325.
— RAUVALA, H., 1977: Carbohyd. Res. **58**, 57—64.
— KRUSIUS, T., RAUVALA, H., HEMMINKI, K., 1977: Eur. J. Biochem. **77**, 319—323.
— — MARGOLIS, R. K., MARGOLIS, R. U., 1979: J. Biol. Chem. **254**, 10295—10300.
FLETCHER, A. P., 1972: In: Glycoproteins, Their Composition, Structure and Function (GOTTSCHALK, A., ed.), 2nd ed., part B, pp. 892—908. Amsterdam: Elsevier.
FOUQUET, J. P., 1971: Comp. Biochem. Physiol. **40 B**, 305—308.
— 1972: J. Reprod. Fertil. **28**, 273—275.
FOURNET, B., FIAT, A. M., ALAIS, C., JOLLÈS, P., 1979: Biochim. Biophys. Acta **576**, 339—346.
FRANCIS, G., THOMAS, J., 1975: Biochem. J. **145**, 299—304.
FRASER, D., CLAMP, J. R., 1975: Clin. Chim. Acta **59**, 301—307.
GALLAGHER, J. T., KENT, P. W., 1975: Biochem. J. **148**, 187—196.
— — PASSATORE, M., PHIPPS, R. J., RICHARDSON, P. S., 1975: Proc. R. Soc. London B, **192**, 49—76.
GHIDONI, R., SONNINO, S., TETTAMANTI, G., WIEGANDT, H., ZAMBOTTI, V., 1976: J. Neurochem. **27**, 511—515.
— — — BAUMANN, N., REUTER, G., SCHAUER, R., 1980: J. Biol. Chem. **255**, 6990—6995.
GIBBONS, R. A., DIXON, S. N., SELLWOOD, R., 1976: Eur. J. Biochem. **66**, 243—250.
GIELEN, W., 1968: Z. Naturforsch. **23 B**, 1598—1601.
GLICK, M. C., FLOWERS, H., 1978: In: The Glycoconjugates (HOROWITZ, M. I., PIGMAN, W., eds.), Vol. 2, pp. 337—384. New York: Academic Press.
GLODE, M. P., LEWIN, E. B., SUTTON, A., LE, C. T., GOTSCHLICH, E. C., ROBBINS, J. B., 1979: J. Infect. Dis. **139**, 52—59.
GOLDITCH, M., BARNES, P. R., SCHNEIR, M., 1974: Arch. Oral Biol. **19**, 65—70.
GOSO, K., HOTTA, K., 1977: Biochem. J. **163**, 169—172.
GOTTSCHALK, A. (ed.), 1960: The Chemistry and Biology of Sialic Acids and Related Substances. Cambridge: Cambridge University Press.
— (ed.), 1966: Glycoproteins, Their Composition, Structure and Function. Amsterdam: Elsevier.
— 1969: Nature **222**, 452—454.
— (ed.), 1972: Glycoproteins, Their Composition, Structure and Function, 2nd ed. Amsterdam: Elsevier.
GRAHAM, E. R. B., 1972: In: Glycoproteins, Their Composition, Structure and Function (GOTTSCHALK, A., ed.), 2nd ed., part A, pp. 717—731. Amsterdam: Elsevier.
GRIMMONPREZ, L., MONTREUIL, J., 1968: Bull. Soc. Chem. Biol. **50**, 843—855.
HAKOMORI, S.-I., SAITO, T., 1969: Biochemistry **8**, 5082—5088.
HANDA, S., YAMAKAWA, T., 1964: Jpn. J. Exp. Med. **34**, 293—299.
HARMON, R. E. (ed.), 1978: Cell Surface Carbohydrate Chemistry. New York: Academic Press.
HARTREE, E. F., 1962: Nature **196**, 483—484.
HATCHER, V. B., SCHWARZMANN, G. O. H., JEANLOZ, R. W., MCARTHUR, J. W., 1977: Biochemistry **16**, 1518—1524.
HAVERKAMP, J., SCHAUER, R., WEMBER, M., FARRIAUX, J.-P., KAMERLING, J. P., VERSLUIS, C., VLIEGENTHART, J. F. G., 1976: Hoppe-Seyler's Z. Physiol. Chem. **357**, 1699—1705.

HAVERKAMP, J., VEH, R. W., SANDER, M., SCHAUER, R., KAMERLING, J. P., VLIEGENTHART, J. F. G., 1977: Hopple-Seyler's Z. Physiol. Chem. **358**, 1609—1612.

— SPOORMAKER, T., DORLAND, L., VLIEGENTHART, J. F. G., SCHAUER, R., 1979: J. Am. Chem. Soc. **101**, 4851—4853.

HAYES, M. L., CASTELLINO, F. J., 1979a: J. Biol. Chem. **254**, 8768—8771.

— — 1979b: J. Biol. Chem. **254**, 8772—8776.

— — 1979c: J. Biol. Chem. **254**, 8777—8780.

HENTSCHEL, H., MÜLLER, M., 1979: Comp. Biochem. Physiol. **64A**, 585—588.

HERP, A., WU, A. M., MOSCHERA, J., 1979: Mol. Cell Biochem. **23**, 27—44.

HERRING, G. M., 1976: Biochem. J. **159**, 749—755.

HERRMANN, W. P., UHLENBRUCK, G., 1972: Z. Klin. Chem. **10**, 363—366.

HERZBERG, M. C., LEVINE, M. J., ELLISON, S. A., TABAK, L. A., 1979: J. Biol. Chem. **254**, 1487—1494.

HILBIG, R., RAHMANN, H., 1980: J. Neurochem. **34**, 236—240.

— RÖSNER, H., RAHMANN, H., 1981: Comp. Biochem. Physiol. **68B**, 301—305.

HIRABAYASHI, Y., TAKI, T., MATSUMOTO, M., 1979: FEBS Lett. **100**, 253—257.

HIRSCHBERG, C. B., ROBBINS, P. W., 1974: Virology **61**, 602—608.

HOLBROOK, I. B., LEAVER, A. G., 1976: Archs. Oral Biol. **21**, 509—512.

HOLDEN, K. G., GRIGGS, L. J., 1977: In: The Glycoconjugates (HOROWITZ, M. I., PIGMAN, W., eds.), Vol. 1, pp. 215—237. New York: Academic Press.

HOROWITZ, M. I., 1977: In: The Glycoconjugates (HOROWITZ, M. I., PIGMAN, W., eds.), Vol. 1, pp. 189—213. New York: Academic Press.

— PIGMAN, W. (eds.), 1977: The Glycoconjugates, Vol. 1. New York: Academic Press.

— — (eds.), 1978: The Glycoconjugates, Vol. 2. New York: Academic Press.

HOSHI, M., NAGAI, Y., 1975: Biochim. Biophys. Acta **388**, 152—162.

HOTTA, K., 1977: Adv. Invertebrate Reprod. **1**, 130—144.

HOUDRET, N., LHERMITTE, M., DEGAND, P., ROUSSEL, P., 1975: Biochimie **57**, 603—608.

HUDSON, M. T., WELLERSON, R., jr., KUPFERBERG, A. B., 1965: J. Reprod. Fertil. **9**, 189—195.

HUGHES, R. C. (ed.), 1976: Membrane Glycoproteins. London: Butterworths.

HUNT, L. A., SUMMERS, D. F., 1976a: J. Virol. **20**, 637—645.

— — 1976b: J. Virol. **20**, 646—657.

HUTTUNEN, J. K., 1966: Ann. Med. Exp. Biol. Fenn **44**, Suppl. 12, 1—60.

— MIETTINEN, T. A., 1969: Anal. Biochem. **29**, 441—458.

INOUE, S., IWASAKI, M., 1980: Biochem. Biophys. Res. Commun. **93**, 162—165.

— MATSUMURA, G., 1979: Carbohyd. Res. **74**, 361—368.

IRWIN, L. N., CHEN, H., BARRACO, R. A., 1976: Develop. Biol. **49**, 29—39.

ISEMURA, M., ZAHN, R. K., SCHMID, K., 1973: Biochem. J. **131**, 509—521.

ISHIZUKA, I., KLOPPENBURG, M., WIEGANDT, H., 1970: Biochim. Biophys. Acta **210**, 299—305.

IUPAC-IUB Commission of Biochemical Nomenclature, 1977: Lipids **12**, 455—468.

IWAMORI, M., NAGAI, Y., 1978a: J. Biochem. **84**, 1609—1615.

— — 1978b: J. Biol. Chem. **253**, 8323—8331.

— — 1979: J. Neurochem. **32**, 767—777.

JAKOBY, R. K., WARREN, L., 1961: Neurology **11**, 232—238.

JASS, J. R., 1980: J. Clin. Pathol. **33**, 801—810.

— FILIPE, M. I., 1980: Histopathology **4**, 271—279.

JÄRNEFELT, J., FINNE, J., KRUSIUS, T., RAUVALA, H., 1978: Trends Biochem. Sci. **3**, 110—114.

JEANLOZ, R. W., CODINGTON, J. F., 1976: In: Biological Roles of Sialic Acid (ROSENBERG, A., SCHENGRUND, C.-L., eds.), pp. 201—238. New York: Plenum Press.

JENNINGS, H. J., BHATTACHARJEE, A. K., BUNDLE, D. R., KENNY, C. P., MARTIN, A., SMITH, I. C. P., 1977: J. Infect. Dis. Suppl. **136**, S78—S83.

Jutisz, M., De la Llosa, P., 1972: In: Glycoproteins, Their Composition, Structure and Function (Gottschalk, A., ed.), 2nd ed., part B, pp. 1019—1068. Amsterdam: Elsevier.
Kamerling, J. P., Vliegenthart, J. F. G., Schauer, R., Strecker, G., Montreuil, J., 1975: Eur. J. Biochem. **56**, 253—258.
— Strecker, G., Farriaux, J.-P., Dorland, L., Haverkamp, J., Vliegenthart, J. F. G., 1979: Biochim. Biophys. Acta **583**, 403—408.
— Schauer, R., Vliegenthart, J. F. G., Hotta, K., 1980: Hoppe-Seyler's Z. Physiol. Chem. **361**, 1511—1516.
— Dorland, L., Van Halbeek, H., Vliegenthart, J. F. G., Messer, M., Schauer, R., 1982a: Carbohyd. Res. **100**, 331—340.
— Makovitzky, J., Schauer, R., Vliegenthart, J. F. G., Wember, M., 1982b: Biochim. Biophys. Acta **714**, 351—355.
Kathan, R. H., Weeks, I. D., 1969: Arch. Biochem. Biophys. **134**, 572—576.
Kaufmann, S. H. E., Schauer, R., Hahn, H., 1981: Immunobiol. **160**, 184—195.
Kawasaki, T., Ashwell, G., 1977: J. Biol. Chem. **252**, 6536—6543.
Kędzierska, B., 1978: Eur. J. Biochem. **91**, 545—552.
Keegstra, K., Sefton, B., Burke, D., 1975: J. Virol. **16**, 613—620.
Keenan, T. W., Powell, K. M., Sasaki, M., Eigel, W. N., Franke, W. W., 1977: Cytobiol. **15**, 96—115.
Kent, P. W., 1973: Exp. Ann. Biochim. Méd. **32**, 97—120.
— 1978: In: CIBA Foundation Symposium 54, Respiratory Tract Mucus, pp. 155—170. Amsterdam: Elsevier.
— Draper, P., 1968: Biochem. J. **106**, 293—299.
Kessler, M. J., Reddy, M. S., Shah, R. H., Bahl, O. P., 1979a: J. Biol. Chem. **254**, 7901—7908.
— Mise, T., Ghai, R. D., Bahl, O. P., 1979b: J. Biol. Chem. **254**, 7909—7914.
Kleinberg, I., Ellison, S. A., Mandel, I. D., 1979: Proceedings Saliva and Dental Caries, Sp. Supp. Microbiology Abstracts, Information Retrieval Inc., New York.
Klenk, E., 1941: Hoppe-Seyler's Z. Physiol. **268**, 50—58.
— Faillard, H., 1954: Hoppe-Seyler's Z. Physiol. Chem. **298**, 230—238.
Kobata, A., 1977: In: The Glycoconjugates (Horowitz, M. I., Pigman, W., eds.), Vol. 1, pp. 423—440. New York: Academic Press.
— 1979: Cell Struct. Funct. **4**, 169—181.
Kochetkov, N. K., Zhukova, I. G., Smirnova, G. P., Bogdanovskaya, T. A., 1970: Dokl. Acad. Nauk USSR **191**, 358—362.
— — — Glukhoded, I. S., 1973: Biochim. Biophys. Acta **326**, 74—83.
— Smirnova, G. P., Chekareva, N. V., 1976: Biochim. Biophys. Acta **424**, 274—283.
Kornfeld, R., Kornfeld, S., 1976: Ann. Rev. Biochem. **45**, 217—237.
Kornfeld, S., Li, E., Tabas, I., 1978: J. Biol. Chem. **253**, 7771—7778.
Köttgen, E., Reutter, W., Gerok, W., 1976: Biochem. Biophys. Res. Commun. **72**, 61—66.
Krantz, M. J., Lee, Y. C., Hung, P. P., 1974: Nature **248**, 684—686.
Krenger, J., Sakoloff, N., Botelho, S. Y., 1976: Invest. Ophthalmol. **15**, 479—481.
Krusius, T., Finne, J., 1978: Eur. J. Biochem. **84**, 395—403.
— — Margolis, R. U., Margolis, R. K., 1978a: Biochemistry **17**, 3849—3854.
— — Rauvala, H., 1978b: Eur. J. Biochem. **92**, 289—300.
Kuhn, N. J., 1972: Biochem. J. **130**, 177—180.
Kuhn, R., 1959: Naturwissenschaften **46**, 43—50.
— Brossmer, R., 1956: Chem. Ber. **89**, 2013—2025.
— Ekong, D., 1963: Chem. Ber. **96**, 683—688.
— Gauhe, A., 1962: Chem. Ber. **95**, 513—517.

KUHN, R., GAUHE, A., 1965: Chem. Ber. **98**, 395—413.
LABAT-ROBERT, J., DECAENS, C., 1979: Path. Biol. **27**, 241—247.
LAMBLIN, G., LHERMITTE, M., DEGAND, P., ROUSSEL, P., 1979: Biochimie **61**, 23—43.
— — BOERSMA, A., ROUSSEL, P., REINHOLD, V., 1980: J. Biol. Chem. **255**, 4595—4598.
LAMBOTTE, R., UHLENBRUCK, G., 1966: Nature **212**, 290—291.
LANDA, C. A., MACCIONI, H. J. F., CAPUTTO, R., 1979: J. Neurochem. **33**, 825—838.
LEDEEN, R. W., 1978: J. Supramol. Struct. **8**, 1—17.
— 1979: In: Complex Carbohydrates of Nervous Tissue (MARGOLIS, R. U., MARGOLIS, R. K., eds.), pp. 1—23. New York: Plenum Press.
— COCHRAN, F. B., YU, R. K., SAMUELS, F. G., HALEY, J. E., 1980: In: Structure and Function of Gangliosides (SVENNERHOLM, L., MANDEL, P., DREYFUS, H., URBAN, P. F., eds.), pp. 167—176. New York: Plenum Press.
LEMOINE, A.-M., 1974: C. R. Soc. Biol. (Paris) **168**, 402—405.
— OLIVEREAU, M., 1973: Acta Zool. **54**, 223—228.
— — 1974: Acta Zool. **55**, 255—261.
LEMONNIER, M., BOURRILLON, R., 1975: Biochem. Biophys. Res. Commun. **64**, 226—231.
— — 1976: Carbohyd. Res. **51**, 99—106.
— FOURNET, B., BOURRILLON, R., 1975: C. R. Acad. Sci. **280 D**, 2705—2707.
— — — 1977: Biochem. Biophys. Res. Commun. **77**, 767—774.
— DERAPPE, C., BOURRILLON, R., 1978: Biomédicine **29**, 146—150.
LENNARZ, W. J. (ed.), 1980: The Biochemistry of Glycoproteins and Proteoglycans. New York: Plenum Press.
LIAO, T.-H., BLUMENFELD, O. O., PARK, S. S., 1979: Biochim. Biophys. Acta **577**, 442—453.
LICHT, P., PAPKOFF, H., 1974: Gen. Comp. Endocrinol. **23**, 415—420.
LIU, S., KALANT, N., 1974: Exp. Mol. Pathol. **21**, 52—62.
LIU, T.-Y., GOTSCHLICH, E. C., DUNNE, F. T., JONSSEN, E. K., 1971: J. Biol. Chem. **246**, 4703—4712.
— — EGAN, W., ROBBINS, J. B., 1977: J. Infect. Dis. Suppl. **136**, S71—S77.
LOHMANDER, S., DE LUCA, S., NILSSON, B., HASCALL, V. C., CAPUTO, C. B., KIMURA, J.-H., HEINEGÅRD, D., 1980: J. Biol. Chem. **255**, 6084—6091.
LOMBART, C., WINZLER, R. J., 1974: Eur. J. Biochem. **49**, 77—86.
LONDESBOROUGH, J. C., HAMBERG, U., 1975: Biochem. J. **145**, 401—403.
LOPES, R. A., VALERI, V., COMPOS, G. M., LOPES, O. V. P., DE FAVIA, R. M., 1973: Ann. Histochim. **18**, 131—139.
— DE OLIVEIRA, C., CAMPOS, M. N. M., CAMPOS, S. M., BIRMAN, E. G., 1974: Acta Zool. **55**, 17—24.
LUNDBLAD, A., 1977: In: The Glycoconjugates (HOROWITZ, M. I., PIGMAN, W., eds.), Vol. 1, pp. 441—458. New York: Academic Press.
— MASSON, P., NORDÉN, N. E., SVENSSON, S., ÖCKERMAN, P. A., PALO, J., 1976: Eur. J. Biochem. **67**, 209—214.
LUNNEY, J., ASHWELL, G., 1976: Proc. Natl. Acad. Sci. USA **73**, 341—343.
MAGHUIN-ROGISTER, G., CLOSSET, J., HENNEN, G., 1975: FEBS Lett. **60**, 263—266.
MANDEL, I. D., 1977: In: The Glycoconjugates (HOROWITZ, M. I., PIGMAN, W., eds.), Vol. 1, pp. 153—179. New York: Academic Press.
MARCHESI, V. T., FURTHMAYR, H., TOMITA, M., 1976: Ann. Rev. Biochem. **45**, 667—698.
MARGOLIS, R. K., MARGOLIS, R. U., 1979: In: Complex Carbohydrates of Nervous Tissue (MARGOLIS, R. U., MARGOLIS, R. K., eds.), pp. 45—73. New York: Plenum Press.
MARSHALL, M. O., 1976: Biochim. Biophys. Acta **455**, 837—848.
MARSHALL, T., ALLEN, A., 1978: Biochem. J. **173**, 569—578.
MARTINEZ, N. R., DE, OLAVARRIA, J. M., 1973: Biochim. Biophys. Acta **320**, 295—300.
— MENDEZ, B. A., OLAVARRIA, J. M., 1975: Comp. Biochem. Physiol. **50 B**, 603—607.

MARTINEZ, N. R., OLAVARRIA, J. M., 1977: Acta Physiol. Latinoam. **27**, 42—48.
MÅRTENSSON, E., RAAL, A., SVENNERHOLM, L., 1958: Biochim. Biophys. Acta **30**, 124—129.
MAURY, P., 1971a: Biochim. Biophys. Acta **252**, 48—57.
— 1971b: Biochim. Biophys. Acta **252**, 472—480.
— 1972: J. Biol. Chem. **247**, 3153—3158.
— 1976: Clin. Chim. Acta **71**, 335—338.
MAYER, F. C., DAM, R., PAZUR, J. H., 1964: Arch. Biochem. Biophys. **108**, 356—357.
MCCLUER, R. H., AGRANOFF, B. W., 1972: J. Neurochem. **19**, 1—9.
MESSER, M., 1974: Biochem. J. **139**, 415—420.
— KERRY, K. R., 1973: Science **180**, 201—203.
MEYER, F. A., 1977: Biochim. Biophys. Acta **493**, 272—282.
MICHALSKI, J.-C., STRECKER, G., FOURNET, B., CANTZ, M., SPRANGER, J., 1977: FEBS Lett. **79**, 101—104.
MILLER-PODRAZA, H., 1979: Biochim. Biophys. Acta **586**, 209—212.
MOMOI, T., ANDO, S., NAGAI, Y., 1976: Biochim. Biophys. Acta **441**, 488—497.
MONTGOMERY, R., 1972: In: Glycoproteins, Their Composition, Structure and Function (GOTTSCHALK, A., ed.), 2nd ed., part A, pp. 518—528. Amsterdam: Elsevier.
MONTREUIL, J., 1975: Pure Appl. Chem. **42**, 431—477.
— 1980: Adv. Carbohyd. Chem. Biochem. **37**, 158—223.
— MULLET, S., 1960: Bull. Soc. Chim. Biol. **42**, 365—377.
MORGAN, I. G., GOMBOS, G., TETTAMANTI, G., 1977: In: The Glycoconjugates (HOROWITZ, M. I., PIGMAN, W., eds.), Vol. 1, pp. 351—383. New York: Academic Press.
MORRISETT, J. D., JACKSON, R. L., GOTTO, A. M., jr., 1975: Ann. Rev. Biochem. **44**, 183—207.
MOSCHERA, J., PIGMAN, W., 1975: Carbohyd. Res. **40**, 53—67.
MURTY, V. L. N., DOWNS, F. J., PIGMAN, W., 1978: Carbohyd. Res. **61**, 139—145.
MÜLLER, H. E., 1975: Zbl. Bakt. Hyg., 1. Abtl. Orig. **A 232**, 365—372.
NAGAI, Y., HOSHI, M., 1975: Biochim. Biophys. Acta **388**, 146—151.
— IWAMORI, M., 1980: Mol. Cell. Biochem. **29**, 81—90.
NEUBERGER, A., RATCLIFFE, W. A., 1972: Biochem. J. **129**, 683—693.
NG, S.-S., DAIN, J., 1976: In: Biological Roles of Sialic Acid (ROSENBERG, A., SCHENGRUND, C.-L., eds.), pp. 59—102. New York: Plenum Press.
NICHOLS, J. H., BEZKOROVAINY, A., 1973: Biochem. J. **135**, 875—880.
— — PAQUE, R., 1975: Biochim. Biophys. Acta **412**, 99—108.
OATES, M. D. G., ROSBOTTOM, A. C., SCHRAGER, J., 1974: Carbohyd. Res. **34**, 115—137.
ODIN, L., 1955: Acta Chem. Scand. **9**, 862—864.
OHASHI, M., 1980: J. Biochem. **88**, 583—589.
OKUMURA, T., LOMBART, C., JAMIESON, G. A., 1976: J. Biol. Chem. **251**, 5950—5955.
ONODERA, K., HIRANO, S., HIYASHI, H., 1966: Agr. Biol. Chem. (Tokyo) **30**, 1170—1176.
ØRSKOV, F., ØRSKOV, I., SUTTON, A., SCHNEERSON, R., LIU, W., EGAN, W., HOFF, G. E., ROBBINS, J. B., 1979: J. Exp. Med. **149**, 669—685.
OSHIMA, G., NAGASAWA, K., KATO, J., 1976: J. Biochem. **80**, 477—483.
PALLAVICINI, J. C., GABRIEL, O., DI SANT'AGNESE, P. A., BUSKIRK, E. R., 1963: Ann. N.Y. Acad. Sci. **106**, 330—338.
PAPE, L., KRISTENSEN, B. I., BENGTSON, O., 1975: Biochim. Biophys. Acta **406**, 516—525.
PATTON, S., KEENAN, T. W., 1975: Biochim. Biophys. Acta **415**, 273—309.
PEPPER, D. S., 1968: Biochim. Biophys. Acta **156**, 317—326.
PFEIL, R., KAMERLING, J. P., KÜSTER, J. M., SCHAUER, R., 1980: Hoppe-Seyler's Z. Physiol. Chem. **361**, 314—315.
PHELPS, C. F., STEVENS, A. M., 1978: In: CIBA Foundation Symposium 54, Respiratory Tract Mucus, pp. 91—107. Amsterdam: Elsevier.

PHIPPS, R. J., RICHARDSON, P. S., CORFIELD, A., GALLAGHER, J. T., JEFFERY, P. K., KENT, P. W., PASSATORE, M., 1977: Phil. Trans Roy. Soc. Lond. **279 B**, 513—543.

PICKERING, A. D., 1976: Comp. Biochem. Physiol. **54 B**, 325—328.

PIGMAN, W., 1977a: In: The Glycoconjugates (HOROWITZ, M. I., PIGMAN, W., eds.), Vol. 1, pp. 131—135. New York: Academic Press.

— 1977b: In: The Glycoconjugates (HOROWITZ, M. I., PIGMAN, W., eds.), Vol. 1, pp. 137—152. New York: Academic Press.

— 1977c: In: The Glycoconjugates (HOROWITZ, M. I., PIGMAN, W., eds.), Vol. 2, pp. 73—77. New York: Academic Press.

POLLIT, R. J., PRETTY, K. M., 1974: Biochem. J. **141**, 141—146.

PORTER, W. H., WINZLER, R. J., 1975: Arch. Biochem. Biophys. **166**, 152—163.

PUTNAM, F. W. (ed.), 1975: The Plasma Proteins, Vols. 1 and 2. New York: Academic Press.

— (ed.), 1977: The Plasma Proteins, Vol. 3. New York: Academic Press.

RAHMANN, H., 1978: Jap. J. Exp. Med. **48**, 85—96.

— 1980: In: Structure and Function of Gangliosides (SVENNERHOLM, L., MANDEL, P., DREYFUS, H., URBAN, P. F., eds.), pp. 505—514. New York: Plenum Press.

— BREER, H., 1976: Roux's Arch. Develop. Biol. **180**, 253—256.

— HILBIG, R., 1981: In: Survival in the Cold (MUSACCHIA, X. J., JANSKY, L., eds.), pp. 177—189. Amsterdam: Elsevier and North-Holland.

RAUVALA, H., 1976: J. Biol. Chem. **251**, 7517—7520.

— FINNE, J., 1979: FEBS Lett. **97**, 1—8.

— KRUSIUS, T., FINNE, J., 1978: Biochim. Biophys. Acta **531**, 266—274.

RAY, P. K., CHATTERJEE, S., 1975: Z. Naturforsch. **30 C**, 551—552.

RENLUND, M., CHESTER, M. A., LUNDBLAD, A., AULA, P., RAIVIO, K. O., AUTIO, S., KOSKELA, S.-L., 1979: Eur. J. Biochem. **101**, 245—250.

REUTER, G., VLIEGENTHART, J. F. G., WEMBER, M., SCHAUER, R., HOWARD, R. J., 1980: Biochem. Biophys. Res. Commun. **94**, 567—572.

ROBERTS, G. P., 1974: Eur. J. Biochem. **50**, 265—280.

ROCHA, M., CABEZAS, M., CABEZAS, J. A., 1975: Abstract 10th FEBS Meeting Paris, no. 959, Soc. Chim. Biol., Paris.

ROGERS, C. M., COOKE, K. B., FILIPE, M. I., 1978: Gut **19**, 587—592.

— CORFIELD, A. P., REUTER, G., FILIPE, M. I., COOKE, K. B., SCHAUER, R., 1979: In: Glycoconjugates (SCHAUER, R., BOER, P., BUDDECKE, E., KRAMER, M. F., VLIEGENTHART, J. F. G., WIEGANDT, H., eds.), pp. 652—653. Stuttgart: G. Thieme.

ROSE, M. C., LYNN, W. S., KAUFMAN, B., 1979: Biochemistry **18**, 4030—4037.

ROSENBERG, A., SCHENGRUND, C.-L. (eds.), 1976: Biological Roles of Sialic Acid. New York: Plenum Press.

ROTHMAN, J. E., LODISH, H. F., 1977: Nature **269**, 775—780.

ROUKEMA, P. A., ODERKERK, C. H., SALKINOJA-SALONEN, M. S., 1976: Biochim. Biophys. Acta **428**, 432—440.

RÖSNER, H., 1975: Brain Res. **97**, 107—116.

— 1981: J. Neurochem. **37**, 993—997.

— RAHMANN, H., 1981: Personal communication.

— WIEGANDT, H., RAHMANN, H., 1973: J. Neurochem. **21**, 655—665.

ROST, K., SCHAUER, R., 1977: Phytochemistry **16**, 1365—1368.

RUDZKI, Z., DELLER, D. J., 1973: Digestion **8**, 35—52.

RYAN, L. C., CARUBELLI, R., CAPUTTO, R., TRUCCO, R. E., 1965: Biochim. Biophys. Acta **101**, 252—258.

SAITO, H., YOSIZAWA, Z., 1975: J. Biochem. **77**, 919—930.

SARRIS, A. H., PALADE, G. E., 1979: J. Biol. Chem. **254**, 6724—6731.

SCANU, A. M., EDELSTEIN, C., KEIM, P., 1975: In: The Plasma Proteins: Structure, Function and General Control (PUTNAM, F. W., ed.), Vol. 1, pp. 318—391. New York: Academic Press.

SCHAUER, R., 1973: Angew. Chem. Int. Edn. **12**, 127—140.

— 1978 a: Methods Enzymol. **50 C**, 64—89.

— 1978 b: Methods Enzymol. **50 C**, 374—386.

— 1982: Adv. Carbohyd. Chem. Biochem. **40**, 131 – 234.

— WEMBER, M., 1973: Hoppe-Seyler's Z. Physiol. Chem. **354**, 1405—1414.

— — FERREIRA DO AMARAL, C., 1972: Hoppe-Seyler's Z. Physiol. Chem. **353**, 883—886.

— BUSCHER, H.-P., CASALS-STENZEL, J., 1974: Biochem. Soc. Symp. **40**, 87—116.

— HAVERKAMP, J., WEMBER, M., VLIEGENTHART, J. F. G., KAMERLING, J. P., 1976: Eur. J. Biochem. **62**, 237—242.

— VEH, R. W., SANDER, M., CORFIELD, A. P., WIEGANDT, H., 1980: In: Structure and Function of Gangliosides (SVENNERHOLM, L., MANDEL, P., DREYFUS, H., URBAN, P. F., eds.), pp. 283—294. New York: Plenum Press.

SCHENK, E. A., SCHWARTZ, R. H., LEWIS, R. A., 1971: Lab. Invest **25**, 92—95.

SCHLOEMER, R. H., WAGNER, R. R., 1974: J. Virol. **14**, 270—281.

— — 1975: J. Virol. **15**, 1029—1032.

SCHWICK, H. G., HEIDE, K., HAUPT, H., 1977: In: The Glycoconjugates (HOROWITZ, M. I., PIGMAN, W., eds.), Vol. 1, pp. 261—321. New York: Academic Press.

SCOTT, F. W., ANASTASSIADIS, P. A., 1978: Comp. Biochem. Physiol. **60 A**, 391—398.

SEED, T. M., AIKAWA, M., STERLING, C., RABBEGE, J., 1974: Infect. Immun. **9**, 750—756.

SEGLER, K., RAHMANN, H., RÖSNER, H., 1978: Biochem. Systemat. Ecol. **6**, 87—93.

SHARON, N. (ed.), 1975: Complex Carbohydrates. Their Chemistry, Biosynthesis and Functions, Reading, Mass.: Addison-Wesley.

SHIPP, D. W., BOWNESS, J. M., 1975: Biochim. Biophys. Acta **379**, 282—294.

SHIRAISHI, H., KOHAMA, T., SHIRACHI, R., ISHIDA, N., 1977: J. Gen. Virol. **36**, 207—210.

SIEFRING, G. E., jr., CASTELLINO, F. J., 1974: J. Biol. Chem. **249**, 7742—7746.

SLOMIANY, A., SLOMIANY, B. L., 1978: J. Biol. Chem. **253**, 7301—7306.

— KOJIMA, K., BANAS-GRUSZKA, Z., SLOMIANY, B. L., 1981 a: Biochem. Biophys. Res. Commun. **100**, 778—784.

— KOJIMA, K., BANAS-GRUSZKA, Z., MURTY, V. L. N., GALICKI, N. I., SLOMIANY, A., 1981 b: Eur. J. Biochem. **119**, 647—650.

SLOMIANY, B. L., MEYER, K., 1972: J. Biol. Chem. **247**, 5062—5070.

— SLOMIANY, A., HERP, A., 1978: Eur. J. Biochem. **90**, 255—260.

— MURTY, V. L. N., SLOMIANY, A., 1980: J. Biol. Chem. **255**, 9719—9723.

SNOW, L. D., COLTON, D. G., CARRAWAY, K. L., 1977: Arch. Biochem. Biophys. **179**, 690—697.

SOUPART, P., NOYES, R. W., 1964: J. Reprod. Fertil. **8**, 251—253.

SPEE-BRAND, R., STROUS, G. J. A. M., KRAMER, M. F., 1980: Biochim. Biophys. Acta **621**, 104—116.

SRIVASTAVA, P. N., BREWER, J. M., MENDICINO, J., 1974: Fed. Proc. Fed. Am. Soc. Exp. Biol. **33**, Abstr. 1299.

STARKEY, B. J., SNARY, D., ALLEN, A., 1974: Biochem. J. **141**, 633—639.

STEPAN, J., FERWERDA, W., 1973: Experientia **29**, 948—949.

STRECKER, G., MONTREUIL, J., 1979: Biochimie **61**, 1199—1246.

— HONDI-ASSAH, T., FOURNET, B., SPIK, G., MONTREUIL, J., MAROTEAUX, P., DURAND, P., FARRIAUX, J.-P., 1976: Biochim. Biophys. Acta **444**, 349—358.

SUGAHARA, K., FUNAKOSHI, S., FUNAKOSHI, I., AULA, P., YAMASHINA, I., 1976: J. Biochem. **80**, 195—201.

SUGITA, M., 1979 a: J. Biochem. **86**, 289—300.

SUGITA, M., 1979b: J. Biochem. **86**, 765—772.

SULLIVAN, C. W., VOLCANI, B. E., 1974: Arch. Biochem. Biophys. **163**, 29—45.

SUZUKI, A., ISHIZUKA, I., YAMAKAWA, T., 1975: J. Biochem. **78**, 947—954.

SVENNERHOLM, L., 1970: In: Handbook of Neurochemistry (LAJTHA, A., ed.), Vol. 3, pp. 425—452. New York: Plenum Press.

— MANDEL, P., DREYFUS, H., URBAN, P. F. (eds.), 1980: Structure and Function of Gangliosides. New York: Plenum Press.

SWAMINATHAN, N., ALADJEM, F., 1976: Biochemistry **15**, 1516—1522.

SWANN, D. A., SOTMAN, S., DIXON, M., BROOKS, C., 1977: Biochem. J. **161**, 473—485.

SWEELEY, C. C., SIDDIQUI, B., 1977: In: The Glycoconjugates (HOROWITZ, M. I., PIGMAN, W., eds.), Vol. 1, pp. 459—540. New York: Academic Press.

TETTAMANTI, G., PIGMAN, W., 1968: Arch. Biochem. Biophys. **124**, 41—50.

— BERTOANA, L., GUALANDI, V., ZAMBOTTI, V., 1965: Istituto Lombardo (Rend. Sc.) **B 99**, 173—180.

— VENERANDO, B., PRETI, A., LOMBARDO, A., ZAMBOTTI, V., 1972: In: Glycolipids, Glycoproteins and Mucopolysaccharides of the Nervous System (ZAMBOTTI, V., TETTAMANTI, G., ARRIGONI, M., eds.), pp. 161—181. New York: Plenum Press.

TUPPY, H., GOTTSCHALK, A., 1972: In: Glycoproteins, Their Composition, Structure and Function (GOTTSCHALK, A., ed.), 2nd ed., pp. 403—449. Amsterdam: Elsevier.

TURUMI, K., SAITO, Y., 1953: Tohoku J. Exp. Med. **58**, 247—249.

UCHIDA, Y., TSUKADA, Y., SUGIMORI, T., 1974: Biochim. Biophys. Acta **350**, 425—431.

UHLENBRUCK, G., SCHMITT, J., 1965: Naturwissenschaften **52**, 163.

UNGER, F., 1981: Adv. Carbohyd. Chem. Biochem. **38**, 324—388.

URBAN, P. F., HARTH, S., FREYSZ, L., DREYFUS, H., 1980: In: Structure and Function of Gangliosides (SVENNERHOLM, L., MANDEL, P., DREYFUS, H., URBAN, P. F., eds.), pp. 149—157. New York: Plenum Press.

VAN DEN EIJNDEN, D. H., EVANS, N. A., CODINGTON, J. F., REINHOLD, V., SILBER, C., JEANLOZ, R. W., 1979: J. Biol. Chem. **254**, 12153—12159.

VAN HALBEEK, H., DORLAND, L., VLIEGENTHART, J. F. G., FIAT, A.-M., JOLLÈS, P., 1980: Biochim. Biophys. Acta **623**, 295—300.

VARKI, A., KORNFELD, S., 1980: J. Exp. Med. **152**, 532—544.

VASKOVSKY, V. E., KOSTETSKY, E. I., SVETASHEV, V. I., ZHUKOVA, I. G., SMIRNOVA, G. P., 1970: Comp. Biochem. Physiol. **34**, 163—177.

VEH, R. W., CORFIELD, A. P., SANDER, M., SCHAUER, R., 1977: Biochim. Biophys. Acta **486**, 145—160.

— SANDER, M., HAVERKAMP, J., SCHAUER, R., 1979: In: Glycoconjugate Research (GREGORY, J. D., JEANLOZ, R. W., eds.), Vol. 1, pp. 557—559. New York: Academic Press.

— MICHALSKI, J.-C., CORFIELD, A. P., SANDER-WEWER, M., GIES, D., SCHAUER, R., 1981: J. Chromatogr. **212**, 313—322.

WALDRON-EDWARD, D., DECAENS, C., BADER, J. P., ROBERT, A. M., ROBERT, L., 1976: Path. Biol. **24**, 531—536.

WARREN, L., 1960: Biochim. Biophys. Acta **44**, 347—351.

— 1963: Comp. Biochem. Physiol. **10**, 153—171.

— 1976: In: Biological Roles of Sialic Acid (ROSENBERG, A., SCHENGRUND, C.-L., eds.), pp. 103—121. New York: Plenum Press.

WATANABE, K., HAKOMORI, S.-I., 1979: Biochemistry **18**, 5502—5504.

— POWELL, M., HAKOMORI, S.-I., 1978: J. Biol. Chem. **253**, 8962—8967.

— — — 1979: J. Biol. Chem. **254**, 8223—8229.

WESSLER, E., WERNER, I., 1957: Acta Chem. Scand. **11**, 1240—1247.

White, B. N., Lockhart, M. S., Schilling, J. A., 1975: Comp. Biochem. Physiol. **50 B**, 413—418.
Wiegandt, H., 1971: Adv. Lipid Res. **9**, 249—289.
— 1973: Hoppe-Seyler's Z. Physiol. Chem. **354**, 1049—1056.
— 1980: In: Structure and Function of Gangliosides (Svennerholm, L., Mandel, P., Dreyfus, H., Urban, P. F., eds.), pp. 3—10. New York: Plenum Press.
Winzler, R. J., 1973: In: Membrane Mediated Information (Kent, P. W., ed.), Vol. 1, pp. 3—19. Lancaster: Medical and Technical Publ. Co.
Wold, J. K., Midtvedt, T., Jeanloz, R. W., 1974: Acta Chem. Scand. B. **28**, 277—284.
— Selset, R., 1977: Comp. Biochem. Physiol. **56 B**, 215—218.
— — 1978: Comp. Biochem. Physiol. **61 B**, 271—273.
Wolf, D., Sokoloski, J. E., Litt, M., 1980: Biochim. Biophys. Acta **630**, 545—558.
Woussen-Colle, M. C., Rasinski, C., De Graef, J., 1975a: Biol. Gastroenterol. (Paris) **8**, 275—282.
— — — 1975b: Biol. Gastroenterol. (Paris) **8**, 283—289.
Wright, P., Mackie, I. A., 1977: Trans. Ophthalmol. Soc. UK **97**, 1—7.
Wu, A. M., Pigman, W., 1977: Biochem. J. **161**, 37—47.
— Wu, J. C., Herp, A., 1978: Biochem. J. **175**, 47—51.
— Slomiany, A., Herp, A., Slomiany, B., 1979: Biochim. Biophys. Acta **578**, 297—304.
Yamakawa, T., Nagai, Y., 1978: Trends Biochem. Sci. **3**, 128—131.
— Suzuki, S., 1951: J. Biochem. **38**, 199—212.
Yeager, H., jr., Massaro, G., Massaro, D., 1971: Am. Rev. Resp. Dis. **103**, 188—197.
Yoshima, H., Furthmayr, H., Kobata, A., 1980: J. Biol. Chem. **255**, 9713—9718.
Yu, R. K., Ando, S., 1980: In: Structure and Function of Gangliosides (Svennerholm, L., Mandel, P., Dreyfus, H., Urban, P. F., eds.), pp. 33—45. New York: Plenum Press.
— Ledeen, R. W., 1970: J. Lipid. Res. **11**, 506—516.

C. Isolation and Purification of Sialic Acids

ROLAND SCHAUER and ANTHONY P. CORFIELD

Biochemisches Institut, Christian-Albrechts-Universität, Kiel, Federal Republic of Germany

Contents

I. Introduction

The preparation of acylneuraminic acids is desirable to provide characterized standards for analytical methods and as a basis for the development of synthetic and new analytical techniques. Several methods exist for this, involving hydrolysis of material containing glycosidically bound sialic acids by dilute aqueous or methanolic acids or by enzymic means. Problems exist with respect to O-acyl-sialic acid preparation due to their lability during acid hydrolysis and their reduced or non-susceptibility towards the action of sialidases. Standard techniques have been developed over the last 15 years in the authors' laboratory and these are discussed in the light of alternative methods and improvements which have appeared in the literature.

II. Acid Hydrolysis

The release of acylneuraminic acids by mild acid hydrolysis has been detailed in the literature (FAILLARD 1959, GOTTSCHALK 1960, BLIX 1962, SVENNERHOLM 1963, SCHAUER *et al.* 1968, TUPPY and GOTTSCHALK 1972, SCHAUER 1973, 1978). Essentially complete release of acylneuraminic acids from complex carbohydrates can be obtained by hydrolysis in aqueous 0.1 M H_2SO_4 or HCl at 80 °C for 50 min

(SCHAUER *et al.* 1975, SCHAUER 1978), gangliosides requiring 60–70 min under these conditions. O-Acetylated sialic acids show a greater resistance to acid hydrolysis compared with the non-O-acetylated derivatives (NEUBERGER and RATCLIFFE 1972). Under these conditions there is a loss of sialic acid amounting to about 10% due to acid- and heat-mediated degradation to chromogens (WIRTZ-PEITZ 1969). In addition there is great loss of O-acyl groups. This is mainly due to the increase in lability of acylneuraminic acids in the free state after release from the glycosidic linkage. Optimal conditions for the release of sialic acids from different complex carbohydrates must first be determined (*e.g.* SCHAUER *et al.* 1975), as variations do occur (O'KENNEDY 1980).

As O-acyl groups in sialic acids are rather labile, considerably milder conditions must be employed to preserve at least a portion of these ester groups. Using 0.01 M HCl or formic acid at pH 2 for 60 min at 70 °C (SCHAUER 1978) only 20–30% of the O-acyl groups are lost. However, release of acylneuraminic acids is incomplete.

The free sialic acids can be removed from high molecular weight material by ultrafiltration or dialysis (~ 10,000 daltons cut off) at 2–4 °C. Depending on the nature of the sample, dialysis may require between 6–24 h and can be monitored using a trace amount of radioactive Neu5Ac. Once again it is advisable to determine optimal dialysis times in each case. The remaining glycosidically bound acylneuraminic acid may be subjected to the mild hydrolysis conditions (0.01 M HCl) once again, or to the stronger conditions with 0.1 M acid, and the low molecular weight fraction prepared as above. Using this technique, the recoveries of total sialic acid are 70–80% and the loss of O-acyl groups is about 40% (SCHAUER 1978).

A simple method for the preparation of gramme amounts of Neu5Ac has been described using edible bird's nest substance ("collocalia mucin") (MARTIN *et al.* 1977). This method involves boiling the pulverized material in water under reflux for 6 h and subsequent ultrafiltration before ion-exchange purification of the liberated sialic acid. The yield is 2.2 g Neu5Ac from 100 g of starting material. In the authors' laboratory edible bird's nest substance is solubilized by pronase or ficin treatment, Neu5Ac is then hydrolyzed in 0.1 M H_2SO_4 from the glycopeptides and purified by ion-exchange chromatography after precipitation of the sulfate by barium hydroxide. Thus, 7 g of crystalline Neu5Ac are obtained from 100 g glycoprotein (unpublished). Neu5Ac can also be obtained from the cell wall homopolysaccharide colominic acid present in some strains of *Escherichia coli* (BARRY and GOEBEL 1957; see chapter B), and directly from the urine of a sialuria patient (MONTREUIL *et al.* 1968). The use of bovine, ovine, porcine and equine submandibular gland mucus glycoproteins, containing 15–30% of sialic acid on a dry weight basis, gives high yields, but mixtures of Neu5Ac, Neu5Gc and their O-acetylated derivatives are obtained when using these sources. For instance, porcine submandibular gland glycoprotein contains over 90% of the total sialic acid as Neu5Gc. Separation of Neu5Ac and Neu5Gc requires additional techniques described below, and residual amounts of each N-acyl derivative are difficult to remove from such mixed preparations.

Neu5Ac or Neu5Gc with O-acetyl groups at positions 7, 8, and 9 may be prepared from bovine submandibular gland mucin. Isolation of the crude mucous glycoprotein before hydrolysis is advisable to improve the resolution of sialic acids at the ion-exchange chromatography step. The 4-O-acetylated Neu5Ac and Neu5Gc derivatives can be prepared from equine submandibular gland glycoprotein and equine erythrocyte membranes (BUSCHER *et al.* 1974, CORFIELD *et al.* 1976, SCHAUER 1978).

An alternative method for the preparation of sialic acids is methanolysis of crude complex carbohydrate preparations with hydrochloric acid, which leads to the formation of neuraminic acid-β-methylglycoside (Neu-β-Me) from all natural N- and O-acyl derivatives of neuraminic acid (SCHAUER and BUSCHER 1974, SCHAUER 1978). This method can easily be employed on a large scale, *e.g.* yielding 8.8 g Neu-β-Me from 100 g of edible bird's nest substance. It allows preparation of various homogeneous N-acyl derivatives of neuraminic acid on re-N-acetylation and acid removal of the β-methyl group, with or without a radioactive label (SCHAUER *et al.* 1970, SCHAUER and BUSCHER 1974).

III. Enzymic Hydrolysis

The use of sialidases for the preparation of acylneuraminic acids has several advantages over the relatively destructive acid hydrolysis techniques. The hydrolysis is carried out under milder conditions of temperature and pH. Low temperatures (0–4 °C) can be employed and even on prolonged incubation (24–48 h) the destruction of released acylneuraminic acids is usually below 5%. The sialic acids are released into aqueous solution at pH 5–6, where they are stable for the duration of the incubation. The use of sialidases is widespread, and several bacterial preparations are available in partially purified form well suited for the experiments outlined here. The most widely available sialidases are those from *Vibrio cholerae*, *Clostridium perfringens* and *Arthrobacter ureafaciens*, and these have sufficiently high specific activities to be used in preparative work. Details of the properties and specificities of these and other sialidases are given in chapter I and in reviews by DRZENIEK (1972, 1973) and CORFIELD *et al.* (1981).

It is important to consider the nature of the desialylation experiment in view of the purity of sialidase preparations. The use of highly purified sialidase can provide additional information related to complex carbohydrate structure and the sialidase specificity while effecting desialylation. This approach also generates sialic acid-free complex carbohydrates for use as *e.g.* sialyltransferase substrates, or, in biological systems, desialylated cell-surface glycoconjugates. Cruder preparations frequently contain proteinase and other glycosidase activities which will also degrade the material. Furthermore, consideration of the buffer system must include those ions necessary for enzyme activity (*e.g. V. cholerae* sialidase requires approx. 4 mM Ca^{2+} for activity).

Problems arising due to the removal of sialidase from substrates or its adherence to cells or isolated membranes can be overcome with immobilized sialidase. The use of *V. cholerae* and *C. perfringens* sialidases immobilized on Sepharose 4 B or controlled-pore glass has been demonstrated with a variety of substrates including cells (PARKER *et al.* 1977, CORFIELD *et al.* 1978, BAZARIAN and WINGARD 1979, DEPIERRE *et al.* 1980). These enzymes can be used repeatedly and if suitable, also at higher temperatures. Complete desialylation, dependent on enzyme specificity, has been demonstrated using small columns of immobilized sialidase in a closed circuit including a dialysis reservoir to collect the liberated sialic acids (CORFIELD *et al.* 1978).

The use of enzymes for the release of acylneuraminic acids imposes some limitations on this method of isolation. The amount of sialic acid which can be prepared in this way is limited by enzyme availability, thus the preparation of gramme quantities of sialic acids is not feasible. In addition, the specificity of the enzymes means that some sialic acid residues are resistant to cleavage. Thus, 4-O-acetylated sialic acids in any complex carbohydrate are resistant to all sialidases tested so far, and the sialic acid linked to the internal galactose in the oligosaccharide moiety of gangliosides (*e.g.* II^3Neu5AcGgOse$_4$Cer) has been found to be completely resistant to the action of some sialidases and to be only slowly cleaved by others (CORFIELD *et al.* 1981). In addition, the use of immobilized sialidase imposes spatial limitations on desialylation of cells and membranes (*e.g.* NORDT *et al.* 1981). Sialic acids O-acylated at positions 7–9 of the carbon skeleton are released at reduced rates, but can be completely removed on prolonged incubation (SCHAUER and FAILLARD 1968, CORFIELD *et al.* 1981).

The sialic acid preparations obtained with the aid of sialidase generally are purer than those obtained by acid hydrolysis. In spite of this, ultrafiltration or dialysis (~ 10,000 daltons cut off) is advisable before further purification.

IV. Lipid Extraction

Complex carbohydrates frequently contain adsorbed lipids, and this is especially the case where tissue, cell and membrane preparations are desialylated. For this reason, two extractions of the supernatant or dialysate containing free sialic acids with one volume each of diethyl ether or n-hexane is advisable, prior to ion-exchange chromatography (SCHAUER *et al.* 1975, SCHAUER 1978). If this extraction is not performed, erroneous results due to non-sialic acid compounds may be obtained in the analytical procedures, especially in the periodic acid/thiobarbituric acid assay of sialic acids (SCHAUER *et al.* 1975). Furthermore, lipid extraction improves sialic acid resolution on the following column chromatographic steps.

V. Ion-Exchange Chromatography

Purification of sialic acids is best effected by ion-exchange chromatography which should be carried out at 0–4 °C. The aqueous acylneuraminic acid fraction obtained after lipid extraction is passed through Dowex 50 H^+, usually 20–50 or 50–100 mesh. This step should be rapid, as prolonged contact of sialic acids, particularly their O-acylated derivatives, may result in degradation. The column is eluted with water and the effluent passed on immediately to the anion-exchange column. The size of the Dowex 50 H^+ column is governed by the total amount of material and the nature of the preparation (*e.g.* acid hydrolysis generates more contaminants than enzymic hydrolysis).

The anion-exchange resin, Dowex 2 × 8, is employed either in the formate or the acetate form. Columns of 3 ml of resin are used for less than 100 μg sialic acid and about 400 ml per mmole of sialic acid (SCHAUER 1978). Batchwise elution is achieved with either 0.8 M formic acid or 0.25 M pyridinium acetate, pH 5.4. As slight variations between batches of resin have been observed, the use of a gradient is recommended to optimize conditions for each new batch of resin with tracer amounts of radioactive sialic acid. Gradient elution can be carried out using at least the four-fold resin bed volume of a linear gradient from 0–2 M formic acid or 0–0.5 M pyridinium acetate. This procedure is recommended mainly for preparative purposes, leading to purer sialic acids. A partial fractionation of acylneuraminic acid mixtures is possible on Dowex 2 × 8 formate using ten column volumes of a shallow linear gradient of formic acid from 0–0.5 M (PFEIL and SCHAUER 1979). The separations are thought to be due to a combination of the differences in the pK values of sialic acids (SVENNERHOLM 1956, SCHEINTHAL and BETTELHEIM 1968), the hydrophobic interactions due to the presence of O-acetyl groups and the structural conformations governed by intramolecular hydrogen bonding (unpublished).

Purification of Neu-β-Me by ion-exchange chromatography is strongly influenced by the free amino group. The aqueous solution of Neu-β-Me (e.g. 50 ml for sialic acids from 30 g submandibular gland glycoprotein) is passed slowly through a column (4 × 70 cm) of Dowex 50 H^+ resin. The column is washed with water and Neu-β-Me eluted with a linear gradient (4 *l* in this example) of 0–2 M HCl. Chloride ions are removed using a column of Dowex 1 × 4 in the formate form, and the effluent and two column volumes of a water wash pooled (SCHAUER 1978).

Acylneuraminic acid and Neu-β-Me fractions are either lyophilized or rotary evaporated to dryness at 30–35 °C and stored at −20 °C. Crystallization of these substances will be described below. Sialic acids obtained by ion-exchange chromatography may still contain contaminants which co-chromatograph on ion-exchange resins. Further separation may be achieved on Sephadex G 10 or G 25 or Biogel P 2 using water as eluent. Alternatively, cellulose chromatography may be used.

VI. Cellulose Chromatography

Resolution of individual sialic acid species can be achieved using partition chromatography on cellulose, depending on the relative amounts of the different sialic acids. Cellulose columns (Cellulose MN 2100 ff from Macherey, Nagel, and Co., Düren, Federal Republic of Germany, is routinely used in the authors' laboratory) are used with butan-1-ol : propan-1-ol : water (1 : 2 : 1, v/v/v) as solvent (SCHAUER and FAILLARD 1968, BUSCHER *et al.* 1974, SCHAUER 1978). Careful removal of fines prior to packing the column yields improved flow rates with no loss of resolution. The chromatography is carried out at 0–4 °C, and 50–100 mg of a sialic acid mixture are applied to a column of 2.8 × 100 cm. The elution sequence is N-acetyl-di-O-acetylneuraminic acids, N-acetyl-mono-O-acetylneuraminic acids, N-glycolyl-mono-O-acetylneuraminic acids, Neu5Ac and Neu5Gc (SCHAUER and FAILLARD 1968, BUSCHER *et al.* 1974, CORFIELD *et al.* 1976, SCHAUER 1978). Little loss of O-acetyl groups occurs during this chromatography. Overlapping of the different sialic acids may occur, depending on the relative amounts in the mixture to be separated.

The solvent must be removed from the sialic acid fractions after elution, and this is usually carried out by freeze-drying or rotary evaporation at 30–35 °C. Prolonged storage in the solvent leads to the formation of butyl and propyl esters, and quantitation of the sialic acids in the solvent is not possible. The isolated sialic acids, including the O-acylated species, can be stored at —20 °C for years, without major decomposition.

VII. Crystallization

Crystallization is not always necessary in sialic acid preparation, as the isolation techniques described above usually yield products of high purity. Furthermore, the conditions used for crystallization result in the loss of O-acetyl groups and the method is not recommended for O-acyl-sialic acids. Crystallization of Neu5Ac and Neu5Gc from water at 0 °C has been described using mixtures of ethanol or methanol, diethyl ether and light petroleum (SVENNERHOLM 1956, BLIX and LINDBERG 1960, MEINDL and TUPPY 1969) or addition of about 9 volumes of acetic acid to the saturated, aqueous sialic acid solution (KUHN and BASCHANG 1962, SCHAUER *et al.* 1970, SCHAUER 1978). Neu-β-Me can be readily crystallized from dioxane : water (1 : 1) adding dioxane dropwise to the compound dissolved in water at 40 °C (SCHAUER and BUSCHER 1974).

VIII. Other Methods

Acylneuraminic acids can be prepared by preparative thin-layer chromatography. Details of various systems are given in chapter E. Localization of the sialic acids on the thin-layer plate can for instance be achieved using a radioactive Neu5Ac sample and the position of other sialic acids calculated from R_f values and eluted from the support with water after scraping from the plate. Large amounts cannot be prepared in this way, but the method is especially valuable in the preparation of radioactive acylneuraminic acids from metabolic studies.

The use of HPLC to separate sialic acids in preparative scale is still in the developmental stage. However, ion-exchange columns in analytical HPLC systems have been successfully applied to the separation of free sialic acids, Neu5Ac9P and CMP-sialic acids (unpublished). Analytical separation of individual sialic acids is reported in chapter E.

Bibliography

BARRY, G. T., GOEBEL, W. F., 1957: Nature **179**, 206.

BAZARIAN, E. R., WINGARD, L. B., jr., 1979: J. Histochem. Cytochem. **27**, 125—127.

BLIX, G., 1962: Methods Carbohydr. Chem. **1**, 246—250.

— LINDBERG, E., 1960: Acta Chem. Scand. **14**, 1809—1814.

BUSCHER, H.-P., CASALS-STENZEL, J., SCHAUER, R., 1974: Eur. J. Biochem. **50**, 71—82.

CORFIELD, A. P., BEAU, J.-M., SCHAUER, R., 1978: Hoppe-Seyler's Z. Physiol. Chem. **359**, 1335—1342.

— FERREIRA DO AMARAL, C., WEMBER, M., SCHAUER, R., 1976: Eur. J. Biochem. **68**, 597—610.

— MICHALSKI, J.-C., SCHAUER, R., 1981: In: Sialidases and Sialidoses, Perspectives in Inherited Metabolic Diseases (TETTAMANTI, G., DURAND, P., DI DONATO, S., eds.), Vol. 4, pp. 3—70. Milano: Edi Ermes.

DEPIERRE, J. W., LAZDINS, J., KARNOVSKY, M. L., 1980: Biochem. J., **192**, 543—550.

DRZENIEK, R., 1972: Curr. Top. Microbiol. Immunol. **59**, 35—74.
— 1973: Histochem. J. **5**, 271—290.
FAILLARD, H., 1959: Hoppe-Seyler's Z. Physiol. Chem. **317**, 257—268.
GOTTSCHALK, A., 1960: The Chemistry and Biology of Sialic Acids and Related Substances. London-New York: Cambridge University Press.
KUHN, R., BASCHANG, G., 1962: Justus Liebig's Ann. Chem. **659**, 156—163.
MARTIN, J. E., TANENBAUM, S. W., FLASHNER, M., 1977: Carbohydr. Res. **56**, 423—425.
MEINDL, P., TUPPY, H., 1969: Monatsh. Chem. **100**, 1295—1306.
MONTREUIL, J., BISERTE, G., STRECKER, G., SPIK, G., FONTAINE, G., FARRIAUX, J.-P., 1968: Clin. Chim. Acta **21**, 61—69.
NEUBERGER, A., RATCLIFFE, W. A., 1972: Biochem. J. **129**, 683—693.
NORDT, F. J., FRANCO, M., CORFIELD, A., SCHAUER, R., RUHENSTROTH-BAUER, G., 1981: Blut **42**, 95—98.
O'KENNEDY, R., 1980: Irish J. Med. Sci. **148**, 92—96.
PARKER, T. L., CORFIELD, A. P., VEH, R. W., SCHAUER, R., 1977: Hoppe-Seyler's Z. Physiol. Chem. **358**, 789—795.
PFEIL, R., SCHAUER, R., 1979: In: Glycoconjugates (SCHAUER, R., BOER, P., BUDDECKE, E., KRAMER, M. F., VLIEGENTHART, J. F. G., WIEGANDT, H., eds.), pp. 44—45. Stuttgart: G. Thieme.
SCHAUER, R., 1973: Angew. Chem. Internat. edit. **12**, 127—138.
— 1978: Methods Enzymol. **50 C**, 64—89.
— BUSCHER, H.-P., 1974: Biochim. Biophys. Acta **338**, 369—373.
— FAILLARD, H., 1968: Hoppe-Seyler's Z. Physiol. Chem. **349**, 961—968.
— CORFIELD, A. P., WEMBER, M., DANON, D., 1975: Hoppe-Seyler's Z. Physiol. Chem. **356**, 1727—1732.
— SCHOOP, H. J., FAILLARD, H., 1968: Hoppe-Seyler's Z. Physiol. Chem. **349**, 645—652.
— WIRTZ-PEITZ, F., FAILLARD, H., 1970: Hoppe-Seyler's Z. Physiol. Chem. **351**, 359—364.
SCHEINTHAL, B. M., BETTELHEIM, F. A., 1968: Carbohydr. Res. **6**, 257—265.
SVENNERHOLM, L., 1956: Acta Soc. Med. Upsal. **61**, 75—85.
— 1963: Methods Enzymol. **6**, 453—459.
TUPPY, H., GOTTSCHALK, A., 1972: In: Glycoproteins: Their Composition, Structure and Function (GOTTSCHALK, A., ed.), 2nd ed., part A, pp. 403—449. Amsterdam: Elsevier.
WIRTZ-PEITZ, F., 1969: Doctoral Thesis, Ruhr-University, Bochum, Federal Republic of Germany.

D. Synthesis of Sialic Acids and Sialic Acid Derivatives

JOHANNES F. G. VLIEGENTHART and JOHANNIS P. KAMERLING

Department of Bio-Organic Chemistry, State University of Utrecht, Utrecht, The Netherlands

Contents

I. Introduction

Sialic acids as such, sialic acid derivatives, analogues, glycosides and sialo-oligosaccharides have been subject of many synthetic investigations. These studies were aimed at a further exploration of the properties of sialic acids, the preparation of substrates and inhibitors for sialidases, sialyltransferases or for sialic acid converting enzymes, and to make compounds accessable for analytical purposes or for studies related to metabolism or biological functions of sialic acids.

In comparison to the synthesis of analogous compounds from simple aldoses or ketoses, the situation for sialic acid is more complicated, due to the occurrence of several different functional groups. Synthetic studies on sialo-compounds have been reviewed by BLIX and JEANLOZ (1969), TUPPY and GOTTSCHALK (1972), HOLMQUIST (1975), and VAN DER VLEUGEL (1981).

II. Glycosides of Sialic Acids

For the synthesis of alkyl and aryl α-glycosides of sialic acids often Koenigs-Knorr-like procedures have been applied. A large number of these glycosides have been obtained by silver carbonate-promoted condensation of 5-acylamino-4,7,8,9-

tetra-O-acetyl-2-chloro-2,3,5-trideoxy-β-D-*glycero*-D-*galacto*-2-nonulopyranosonic acids or the corresponding methyl esters with the appropriate alcohol, followed by removal of protecting groups. This reaction sequence is summarized in Scheme 1. The yields vary from 20–60%. Low reaction rates influence the yield unfavourably, since concomitantly the elimination of hydrogen chloride from the sialic acid 2-chloride derivative takes place, which becomes then quantitatively

Scheme 1

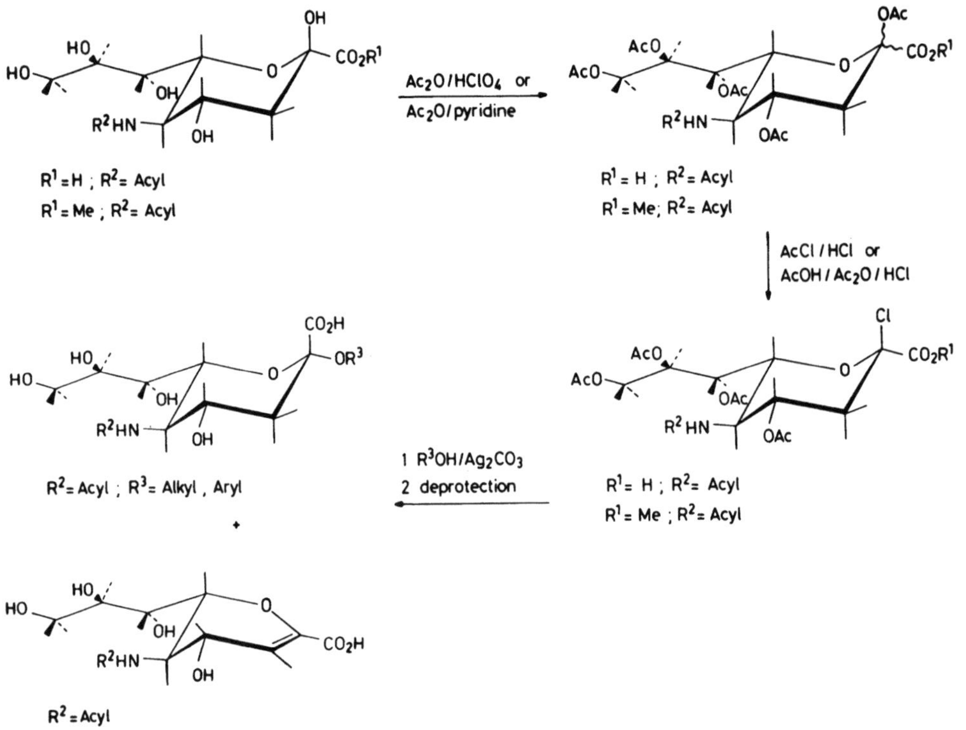

more important (see section IV). Although the replacing of silver carbonate by silver oxide or mercury(II) cyanide (MEINDL and TUPPY 1965 a) does not lead to a significant improvement in the outcome of the reaction, the nature of the condensation promoting agent is important. The 2-pyridyl α-glycoside of N-acetylneuraminic acid has been prepared by condensation of 4,7,8,9-tetra-O-acetyl-2-chloro-2-deoxy-β-N-acetylneuraminic acid methyl ester with the silver salt of 2-hydroxypyridine. The aglycon in this reaction acts also as a promotor. The nucleophilicity of the oxygen atom of 2-hydroxypyridine is increased by substitution of silver for hydrogen (HOLMQUIST and BROSSMER 1972 b, HOLMQUIST 1975). Higher yields in the preparation of α-glycosides were obtained by ESCHENFELDER *et al.* (1975) and ESCHENFELDER and BROSSMER (1979, 1980), by application of polymeric promotors like silver polymaleate or silver polymethacrylate. The underlying principle of this catalyst is the observation that glycosyl halides react readily with alcohols in the presence of insoluble silver salts

Table 1. *Survey of alkyl (aryl) α-glycosides of N-acylneuraminic acid(s) (methyl esters)*

N-Acyl group	α-Glycoside	References
Formyl	benzyl	BROSSMER and NEBELIN 1969
Acetyl	methyl	KUHN *et al.* 1966(*), MEINDL and TUPPY 1965a, 1966b*, VAN DER VLEUGEL *et al.* 1982a*, YU and LEDEEN 1969*, BEAU and SCHAUER 1979
Acetyl	ethyl	ESCHENFELDER and BROSSMER 1979(*), VAN DER VLEUGEL *et al.* 1982a*
Acetyl	isopropyl	VAN DER VLEUGEL *et al.* 1982a*
Acetyl	pentyl	MEINDL and TUPPY 1965a, b
Acetyl	neopentyl	VAN DER VLEUGEL *et al.* 1982a*
Acetyl	hexyl	MEINDL and TUPPY 1965a, b
Acetyl	decyl	MEINDL and TUPPY 1965a
Acetyl	allyl	BROSSMER *et al.* 1974
Acetyl	carboxymethyl	HOLMQUIST and BROSSMER 1972a
Acetyl	2-aminoethyl	ESCHENFELDER and BROSSMER 1980, HOLMQUIST and BROSSMER 1972b
Acetyl	2-aminoethylamino-carbonylmethyl	HOLMQUIST 1974
Acetyl	2-hydroxyethyl	BROSSMER *et al.* 1974
Acetyl	3-hydroxypropyl	MEINDL and TUPPY 1965a
Acetyl	4-hydroxybutyl	BROSSMER *et al.* 1974
Acetyl	5-hydroxypentyl	MEINDL and TUPPY 1965a
Acetyl	cyclohexyl	MEINDL and TUPPY 1965a
Acetyl	cyclohexylmethyl	MEINDL and TUPPY 1965a
Acetyl	phenyl	MEINDL and TUPPY 1967(*)
Acetyl	*m*-methoxyphenyl	TUPPY and PALESE 1969
Acetyl	*p*-nitrophenyl	PRIVALOVA and KHORLIN 1969
Acetyl	benzyl	FAILLARD *et al.* 1966, MEINDL and TUPPY 1965a, 1966b(*)
Acetyl	*p*-methoxybenzyl	MEINDL and TUPPY 1965a
Acetyl	*m*-nitrobenzyl	MEINDL and TUPPY 1965a
Acetyl	*p*-nitrobenzyl	GIELEN and UHLENBRUCK 1969
Acetyl	*m*-chlorobenzyl	MEINDL and TUPPY 1965a
Acetyl	*m*-bromobenzyl	MEINDL and TUPPY 1965a
Acetyl	*m*-iodobenzyl	MEINDL and TUPPY 1965a
Acetyl	4-methylumbelliferyl	MYERS *et al.* 1980, POTIER *et al.* 1979, THOMAS *et al.* 1978, WARNER and O'BRIEN 1979
Acetyl	2-pyridyl	HOLMQUIST and BROSSMER 1972b
Acetyl	5-bromo-3-indolyl	GOSSRAU *et al.* 1977
Glycolyl	methyl	MEINDL and TUPPY 1966a, b(*)
Glycolyl	pentyl	MEINDL and TUPPY 1966a
Glycolyl	decyl	MEINDL and TUPPY 1966a
Glycolyl	benzyl	MEINDL and TUPPY 1966a
Glycolyl	*m*-nitrobenzyl	MEINDL and TUPPY 1966a
Propionyl	benzyl	MEINDL and TUPPY 1966c

Table 1 (continued)

N-Acyl group	α-Glycoside	References
Propionyl	2-aminoethylamino-carbonylmethyl	HOLMQUIST and NILSSON 1979
Butyryl	benzyl	MEINDL and TUPPY 1966c
Succinyl	benzyl	BROSSMER and NEBELIN 1969
Benzoyl	methyl	MEINDL and TUPPY 1966c(*)
Benzoyl	benzyl	MEINDL and TUPPY 1966c
Benzyloxycarbonyl	benzyl	FAILLARD *et al.* 1969

* Methyl ester instead of free acid.

of dicarboxylic acids or hydroxycarboxylic acids (HELFERICH and MÜLLER 1970, WULFF *et al.* 1970, 1972, WULFF and RÖHLE 1974). The preparation of alkyl α-glycosides was also conveniently carried out with silver salicylate as promotor, affording high yields of sialic acid derivatives (VAN DER VLEUGEL *et al.* 1982a). Table 1 summarizes a series of alkyl (aryl) α-glycosides of N-acylneuraminic acid(s) (methyl esters), prepared by several investigators.

β-Linked simple glycosides of sialic acid can be obtained by heating of a solution of sialic acid in the appropriate alcohol in the presence of an acid catalyst, followed by saponification of the formed ester (Scheme 2). This procedure is an

Scheme 2

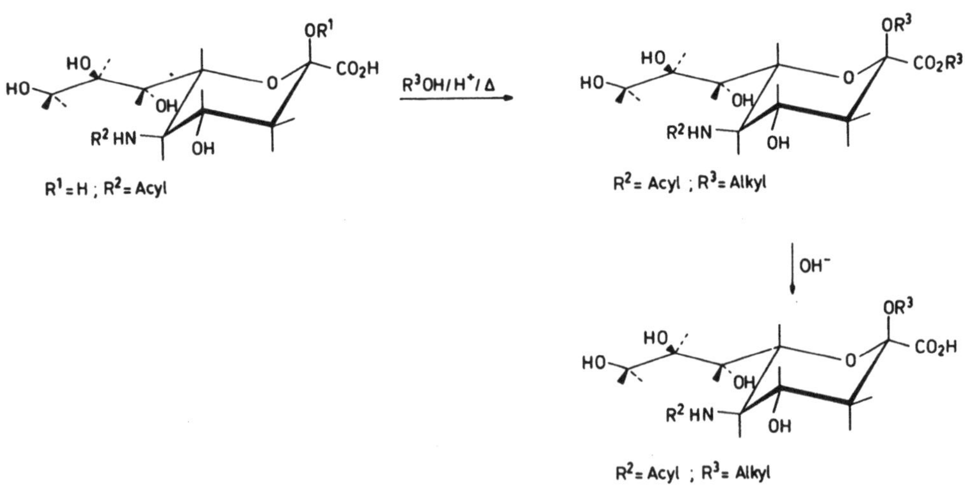

example of alcoholysis according to the FISCHER method (see BOCHKOV and ZAIKOV 1979). Under the thermodynamically controlled reaction conditions only a few percent of the α-anomer is formed. The method is limited to sialic acids which are

Table 2. *Survey of alkyl (aryl) β-glycosides of N-acylneuraminic acid(s) (methyl esters)*

N-Acyl group	β-Glycoside	References
Free amino	methyl	Böhm and Baumeister 1955, Gielen 1965, 1967, Klenk and Lauenstein 1952
Acetyl	methyl	Blix *et al.* 1956*, Kuhn *et al.* 1966(*), McGuire and Binkley 1964*, Wirtz-Peitz *et al.* 1969, Yu and Ledeen 1969*
Acetyl	ethyl	Eschenfelder and Brossmer 1979(*)
Acetyl	pentyl	Meindl and Tuppy 1965 b
Acetyl	hexyl	Meindl and Tuppy 1965 b
Acetyl	carboxymethyl	Holmquist and Brossmer 1972 a
Acetyl	phenyl	Meindl and Tuppy 1967
Acetyl	benzyl	Faillard *et al.* 1966
Acetyl	2-pyridyl	Holmquist and Brossmer 1972 b
Acetyl	CMP	Kean and Roseman 1966, Schauer *et al.* 1972, Schauer and Wember 1973, Corfield *et al.* 1979 b, Haverkamp *et al.* 1979 a, b
Acetyl	2-aminoethylamino-carbonylmethyl	Holmquist and Östman 1975
Fluoroacetyl	methyl	Schauer *et al.* 1970
Trifluoroacetyl	methyl	Klenk *et al.* 1956
Chloroacetyl	methyl	Schauer *et al.* 1970
Glycolyl	methyl	Blix *et al.* 1956*, Wirtz-Peitz *et al.* 1969, Yu and Ledeen 1970*
Glycolyl	CMP	Kean and Roseman 1966, Schauer *et al.* 1972, Schauer and Wember 1973, Corfield *et al.* 1979 b
Glycyl	methyl	Derevitskaya *et al.* 1965 a*
N-Benzyloxy-carbonylglycyl	methyl	Derevitskaya *et al.* 1965 a*
Benzoyl	methyl	Meindl and Tuppy 1966 c(*)
Benzyloxycarbonyl	methyl	Gielen 1965, 1967(*)

* Methyl ester instead of free acid.

stable under the acid conditions. Glycosides from complex alcohols cannot be prepared along this route. Acid catalysts in this reaction can be e.g. Dowex 50 (H^+ form) (20-48 h; 60-70 °C), or 0.1 N HCl (2-3 h; 60-70 °C). In the latter case long reaction times may lead to extensive N-deacylation. It should be noted that the methanolysis procedure which is currently used for sugar analysis of carbohydrates or glycoconjugates containing sialic acid, leads to complete N-deacylation (see chapter F). However, the removal of the N-acyl group followed by N-reacylation opens the way for preparing sialic acids bearing different N-acyl groups in the free as well as in the glycosidically bound form (see section IX). The benzyl β-glycoside of N-acetylneuraminic acid has been obtained by Faillard *et al.* (1966) in 40% yield (besides 4% of the α-anomer) by heating 2,4,7,8,9-penta-O-acetyl-N-acetylneuraminic acid with benzyl alcohol in the presence of $ZnCl_2$, followed by O-deacetylation. β-Glycosides of sialic acids have sometimes been obtained as more or less important side-products in syntheses aimed at the preparation of α-anomers. The β-glycosides have been summarized in Table 2.

Besides the O-glycosides mentioned so far, also attention has been paid to the synthesis of N- and S-analogues. The *p*-nitrophenyl N-glycoside of α-N-acetylneuraminic acid has been prepared by silver carbonate-promoted reaction of 4,7,8,9-tetra-O-acetyl-2-chloro-2-deoxy-β-N-acetylneuraminic acid methyl ester with *p*-nitroaniline and subsequent deprotection (Privalova and Khorlin 1969). Treatment of the same precursor with sodium azide, followed by de-esterification, led to the formation of 2-azido-2-deoxy-α-N-acetylneuraminic acid (Supp *et al.* 1980). The same authors reported also the preparation of the β-form. The azides can readily be converted into the 2-amino derivatives of N-acetylneuraminic acid. Subsequently, the amino function has been benzoylated. Data on the syntheses of the methyl and *p*-nitrophenyl S-glycosides of α-N-acetylneuraminic acid have been presented by Privalova and Khorlin (1969), whereas Ponpipom *et al.* (1980) have described a method for the synthesis of 5-acetamido-2-S-[6-(5-cholesten-3β-yloxy)hexyl]-3,5-dideoxy-2-thio-β-**D**-*glycero*-**D**-*galacto*-2-nonulopyranosonic acid.

III. Sialodisaccharides

In the synthesis of sialodisaccharides most studies were directed to α-glycosidically linked compounds, because virtually all naturally occurring sialo-compounds contain sialic acid in α-glycosidic linkage. It is clear that the coupling reaction is more complicated than for the preparation of simple glycosides. By consequence side-reactions may become more important, leading to rather low yields.

The compounds N-acetyl-**D**-neuraminyl-α(2→6)-**D**-glucose, N-acetyl-**D**-neuraminyl-α(2→6)-**D**-galactose, N-acetyl-**D**-neuraminyl-α(2→6)-N-acetyl-**D**-glucosamine, N-acetyl-**D**-neuraminyl-α(2→3)-**D**-glucose, and N-acetyl-**D**-neuraminyl-α(2→3)-N-acetyl-**D**-glucosamine have been obtained in 8–18% yield by coupling methyl 5-acetamido-4,7,8,9-tetra-O-acetyl-2-chloro-2,3,5-trideoxy-β-**D**-*glycero*-**D**-*galacto*-2-nonulopyranosonate with the appropriately protected derivatives of **D**-glucose, **D**-galactose and N-acetyl-**D**-glucosamine under silver carbonate-promoted Koenigs-Knorr conditions (Khorlin *et al.* 1971). It has to be emphasized that the choice of the protecting groups is an essential part of the synthesis strategy. Removal of such groups must be possible without disruption of the newly formed glycosidic bond.

The disaccharides N-acetyl-**D**-neuraminyl-α(2→6)-1,2 : 3,4-di-O-isopropylidene-α-**D**-galactopyranose, methyl N-acetyl-**D**-neuraminyl-α(2→6)-α-**D**-galactopyranoside, N-acetyl-**D**-neuraminyl-α(2→6)-**D**-glucose, N-acetyl-**D**-neuraminyl-α(2→2)-N-acetyl-α-**D**-neuraminide, and N-acetyl-**D**-neuraminyl-α(2→2)-N-acetyl-β-**D**-neuraminide have been prepared by condensation of the abovementioned

N-acetylneuraminic acid synthon with suitable protected derivatives of **D**-galactose, **D**-glucose and N-acetylneuraminic acid, respectively, in the presence of a polymer-bound silver salt as promotor. The latter catalyst afforded sialodisaccharides in yields of 10–40%, thereby showing the influence of the type of catalyst (BROSSMER *et al.* 1978 a, b).

Condensation of the same N-acetylneuraminic acid synthon with benzyl 2,3,4-tri-O-benzyl-β-**D**-galactopyranoside in the presence of silver salicylate afforded after deprotection the potassium salt of N-acetyl-**D**-neuraminyl-α(2→6)-**D**-galactose in 48% yield (VAN DER VLEUGEL *et al.* 1982 b). During the coupling reaction also a few percent of the β-isomer was formed.

VAN DER VLEUGEL *et al.* (1982 c) also reported on the synthesis of a β-linked sialodisaccharide, namely, N-acetyl-**D**-neuraminyl-β(2→6)-N-acetyl-**D**-glucosamine. This product was obtained by the silver triflate-promoted coupling of the already mentioned N-acetylneuraminic acid synthon with benzyl 2-acetamido-2-deoxy-3,4-O-(1,1,3,3-tetraisopropyldisiloxane-1,3-diyl)-α-**D**-glucopyranoside followed by removal of the protecting groups. It has to be noted that after the condensation reaction the fully protected β-isomer was contaminated, among other products, with the corresponding α-isomer (ratio β : α = 4 : 1). When silver salicylate was used instead of silver triflate, the fully protected 2-O-salicyloyl derivative of N-acetylneuraminic acid was formed almost exclusively.

The aforementioned results demonstrate clearly that for the synthesis of sialo-oligosaccharides a generally applicable route has still to be developed.

Also attention has been paid to the synthesis of sialodisaccharides, having sialic acid in reducing position. By condensation of tetra-O-acetyl-α-**D**-glucopyranosyl bromide or tetra-O-acetyl-α-**D**-galactopyranosyl bromide with methyl 5-acetamido-2,4,7,8-tetra-O-acetyl-3,5-dideoxy-9-O-trityl-**D**-*glycero*-**D**-*galacto*-2-nonulopyranosonate in nitromethane under the influence of silver perchlorate, followed by O-deacetylation, the methyl esters of the corresponding β(1→9) disaccharides were obtained in 67 and 46% yield, respectively (KHORLIN and PRIVALOVA 1968).

IV. 2-Deoxy-2,3-dehydro-N-acylneuraminic Acids

2-Deoxy-2,3-dehydro-N-acylneuraminic acids have been applied in several competitive inhibition studies with neuraminidases from different sources (see for instance: MEINDL and TUPPY 1969 b, MEINDL *et al.* 1971, 1974, PALESE and COMPANS 1976, VEH and SCHAUER 1978, KUMAR *et al.* 1981).

2-Deoxy-2,3-dehydro-N-acetylneuraminic acid has been synthesized by treatment of 4,7,8,9-tetra-O-acetyl-2-chloro-2-deoxy-β-N-acetylneuraminic acid (MEINDL and TUPPY 1965 a) with triethylamine (10 min, 20 °C) or silver carbonate (60–90 min, 80–90 °C) in dioxan or acetone, and subsequent alkaline O-deacetylation (MEINDL and TUPPY 1969 a). This compound has also been detected frequently as a by-product in the synthesis of α-glycosides (MEINDL and TUPPY 1965 a) and of α-linked sialodisaccharides (VAN DER VLEUGEL *et al.* 1982 b, 1982 c).

The unsaturated sialic acid could also be obtained by prolonged heating (5 h, 90 °C) of 2,4,7,8,9-penta-O-acetyl-N-acetylneuraminic acid in dioxan, followed by O-deacetylation (MEINDL and TUPPY 1969 a).

Using the two routes mentioned above, also a series of other unsaturated sialic acids with different N-acyl groups have been prepared (MEINDL and TUPPY 1969 a). Methyl esters, prepared by esterification with diazomethane, can be hydrogenated very easily with a Pd catalyst (MEINDL and TUPPY 1969 a). An additional series of unsaturated N-acylneuraminic acids was synthesized using 2-deoxy-2,3-dehydroneuraminic acid (obtained from the N-benzyloxycarbonyl-analogue by hydrogenolytic cleavage with $Pd/BaSO_4/H_2$) or its benzhydryl ester as precursors (MEINDL and TUPPY 1973). The various N-acyl groups reported by MEINDL and TUPPY (1969 a, 1973) are summarized in Table 3; for recent analytical data of 2-deoxy-2,3-dehydro-N-acetylneuraminic acid, see chapters E, F, and G.

Table 3. *Survey of N-acyl groups of 2-deoxy-2,3-dehydro-N-acylneuraminic acid(s) (methyl esters)* (MEINDL and TUPPY 1969 a, 1973)

Formyl	aminoacetyl	β-(N′-benzyloxycarbonyl)aminopropionyl
Acetyl	acetylaminoacetyl	β-carboxypropionyl
Fluoroacetyl	dimethylaminoacetyl	butyryl
Difluoroacetyl	mercaptoacetyl	carbamoyl
Trifluoroacetyl	glycolyl	β-carboxyacrylyl
Chloroacetyl	propionyl	thiodiacetyl
Iodoacetyl	β-aminopropionyl	benzoyl
Cyanoacetyl	β-acetylaminopropionyl	benzyloxycarbonyl

Another approach for the synthesis of 2-deoxy-2,3-dehydro-N-acetylneuraminic acid has been reported by BEAU and SCHAUER (1979). 4,7,8,9-Tetra-O-*p*-nitrobenzoyl-2-bromo-2-deoxy-N-acetylneuraminic acid methyl ester could be converted quantitatively into the unsaturated analogue using triethylamine or molecular sieves 4 Å in dichloromethane. After deprotection of the latter derivative the free unsaturated N-acetylneuraminic acid was obtained. These authors also mentioned the synthesis of a 2-deoxy-2,3-dehydro-N-acetylneuraminic acid methyl ester, in which the primary hydroxyl group had been replaced by a Cl-atom.

When N-acetylneuraminic acid methyl ester was treated with sulfuric acid and acetic anhydride, after O-deacetylation a mixture of 2-deoxy-2,3-dehydro-N-acetylneuraminic acid methyl ester and 2-deoxy-2,3-dehydro-4-epi-N-acetylneuraminic acid methyl ester was obtained. The ratio of the two compounds was influenced by the reaction temperature. A minor by-product of the acetylation reaction showed to be 2-methyl-(methyl 7,8,9-tri-O-acetyl-2,6-anhydro-2,3,5-trideoxy-**D**-*glycero*-**D**-*talo*-non-2-enonate)-[4,5-*d*]-2-oxazoline (KUMAR *et al.* 1981).

V. Methylated Sialic Acids

Condensation of 2-acetamido-3-O-benzyl-4,6-O-benzylidene-2-deoxy-**D**-mannopyranose with potassio-di-*tert*-butyloxaloacetate, followed by methylation with methyl iodide/silver oxide gave 6-O-benzyl-7,9-O-benzylidene-8-O-methyl-N-acetylneuraminic acid γ-lactone. After removal of the benzyl and benzylidene groups, and opening of the lactone ring, 8-O-methyl-N-acetylneuraminic acid was obtained (KHORLIN and PRIVALOVA 1970). This sialic acid has been found to occur in glycolipids (KOCHETKOV *et al.* 1973, SUGITA 1979).

Reaction of 3-acetamido-3-deoxy-4,5:6,7-di-O-isopropylidene-2-O-methyl-*aldehydo*-**D**-*glycero*-**D**-*galacto*-heptose with [ethoxy(ethoxycarbonyl)methylene]-triphenylphosphorane, followed by an ethoxymercuration-demercuration reaction and acid hydrolysis resulted in the formation of the ethyl esters of 4-O-methyl-N-acetylneuraminic acid and 4-O-methyl-4-epi-N-acetylneuraminic acid (BEAU *et al.* 1978). By oxymercuration of ethyl 5-acetamido-3,5-dideoxy-2-O-ethyl-4-O-methyl-**D**-*glycero*-**D**-*galacto*-non-2-enonate with mercury(II) trifluoroacetate, followed by borohydride-demercuration, 4-O-methyl-N-acetylneuraminic acid ethyl β-glycoside was obtained (BEAU *et al.* 1980). 4-O-Methyl-N-acetylneuraminic acid has been used in metabolism studies, as reported by BEAU and SCHAUER (1980). Using tritiated sodiumborohydride, the corresponding 4-O-methyl-N-acetyl-[3-^{3}H]neuraminic acid was obtained (BEAU and SCHAUER 1980).

A set of partially O-methylated derivatives of methyl N,N-acetyl,methyl-β-**D**-neuraminate methyl glycoside for methylation analyses studies have been prepared by VAN HALBEEK *et al.* (1978) and BRUVIER *et al.* (1981) (see chapter F). VAN HALBEEK *et al.* (1978) started from well-defined partially O-acetylated methyl N-acetyl-β-**D**-neuraminate methyl glycosides. After treatment with methyl vinyl ether and methyl iodide/sodium methylsulfinylmethanide/methylsulfoxide, removal of the 1-methoxyethyl groups yielded the partially O-methylated derivatives (N-methylated). BRUVIER *et al.* (1981) carried out a partial methylation of methyl N-acetyl-β-**D**-neuraminate methyl glycoside using methyl iodide/silver oxide/N,N-dimethylformamide. For the preparation of the 4,7,8,9-tetra-O-methyl derivatives of the methyl esters of N-acetylneuraminic acid benzyl α- and β-glycoside, see LUTZ *et al.* (1968).

VI. Isotopically Labelled Sialic Acids

For the study of biochemical processes several labelled sialic acids and sialic acid derivatives have been prepared.

Several approaches have been reported for the synthesis of N-[1-^{14}C]acetyl-neuraminic acid: i. Hydrogenolysis of N-benzyloxycarbonylneuraminic acid in the presence of [1-^{14}C]acetic anhydride (WESEMANN and ZILLIKEN 1966); ii. Acylation of neuraminic acid methyl β-glycoside with [1-^{14}C]acetic anhydride followed by mild acid hydrolysis (WIRTZ-PEITZ *et al.* 1969); iii. Acylation of neuraminic acid methyl ester methyl β-glycoside with [1-^{14}C]acetic anhydride and subsequent removal of protecting groups (SCHAUER and BUSCHER 1974). N-[1-^{14}C]Glycolyl-neuraminic acid has been prepared by acylation of neuraminic acid methyl β-glycoside with [4-^{14}C]1,3-dioxolan-2,4-dione followed by mild acid hydrolysis (SCHAUER *et al.* 1970). N-[1-^{14}C]acetyl-neuraminic acid has been obtained also enzymatically by condensation of N-[1-^{14}C]acetyl-mannosamine and phospho-enolpyruvate (WARREN and GLICK 1966), and by incubations of surviving slices of submaxillary salivary glands of ox and horse with [1-^{14}C]acetate followed by isolation of the glycoprotein fraction and mild acid hydrolysis (SCHAUER 1970, SCHAUER *et al.* 1972). Using the latter approach also O-[1-^{14}C]acetylated sialic acids were obtained and N-[1-^{14}C]glycolyl-neuraminic acid. See also BRUNETTI *et al.* (1962), WARREN and FELSENFELD (1962), and BLACKLOW and WARREN (1962).

Reaction of neuraminic acid methyl ester methyl β-glycoside with [^{3}H]acetic anhydride followed by release of the methyl groups led to the formation of N-[^{3}H]acetyl-neuraminic acid (SCHAUER and BUSCHER 1974). Starting from 2-acetamido-2-deoxy-**D**-mannose and using $Na^{13}CN$, BENZING-NGUYEN and PERRY (1978) reported a stepwise synthesis of N-acetyl-[1-^{13}C]neuraminic acid. N-acetyl-[1-^{14}C]neuraminic acid or N-glycolyl-[1-^{14}C]neuraminic acid have been prepared also enzymatically using [1-^{14}C]pyruvate and N-acetylmannosamine or N-glycolylmannosamine, respectively (KEAN and ROSEMAN 1966). Enzymatic coupling of N-acetyl-[1-^{14}C]mannosamine and phosphoenolpyruvate led to the formation of N-acetyl-[4-^{14}C]neuraminic acid (WARREN and GLICK 1966). N-acetyl-[3-^{3}H]neuraminic acid can easily be prepared by treatment of N-acetyl-neuraminic acid with alkaline $^{3}H_2O$. Depending on the pH only H3ax or both H3eq and H3ax are exchanged (FRIEBOLIN *et al.* 1981, DORLAND *et al.* 1982, see also chapter G). The synthesis of 4-O-methyl-N-acetyl-[3-^{3}H]-neuraminic acid (BEAU and SCHAUER 1980) has already been mentioned in section V. The same authors reported also the acylation of 2-deoxy-2,3-dehydro-neuraminic acid with [1-^{14}C]acetic anhydride, yielding 2-deoxy-2,3-dehydro-N-[1-^{14}C]acetyl-neuraminic acid (BEAU and SCHAUER 1979).

The doubly labelled sialic acids N-acetyl-[2-^{14}C,9-^{3}H]neuraminic acid (NÖHLE and SCHAUER 1981) and N-glycolyl-[2-^{14}C,9-^{3}H]neuraminic acid (NÖHLE *et al.* 1982) were prepared from sodium[2-^{14}C]pyruvate and either N-acetyl-[6-^{3}H]mannosamine or N-glycolyl-[6-^{3}H]mannosamine, with the aid of the N-acetylneuraminate lyase from *Clostridium perfringens*. The metabolic fate of these compounds was studied after oral and intravenous application to mice and rats.

A number of these labelled sialic acids have been converted into their CMP-analogues, and subsequently incorporated into glycoconjugates (chapter I). Sialoglycoconjugates labelled in sialic acid can also be prepared by incubations of surviving tissue slices with isotopically labelled precursors of sialic acid, as N-[^{3}H]acetyl-mannosamine.

By using periodate oxidation/tritiated borohydride reduction (see also section VIII), sialoglycoconjugates can be labelled very easily. VAN LENTEN and ASHWELL (1972) and SCHAUER *et al.* (1976) reported the use of this technique for the modification of sialoglycoproteins. VEH *et al.* (1977) have described this degradation for gangliosides; see also PFANNSCHMIDT and SCHAUER (1980) for additional data.

VII. Substrates for Sialidase Investigations

For the tracing of sialidase activities in biological materials, a great variety of natural and synthetic sialyl-substrates have been reported. After incubation two approaches have been worked out: i. Determination of the liberated (modified) sialic acid (e.g. spectrofotometrically or radiochemically); ii. Determination of the released aglycon. Concerning the second approach, besides the use of radio-labelled substrates of sialo-oligosaccharides (BHAVANANDAN *et al.* 1975, STRECKER *et al.* 1976) or gangliosides (SCHRAVEN *et al.* 1977, GHIDONI *et al.* 1981) obtained by treatment with tritiated borohydride, also some useful synthetic glycosides of α-N-acetylneuraminic acid have been developed.

PRIVALOVA and KHORLIN (1969) reported the synthesis of the *p*-nitrophenyl glycoside of α-N-acetylneuraminic acid. The substrate was obtained by glycosylation of *p*-nitrophenol with 4,7,8,9-tetra-O-acetyl-2-chloro-2-deoxy-β-N-acetylneuraminic acid methyl ester in chloroform in the presence of silver carbonate and Drierite, followed by O-deacetylation and saponification. In enzymatic reactions, released *p*-nitrophenol can be determined spectrofotometrically.

The chromogenic substrate α-N-acetylneuraminic acid 3-methoxyphenyl glycoside has been described by TUPPY and PALESE (1969). This substrate was synthesized by reaction of 4,7,8,9-tetra-O-acetyl-2-chloro-2-deoxy-β-N-acetylneuraminic acid (MEINDL and TUPPY 1965a) with 3-methoxyphenol in the presence of silver carbonate, followed by O-deacetylation. For the determination of the liberated 3-methoxyphenol in enzymatic reactions two methods have been described, namely, coupling with the diazonium salt of 4-amino-2,5-dimethoxy-4′-nitroazobenzene (TUPPY and PALESE 1969) yielding a red coloured product, and, coupling with 4-aminoantipyrine in the presence of the oxidizing agent potassium ferricyanide yielding a coloured quinone (SANTER *et al.* 1978).

The synthesis of a fluorogenic substrate, namely, the 4-methylumbelliferyl glycoside of α-N-acetylneuraminic acid has been reported by four groups. In all approaches a similar main strategy was followed. THOMAS *et al.* (1978) described the coupling of 4,7,8,9-tetra-O-acetyl-2-chloro-2-deoxy-β-N-acetylneuraminic acid methyl ester with 4-methylumbelliferone in distilling toluene using $CdCO_3$ as a catalyst. The substrate was obtained after O-deacetylation and saponification, $[\alpha]_D^{25} = -9.7°$ (H_2O). WARNER and O'BRIEN (1979) used the sodium salt of 4-methylumbelliferone, whereas the coupling was carried out in acetonitrile in the presence of silver carbonate. The ammonium salt of the substrate obtained after deblocking had $[\alpha]_D^{22} = +59.8°$ (H_2O, pH 5.0). POTIER *et al.* (1979) advised the use of 4-methylumbelliferone with acetonitrile in the presence of silver carbonate and activated molecular sieves. The obtained sodium salt had $[\alpha]_D^{20} = +51°$ (c 0.99, H_2O). Finally, MYERS *et al.* (1980) used N,N-dimethylformamide as a solvent for the coupling of the sialic acid derivative and the sodium salt of 4-methylumbelliferone.

VIII. C_7- and C_8-Analogues of Sialic Acids

For the preparation of 5-acetamido-3,5-dideoxy-**D**-*galacto*-octulosonic acid (C_8-Neu5Ac) two approaches have been developed. The compound has been prepared by alkaline condensation of 2-acetamido-2-deoxy-**D**-lyxose with potassio-di-*tert*-butyloxaloacetate in methanol (MCLEAN and BEIDLER 1969, MCLEAN *et al.* 1971). Furthermore, periodate oxidation/borohydride reduction of the glycerol side-chain of sialic acids has been applied. Using varying amounts of periodate, C_8-Neu5Ac as well as 5-acetamido-3,5-dideoxy-**L**-*arabino*-heptulosonic acid (C_7-Neu5Ac) are obtained. Although in principle many sialoglycoconjugates can be used, procedures have been worked out for the mild periodate oxidation followed by borohydride reduction of *Collocalia* mucoid. After mild acid hydrolysis of the degraded mucin, C_8-Neu5Ac and C_7-Neu5Ac can be isolated via column chromatography (MCLEAN *et al.* 1971, VEH *et al.* 1977). It is also possible to start with N-acetylneuraminic acid methyl ester methyl β-glycoside. In this case, to prevent reduction of the ester function of the periodate-oxidized product, Dowex 1/borohydride was used as a reducing agent (MCLEAN *et al.* 1971). For sialic acid-modifications in glycoproteins, glycolipids and oligosaccharides, and their use in biological studies, see for instance SUTTAJIT and WINZLER (1971), SUTTAJIT *et al.* (1971), VEH *et al.* (1977), and PFANNSCHMIDT and SCHAUER (1980).

The periodate oxidation/reduction approach has also been used to differentiate between the methyl α- and β-glycosides of N-acetylneuraminic acid. Only the C_7-analogue of the β-anomer could be lactonized (YU and LEDEEN 1969). The preparation of the C_7-analogue of 2-deoxy-2,3-dehydro-N-acetylneuraminic acid using periodate oxidation/borohydride reduction was reported by MEINDL and TUPPY (1970).

IX. Miscellaneous Sialic Acids and Sialic Acid Derivatives

The preparation of sialic acids with different N-acyl groups has already got attention in the sections II (including Tables 1 and 2), IV and VI. More details can be obtained from the review of TUPPY and GOTTSCHALK (1972). Surveys of the known reactions involving the carboxyl group and the alcoholic groups of sialic acid have been presented by TUPPY and GOTTSCHALK (1972) and HOLMQUIST (1975). See also HOLMQUIST (1971), BROSSMER and HOLMQUIST (1971), BROSSMER and HOLMQUIST (1974), and BROSSMER *et al.* (1974). For studies on the stability of simple derivatives of sialic acid, see KARKAS and CHARGAFF (1964).

The 9-azido-9-deoxy derivative of N-acetylneuraminic acid has been prepared both chemically and enzymatically. For the chemical synthesis, N-acetylneuraminic acid methyl ester methyl α-glycoside was converted into the corresponding 9-O-tosyl derivative as intermediate and then in the 9-azido-9-deoxy compound by reaction with NaN_3. After removal of methyl groups, 5-acetamido-9-azido-3,5,9-trideoxy-**D**-*glycero*-**D**-*galacto*-2-nonulosonic acid was obtained (BROSSMER *et al.* 1979). The enzymatic preparation was carried out by condensing phosphoenolpyruvate and 6-azido-6-deoxy-N-acetylmannosamine in the presence of N-acetylneuraminate synthase (BROSSMER *et al.* 1980). In principle the 9-azido-9-deoxy derivative of N-acetyl-[1-^{14}C]neuraminic acid can be prepared using [1-^{14}C]phosphoenolpyruvate. Also the CMP-analogue has been reported (BROSSMER *et al.* 1979). For another synthetic approach of azido-derivatives, see BRANDSTETTER *et al.* (1982). Furthermore, the syntheses of the 9-amino-9-deoxy and 9-iodo-9-deoxy derivatives of N-acetylneuraminic acid have been mentioned (SUPP *et al.* 1980).

The synthesis of the methyl ester of 9-O-glycyl-N-acetylneuraminic acid has been reported by DEREVITSKAYA *et al.* (1965b). The methyl ester of N-acetylneuraminic acid was coupled with N-benzyloxycarbonylglycine in the presence of dicyclohexylcarbodiimide; subsequently the benzyloxycarbonyl group was removed by hydrogenolysis. The 9-phosphate esters of N-acetylneuraminic acid and N-glycolylneuraminic acid were prepared enzymatically (ROSEMAN *et al.* 1961, WARREN and FELSENFELD 1961, WATSON *et al.* 1966a, b). Also a chemical synthesis has been mentioned (BROSSMER *et al.* 1974). The preparation of partially O-acetylated derivatives of the methyl ester methyl β-glycoside has been reported by HAVERKAMP *et al.* (1975) and VAN HALBEEK *et al.* (1978).

N-acetyl-[3-OH]neuraminic acid was synthesized by reaction of N-acetylmannosamine with bromopyruvate or hydroxypyruvate (DEVRIES and BINKLEY 1972). The synthesis of N-acetyl-[3-F]neuraminic acid by condensing fluoropyruvic acid and N-acetylglucosamine or N-acetylmannosamine in alkaline solution has been published by GANTT *et al.* (1964).

Scheme 3

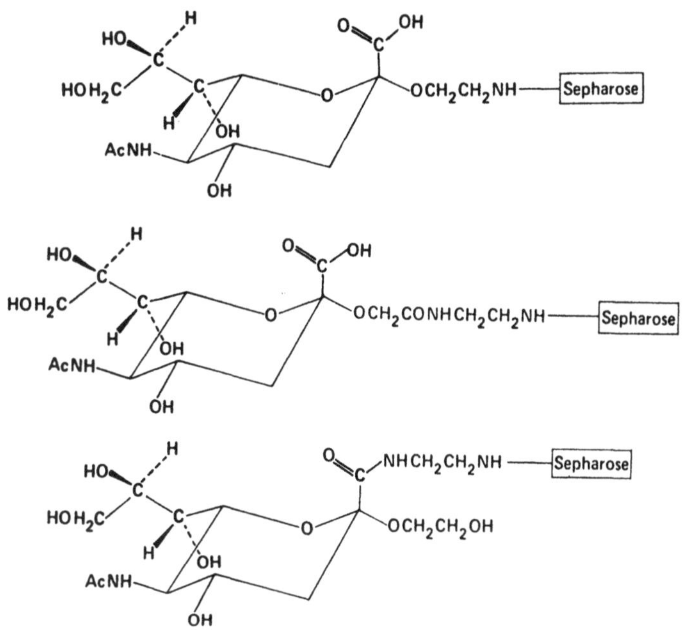

A method for the incorporation of fluorescent probes in sialoglycoproteins has been described by INGHAM and BREW (1981). After mild periodate oxidation the resulting aldehyde functions of the degraded side-chains of sialic acids were condensed with dansylhydrazine, dansylethylenediamine or fluoresceinamine, followed by reduction. Compared with the other derivatives, dansylhydrazine conjugates are relatively unstable. ABRAHAM and LOW (1980) used fluorescent probes for the investigation of human erythrocyte membranes.

To study glycoconjugates by e.p.r. spectroscopy, techniques have been reported for the spin-labelling of sialic acids. For the attachment of the spin-label 4-amino-2,2,6,6-tetramethylpiperidin-1-oxyl, the carboxyl group (APLIN *et al.* 1979b) as well as the periodate-oxidized side-chain (APLIN *et al.* 1979a, FEIX and BUTTERFIELD 1980) of sialic acid were chosen. See also DAVOUST *et al.* (1981).

Sialic acids have also been coupled to proteins. To investigate the haptenic properties of sialic acid, the *p*-nitrobenzyl α-glycoside of N-acetylneuraminic acid has been reduced to the corresponding *p*-aminobenzyl glycoside, which was diazotized and coupled to ovalbumin (GIELEN and UHLENBRUCK 1969).

It has been demonstrated that α-glycosides of sialic acid(s) (derivatives), when immobilized on solid supports, can be used for affinity chromatography of sialidases. HOLMQUIST (1974) reported the preparation of the 2-aminoethyl and the 2-aminoethylaminocarbonylmethyl α-glycosides of N-acetylneuraminic acid, and of the 2-aminoethylamide of the 2-hydroxyethyl α-glycoside of N-acetylneuraminic acid. These derivatives were coupled to cyanogen bromide-activated Sepharose 2B or 4B via the amino group of the α-glycoside (N-acetylneuraminic acid-Sepharose) or the amide (N-acetylneuraminamide-Sepharose) (Scheme 3). The matrix-bound derivatives were also treated with periodate/borohydride to afford the 7-carbon analogues. It was demonstrated that especially N-acetylneuraminamide-Sepharose did not significantly adsorb *Vibrio cholerae* sialidase. In this context also the behaviour of the 2-aminoethylaminocarbonylmethyl β-glycoside of N-acetylneuraminic acid coupled to Sepharose has been studied (HOLMQUIST and ÖSTMAN 1975). For additional data, see HOLMQUIST and NILSSON (1979). Promising results have also been obtained using other sialic acid-Sepharose coupling products (CORFIELD and SCHAUER, unpublished results). Sepharose 4B was activated by cyanogen bromide or periodate. To connect the sialic acids with the support material, two types of spacers were studied, namely, adipic acid dihydrazide and polymethylacrylic acid hydrazide. Binding of neuraminic acid methyl β-glycoside and 2-deoxy-2,3-dehydro-neuraminic acid was carried out via the free amino group after activation of the matrix with HNO_2. Binding of the N-acetyl analogues was performed via the side-chain, after periodate oxidation, in the presence of cyanoborohydride. For the coupling of sialoglycoconjugates to insoluble supports, see CORFIELD *et al.* (1979 a).

Acknowledgement

The studies from the authors' laboratory were supported by the Netherlands Foundation for Chemical Research (SON) with financial aid from the Netherlands Organization for the Advancement of Pure Research (ZWO).

Bibliography

ABRAHAM, G., LOW, P. S., 1980: Biochim. Biophys. Acta **597**, 285—291.

APLIN, J. D., BERNSTEIN, M. A., CULLING, C. F. A., HALL, L. D., REID, P. E., 1979 a: Carbohydr. Res. **70**, C9–C12.

— BROOKS, D. E., CULLING, C. F. A., HALL, L. D., REID, P. E., 1979 b: Carbohydr. Res. **75**, 11—16.

BEAU, J.-M., SCHAUER, R., 1979: Proc. 5th Int. Symp. Glycoconjugates (SCHAUER, R., *et al.*, eds.), pp. 356—357. Stuttgart: G. Thieme.

BEAU, J.-M., SCHAUER, R., 1980: Eur. J. Biochem. **106**, 531—540.
— — HAVERKAMP, J., DORLAND, L., VLIEGENTHART, J. F. G., SINAŸ, P., 1980: Carbohydr. Res. **82**, 125—129.
— SINAŸ, P., KAMERLING, J. P., VLIEGENTHART, J. F. G., 1978: Carbohydr. Res. **67**, 65—77.
BENZING-NGUYEN, L., PERRY, M. B., 1978: J. Org. Chem. **43**, 551—554.
BHAVANANDAN, V. P., YEH, A. K., CARUBELLI, R., 1975: Anal. Biochem. **69**, 385—394.
BLACKLOW, R. S., WARREN, L., 1962: J. Biol. Chem. **237**, 3520—3526.
BLIX, G., JEANLOZ, R. W., 1969: In: The Amino Sugars, the Chemistry and Biology of Compounds Containing Amino Sugars, Vol. IA (JEANLOZ, R. W., ed.), pp. 213–247. New York-London: Academic Press.
— LINDBERG, E., ODIN, L., WERNER, I., 1956: Acta Soc. Med. Upsalien. **61**, 1—25.
BOCHKOV, A. F., ZAIKOV, G. E., 1979: Chemistry of the O-Glycosidic Bond. – Oxford: Pergamon Press.
BÖHM, P., BAUMEISTER, L., 1955: Hoppe-Seyler's Z. Physiol. Chem. **300**, 153—156.
BRANDSTETTER, H. H., ZBIRAL, E., SCHULZ, G., 1982: Liebigs Ann. Chem. 1—13.
BROSSMER, R., HOLMQUIST, L., 1971: Hoppe-Seyler's Z. Physiol. Chem. **353**, 1715—1719.
— — 1974: FEBS Lett. **40**, 250—252.
— NEBELIN, E., 1969: FEBS Lett. **4**, 335—336.
— FRIEBOLIN, H., KEILICH, G., LÖSER, B., SUPP, M., 1978a: Hoppe-Seyler's Z. Physiol. Chem. **359**, 1064.
— — LÖSER, B., SUPP, M., 1978b: Abstr. 9th Int. Symp. Carbohydr. Chem., pp. 91—92.
— BÜRK, G., ESCHENFELDER, V., HOLMQUIST, L., JÄCKH, R., NEUMANN, B., ROSE, U., 1974: Behring Inst. Mitt. **55**, 119—123.
— ROSE, U., KASPER, D., SMITH, T. L., GRASMUK, H., UNGER, F. M., 1980: Biochem. Biophys. Res. Commun. **96**, 1282—1289.
— — UNGER, F. M., GRASMUK, H., 1979: Proc. 5th Int. Symp. Glycoconjugates (SCHAUER, R., *et al.*, eds.), pp. 242—243. Stuttgart: G. Thieme.
BRUNETTI, P., JOURDIAN, G. W., ROSEMAN, S., 1962: J. Biol. Chem. **237**, 2447—2453.
BRUVIER, C., LEROY, Y., MONTREUIL, J., FOURNET, B., KAMERLING, J. P., 1981: J. Chromatogr. **210**, 487—504.
CORFIELD, A. P., PARKER, T. L., SCHAUER, R., 1979a: Anal. Biochem. **100**, 221—232.
— SCHAUER, R., WEMBER, M., 1979b: Biochem. J. **177**, 1—7.
DAVOUST, J., MICHEL, V., SPIK, G., MONTREUIL, J., DEVAUX, P. F., 1981: FEBS Lett. **125**, 271—276.
DEREVITSKAYA, V. A., KALINEVICH, V. M., KOCHETKOV, N. K., 1965a: Khim. Prirodn. Soedin., 241.
— — — 1965b: Dokl. Akad. Nauk. SSSR **160**, 596—599.
DEVRIES, G. H., BINKLEY, S. B., 1972: Arch. Biochem. Biophys. **151**, 243—250.
DORLAND, L., HAVERKAMP, J., SCHAUER, R., VELDINK, G. A., VLIEGENTHART, J. F. G., 1982: Biochem. Biophys. Res. Commun. **104**, 1114—1119.
ESCHENFELDER, V., BROSSMER, R., 1979: Hoppe-Seyler's Z. Physiol. Chem. **360**, 1253—1256.
— — 1980: Carbohydr. Res. **78**, 190—194.
— — WACHTER, M., 1975: Angew. Chem. **87**, 747—748.
FAILLARD, H., FERREIRA DO AMARAL, C., BLOHM, M., 1969: Hoppe-Seyler's Z. Physiol. Chem. **350**, 798—802.
— KIRCHNER, G., BLOHM, M., 1966: Hoppe-Seyler's Z. Physiol. Chem. **347**, 87—93.
FEIX, J. B., BUTTERFIELD, D. A., 1980: FEBS Lett. **115**, 185—188.
FRIEBOLIN, H., SCHMIDT, H., SUPP, M., 1981: Tetr. Lett. **22**, 5171—5174.
GANTT, R., MILLNER, S., BINKLEY, S. B., 1964: Biochemistry **3**, 1952—1960.
GHIDONI, R., SONNINO, S., MASSERINI, M., ORLANDO, P., TETTAMANTI, G., 1981: J. Lip. Res. **22**, 1286—1295.

GIELEN, W., 1965: Hoppe-Seyler's Z. Physiol. Chem. **342**, 170—171.
— 1967: Hoppe-Seyler's Z. Physiol. Chem. **348**, 378—380.
— UHLENBRUCK, G., 1969: Hoppe-Seyler's Z. Physiol. Chem. **350**, 672—673.
GOSSRAU, R., ESCHENFELDER, V., BROSSMER, R., 1977: Histochem. **53**, 189—192.
HAVERKAMP, J., BEAU, J.-M., SCHAUER, R., 1979a: Hoppe-Seyler's Z. Physiol. Chem. **360**, 159—166.
— SCHAUER, R., WEMBER, M., KAMERLING, J. P., VLIEGENTHART, J. F. G., 1975: Hoppe-Seyler's Z. Physiol. Chem. **356**, 1575—1583.
— SPOORMAKER, T., DORLAND, L., VLIEGENTHART, J. F. G., SCHAUER, R., 1979b: J. Am. Chem. Soc. **101**, 4851—4853.
HELFERICH, B., MÜLLER, W. M., 1970: Chem. Ber. **103**, 3350—3352.
HOLMQUIST, L., 1971: Acta Chem. Scand. **25**, 712—716.
— 1974: Acta Chem. Scand. **B 28**, 1065—1068.
— 1975: FOA Reports **9**, no. 3.
— BROSSMER, R., 1972a: FEBS Lett. **22**, 46—48.
— — 1972b: Hoppe-Seyler's Z. Physiol. Chem. **353**, 1346—1350.
— NILSSON, G., 1979: Acta Path. Microbiol. Scand. **B 87**, 129—135.
— ÖSTMAN, B., 1975: FEBS Lett. **60**, 327—330.
INGHAM, K. C., BREW, S. A., 1981: Biochim. Biophys. Acta **670**, 181—189.
KARKAS, J. D., CHARGAFF, E., 1964: J. Biol. Chem. **239**, 949—957.
KEAN, E. L., ROSEMAN, S., 1966: Meth. Enzymol. **8**, 208—215.
KHORLIN, A. YA., PRIVALOVA, I. M., 1968: Bull. Acad. Sci. USSR., Div. Chem. Sci., 217—219.
— — 1970: Carbohydr. Res. **13**, 373—377.
— — BYSTROVA, I. B., 1971: Carbohydr. Res. **19**, 272—275.
KLENK, E., LAUENSTEIN, K., 1952: Hoppe-Seyler's Z. Physiol. Chem. **291**, 147—152.
— FAILLARD, H., WEYGAND, F., SCHÖNE, H. H., 1956: Hoppe-Seyler's Z. Physiol. Chem. **304**, 35—52.
KOCHETKOV, N. K., CHIZHOV, O. S., KADENTSEV, V. I., SMIRNOVA, G. P., ZHUKOVA, I. G., 1973: Carbohydr. Res. **27**, 5—10.
KUHN, R., LUTZ, P., McDONALD, D. L., 1966: Chem. Ber. **99**, 611—617.
KUMAR, V., KESSLER, J., SCOTT, M. E., PATWARDHAN, B. H., TANENBAUM, S. W., FLASHNER, M., 1981: Carbohydr. Res. **94**, 123—130.
LUTZ, P., LOCHINGER, W., TAIGEL, G., 1968: Chem. Ber. **101**, 1089—1094.
McGUIRE, E. J., BINKLEY, S. B., 1964: Biochemistry **3**, 247—251.
McLEAN, R. L., BEIDLER, J., 1969: J. Am. Chem. Soc. **91**, 5388.
— SUTTAJIT, M., BEIDLER, J., WINZLER, R. J., 1971: J. Biol. Chem. **246**, 803—809.
MEINDL, P., TUPPY, H., 1965a: Monatsh. Chem. **96**, 802—815.
— — 1965b: Monatsh. Chem. **96**, 816—827.
— — 1966a: Monatsh. Chem. **97**, 654—661.
— — 1966b: Monatsh. Chem. **97**, 990—999.
— — 1966c: Monatsh. Chem. **97**, 1628—1647.
— — 1967: Monatsh. Chem. **98**, 53—60.
— — 1969a: Monatsh. Chem. **100**, 1295—1306.
— — 1969b: Hoppe-Seyler's Z. Physiol. Chem. **350**, 1088—1092.
— — 1970: Monatsh. Chem. **101**, 639—647.
— — 1973: Monatsh. Chem. **104**, 402—414.
— BODO, G., LINDNER, J., TUPPY, H., 1971: Z. Naturforsch. **26b**, 792—797.
— — PALESE, P., SCHULMAN, J., TUPPY, H., 1974: Virology **58**, 457—463.
MYERS, R. W., LEE, R. T., LEE, Y. C., THOMAS, G. H., REYNOLDS, L. W., UCHIDA, Y., 1980: Anal. Biochem. **101**, 166—174.

Nöhle, U., Schauer, R., 1981: Hoppe-Seyler's Z. Physiol. Chem. **362**, 1495—1506.
— Beau, J.-M., Schauer, R., 1982: Eur. J. Biochem. **126**, 543–548.
Palese, P., Compans, R. W., 1976: J. Gen. Virol. **33**, 159–163.
Pfannschmidt, G., Schauer, R., 1980: Hoppe-Seyler's Z. Physiol. Chem. **361**, 1683—1695.
Ponpipom, M. M., Bugianesi, R. L., Shen, T. Y., 1980: Can. J. Chem. **58**, 214—220.
Potier, M., Mameli, L., Bélisle, M., Dallaire, L., Melançon, S. B., 1979: Anal. Biochem. **94**, 287—296.
Privalova, I. M., Khorlin, A. Ya., 1969: Izv. Akad. Nauk. SSSR, Ser. Khim. 2785—2792.
Roseman, S., Jourdian, G. W., Watson, D., Rood, R., 1961: Proc. Nat. Acad. Sci. U.S.A. **47**, 958.
Santer, U. V., Yee-Foon, J., Glick, M. C., 1978: Biochim. Biophys. Acta **523**, 435—442.
Schauer, R., 1970: Hoppe-Seyler's Z. Physiol. Chem. **351**, 595—602.
— Buscher, H.-P., 1974: Biochim. Biophys. Acta **338**, 369—373.
— Wember, M., 1973: Hoppe-Seyler's Z. Physiol. Chem. **354**, 1405—1414.
— — Ferreira do Amaral, C., 1972: Hoppe-Seyler's Z. Physiol. Chem. **353**, 883—886.
— Wirtz-Peitz, F., Faillard, H., 1970: Hoppe-Seyler's Z. Physiol. Chem. **351**, 359—364.
— Veh, R. W., Wember, M., Buscher, H.-P., 1976: Hoppe-Seyler's Z. Physiol. Chem. **357**, 559—566.
Schraven, J., Čáp, C., Nowoczek, G., Sandhoff, K., 1977: Anal. Biochem. **78**, 333—339.
Strecker, G., Michalski, J.-C., Montreuil, J., Farriaux, J.-P., 1976: Biomedicine **25**, 238—239.
Sugita, M., 1979: J. Biochem. **86**, 765—772.
Supp, M., Rose, U., Brossmer, R., 1980: Hoppe-Seyler's Z. Physiol. Chem. **361**, 338.
Suttajit, M., Winzler, R. J., 1971: J. Biol. Chem. **246**, 3398—3404.
— Reichert, L. E., Winzler, R. J., 1971: J. Biol. Chem. **246**, 3405—3408.
Thomas, J. J., Folger, E. C., Nist, D. L., Thomas, B. J., Jones, R. H., 1978: Anal. Biochem. **88**, 461—467.
Tuppy, H., Gottschalk, A., 1972: In: Glycoproteins, Their Composition, Structure and Function (Gottschalk, A., ed.), pp. 401—449. Amsterdam-London-New York: Elsevier.
— Palese, P., 1969: FEBS Lett. **3**, 72—75.
Van der Vleugel, D. J. M., 1981: Thesis Ph. D., State University of Utrecht (The Netherlands), pp. 9—35.
— Van Heeswijk, W. A. R., Vliegenthart, J. F. G., 1982a: Carbohydr. Res. **102**, 121—130.
— Wassenburg, F. R., Zwikker, J. W., Vliegenthart, J. F. G., 1982b: Carbohydr. Res. **104**, 221—233.
— Zwikker, J. W., Vliegenthart, J. F. G., Van Boeckel, S. A. A., Van Boom, J. H., 1982c: Carbohydr. Res. **105**, 19—31.
Van Halbeek, H., Haverkamp, J., Kamerling, J. P., Vliegenthart, J. F. G., Versluis, C., Schauer, R., 1978: Carbohydr. Res. **60**, 51—62.
Van Lenten, L., Ashwell, G., 1972: Meth. Enzymol. **28**, 209—211.
Veh, R. W., Schauer, R., 1978: In: Enzymes of Lipid Metabolism (Gatt, S., *et al.*, eds.), pp. 447—462. New York-London: Plenum.
— Corfield, A. P., Sander, M., Schauer, R., 1977: Biochim. Biophys. Acta **486**, 145—160.
Warner, T. G., O'Brien, J. S., 1979: Biochemistry **18**, 2783—2787.
Warren, L., Felsenfeld, H., 1961: Biochem. Biophys. Res. Commun. **5**, 185—190.
— — 1962: J. Biol. Chem. **237**, 1421—1431.
— Glick, M. C., 1966: Meth. Enzymol. **8**, 131—133.
Watson, D. R., Jourdian, G. W., Roseman, S., 1966a: J. Biol. Chem. **241**, 5627—5636.
— — — 1966b: Meth. Enzymol. **8**, 201—205.

WESEMANN, W., ZILLIKEN, F., 1966: Liebigs Ann. Chem. **695**, 209—216.

WIRTZ-PEITZ, F., SCHAUER, R., FAILLARD, H., 1969: Hoppe-Seyler's Z. Physiol. Chem. **350**, 111—115.

WULFF, G., RÖHLE, G., 1974: Angew. Chem. **86**, 173—187.

— — KRÜGER, W., 1970: Angew. Chem. **82**, 480.

— — — 1972: Chem. Ber. **105**, 1097—1110.

YU, R. K., LEDEEN, R. W., 1969: J. Biol. Chem. **244**, 1306—1313.

— — 1970: J. Lip. Res. **11**, 506—516.

E. Colorimetry and Thin-Layer Chromatography of Sialic Acids

ROLAND SCHAUER and ANTHONY P. CORFIELD

Biochemisches Institut, Christian-Albrechts-Universität, Kiel, Federal Republic of Germany

With 1 Figure

Contents

I. Introduction

Analysis and quantitation of the sialic acids are carried out on a vast range of molecules where they occur. Colorimetric analysis has been used since the discovery of the sialic acids as a routine method for detection and, in some cases, accurate quantitation of these monosaccharides. Several factors have contributed to the difficulties encountered using these methods. These include non-identical reactions of different sialic acids in the same assay, the non-specificity of the reactions for the sialic acids, and the need to allow for the contamination of samples with compounds interfering with the assay. In this light it is perhaps not surprising that errors have arisen due to one or more of these factors. Reviews of the colorimetric assays available for sialic acids have pointed out these problems (TUPPY and GOTTSCHALK 1972, LEDEEN and YU 1976, SCHAUER 1978) and they will again be emphasized below. Greatest problems are encountered using cells or tissue extracts, as the level of contamination is inevitably higher and purification of the sialic acids as described in chapter C is necessary. In many enzyme assays the addition of a pure substrate enables accurate quantitation using colorimetry. However, cases may arise where purification is still required (e.g. the assay of tissue and cell homogenates). Colorimetry provides a routine and accurate assay for sialic acids if properly controlled.

Thin-layer chromatography is a further valuable technique easily adapted for routine analysis, and one which can be used to obtain much information about the nature of sialic acids. As with the colorimetric methods, care must be exercised in analysis, and control samples are required if meaningful data are to be obtained.

These methods have their value in the relative ease of use and adaptation to large sample numbers. However, although tentative assignment of N- and O-acyl substitutions can be made, the unequivocal determination of the structure of sialic acids must be made with mass spectrometry and n.m.r. spectroscopy as described in chapters F and G.

II. Colorimetry and Fluorimetry

The initial attempts to develop a quantitative assay for sialic acids were colorimetric. The majority of these early methods were unsuitable because of their lack of specificity and/or sensitivity required for sialic acid analysis in biological materials or for enzymic assays. Some of these methods included diphenylamine (AYALA *et al.* 1951), tryptophane/perchloric acid (SEIBERT *et al.* 1948), sulphuric acid/acetic acid (HESS *et al.* 1957), hydrochloric acid (FOLCH *et al.* 1951) and the "Ehrlich" reagent, dimethyl-aminobenzaldehyde (WERNER and ODIN 1952).

Those colorimetric methods developed at this time and found to be suitably sensitive and specific are the orcinol and resorcinol methods (see II.1 below) and the periodic acid/thiobarbituric acid assay (see II.2). These methods together with a recently introduced assay using methyl-3-benzothiazolinone-2-hydrazone and fluorimetric methods (II.3 and II.4, respectively) form the basis of currently used techniques.

1. Diphenol Reagent Assays

a) Orcinol

The orcinol reaction, involving heating of the sample in orcinol/Fe^{3+} in strong aqueous hydrochloric acid and known as the "Bial" reaction, was introduced by KLENK and LANGERBEINS (1941) and has been in continual use since then (BÖHM *et al.* 1954, SVENNERHOLM 1957a, 1963a, VEH *et al.* 1977, SCHAUER 1978). The assay has been used routinely in the authors' laboratory for several years. The chromophore formed after heating at 100 °C and believed to be a methine dye (WIRTZ-PEITZ 1969), is extracted with isoamyl alcohol after cooling and measured at 572 nm. Molar extinction coefficients for the different sialic acids are given in Table 1. Using microadaptations of the originally described technique, minimum amounts of 2–3 μg sialic acids can be determined (SCHAUER 1978). The method allows determination of total sialic acids (i.e. bound and free) including the O-acetylated species. Due to the strong acid conditions O-acyl groups are released and the extinction coefficient of the parent sialic acid (Neu5Ac or Neu5Gc) is found. A shortening of the carbon skeleton to the C-8 or C-7 derivatives of Neu5Ac results in an increase in the molar extinction coefficient, two-fold for the C-8 derivative and slightly less for the C-7 derivative, which is also accompanied by changes in absorption maxima from 572 nm to 562 nm and 602 nm, respectively (VEH *et al.* 1977).

As the assay does not discriminate between bound and free sialic acids, it is widely used to monitor the presence of sialic acids in either form during fractionation of biological material. Interference by other monosaccharides,

Table 1. *Colorimetric analysis of the sialic acids. Molar absorption coefficients*

The molar absorption coefficients are noted for the different sialic acids using the commonly used colorimetric assays. The chromophore was measured at the absorption maximum, noted at the head of each column. Abbreviations of the sialic acids are those used in chapter B, Table 1

Compounds	Diphenol assays			Periodic acid/thiobarbituric acid assay	
	orcinol/Fe^{3+} 572 nm[a]	resorcinol/Cu^{2+} 580 nm[b]	periodate/resorcinol/Cu^{2+} 630 nm[c]	AMINOFF 549 nm[d]	WARREN 549 nm[e]
Neu5Ac	5,300	6,900	27,900	70,000	61,000
Neu4,5Ac$_2$	5,300	—	—	65,000	—
Neu5Ac4Me	—	—	—	74,000	—
Neu5,7Ac$_2$	5,300	—	—	3,500	—
Neu5,9Ac$_2$	5,300	—	—	30,000	—
Neu5,7,9Ac$_3$	5,300	—	—	0	—
C-8-Neu5Ac	10,300[f]	—	—	69,500[i]	61,000
C-7-Neu5Ac	8,000[g]	17,900[h]	—	69,000[i]	61,000
Neu5Gc	6,300	8,200	27,300	51,000	50,000
Neu4Ac5Gc	6,300	—	—	51,000	—
Neu9Ac5Gc	6,300	—	—	23,000	—
Neu5Gc8Me	6,300	—	—	—	6,000
Neu5Ac2en	4,900	—	—	0	0
Neu-β-Me	6,780	—	—	0	0

[a] BÖHM *et al.* 1954, WARREN 1964, SCHAUER *et al.* 1970, VEH *et al.* 1977, SCHAUER 1978; [b] SVENNERHOLM 1957 a, b, SCHAUER 1978; [c] JOURDIAN *et al.* 1971, SCHAUER 1978; [d] AMINOFF 1961, SCHAUER and FAILLARD 1968, VEH *et al.* 1977, SCHAUER 1978; [e] WARREN 1959, 1964, SCHAUER 1978; [f] Maximum absorbance at 562 nm, VEH *et al.* 1977; [g] Maximum absorbance at 602 nm, VEH *et al.* 1977; [h] PETERS and ARONSON 1976; [i] McLEAN *et al.* 1971.—Not determined.

especially pentoses, hexoses and uronic acids, limit the quantitative value of the assay, markedly when only low amounts of sialic acid are present (BÖHM *et al.* 1954, SVENNERHOLM 1963b, SPIRO 1966, TUPPY and GOTTSCHALK 1972). The purification of liberated sialic acid from samples where excessive interference prevents quantitation enables more accurate measurements in both this and the resorcinol/Cu^{2+} methods and is discussed under II.10.

b) Resorcinol

α) *Resorcinol*/Cu^{2+} *Assay*

This assay was introduced by SVENNERHOLM (1957b) and proved to be more sensitive (30-50%) and specific than the orcinol/Fe^{3+} assay (see Table 1). The use of Cu^{2+} in place of Fe^{3+} gave the highest extinction coefficient for Neu5Ac (SVENNERHOLM 1957b). Recent work on the mechanism of chromogen formation has led to the proposal that heterocyclic breakdown products of Neu5Ac and Neu5Gc containing a carbonyl group formed in the reaction under the influence of Cu^{2+} are the major substances involved in the condensation with resorcinol (PETERS and ARONSON 1976). On heating in strong aqueous acid solution the sialic acids are released from glycosidic linkage, their carboxyl groups are removed as CO_2 and N-deacetylation takes place. The result of the complex transformations are heterocyclic derivatives (BERGGÅRD and ODIN 1958). PETERS and ARONSON (1976) have proposed the formation of a tetrahydroxybutylpyrrole, which is dehydrated to give a carbonyl group reactive with resorcinol as follows:

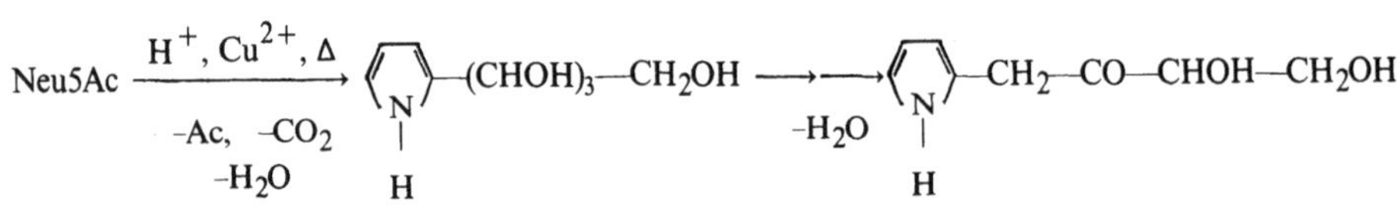

The principal chromogen gives rise to a chromophore with an absorption maximum at 580-585 nm (SVENNERHOLM 1957b, PETERS and ARONSON 1976), when extracted into butan-1-ol. Ketohexoses, pentoses and 2-deoxyhexoses all give interference. Correction for ketohexoses can be made by measuring the absorption at a second wavelength, 450 nm (SVENNERHOLM 1957b, 1963a, b). Extraction of the chromophore with a butylacetate: butan-1-ol (85:15) mixture gives some improvement on the interference (MIETTINEN and TAKKI-LUUKKAINEN 1959).

As with the orcinol/Fe^{3+} assay, both free and bound sialic acids are measured by this method and no discrimination can be made between these forms.

β) *Periodate/Resorcinol*/Cu^{2+} *Assay*

The sensitivity of the resorcinol/Cu^{2+} method can be increased significantly (between 3- to 6-fold), if periodate oxidation is carried out prior to heating with the resorcinol/Cu^{2+} strong acid reagent (JOURDIAN *et al.* 1971; Table 1). The chromogen formed with sialic acid glycosides under the influence of periodate is stable to further oxidation at 37 °C and leads to a chromophore with an extinction maximum at 630 nm. The chromogen formed with free sialic acids is destroyed at this temperature, but is stable at 0 °C. These characteristics were employed to allow measurement of total, free or bound sialic acid concentrations in a sample, previously not possible with the resorcinol/Cu^{2+} assay.

An explanation of the increase in molar extinction coefficient on periodate oxidation has been put forward based on the reactivity of the C-7 aldehyde of sialic acid in the assay (PETERS and ARONSON 1976). The synthesis of intermediates in the oxidation reaction, including the C-7 aldehyde derivative of Neu5Ac, confirmed this proposal. The change in absorption maximum from 585 nm to 630 nm and the increase in colour yield were also shown (PETERS and ARONSON 1976). Further evidence in favour of such a mechanism has come from the low colour formation due to any substitution of the C-7 to C-9 side chain of Neu5Ac thus blocking the periodate oxidation (NEUBERGER and RATCLIFFE 1973). In addition, the presence of disialyl groups (Neu5Acα(2-8)Neu5Ac-) yields a result equivalent only to one mole instead of two moles of Neu5Ac, as the α(2-8)-glycosidic linkage prevents periodate attack on the polyhydroxy side chain of the internal sialic acid moiety.

The method has advantages over the resorcinol and orcinol assays by virtue of its sensitivity and flexibility in determining free and bound sialic acid, and also in the lower level of interference from other monosaccharides and other substances (JOURDIAN *et al.* 1971).

2. *The Periodic Acid/Thiobarbituric Acid Assay*

a) The WARREN and AMINOFF Methods

The periodic acid/thiobarbituric acid assay for sialic acid is probably the most widely used for quantitative analysis. This is due to its sensitivity and specificity.

The method was introduced for sialic acids simultaneously by WARREN (1959, 1963) and AMINOFF (1959, 1961), and was earlier employed for deoxyribose measurement (WARAVDEKAR and SASLAW 1957, 1959). The two methods are similar and generate the same chromogen which condenses with thiobarbituric acid to give a red pigment with an absorption maximum at 549 nm. Differences in the methods lie in the acidity of the initial periodate oxidation and the extraction of the pigment into cyclohexanone (WARREN 1959) or acidic butan-1-ol (AMINOFF 1961).

The oxidation of free sialic acids by periodic acid under the strongly acidic conditions of the assay leads to the formation of a prechromogen, a six carbon aldehyde (I), which then yields the chromogen β-formyl pyruvic acid (III) by aldol cleavage between the carbon atoms 4 and 5 (PAERELS and SCHUT 1965) (Fig. 1). β-Formyl pyruvic acid reacts with thiobarbituric acid to give the red chromophore. A study of Neu5Ac4Me demonstrated a stronger reaction in the WARREN assay than with Neu5Ac, and a lactone form (II, Fig. 1) of the prechromogen was proposed as an explanation for this difference (BEAU and SCHAUER 1980). This 1-4 lactone would be formed under the acidic conditions of the assay only from Neu5Ac having a free hydroxyl group at C-4; it gives no colour with thiobarbituric acid. The AMINOFF (1961) method employs less acidic conditions leading to smaller amounts of the lactone (BEAU and SCHAUER 1980) and shows a higher molar absorption coefficient for Neu5Ac relative to the Warren (1959) method (Table 1).

The presence of O-acyl substituents in the C-7 to C-9 side chain of sialic acid will influence the periodate oxidation and thus the chromogen formation in the assay. The low molar extinction coefficients relative to Neu5Ac (Table 1) indicate that such sialic acids are not de-O-acylated completely in the assay (Beau and Schauer 1980). The presence of O-acetyl groups in the side chain prevents periodate oxidation (Haverkamp *et al.* 1975) and precludes the formation of compound I (Fig. 1). This effect is complete with O-substitution at C-7 or C-8, while C-9 substitution leads to partial blocking of oxidation. An analysis of total sialic acid

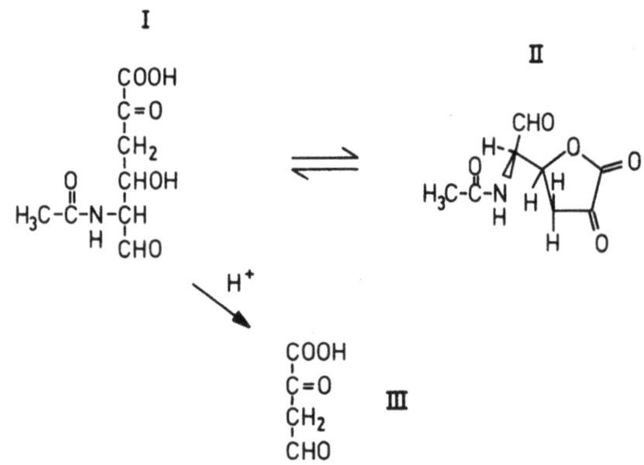

Fig. 1. Compounds formed during the oxidation step of the periodic acid/thiobarbituric acid assay. The six carbon aldehyde (*I*) and its hypothetical lactone (*II*) give rise to the chromogen β-formyl pyruvic acid (*III*)

should therefore include a saponification step to ensure that all O-acyl substituents are removed prior to analysis by this method (0.05 M NaOH for 45 min at 0–4 °C is sufficient).

Sialic acids which do not react in the periodic acid/thiobarbituric acid assay are Neu5Ac2en and Neu-β-Me, neither of these sialic acids yield chromogen I (Fig. 1) on periodate oxidation.

As mentioned above, the periodic acid/thiobarbituric acid methods only function with free sialic acids. This means that prior release of glycosidically bound sialic acids must be undertaken before analysis. Care should be taken in the case of sialic acid glycosides known to break down under the conditions of the assay, e.g. CMP-β-Neu5Ac (Aminoff 1961) and various synthetic ketosides of sialic acid (Meindl and Tuppy 1967, Khorlin *et al.* 1970). This inevitably leads to erroneous values of free sialic acid and involves other parameters which need consideration. These are discussed below in II.10.

In contrast with the requirement for free sialic acids in this assay, several reports have appeared in the literature which describe the occurrence of a non-dialysable periodic acid/thiobarbituric acid positive material (Eichberg and Karnovsky 1966, Brown *et al.* 1970, Srivastava *et al.* 1970, Srivastava and Abou-Issa 1977).

A satisfactory explanation for this phenomenon has, however, not been given by the authors.

The extraction of the pigment formed in aqueous solution into an organic solvent serves to stabilize and intensify the chromophore for measurement. The methods of AMINOFF and WARREN employed acidified butan-1-ol and cyclohexanone, respectively, to achieve this. Several alternative organic solvents for extraction have been put forward including dimethylsulphoxide (DMSO) (SKOZA and MOHOS 1976), HCl/acetone or methyl cellosolve (SAIFER and GERSTENFELD 1962, UCHIDA *et al.* 1977). These alternative extraction solvents have the advantage of increased pigment stability in comparison with the original methods.

b) Interference

A large number of compounds giving interference with the periodic acid/thiobarbituric acid assays have been identified (LEDEEN and YU 1976, KUWAHARA 1980). Those causing the most serious errors are 2-deoxyribose, 2-keto-3-deoxyaldonic acids, glycosides such as lactose, maltose and α-methyl-galactoside, and unsaturated fatty acids (UNGER 1981). These compounds lead to the formation of either similar pigments which cause significant absorption at the true chromophore absorption maximum, or of the same pigment due to the presence or generation of β-formylpyruvic acid (LEDEEN and YU 1976, UNGER 1981). The greatest errors arise in the quantitation of sialic acids from cellular extracts or homogenates containing membrane and nucleic acid material (SCHAUER 1978). As a result, several approaches have been adopted to eliminate interference and allow accurate sialic acid quantitation.

The use of ion-exchange chromatography to purify sialic acids after mild acid hydrolysis and calibration of losses using a standard sialic acid sample has been suggested (SVENNERHOLM 1958, 1963 b, SCHAUER 1978, CAIMI *et al.* 1979) and is routinely carried out in the authors' laboratory. Comparison of this technique with others designed to eliminate interference has underlined its value (e.g. SARRIS and PALADE 1979, ROBOZ *et al.* 1981).

Combination of ion-exchange purification with ether or hexane extraction has proved to be necessary in quantitation of sialic acids released by mild acid hydrolysis from membrane preparations (SCHAUER *et al.* 1975, SCHAUER 1978; see also chapter C). The extraction removes lipid material which would otherwise yield similar pigments in the assay, and which is not always removed after ion-exchange chromatography.

An alternative method for elimination of 2-deoxyribose interference put forward by WARREN (1959) is the measurement of the pigment at other wavelengths and use of a formula to correct the result for sialic acid only. The additional wavelengths were 532 nm, the maximum for malonaldehyde generated from 2-deoxyribose, and 562 nm in cases where the level of contamination was high (WARREN 1959). In cases where the genuine sialic acid pigment is generated the ratio of optical densities at 549 nm : 532 nm has been found to be 2.75, and this value decreases with increasing malonaldehyde contamination.

Methods aimed at removing the 532 nm absorption using alternative extraction procedures have been put forward by Warren (1963) and Roboz *et al.* (1981). The latter authors reviewed the use of isoamyl alcohol to remove the deoxyribose interference, as suggested by Warren (1963), but favoured the use of a pH-dependent extraction in cyclohexanone for this correction. The deoxyribose pigment could be completely removed with cyclohexanone at pH 5.6–6.0. The sialic acid pigment remains in the aqueous phase and can be quantitatively extracted with cyclohexanone after readjustment of the pH to 1.7–1.9 (Roboz *et al.* 1981).

Other methods of removing interference include the use of glucose to suppress deoxyribose pigment formation (Smith *et al.* 1975). The suppression could be augmented with NaCl and trichloracetic acid and deoxyribose interference at concentrations up to 75 μM completely eliminated. However, the method leads to decreases (up to 25%) in sialic acid reactivity.

Interference due to protein, especially haemoglobin, results in a reduction of sensitivity, and proteins should therefore be removed prior to analysis of sialic acid. A method using ethanol/chloroform precipitation has been put forward (Rivetz *et al.* 1980), but such deproteination methods still require additional correction for lipid, and the ion-exchange purification and lipid extraction should be part of the procedure to achieve reproducible and accurate quantitation (Ledeen and Yu 1976, Schauer 1978, 1982).

3. *The Periodic Acid/Methyl-3-Benzothiazolinone-2-Hydrazone Method*

A new colorimetric method for the measurement of free and glycosidically bound sialic acid was introduced by Durand *et al.* (1974) and Massamiri *et al.* (1978, 1979), based on the reaction of methyl-3-benzothiazolinone-2-hydrazone (MBTH) with the formaldehyde liberated on periodate oxidation of the sialic acid side chain. The molar extinction coefficient for Neu5Ac is 67,000 at 625 nm, and the method is thus similar in sensitivity to the periodic acid/thiobarbituric acid methods.

Periodate oxidation is optimal for free sialic acid at a ratio of $NaIO_4$: Neu5Ac of 1.5 : 1. At higher ratios (e.g. 50 : 1) β-formyl pyruvic acid is generated and this leads to error by formation of an alternative chromogen. This drawback does not arise with glycosidically linked sialic acid, although some interference due to protein was observed and deproteination recommended (Massamiri *et al.* 1978, 1979). Although this method has been applied to the measurement of erythrocyte membrane sialic acid (Massamiri *et al.* 1979), interference due to other membrane components presented a major problem with some species and no satisfactory measurement could be made (Shukla and Schauer 1981; see also below under fluorimetric assays).

Where applicable, the MBTH-method is of value for determining total sialic acid concentration without prior hydrolysis or sialidase treatment. However, the problems arising with O-acyl-sialic acids and periodate oxidation also apply here, and saponification may be necessary for the complete assay of sialic acids.

On the other hand, in this assay and also in the fluorimetric adaptation described below, the presence of O-acyl groups in the C-7—C-9 side chain of sialic acids can be detected, as the periodate oxidation is blocked as described in section

II.2.a). Here, mild conditions are necessary to avoid de-O-acylation of 8- or 9-mono-O-acyl-sialic acids and 7,9- or 8,9-di-O-acyl-sialic acids, where no formaldehyde is formed and thus no chromophore. Substitution at C-7 or C-4 will not influence the formaldehyde formation. Comparison of total sialic acid concentrations, measured either by other methods or after saponification using the MBTH assay, with the results of a second MBTH assay without prior saponification, will give a measure of the proportion of sialic acids containing O-acyl groups in positions as described above. The measurement of glycosidically bound sialic acids by these methods gives an advantage over the thiobarbituric acid assay, as O-acyl groups are lost during mild acid hydrolysis, introducing an unknown error into the method. Using these concepts the same estimations can be made with the periodic acid/resorcinol method described in section II.1.b)β).

4. Fluorimetric Assays

A number of fluorimetric assays for sialic acid have been developed. Heating sialic acid with 3,5-diaminobenzoic acid in dilute HCl produces an intense green fluorescence which was used to detect amounts below 0.3 μg of free or bound sialic acid (HESS and ROLDE 1964). Some interference from lipids and solvents was reported (HESS and ROLDE 1964, HARRIS and KLINGMAN 1972), although other carbohydrates showed little or no reaction in the assay (HESS and ROLDE 1964).

An alternative method for free sialic acids only was introduced by MURAYAMA *et al.* (1976), in which reaction with pyridoxamine after heating at 70 °C for 45 min is followed by measurement of the fluorescence with 395 nm for excitation and 470 nm for emission. The method readily allows detection of 0.1 μg of sialic acid and is not influenced by deoxyribose. However, α-keto acids develop the same fluorescence, and care must be taken to eliminate these compounds before using the assay (MURAYAMA *et al.* 1976).

A modification of the periodic acid/thiobarbituric acid method has been described by excitation of the usual chromophore at 550 nm and measurement of the emission at 570 nm (HAMMOND and PAPERMASTER 1976). The assay is 500-fold more sensitive than the conventional spectrophotometric procedures and can detect 10 ng sialic acid. Deoxyribose remains a problem with this assay and the precautions detailed for the periodic acid/thiobarbituric acid assay are required if accurate quantitation is to be made (HAMMOND and PAPERMASTER 1976).

Developments in the authors' laboratory led to the introduction of a new fluorimetric method, detecting the formaldehyde formed after mild periodate oxidation, with acetylacetone. The slightly yellow chromophore thus formed is measured fluorimetrically (F 410/510 nm), and free or glycosidically linked sialic acid can be detected in quantities as low as 1 nmole (0.3 μg) (SHUKLA and SCHAUER 1981). The method is specific, no interfering substances including triglycerides are known. According to the outlines in subsection II.3 the percentage of sialic acids O-acylated at C-8 or C-9 can be determined. Using this method, the total amount of sialic acids and the quantity of derivatives O-acetylated at the positions mentioned have been determined in erythrocyte membranes from various species. The sialic acid values are in agreement with those found with the classical analytical tools.

5. *Enzymic Assays*

The most specific method for sialic acid determination is possible with enzymes. Using acylneuraminate pyruvate-lyase (EC 4.1.3.3), sialic acids are cleaved to acylmannosamines and pyruvate. By coupling this reaction to lactate dehydrogenase the reduction of pyruvate allows monitoring of the pyruvate concentration and drives the lyase reaction, otherwise an equilibrium reaction, to completion (Brunetti *et al.* 1963). The concentration of sialic acid can be measured by spectrophotometry or fluorimetry of the amount of NADH oxidized on completion of the reaction, or by measurement of the initial rate of NADH oxidation. A less sensitive alternative is determination of N-acylmannosamine using colorimetric methods. Brunetti *et al.* (1962) utilized the Morgan-Elson reaction with dimethyl-aminobenzaldehyde for this purpose.

Commercial sources of the enzyme are available, and the assay has been modified to use smaller amounts (Gantt *et al.* 1964, Schauer *et al.* 1971, Ferwerda *et al.* 1981). Thus, 10 mU/ml of the lyase are required. Care must be taken where O-acyl-sialic acids are to be estimated. The presence of O-acyl groups at C-7 or C-9 of the sialic acid molecule leads to a reduction in the reaction rate of 30–50%, while O-acetylation of C-4 gives 90% reduction (Schauer *et al.* 1971, Buscher *et al.* 1974, Corfield *et al.* 1976).

Saponification prior to analysis, as described previously (section II.2.a)), eliminates this problem (Ledeen and Yu 1976, Schauer 1978). Interference may arise from endogeneous pyruvate in tissue samples, or from NADH oxidase activity.

A modification of the acylneuraminate pyruvate-lyase assay using pyruvate oxidase to generate H_2O_2, and measurement of this with peroxidase and p-chlorophenol-4-aminoantipyrine has been described (Sugahara *et al.* 1980). The measurement of the chromophore is at 505 nm and a molar extinction coefficient of 1.14×10^4 was found at this wavelength. This test has been used in conjunction with sialidase for estimation of serum sialic acid (Sugahara *et al.* 1980), but remains less sensitive than the method coupled with lactate dehydrogenase.

6. *O-Acyl Group Assay*

Quantitative estimation of the O-acyl content of sialic acid preparations can be made using the hydroxamic acid reaction as described by Ludowieg and Dorfman (1960), modified from the method of Hestrin (1949). Calibration with ethyl acetate or other ester standards provides a simple quantitative assay, measuring the chromophore at 520 nm. Volume reductions in the assay procedure allow quantitation of 0.05 μmole of sialic acid acyl ester (Schauer 1978). Qualitative identification of different acylhydroxamates can be made on thin-layer chromatography (see section III.3.b)).

The level of O-acylation of sialic acids can also be quantitated by difference measurements with the periodic acid/thiobarbituric acid method (Neuberger and Ratcliffe 1972, Skoza and Mohos 1976, Sarris and Palade 1979, see section II.2), with the periodate/resorcinol (II.1.b)β)) and the MBTH method (II.3), and a fluorimetric method (Shukla and Schauer 1981, II.4). Using such assays care should be taken to run suitable controls for the saponification step and to consider the increased resistance to mild acid hydrolysis of O-acyl-sialic acids (NEUBERGER and RATCLIFFE 1972).

The detection of O-lactyl groups in some sialic acids (HAVERKAMP *et al.* 1976, SCHAUER *et al.* 1976, REUTER *et al.* 1980) was quantitated, after saponification and neutralization, with the aid of L-lactate dehydrogenase, NAD^+ and hydrazine (HOHORST 1970, HAVERKAMP *et al.* 1976).

O-Glycolyl groups have been reported in sialic acids (BUSCHER *et al.* 1974, CORFIELD *et al.* 1976), although no mass spectrometric data have been obtained to support this. Both O- and N-glycolyl groups can be quantitated after conversion to ethyl glycolate using p-toluene sulphonic acid in ethanol and distillation *in vacuo* before measurement as glycolic acid using EEGRIWE's reagent (2,7-dihydroxynaphthalene in conc. H_2SO_4; SCHOOP and FAILLARD 1967). The chromophore is measured at 546 nm, and minimum amounts determined by this method are 0.5 μg glycolic acid (SCHOOP and FAILLARD 1967, SCHAUER 1978). Although tedious, the inclusion of the distillation step greatly increases the specificity of the assay.

7. *Automatic Procedures*

Several laboratories have reported automated procedures for the quantitation of sialic acid in biological fluids, column effluents or for miscellaneous samples. The periodic acid/resorcinol assay has been adapted for use in a Technicon autoanalyser for routine assay of serum sialic acid (REY *et al.* 1975). The method is based on the assay developed by JOURDIAN *et al.* (1971) and does not involve the organic solvent extraction. As with the original assay, the automated procedure can be adapted to measure free or glycosidically bound sialic acid (REY *et al.* 1975). Modification of this system allows measurement of erythrocyte membrane sialic acid simultaneously with protein (GERBAUT *et al.* 1978).

The periodic acid/thiobarbituric acid assay has been automated by several authors for use with autoanalysers. For instance, KENDALL (1968) and FIDGEN (1973) employed the conditions of the AMINOFF (1961) assay including organic solvent extraction. This assay was also used by GERBAUT *et al.* (1973), but without extraction, for measurement of sialic acid in serum. Adaptation of the WARREN assay to the autoanalyser was reported by DELMOTTE (1968), who also dispensed with extraction of the chromophore. This method does not correct for 2-deoxyribose contamination (DELMOTTE 1968) and has subsequently been adapted to analyze both sialic acid and 2-deoxyribose in the same samples from blood and tissues (ENGEN *et al.* 1974) by measuring the optical density at 550 and 530 nm.

An automated assay based on the AMINOFF method, including acidic butan-1-ol and combining the sensitivity of the assay with ion-exchange purification, has been put forward by KRANTZ and LEE (1975) and remains the most sensitive of all automated methods. It can detect 1.5 nmoles of sialic acid (~ 0.5 μg), eliminates major interference due to 2-deoxyribose and precludes any corrections for the presence of such interference.

8. *High Performance Liquid Chromatography*

Individual sialic acids (for example Neu5Ac, Neu5Gc, $Neu4,5Ac_2$ and $Neu5,9Ac_2$) have been separated on an analytical scale using a strongly basic anion-exchange resin in borate buffer and monitoring the eluting sialic acids at 570 nm after reaction with 4,4′-dicarboxy-2,2′-biquinoline sodium salt (SHUKLA *et al.* 1982a). Modification of this method and direct estimation of the eluting sialic acids at 195 or 215 nm led to qualitative and quantitative analysis of different sialic acids including Neu5Ac2en in a highly sensitive manner allowing the determination of minimum amounts of 6 ng (20 pmol) of sialic acids (SHUKLA *et al.* 1982b, SHUKLA and SCHAUER 1982). HPLC has also been adopted to quantitate Neu5Ac in serum (SILVER *et al.* 1981) and for the identification of the reaction products of sialyltransferase experiments (BERGH *et al.* 1981).

9. Other Methods

Alternative methods of sialic acid determination have been described, but remain largely unused due to their poor sensitivity or difficulty in application relative to the assays described in these sections. These include the sulpho/phospho/vanillin method, originally used with gangliosides (SAIFER and FELDMAN 1971), and the 1,10-phenanthroline method of DIMITROV (1973), which was subsequently refuted as a valid method by SNYDER *et al.* (1974). Of the remaining methods, already detailed at the beginning of section II, only the direct EHRLICH reaction has occasionally been employed where great sensitivity was of secondary importance (WERNER and ODIN 1952, ONODERA *et al.* 1965), or in cases where an additional colorimetric assay for identification of sialic acid was required (ONODERA *et al.* 1965, CABEZAS 1973, SCHAUER 1978).

10. Quantitative Problems in Sialic Acid Release

A number of problems arise when sialic acids are to be released prior to quantitative analysis, in general by sialidase digestion or mild acid hydrolysis, as described in chapter C.

Sialidase digestion frequently leads to incomplete release of sialic acid, thus giving false values for total contents in sialoglycoconjugates. The reasons for the incomplete release are often linked with the nature of the substance as has been discussed in section I.

Mild acid hydrolysis is less specific than sialidase digestion and also less expensive. Attention should be paid to the different conditions of hydrolysis required for various glycoconjugates and the slower release of O-acylated sialic acids relative to Neu5Ac (LEDEEN and YU 1976) (see chapter C). To obtain optimal quantitation, a release curve for sialic acid should be prepared (SCHAUER *et al.* 1975, O'KENNEDY 1979). In addition to this precaution, the destruction of sialic acid under the conditions of the hydrolysis should be assessed (e.g. SCHAUER *et al.* 1975, O'KENNEDY 1979). The inclusion of a sialic acid standard under the conditions of hydrolysis will allow assessment of the level of destruction (SCHAUER *et al.* 1975, SCHAUER 1978, CAIMI *et al.* 1979).

III. Thin-Layer Chromatography

1. Chromatographic Systems

A variety of chromatographic systems for the separation of sialic acids on either cellulose or silica gel thin-layers have been described. Details of these are given in

Table 2. *Solvents for the chromatography of the sialic acids on thin-layers and paper*
Chromatography on cellulose or silica gel thin-layers and by ascending (a) or descending (d) chromatography on Whatman No 1 or 3 MM (x) chromatography paper

Solvent systems		Volume ratios	References
Thin-layer chromatography			
Cellulose			
Butan-1-ol : propan-1-ol : 0.1 M HCl		1 : 2 : 1	Svennerholm and Svennerholm 1958
Butan-1-ol : propan-1-ol : H_2O		1 : 2 : 1	
Butan-1-ol : acetic acid : H_2O		4 : 1 : 5	Crumpton 1959
Butan-1-ol : pyridine : H_2O		6 : 4 : 3	Crumpton 1959
Silica Gel			
Propan-1-ol : H_2O		7 : 3	Granzer 1962
Propan-1-ol : H_2O		6 : 3	Öhman 1967
Propan-1-ol : 14 M NH_4OH : H_2O		6 : 1 : 2	Öhman 1967
Propan-1-ol : ethyl acetate : H_2O		5 : 1 : 4	Gal 1968
Ethyl acetate : acetic acid : H_2O		5 : 2 : 2	Beau and Schauer 1980
2-Butanone : acetic acid : methanol : H_2O		10 : 4 : 3 : 1	Yohe and Yu 1981
Chloroform : methanol		17 : 3	Beau and Schauer 1980
Paper chromatography			
Ethyl acetate : pyridine : H_2O	d	10 : 4 : 3	Blix and Lindberg 1960
Ethyl acetate : pyridine : acetic acid : H_2O	d	5 : 5 : 1 : 3	Maury 1971
Butan-1-ol : acetic acid : H_2O	d	4 : 1 : 5	Berggård and Odin 1958
Butan-1-ol : acetic acid : H_2O x	d	5 : 2 : 2	Liao *et al.* 1973
Butan-1-ol : pyridine : H_2O	d	6 : 4 : 3	Blix *et al.* 1956
Butan-1-ol : propan-1-ol : 0.1 M HCl x	d/a	1 : 2 : 1	Liao *et al.* 1973, Yohe and Yu 1981
Butan-1-ol : pyridine : 1 M HCl	d/a	5 : 3 : 2	McLean *et al.* 1971
1-Butyl acetate : acetic acid : H_2O	d	3 : 2 : 1	Liao *et al.* 1973
95% ethanol : H_2O : NH_4OH	d	170 : 30 : 2	Gold and Miller 1978

Table 2. Many of these systems were developed during the early work on sialic acids (see GOTTSCHALK 1960). The introduction of commercial thin-layers (0.1 mm) of cellulose and silica gel has greatly improved reproducibility and led to greater resolution. This has been especially noticeable in the two-dimensional techniques noted below (III.3). In all cases removal of cations from the sialic acid samples on ion-exchange chromatography is necessary to yield good separations.

Table 3. *R_f values of sialic acids on cellulose and silica gel thin-layers (0.1 mm)*
Solvent systems used were: cellulose, butan-1-ol : propan-1-ol : 0.1 M HCl (1 : 2 : 1, v/v) and silica gel, propan-1-ol : water (7 : 3, v/v)

Sialic acids	Cellulose	Silica gel
Neu5Ac	0.45	0.39
Neu4,5Ac$_2$	0.60	0.61
Neu5,7Ac$_2$	0.54	0.61
Neu5,9Ac$_2$	0.63	0.61
Neu5,7,9Ac$_3$	0.70	0.73
Neu5,8,9Ac$_3$	0.75	—
Neu5,7,8,9Ac$_4$	0.80	—
Neu5Ac9Lt	0.56	0.61
Neu5Ac2en	0.55	0.70
C-8-Neu5Ac	0.54	—
C-7-Neu5Ac	0.58	—
Neu5Gc	0.35	0.39
Neu4Ac5Gc	0.65	0.61
Neu9Ac5Gc	0.55	0.61
Neu7,9Ac$_2$5Gc	0.70	—
C-8-Neu5Gc	0.38	—
C-7-Neu5Gc	0.42	—

Abbreviations of the sialic acids are those used in chapter B, Table 1.—Not determined.

In the authors' laboratory most experience has been obtained using cellulose thin-layer plates, prewashed in 0.1 M HCl, and the butan-1-ol : propan-1-ol : 0.1 M HCl (1 : 2 : 1, v/v/v) solvent. This system gives the best resolution and reproducibility. Silica gel thin-layers run in propan-1-ol : H_2O (7 : 3, v/v) have also frequently been used. The R_f values of different sialic acids for these systems are shown in Table 3. The cellulose support gives better separations of mixtures containing both Neu5Ac and Neu5Gc or their derivatives. The presence of O-acyl groups leads to greater hydrophobicity of the sialic acid and to faster migration in both systems, while on cellulose the presence of the N-glycolyl group results in lower R_f values relative to the N-acetyl derivatives. The butan-1-ol : propan-1-ol : 0.1 M HCl solvent and cellulose thin-layers can also be used to separate the acylmannosamines formed after the degradation of sialic acids by acylneuraminate pyruvate-lyase (BUSCHER *et al.* 1974, SCHAUER 1978).

2. *Visualization*

Several reagents have been described which are suitable for the detection of sialic acid-containing spots on thin-layer chromatograms. These are based on the colorimetric methods described in section II.1 and II.2. The orcinol/Fe^{3+} reagent is diluted 3:1 with water, and the sprayed plate sandwiched between glass plates before incubation for 15 min at 120 °C. The purple-blue colour is typical for sialic acids, other carbohydrates give brown, green or yellow spots. The resorcinol/Cu^{2+} reagent is sprayed after a periodate spray has oxidized the sialic acids. This method is an adaptation of the method of JOURDIAN *et al.* (1971). Finally, the thiobarbituric acid method has also been adapted for thin-layer plates using successively periodic acid, arsenite and thiobarbituric acid sprays (WARREN 1960). The red spots are developed after incubation at 100 °C for 15 min under glass plates. Care must be taken using this staining technique, as O-acyl sialic acids may block colour formation; therefore, saponification is necessary if such sialic acids are expected.

3. *Two-Dimensional Techniques*

a) Individual Sialic Acid Determination

Identification of sialic acids can be aided using thin-layer chromatography in two dimension with intermediate saponification (BUSCHER *et al.* 1974, SCHAUER 1978). A mixture of partially O-acylated sialic acids is applied as a spot at the corner of a cellulose thin-layer plate and developed in the butan-1-ol : propan-1-ol : 0.1 M HCl system. Then, the plate is placed in ammonia vapour to saponify the O-acyl groups and subsequently run in the second dimension in the same solvent. Standards can be run in each dimension to calibrate the plate. The saponification generates the parent sialic acids Neu5Ac and Neu5Gc from their O-acyl derivatives, thus enabling identification of sialic acid mixtures by this method. Measurement of R_f values in the first dimension will give information on the possible nature of the spots, while the second dimension will indicate whether each spot contains one or both of the parent sialic acids.

b) O-Acyl Group Identification

Using the same approach as above a separation of sialic acids in the first dimension is achieved. The plate is then dried and masked, except for the sialic acids which have migrated, and this area is sprayed with alkaline hydroxylamine (equal volumes of 0.35 M hydroxylamine hydrochloride and 2.5 M ammonium hydroxide in water and methanol [1 : 1, v/v]). The plate is covered to prevent evaporation and left for 60 min. Chromatography in the second dimension together with hydroxamate standards is in propan-1-ol : 10% aqueous ammonium carbonate : 5 M ammonium hydroxide (6 : 2 : 1, v/v/v) (BUSCHER *et al.* 1974, SCHAUER 1978). The acylhydroxamates are visualized using a spray of 10% $FeCl_3$ solution. The method is specific but not sensitive, at least 20–30 μg of O-acyl sialic acid in any one spot is required to yield a useful result (SCHAUER 1978).

c) Sialic Acid Identification in Oligosaccharides

A further use of two-dimensional thin-layer chromatography for the identification of sialic acids in oligosaccharide fractions has been demonstrated

(VEH *et al.* 1981). The separation is carried out on cellulose plates in butan-1-ol : propan-1-ol : 0.1 M HCl (1 : 2 : 1, v/v/v), followed by spraying of the plate with 0.1 M aqueous HCl and incubation at 80 °C for 60 min. After drying, chromatography in the second dimension is carried out in the same solvent, and the spots are visualized. The migration of standards in both dimensions allows identification of the sialic acids present in each oligosaccharide band (VEH *et al.* 1981).

4. Paper Chromatography

Paper chromatography was used in the early days of sialic acid detection (BLIX and LINDBERG 1960, GOTTSCHALK 1960) and has now been superceded by the more sensitive thin-layer chromatographic methods. Some paper chromatographic systems and the corresponding references are noted in Table 2.

Bibliography

AMINOFF, D., 1959: Virology **7**, 355—357.
— 1961: Biochem. J. **81**, 384—392.
AYALA, W., MOORE, L. V., HESS, E. L., 1951: J. Clin. Invest. **30**, 781—785.
BEAU, J.-M., SCHAUER, R., 1980: Eur. J. Biochem. **106**, 531—540.
BERGGÅRD, I., ODIN, L., 1958: Arkiv Kemi **12**, 581—595.
BERGH, M. L. E., KOPPEN, P., VAN DEN EIJNDEN, D., 1981: Carbohydr. Res. **94**, 225—229.
BLIX, G., LINDBERG, E., 1960: Acta Chem. Scand. **14**, 1809—1814.
— — ODIN, L., WERNER, I., 1956: Acta Soc. Med. Upsalien. **61**, 1—25.
BÖHM, P., DAUBER, ST., BAUMEISTER, L., 1954: Klin. Wochenschr. **32**, 289—292.
BROWN, C. R., SRIVASTAVA, P. N., HARTREE, E. F., 1970: Biochem. J. **118**, 123—133.
BRUNETTI, P., JOURDIAN, G. W., ROSEMAN, S., 1962: J. Biol. Chem. **237**, 2447—2453.
— SWANSON, A., ROSEMAN, S., 1963: Methods Enzymol. **6**, 465—473.
BUSCHER, H.-P., CASALS-STENZEL, J., SCHAUER, R., 1974: Eur. J. Biochem. **50**, 71—82.
CABEZAS, J. A., 1973: Rev. Esp. Fisiol. **29**, 307—322.
CAIMI, L., LOMBARDO, A., PRETI, A., WIESMANN, U., TETTAMANTI, G., 1979: Biochim. Biophys. Acta **571**, 137—146.
CORFIELD, A. P., FERREIRA DO AMARAL, C., WEMBER, M., SCHAUER, R., 1976: Eur. J. Biochem. **68**, 597—610.
CRUMPTON, M. J., 1959: Biochem. J. **72**, 479—486.
DELMOTTE, P., 1968: Z. Klin. Chem. Klin. Biochem. **6**, 46—48.
DIMITROV, D. G., 1973: Hoppe-Seyler's Z. Physiol. Chem. **354**, 121—124.
DURAND, G., FÉGER, J., COIGNOUX, M., AGNERAY, J., PAYS, M., 1974: Anal. Biochem. **61**, 232—236.
EICHBERG, J., KARNOVSKY, M. L., 1966: Biochim. Biophys. Acta **124**, 118—124.
ENGEN, R. L., ANDERSON, A., ROUZE, L. L., 1974: Clin. Chem. **20**, 1125—1127.
FERWERDA, W., BLOK, C. M., HEIJLMAN, J., 1981: J. Neurochem. **36**, 1492—1499.
FIDGEN, K. J., 1973: Anal. Biochem. **54**, 379—385.
FOLCH, J., ARSOVE, S., MEATH, J. A., 1951: J. Biol. Chem. **191**, 819—831.
GAL, A. E., 1968: Anal. Biochem. **24**, 452—461.
GANTT, R., MILLNER, S., BINKLEY, S. B., 1964: Biochemistry **3**, 1952—1960.
GERBAUT, L., REY, E., LOMBART, C., 1973: Clin. Chem. **19**, 1285—1287.
— DE LAUTURE, D., OLIVE, AG. G., PEQUIGNOT, H., 1978: Clin. Chem. **24**, 1287.
GOLD, D. V., MILLER, F., 1978: Cancer Res. **38**, 3204—3211.

GOTTSCHALK, A., 1960: The Chemistry and Biology of Sialic Acids and Related Substances. London-New York: Cambridge University Press.
GRANZER, E., 1962: Hoppe-Seyler's Z. Physiol. Chem. **328**, 277—279.
HAMMOND, K. S., PAPERMASTER, D. S., 1976: Anal. Biochem. **74**, 292—297.
HARRIS, J. U., KLINGMAN, J. D., 1972: J. Neurochem. **19**, 1267—1278.
HAVERKAMP, J., SCHAUER, R., WEMBER, M., KAMERLING, J. P., VLIEGENTHART, J. F. G., 1975: Hoppe-Seyler's Z. Physiol. Chem. **356**, 1575—1583.
— — — FARRIAUX, J.-P., KAMERLING, J. P., VERSLUIS, C., VLIEGENTHART, J. F. G., 1976: Hoppe-Seyler's Z. Physiol. Chem. **357**, 1699—1705.
HESS, E. L., COBURN, A. F., BATES, R. C., MURPHY, P., 1957: J. Clin. Invest. **36**, 449—455.
HESS, H. H., ROLDE, E., 1964: J. Biol. Chem. **239**, 3215—3220.
HESTRIN, S., 1949: J. Biol. Chem. **180**, 249—261.
HOHORST, H. J., 1970: In: Methoden der enzymatischen Analyse (BERGMEYER, H. U., ed.), Vol. II, pp. 1425—1429. Weinheim: Verlag Chemie.
JOURDIAN, G. W., DEAN, L., ROSEMAN, S., 1971: J. Biol. Chem. **246**, 430—435.
KENDALL, A. P., 1968: Anal. Biochem. **23**, 150—155.
KHORLIN, A. YA., PRIVALOVA, I. M., ZAKSTELSKAYA, L. YA., MOLIBOG, E. V., EVSTIGNEEVA, N. A., 1970: FEBS Lett. **8**, 17—19.
KLENK, E., LANGERBEINS, H., 1941: Hoppe-Seyler's Z. Physiol. Chem. **270**, 185—193.
KRANTZ, M. J., LEE, Y. C., 1975: Anal. Biochem. **63**, 464—469.
KUWAHARA, S. S., 1980: Anal. Biochem. **101**, 54—60.
LEDEEN, R. W., YU, R. K., 1976: In: Biological Roles of Sialic Acid (ROSENBERG, A., SCHENGRUND, C.-L., eds.), pp. 1—57. New York: Plenum Press.
LIAO, T.-H., GALLOP, P. M., BLUMENFIELD, O. O., 1973: J. Biol. Chem. **248**, 8247—8253.
LUDOWIEG, J., DORFMAN, A., 1960: Biochim. Biophys. Acta **38**, 212—218.
MASSAMIRI, Y., BELJEAN, M., DURAND, G., FÉGER, J., PAYS, M., AGNERAY, J., 1978: Anal. Biochem. **91**, 618—625.
— DURAND, G., RICHARD, A., FÉGER, J., AGNERAY, J., 1979: Anal. Biochem. **97**, 346—351.
MAURY, P., 1971: Biochim. Biophys. Acta **252**, 48—57.
MCLEAN, R. L., SUTTAJIT, M., BEIDLER, J., WINZLER, R. J., 1971: J. Biol. Chem. **246**, 803—809.
MEINDL, P., TUPPY, H., 1967: Monatsh. Chem. **98**, 53—60.
MIETTINEN, T., TAKKI-LUUKKAINEN, I. T., 1959: Acta Chem. Scand. **13**, 856—858.
MURAYAMA, J.-I., TOMITA, M., TSUJI, A., HAMADA, A., 1976: Anal. Biochem. **73**, 535—538.
NEUBERGER, A., RATCLIFFE, W. A., 1972: Biochem. J. **129**, 683—693.
— — 1973: Biochem. J. **133**, 623—628.
PAERELS, G. B., SCHUT, J., 1965: Biochem. J. **96**, 787—792.
PETERS, B. P., ARONSON, N. N., JR., 1976: Carbohydr. Res. **47**, 345—353.
ÖHMAN, R., 1967: Acta Chem. Scand. **21**, 1670—1672.
O'KENNEDY, R., 1979: Irish J. Med. Sci. **148**, 92—96.
ONODERA, K., HIRANO, S., HAYASHI, H., 1965: Carbohydr. Res. **1**, 44—51.
REUTER, G., PFEIL, R., KAMERLING, J. P., VLIEGENTHART, J. F. G., SCHAUER, R., 1980: Biochim. Biophys. Acta **630**, 306—310.
REY, E., GERBAUT, L., LOMBART, C., 1975: Clin. Chem. **21**, 412—414.
RIVETZ, B., LIPKIND, M. A., SHICHMANTER, E., BOGIN, E., 1980: Experientia **36**, 370—371.
ROBOZ, J., SUTTAJIT, M., BEKESI, J. G., 1981: Anal. Biochem. **110**, 380—388.
SAIFER, A., FELDMAN, N. I., 1971: J. Lipid. Res. **12**, 112—115.
— GERSTENFELD, S., 1962: Clin. Chim. Acta **7**, 467—475.
SARRIS, A. H., PALADE, G. E., 1979: J. Biol. Chem. **254**, 6724—6731.
SCHAUER, R., 1978: Methods Enzymol. **50 C**, 64—89.
— 1982: Adv. Carbohyd. Chem. Biochem. **40**, 131 – 234.

Schauer, R., Faillard, H., 1968: Hoppe-Seyler's Z. Physiol. Chem. **349**, 961—968.
— Wirtz-Peitz, F., Faillard, H., 1970: **351**, 359—364.
— Wember, M., Wirtz-Peitz, F., Ferreira do Amaral, C., 1971: Hoppe-Seyler's Z. Physiol. Chem. **350**, 111—115.
— Corfield, A. P., Wember, M., Danon, D., 1975: Hoppe-Seyler's Z. Physiol. Chem. **356**, 1727—1732.
— Haverkamp, J., Wember, M., Vliegenthart, J. F. G., Kamerling, J. P., 1976: Eur. J. Biochem. **62**, 237—242.
Schoop, H. J., Faillard, H., 1967: Hoppe-Seyler's Z. Physiol. Chem. **348**, 1509—1517.
Seibert, F. B., Pfaff, M. L., Seibert, M. V., 1948: Arch. Biochem. Biophys. **18**, 279—295.
Shukla, A. K., Schauer, R., 1981: Hoppe-Seyler's Z. Physiol. Chem. **362**, 236—237.
— Scholz, N., Reimerdes, E. H., Schauer, R., 1982a: Anal. Biochem. **122**, 78–82.
— Reuter, G., Schauer, R., 1982b: Fresenius Z. Anal. Chem. **311**, 376.
— Schauer, R., 1982: J. Chromatogr. **244**, 80—89.
Silver, H. K. B., Karim, K. A., Gray, M. J., Salinas, F. P., 1981: J. Chromatog. **224**, 381–388.
Skoza, L., Mohos, S., 1976: Biochem. J. **159**, 457—462.
Smith, C. H., Donohue, T. M., Depper, R., 1975: Anal. Biochem. **67**, 290—297.
Snyder, S. L., Mathewson, N. S., Sobocinski, P. Z., 1974: Clin. Chem. **20**, 387—388.
Spiro, R. G., 1966: Methods Enzymol. **8**, 3—26.
Srivastava, P. N., Abou-Issa, H., 1977: Biochem. J. **161**, 193—200.
— Zanefeld, L. J. D., Williams, W. L., 1970: Biochem. Biophys. Res. Commun. **39**, 575—582.
Sugahara, K., Sugimoto, K., Nomura, O., Usui, T., 1980: Clin. Chim. Acta **108**, 493—498.
Svennerholm, E., Svennerholm, L., 1958: Nature **181**, 1154—1155.
Svennerholm, L., 1957a: Arkiv Kemi **10**, 577—596.
— 1957b: Biochim. Biophys. Acta **24**, 604—611.
— 1958: Acta Chem. Scand. **12**, 547—554.
— 1963a: Methods Enzymol. **6**, 453—459.
— 1963b: Methods Enzymol. **6**, 459—462.
Tuppy, H., Gottschalk, A., 1972: In: Glycoproteins: Their Composition, Structure and Function (Gottschalk, A., ed.), 2nd ed., pp. 403—449. Amsterdam: Elsevier.
Uchida, Y., Tsukada, Y., Sugimori, T., 1977: J. Biochem. **82**, 1425—1433.
Unger, F., 1981: Adv. Carbohydr. Chem. Biochem. **38**, 324—388.
Veh, R. W., Corfield, A. P., Sander, M., Schauer, R., 1977: Biochim. Biophys. Acta **486**, 145—160.
— Michalski, J.-C., Corfield, A. P., Sander-Wewer, M., Gies, D., Schauer, R., 1981: J. Chromatog. **212**, 313—322.
Waravdekar, V. S., Saslaw, L. D., 1957: Biochim. Biophys. Acta **24**, 439.
— — 1959: J. Biol. Chem. **234**, 1945—1950.
Warren, L., 1959: J. Biol. Chem. **234**, 1971—1975.
— 1960: Nature **186**, 237.
— 1963: Methods Enzymol. **6**, 463—465.
— 1964: Biochim. Biophys. Acta **83**, 129—132.
Werner, I., Odin, L., 1952: Acta Soc. Med. Upsalien. **57**, 230—241.
Wirtz-Peitz, F., 1969: Doctoral thesis, Ruhr-University Bochum, Federal Republic of Germany.
Yohe, H. C., Yu, R. K., 1981: Carbohydr. Res. **93**, 1—9.

F. Gas-Liquid Chromatography and Mass Spectrometry of Sialic Acids

JOHANNIS P. KAMERLING and JOHANNES F. G. VLIEGENTHART

Department of Bio-Organic Chemistry, State University of Utrecht, Utrecht, The Netherlands

With 27 Figures

Contents

I. Introduction

For the structural analysis of the carbohydrate chains of glycoconjugates, oligosaccharides and polysaccharides, the introduction of gas-liquid chromatography (g.l.c.) and later on of gas-liquid chromatography combined with mass spectrometry (g.l.c./m.s.) has proved to be of great importance. As has been summarized in several comprehensive reviews, many protecting groups are in use to prepare volatile sugar derivatives (BISHOP 1964, CLAMP *et al.* 1971, DUTTON 1973, 1974).

Among the monosaccharides which occur as constituents of glycoconjugates and polysaccharides, sialic acid is a complicated compound. From a chemical point of view it is a C_9-monosaccharide with several different functional groups, which have each to be taken into account when developing derivatization techniques. One can discern a carboxyl group, a keto-acetal function, an acetylated or glycolylated amino function, hydroxyl groups with different reactivities, and a deoxy-group adjacent to the anomeric centre. One or more of the hydroxyl groups can be acetylated, lactylated, methylated or sulfated. In water, sialic acids exist predominantly in the β-anomeric form (equatorial carboxyl group in the stable $^2C_5$-conformation) (HAVERKAMP *et al.* 1982). Methyl ester methyl glycosides of sialic acids obtained by treatment with methanolic HCl have also mainly the β-configuration.

In glycoproteins and glycolipids, sialic acids occupy generally terminal positions in the carbohydrate chain; they are coupled via α-glycosidic linkages (MACHER and SWEELEY 1978, MONTREUIL 1980). Sometimes oligomeric chains of sialic acid residues are attached to the carbohydrate backbone of these molecules (FINNE *et al.* 1977, MACHER and SWEELEY 1978, INOUE and MATSUMURA 1980). Also internal sialic acid residues have been detected, even as branching points (HOTTA *et al.* 1973, SUGITA 1979a, b, SMIRNOVA and KOCHETKOV 1980, VAN DER MEER *et al.* 1982). Furthermore, sialic acids have been found as constituents of homo- and hetero-polysaccharides (BHATTACHARJEE and JENNINGS 1976).

II. Gas-Liquid Chromatography

Protection of sialic acids by trimethylsilylation is widely applied in g.l.c. analysis. Several variants have been described: trimethylsilylation of sialic acid methyl esters (SWEELEY *et al.* 1963, SCHAUER *et al.* 1976), of sialic acid methyl ester methyl glycosides (SWEELEY and WALKER 1964, YU and LEDEEN 1970, HAVERKAMP *et al.* 1975, KAMERLING *et al.* 1975a, MONONEN and KÄRKKÄINEN 1975), of free sialic acids (CRAVEN and GEHRKE 1968, CASALS-STENZEL *et al.* 1975, ROBOZ *et al.* 1978), and of deaminated sialic acid methyl ester methyl glycoside (MONONEN 1981). Furthermore, the analyses of trifluoroacetylated (ZANETTA *et al.* 1972, YOHE and YU 1981), permethylated and partially O-methylated (BHATTACHARJEE and JENNINGS 1976, RAUVALA and KÄRKKÄINEN 1977, VAN HALBEEK *et al.* 1978, INOUE and MATSUMURA 1979, BRUVIER *et al.* 1981) sialic acid methyl ester methyl glycosides have been reported. The latter derivatives are analyzed after subsequent trimethylsilylation or acetylation of free hydroxyl groups.

The release of monosaccharides from glycoconjugates and polysaccharides can be carried out via hydrolysis or methanolysis. The stabilities of free monosaccharides to aqueous acid vary greatly, necessitating different hydrolysis conditions optimal for each type of sugar. On the other hand, it has been reported that methanolysis is as efficient as hydrolysis in cleaving glycosidic bonds, and causes less destruction of sugars than does aqueous acid. This observation made it possible to develop single step methods for the qualitative and quantitative analyses of commonly occurring monosaccharides as (methyl ester) methyl glycoside derivatives.

Generally used methanolysis conditions are 0.5–1.0 N methanolic HCl (16–24 h, 80–85 °C) (CHAMBERS and CLAMP 1971, CLAMP *et al.* 1971, ZANETTA *et al.* 1972, KAMERLING *et al.* 1975a). Although the glycosidic bonds of the sialic acids are cleaved in a good yield and destruction is minimized, these conditions lead to complete N,O-deacylation. Therefore, most authors apply a (re-)N-acetylation step in the derivatization procedure, resulting in the formation of Neu5Ac methyl ester methyl glycoside (CHAMBERS and CLAMP 1971, CLAMP *et al.* 1971, KAMERLING *et al.* 1975a). After trimethylsilylation this approach makes it possible to determine the sialic acids in the form of Neu5Ac. Details including g.l.c. are given in section II.1. When trifluoroacetylation is applied, the (re-)N-acetylation step can be omitted; sialic acids are analyzed in the form of N-trifluoroacetyl-neuraminic acid (ZANETTA *et al.* 1972).

Special attention has been paid to the protection of N- and O-acyl substituents against methanolic HCl. YU and LEDEEN (1970) proposed a mild methanolysis procedure (0.05 N methanolic HCl; 1 h, 80 °C), which would supposedly lead to only a small degree of N-deacylation. Based on this study, they developed a quantitative g.l.c. method for the combined determination of Neu5Ac and Neu5Gc. It should be noted that O-deacylation occurs even when 0.02 N methanolic HCl is employed (CASALS-STENZEL *et al.* 1975). Application of the latter conditions has the disadvantage of incomplete release of sialic acid as methyl ester methyl glycoside.

Terminal sialic acids are released in a good yield under mild acidic conditions like formic acid, pH 2 (1 h, 70 °C), 0.1 N H_2SO_4 or 0.1 N HCl (1 h, 80 °C) (SCHAUER 1978). These conditions do not lead to N-deacylation; however, O-deacylation occurs to an extent of about 50%. Mild acid hydrolysis has frequently been chosen to release N,O-acylneuraminic acids from biological materials for identification by g.l.c./m.s. Derivatization procedures are presented in sections II.3 and II.4. For g.l.c./m.s., see section III.2. Quantitative g.l.c. procedures after pertrimethylsilylation (section II.4) have been worked out by CASALS-STENZEL *et al.* (1975). Because of partial O-deacylation during hydrolysis and subsequent isolation, quantitative analysis of the various N,O-acylneuraminic acids present in *native* biological material is still a serious problem. The nonavailability of pure N,O-acylneuraminic acids as standard compounds makes it also difficult to determine reliable molar adjustment factors for g.l.c. analysis.

SUGITA (1979 a) developed a different approach for the combined determination of Neu5Ac and Neu5Gc, by applying solvolysis with 1.0 N *n*-butanolic HCl. The formed *n*-butyl acetate and *n*-butyl glycolate are analyzed by g.l.c. (section II.5).

1. Quantitative Sugar Analysis, Including Sialic Acid
(CHAMBERS and CLAMP 1971, CLAMP *et al.* 1971, KAMERLING *et al.* 1975 a)

In an ampoule the sialoglycoconjugate or polysaccharide (0.5-2 mg) is mixed with a mannitol solution (internal standard; 25-100 nmol). After lyophilization and drying over P_2O_5 in a vacuum desiccator, the residue is dissolved in 1.0 N methanolic HCl (0.5 ml). Nitrogen is bubbled through the solution for 30 sec, and then the ampoule is sealed. The solution is heated for 24 h at 85 °C; subsequently, neutralization is carried out by the addition of solid silver carbonate (pH-paper). (Re-)N-acetylation is performed by the addition of acetic anhydride (10-50 μl). The obtained mixture is kept at room temperature for 24 h in the dark. The precipitate is then triturated thoroughly and after centrifugation, the supernatant is collected. The residue of silver salts is washed twice with 0.5 ml dry methanol. The pooled supernatants are evaporated under reduced pressure at 35 °C. The final residue is dried for 12 h in a vacuum desiccator over P_2O_5. Before g.l.c. analysis, the sample is trimethylsilylated with a mixture of pyridine-hexamethyldisilazane-chlorotrimethylsilane, 5 : 1 : 1 (100 μl) for 30 min at room temperature.

The quantitative sugar analysis is carried out on a CPsil5 WCOT fused silica capillary column (25 m × 0.32 mm, i.d.) using flame-ionization detection. The carrier gas nitrogen flow-rate is 1.5 ml/min and the make-up gas nitrogen flow-rate 35 ml/min. The split-ratio amounts 1 : 10. The injection-port temperature is 210 °C, and the detector temperature 230 °C. The oven temperature is programmed from 130 to 220 °C at 2 °C/min and kept isothermally at 220 °C for 1 min. A typical gas chromatogram of a standard mixture is presented in Fig. 1.

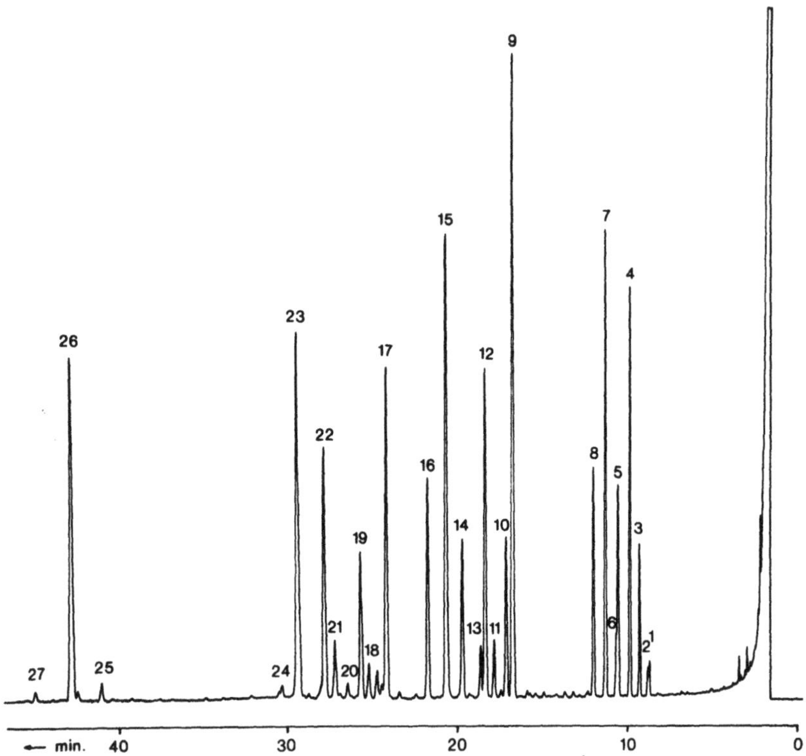

Fig. 1. Gas chromatogram of trimethylsilylated (methyl ester) methyl glycosides on a CPsil5 WCOT fused silica capillary column (25 m × 0.32 mm). Oven temperature program: 130 to 220 °C at 2 °C/min; 1 min at 220 °C. The peaks are numbered in their order of elution and are assigned as follows: *1* Xylose (β-*f*); *2* Xylose (α-*f*); *3* Fucose (β-*f*); *4* Fucose (α-*p*); *5* Fucose (β-*p*); *6* Fucose (α-*f*); *7* Xylose (α-*p*); *8* Xylose (β-*p*); *9* Mannose (α-*p*); *10* Galactose (β-*f*); *11* Mannose (β-*p*); *12* Galactose (α-*p*); *13* Galactose (α-*f*); *14* Galactose (β-*p*); *15* Glucose (α-*p*); *16* Glucose (β-*p*); *17* Mannitol (internal standard); *18* N-acetylglucosamine (α-*f*); *19* N-acetylgalactosamine (α, β-*f*); *20* Mono-O-acetylmannitol; *21* N-acetylglucosamine (β-*p*); *22* N-acetylgalactosamine (α, β-*p*); *23* N-acetylglucosamine (α-*p*); *24* N-acetylglucosamine (α, β-*p*; no methyl glycoside); *25* Neu5Ac (α); *26* Neu5Ac (β); *27* Neu5,9Ac$_2$. *f* furanoside; *p* pyranoside.

Remarks: 1. The amount of added internal standard depends on the carbohydrate-content in the sialoglycoconjugate; 2. When too much acetic anhydride is added, the primary hydroxyl functions of mannitol are O-acetylated giving rise to an additional small peak in the gas chromatogram. The same holds for the primary hydroxyl function of Neu5Ac methyl ester methyl glycoside; 3. Molar adjustment factors of monosaccharides except Neu5Ac are determined by application of the methanolysis procedure on standard mixtures of sugars and internal standard. For Neu5Ac the molar adjustment factor is determined using a methanolyzed known sialooligosaccharide; 4. Under the conditions used, the GlcNAc—Asn linkage is hardly cleaved. This has to be taken into account when calculating ratios.

2. *Hydrolytic Release of Sialic Acids from Sialobiopolymers*

These data are discussed in detail in chapter C.

3. *Preparation and Analysis of Trimethylsilylated N,O-Acylneuraminic Acid Methyl Esters*

(KAMERLING *et al.* 1975c, SCHAUER *et al.* 1976)

Sialic acid dried over P_2O_5 (50–100 μg) dissolved in dry methanol (0.5 ml) is treated with diazomethane in ether until a faint-yellow colour is obtained. The solution is immediately evaporated under reduced pressure, and the residue dissolved in pyridine (1 ml). Subsequently, hexamethyldisilazane (0.2 ml) and chlorotrimethylsilane (0.1 ml) are added. After 2 h at room temperature, chloroform (2 ml) and water (2 ml) are added to the turbid mixture. The chloroform layer is dried over anhydrous Na_2SO_4, and evaporated under reduced pressure. The g.l.c. analysis is carried out on a glass column (2 m × 4 mm, i.d.) packed with 3.8% SE-30 on Chromosorb W HP, 80–100 mesh, using flame-ionization detection. The oven temperature is 215 °C and the carrier gas nitrogen flow-rate 40 ml/min. Retention times R_{Neu5Ac} are given relative to the trimethylsilyl derivative of Neu5Ac methyl ester (see Table 2 on page 104).

4. *Preparation and Analysis of Pertrimethylsilylated N,O-Acylneuraminic Acids*

(CASALS-STENZEL *et al.* 1975)

Sialic acid dried over P_2O_5 (10–100 μg) is mixed with N-trimethylsilylimidazole (25 μl) in a test tube. The tube is flushed with nitrogen and whirled for either

Table 1. *Gas-liquid chromatography of pertrimethylsilylated N,O-acylneuraminic acids. The R_{Neu5Ac}-values on 3.5% OV-17 (temperature program: 200 to 280 °C at 2 °C/min) are given relative to pertrimethylsilylated Neu5Ac*

Sialic acid	R_{Neu5Ac} (β-anomers)
Neu5Ac	1.00
Neu4,5Ac_2	1.37
Neu5,7Ac_2	1.20
Neu5,9Ac_2	1.28
Neu4,5,9Ac_3	1.60
Neu5,7,9Ac_3	1.53
Neu5,8,9Ac_3	1.64
Neu5,7,8,9Ac_4	1.87
Neu5Gc	1.43
Neu9Ac5Gc	1.75
Neu7,9$Ac_2$5Gc	2.06
Neu2en5Ac	1.32

15 min at room temperature or 5 min at 60 °C in a heating block to complete the silylation reaction. For the trimethylsilylation, a mixture of pyridine-hexamethyldisilazane-chlorotrimethylsilane, 5 : 1 : 1, can also be used (reaction conditions: 3 h at room temperature). The g.l.c. analysis is carried out on a glass column (2.2 m × 2 mm, i.d.) packed with 3.5% OV-17 on Chrom GAW-DMCS, 80–100 mesh, using flame-ionization detection. The oven temperature is programmed from 200 to 280 °C at 2 °C/min. The carrier gas nitrogen flow-rate is 30 ml/min. The derivatives are stable in the refrigerator. Table 1 gives a survey of the relative retention times of a series of pertrimethylsilylated N,O-acylneuraminic acids.

5. *Analysis of Acetyl and/or Glycolyl Residues of Sialic Acid* (SUGITA 1979 a)

Glycoconjugate dried over P_2O_5 (1 mg) is treated with 1.0 N *n*-butanolic HCl for 3 h at 100 °C. After cooling, the acidic *n*-butanol solution is neutralized with silver carbonate (pH-paper). An aliquot of the *n*-butanol solution is analyzed for *n*-butyl acetate and/or *n*-butyl glycolate by g.l.c. on 3% diethylene glycol succinate on Shimalite, 80–100 mesh, at an oven temperature of 40 °C (column: 2 m × 3 mm, i.d.).

III. Gas-Liquid Chromatography/Mass Spectrometry

In structure analyses of sialic acids, the combination of g.l.c. and m.s. is almost indispensable. By careful studies of the mass spectra of the trimethylsilylated methyl esters of Neu5Ac and Neu5Gc and of some related derivatives, a general electron impact (e.i.) mass spectrometric micromethod has been developed for the identification of N,O-acylneuraminic acids isolated from biological material (KAMERLING *et al.* 1974, 1975 c, 1978). The method has also proved to be useful for the analysis of other isolated sialic acids, of (partially) O-methylated sialic acid methyl ester methyl glycosides as obtained in methylation analyses, and of synthetic sialic acid(s) (derivatives).

1. *Mass Spectrometric Identification Procedure*

To obtain volatile sialic acid derivatives, free carboxyl groups are converted into methyl esters and free hydroxyl groups into methyl ethers, trimethylsilyl ethers or acetyl esters. In methylation analysis studies, the acetylated or glycolylated amino function is also methylated. Figs. 2 and 3 present the e.i. mass spectra of the trimethylsilylated methyl ester of Neu5Ac, and of the permethylated methyl ester methyl glycoside of Neu5Ac, respectively. In Fig. 4, a schematic survey is depicted showing the selected fragment ions *A*–*H*, which furnish the information (abundances and *m*/*e* values of the ions) necessary to deduce the complete structure of the sialic acids. Although it has not been checked in detail, it is highly probable that the use of trimethylsilyl esters instead of methyl esters will not change the identification procedure. Fragments *A* and *B* indicate the molecular weight of the sialic acid derivatives and thereby the number and type of substituents. Fragments *C*–*H* contain the information concerning the position of the different substituents.

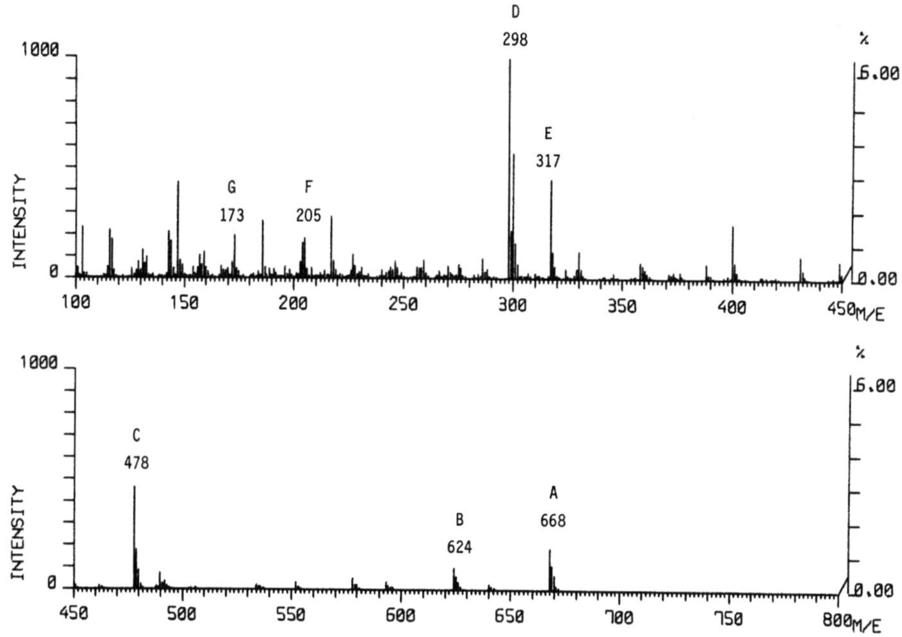

Fig. 2. E.i.-mass spectrum (70 eV) of the trimethylsilylated methyl ester of Neu5Ac (β-form).

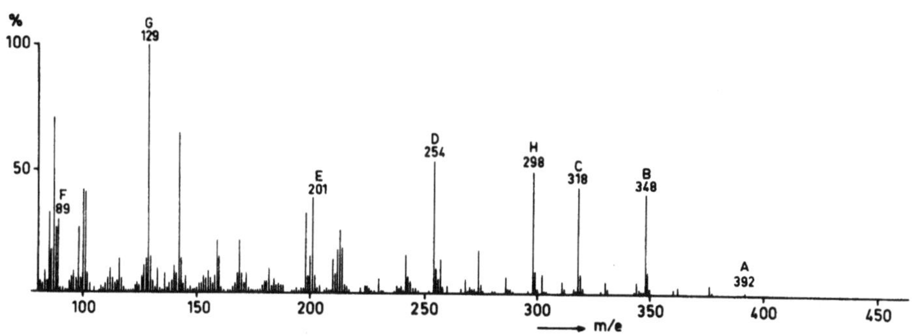

Fig. 3. E.i.-mass spectrum (70 eV) of the permethylated methyl ester β-methyl glycoside of Neu5Ac.

Fragment *A* is formed by elimination of a methyl group from the molecular ion *M*. In trimethylsilylated (O-acylated) N-acylneuraminic acid derivatives, the methyl group originates from a trimethylsilyl substituent, whereas in acetylated partially O-methylated N,N-acetyl,methyl-neuraminic acid methyl ester methyl glycosides the N,N-acetyl,methyl group is responsible. Trimethylsilylated partially O-methylated N,N-acetyl,methyl-neuraminic acid methyl ester methyl glycosides give rise to both possibilities, but the elimination from a trimethylsilyl group dominates (VAN HALBEEK *et al.* 1978).

Fragment *B* is obtained by release of the C-1 part of the sialic acid molecule. Eliminations of $OCOCH_3$ in O-acetylated neuraminic acid derivatives and of NH_2COCH_3 in Neu5Ac derivatives, which in principle give rise to the same *m/e* value as fragment *B* in the case of $R_1 = CH_3$, contribute little to the abundancy of this ion.

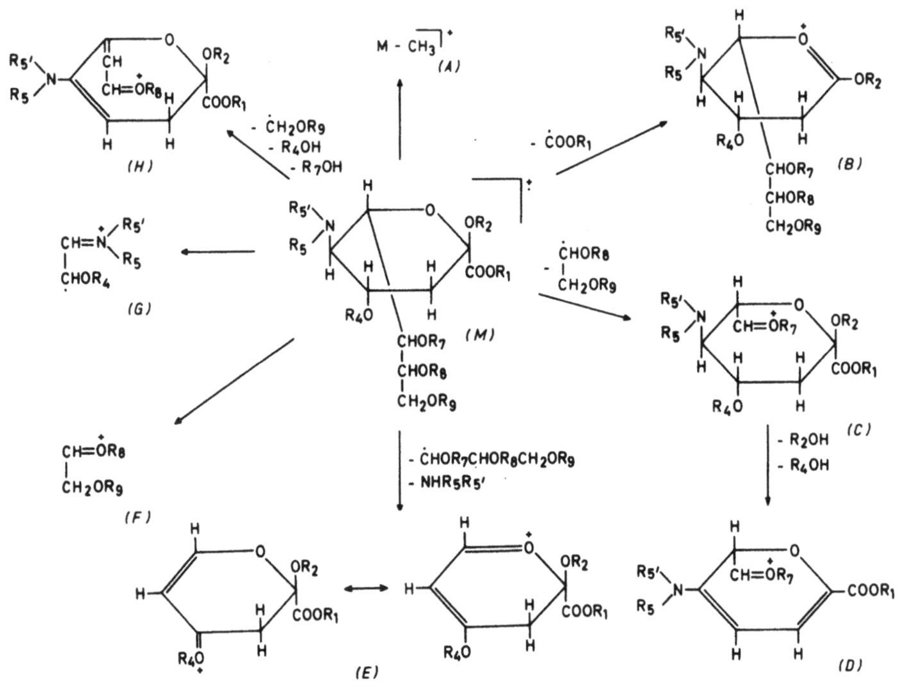

$R_1 = CH_3$

$R_2 = CH_3$ or $Si(CH_3)_3$

$R_4, R_7, R_8, R_9 = CH_3$, $Si(CH_3)_3$, $COCH_3$ and / or $COCH[OSi(CH_3)_3]CH_3$

$R_5 = COCH_3$ or $COCH_2OSi(CH_3)_3$

$R_{5'} = H$ or CH_3

Fig. 4. Survey of the selected fragment ions *A*–*H* worked out for trimethylsilylated (O-acylated) N-acylneuraminic acid methyl ester(s) (methyl glycosides) and for (partially) O-methylated N,N-acetyl,methyl-neuraminic acid methyl ester methyl glycosides. Partially O-methylated compounds are trimethylsilylated or acetylated.

Fragment *C* is formed by elimination of the C-8,9 part, with localization of the charge on position 7. For cleavage of partially methylated alditol acetates, it has been demonstrated that the charge is preferentially located on an ether oxygen instead of on an ester oxygen (BJÖRNDAL *et al.* 1970, LÖNNGREN and SVENSSON 1974). Therefore, in general, cleavage occurs between two alkoxylated (methoxylated or trimethylsiloxylated) carbon atoms, or between an acetoxylated and an alkoxylated carbon atom, rather than between two acetoxylated carbon atoms. Fragment *C* has only significant abundance if C-7 bears an ether group. In case when at C-7 an ester group is present, this fragment ion is absent or hardly observable.

Fragment *D* is formed from fragment *C* by consecutive eliminations of R_2OH and R_4OH. It is evident that the occurrence of this fragment ion is dependent on the presence of *C*.

Fragment *E* is formed by elimination of the side-chain C-7,8,9 and the substituent at C-5. This fragment ion is not observed if an O-acetyl group is attached to C-4, illustrating that the transition state in the McLafferty rearrangement is disfavored when the substituent at C-4 is an ester group rather than an ether group.

Fragment *F* contains the C-8,9 part. Based on the same fragmentation rules as mentioned above for fragments *C* and *D*, this ion can only readily be formed if an ether group is connected at C-8.

Fragment *G* consists of the C-4,5 part of the sialic acid molecule.

Fragment *H* is formed by elimination of the C-9 part of the molecule, followed by elimination of R_4OH and R_7OH. This fragment ion is useful for discriminating between an O-trimethylsilyl group at C-8 or C-9 in trimethylsilylated partially methylated N-acylneuraminic acid methyl ester(s) (methyl glycosides). For the identification of trimethylsilylated O-acylated N-acylneuraminic acid derivatives and acetylated partially methylated N-acylneuraminic acid derivatives, it is not necessary to take fragment *H* into consideration.

2. Analysis of (O-Acylated) N-Acylneuraminic Acids Isolated from Biological Material

In Table 2 a series of (O-acylated) N-acylneuraminic acids isolated from various biological sources is presented (see also chapter B and C). These sialic acids have been analyzed by g.l.c./m.s. after esterification with diazomethane and trimethylsilylation (section II.3): the Table includes relative retention times of the derivatives on g.l.c. and *m/e* values of the fragment ions *A–G*. As was mentioned earlier, sialic acids predominantly occur in the β-anomeric form. However, on 3.8% SE-30 the α-anomer could sometimes be detected separately from the β-anomer (in most cases as a small shoulder). Mass spectra are depicted in Figs. 2 and 5–17.

From Table 2, it is evident that the replacement of one O-trimethylsilyl group by one O-acetyl group causes in involved fragments a negative shift of 30 a.m.u. For two O-acetyl groups a shift of 60 a.m.u. is observed, etc. In the same way, the presence of a trimethylsilylated O-lactyl group instead of an O-trimethylsilyl group causes a positive shift of 72 a.m.u. Replacement of an N-acetyl group by a trimethylsilylated N-glycolyl group shows a positive shift of 88 a.m.u. In conclusion, each sialic acid derivative gives rise to a unique series of fragment ions *A–G*.

Fragment *G* has to be discussed in more detail (KAMERLING *et al.* 1975c, 1978). The occurrence of an O-acetyl group instead of an O-trimethylsilyl group at C-4 as in the derivatives of $Neu4,5Ac_2$, $Neu4,5,9Ac_3$, $Neu4,5Ac_29Lt$, and Neu4Ac5Gc leads to a negative shift of 30 a.m.u. for this fragment ion. Therefore, in the mass spectra of 4-O-acetylated N-acetylneuraminic acids the peak at *m/e* 173 is absent:

Table 2. *G.l.c. and m.s. data of trimethylsilylated (O-acylated) N-acylneuraminic acid methyl esters. The R_{Neu5Ac}-values on 3.8% SE-30 at 215 °C are given relative to the trimethylsilylated methyl ester of Neu5Ac (β-anomeric form). For the esterification and trimethylsilylation procedure and additional analysis conditions, see section II.3. For an explanation of the — signs, see sections III.1 and III.2*

Sialic acid	R_{Neu5Ac} β-anomer	R_{Neu5Ac} α-anomer	*m/e* values							References
			A	*B*	*C*	*D*	*E*	*F*	*G*	
Neu5Ac	1.00	1.05	668	624	478	298	317	205	173	KAMERLING *et al.* 1974, 1975c, SCHAUER *et al.* 1976
Neu4,5Ac$_2$	1.18	1.21	638	594	448	298	—	205	143	KAMERLING *et al.* 1975c, 1982a, REUTER *et al.* 1980a
Neu5,7Ac$_2$	1.04	1.00	638	594	—	—	317	205	173	REUTER *et al.* 1982, SCHAUER *et al.* 1976
Neu5,8Ac$_2$	1.05		638	594	478	298	317	—	173	REUTER *et al.* 1982
Neu5,9Ac$_2$	1.13		638	594	478	298	317	175	173	GHIDONI *et al.* 1980, HAVERKAMP *et al.* 1976, 1977b, KAMERLING *et al.* 1975c, 1982b, REUTER *et al.* 1980b, SCHAUER *et al.* 1976
Neu4,5,9Ac$_3$	1.31		608	564	448	298	—	175	143	KAMERLING *et al.* 1975c, REUTER *et al.* 1980a
Neu5,7,9Ac$_3$	1.14	1.07	608	564	—	—	317	175	173	KAMERLING *et al.* 1975c, REUTER *et al.* 1982, SCHAUER *et al.* 1976
Neu5,8,9Ac$_3$	1.19		608	564	478	298	317	—‡	173	REUTER *et al.* 1982
Neu5,7,8,9Ac$_4$	1.15		578	534	—	—	317	—	173	REUTER *et al.* 1982
Neu5Ac9Lt†	2.55		740	696	478	298	317	277	173	HAVERKAMP *et al.* 1976, SCHAUER *et al.* 1976
Neu4,5Ac$_2$9Lt	3.01		710	666	448	298	—	277	143	REUTER *et al.* 1980a
Neu5Gc	1.81	1.90	756	712	566	386	317	205	261	KAMERLING *et al.* 1974, 1975c, SCHAUER *et al.* 1976
Neu4Ac5Gc	2.02		726	682	536	386	—	205	231	KAMERLING *et al.* 1975c
Neu7Ac5Gc	1.83		726	682	—	—	317	205	261	REUTER *et al.* 1982
Neu9Ac5Gc	2.04		726	682	566	386	317	175	261	KAMERLING *et al.* 1975c, 1980, SCHAUER *et al.* 1976
Neu7,9Ac$_2$5Gc	2.01		696	652	—	—	317	175	261	REUTER *et al.* 1982
Neu8,9Ac$_2$5Gc	1.99		696	652	566	386	317	—‡	261	REUTER *et al.* 1982
Neu7,8,9Ac$_3$5Gc	1.93		666	622	—	—	317	—	261	REUTER *et al.* 1982

† The absolute configuration of the lactyl (Lt) substituent was found to be **L** (SCHAUER *et al.* 1976).

‡ The small peak at *m/e* 145, also present in the mass spectra of other sialic acids, has not been checked by exact mass measurements (see also footnote in Table 5).

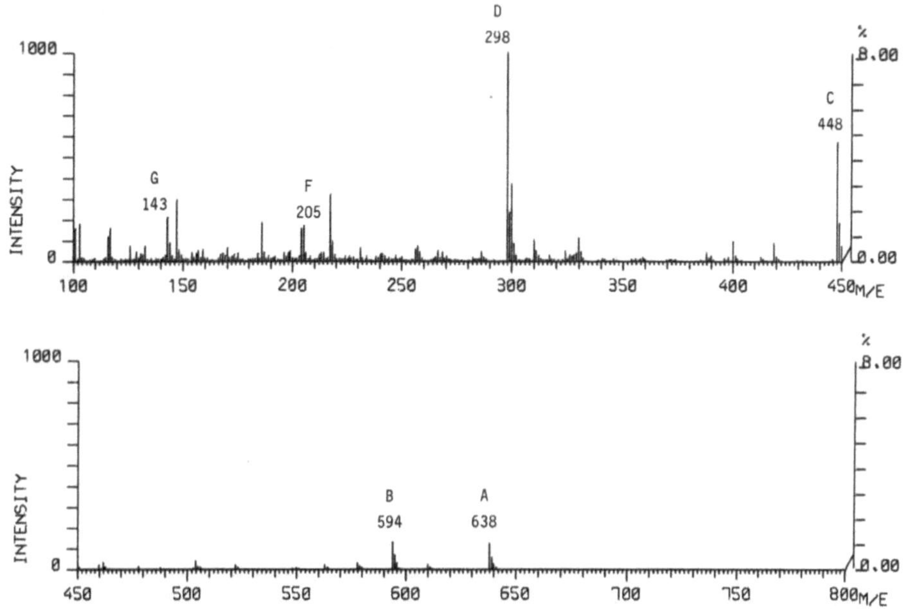

Fig. 5. E.i.-mass spectrum (70 eV) of the trimethylsilylated methyl ester of Neu4,5Ac_2 (β-form).

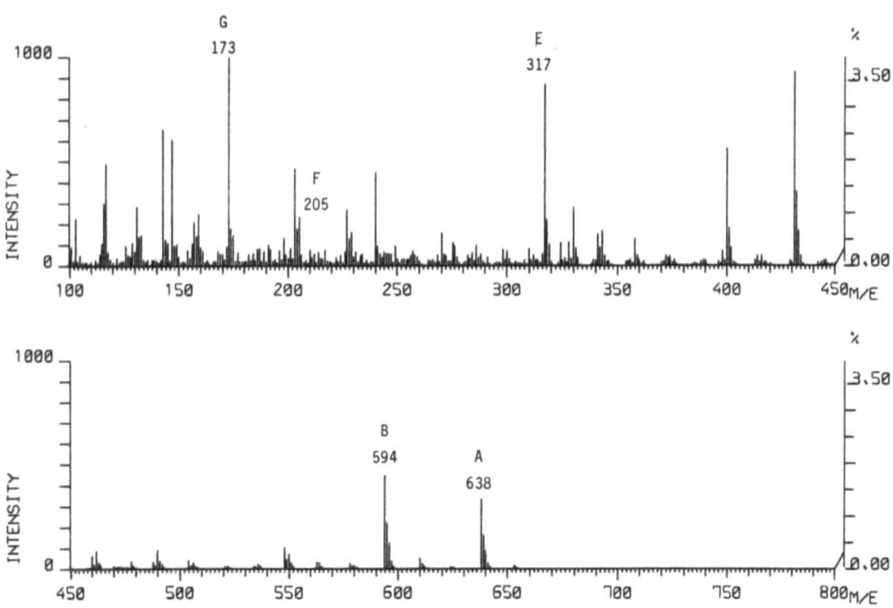

Fig. 6. E.i.-mass spectrum (70 eV) of the trimethylsilylated methyl ester of Neu5,7Ac_2 (β-form).

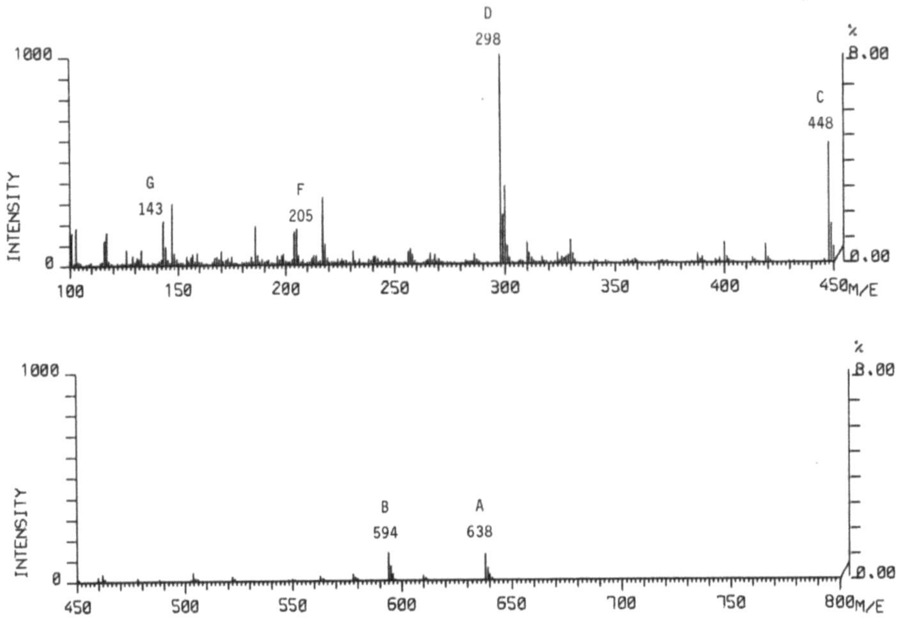

Fig. 7. E.i.-mass spectrum (70 eV) of the trimethylsilylated methyl ester of Neu5,9Ac$_2$.

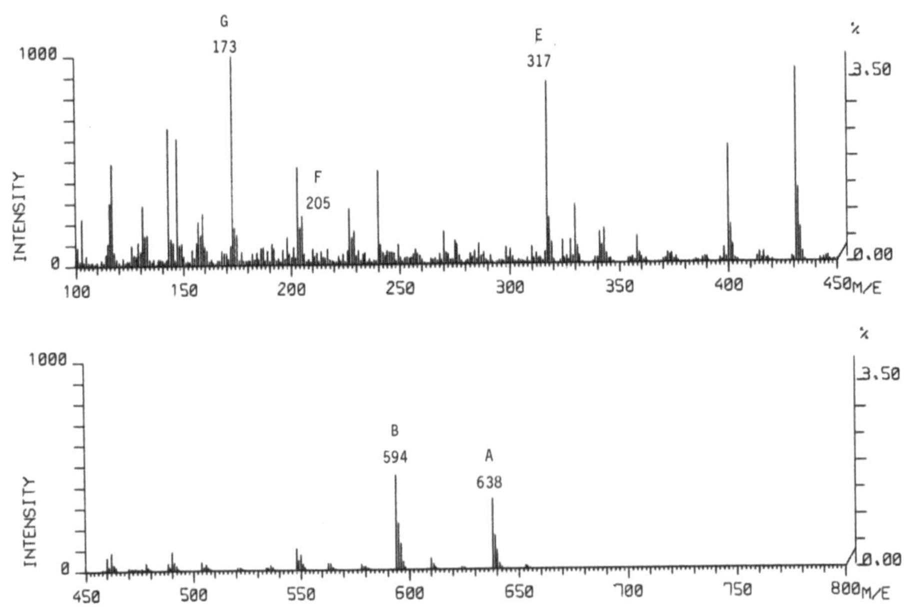

Fig. 8. E.i.-mass spectrum (70 eV) of the trimethylsilylated methyl ester of Neu4,5,9Ac$_3$.

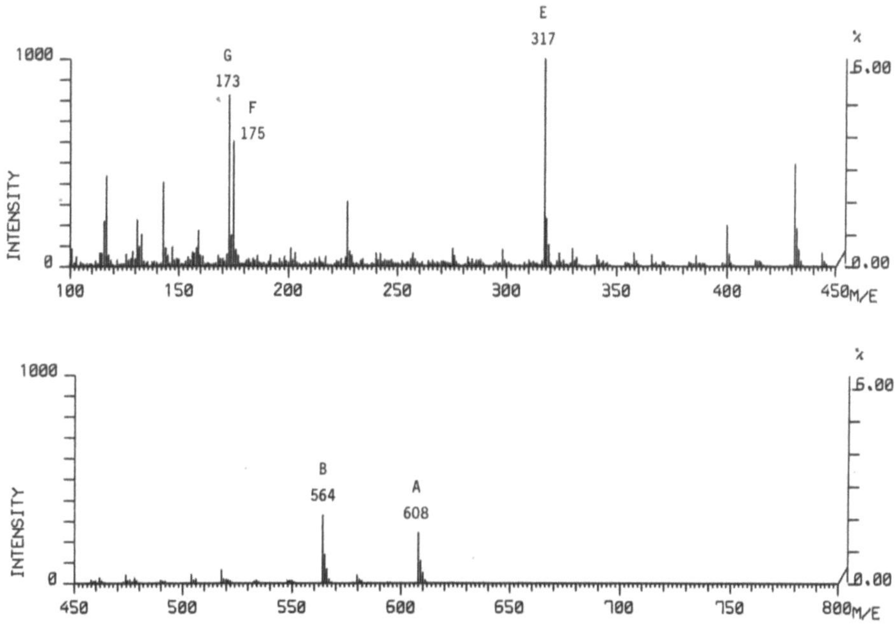

Fig. 9. E.i.-mass spectrum (70 eV) of the trimethylsilylated methyl ester of Neu5,7,9Ac$_3$ (β-form).

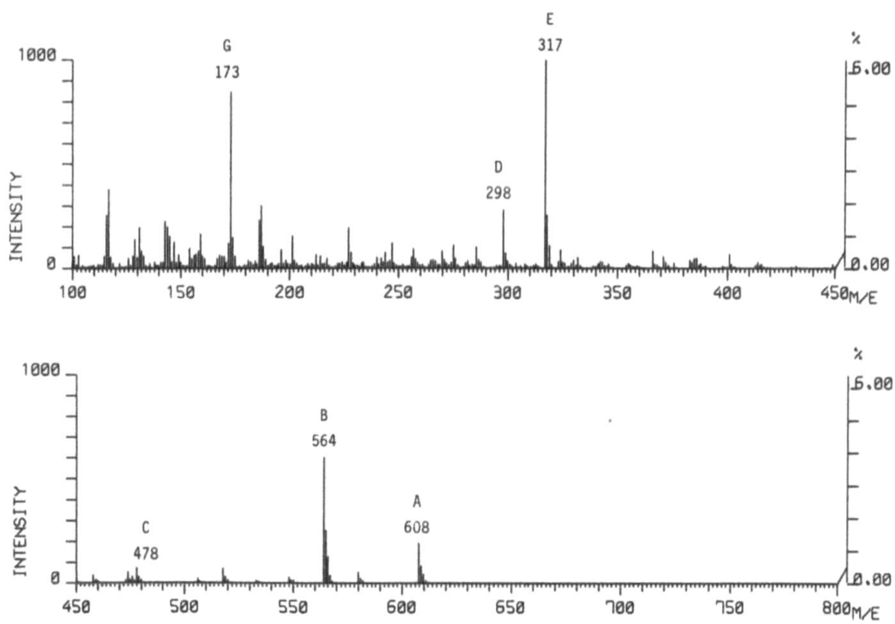

Fig. 10. E.i.-mass spectrum (70 eV) of the trimethylsilylated methyl ester of Neu5,8,9Ac$_3$.

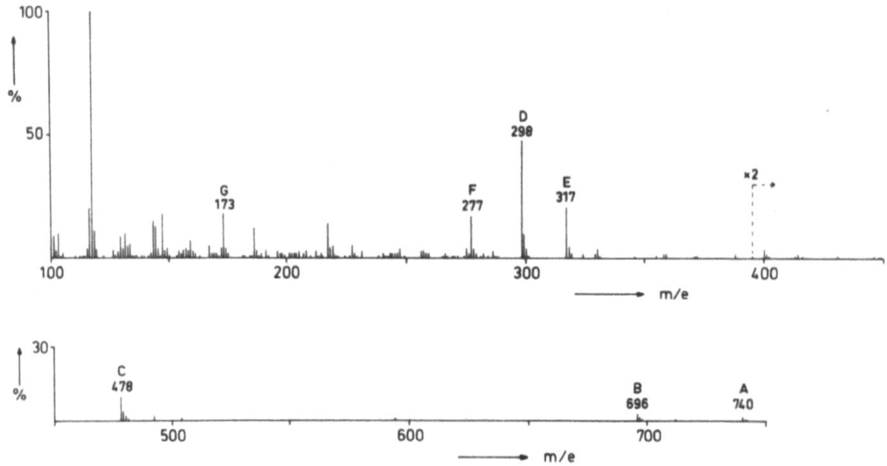

Fig. 11. E.i.-mass spectrum (70 eV) of the trimethylsilylated methyl ester of Neu5Ac9Lt (Lt, lactyl).

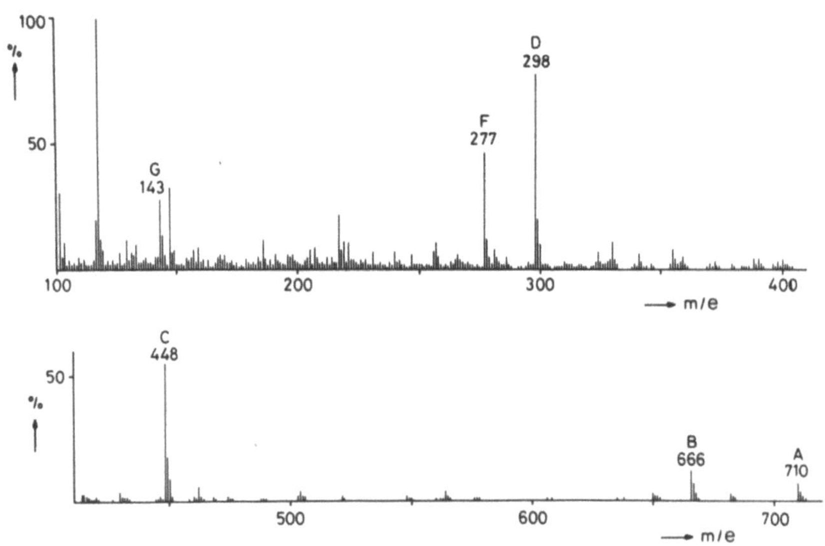

Fig. 12. E.i.-mass spectrum (70 eV) of the trimethylsilylated methyl ester of Neu4,5Ac$_2$9Lt.

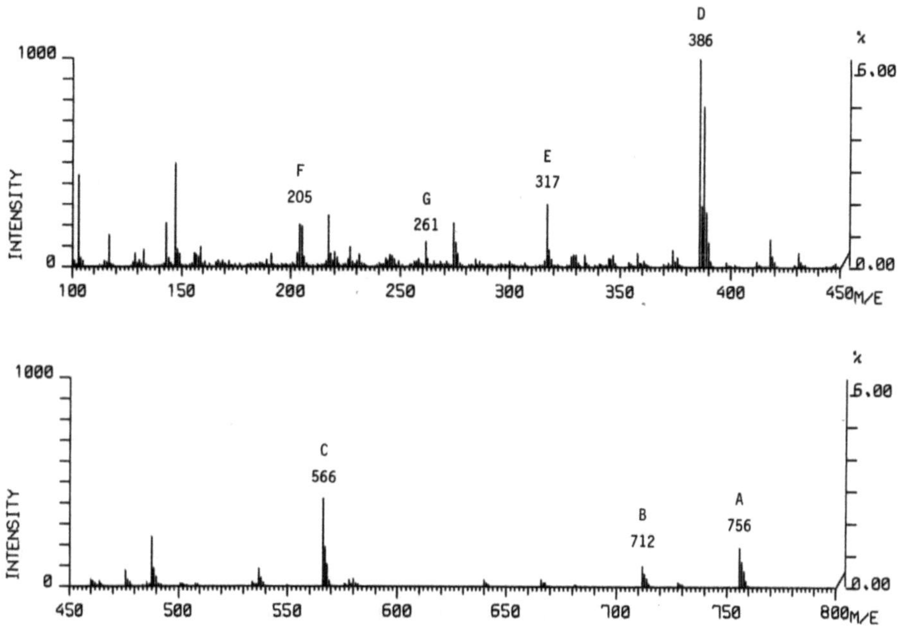

Fig. 13. E.i.-mass spectrum (70 eV) of the trimethylsilylated methyl ester of Neu5Gc (β-form).

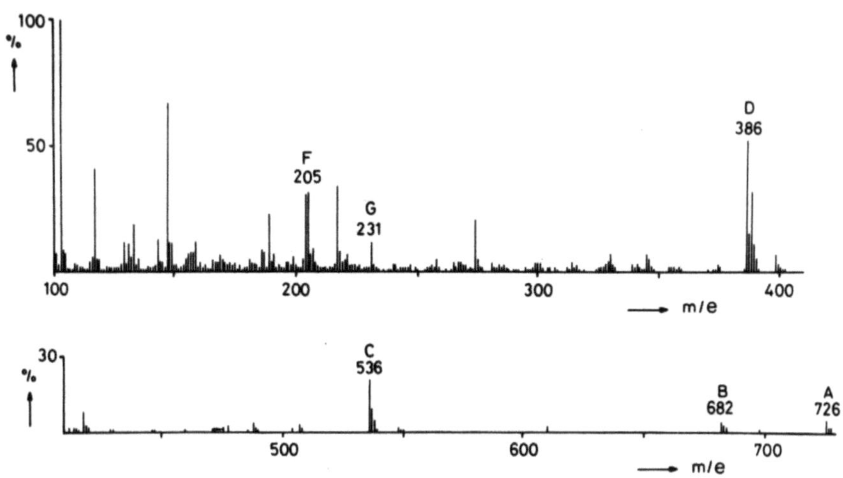

Fig. 14. E.i.-mass spectrum (70 eV) of the trimethylsilylated methyl ester of Neu4Ac5Gc.

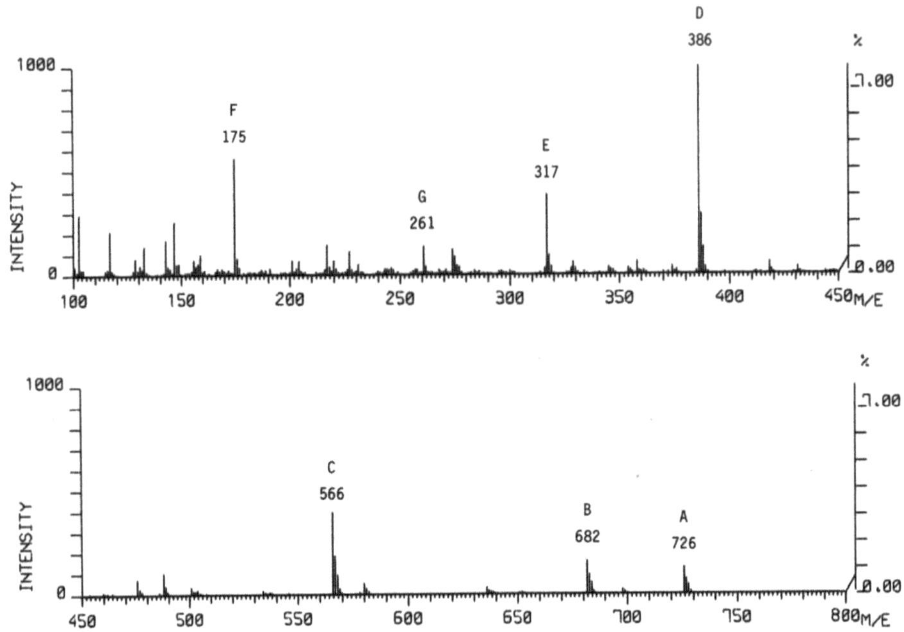

Fig. 15. E.i.-mass spectrum (70 eV) of the trimethylsilylated methyl ester of Neu9Ac5Gc.

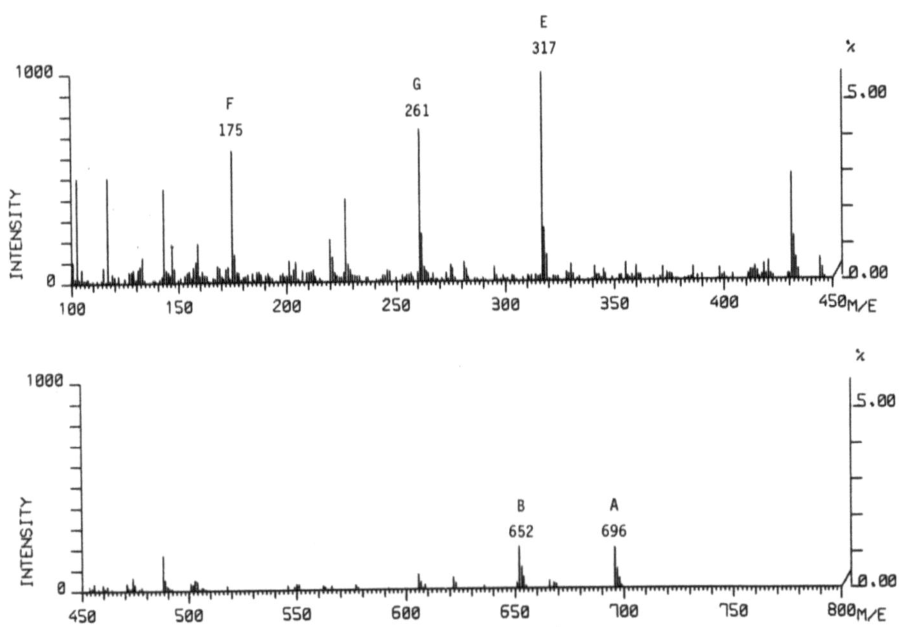

Fig. 16. E.i.-mass spectrum (70 eV) of the trimethylsilylated methyl ester of Neu7,9Ac$_2$5Gc.

m/e 173 → *m/e* 143. However, in all mass spectra a peak at *m/e* 143 with a general formula $C_6H_{11}O_2Si$ (143.0528) is observed. But in the mass spectra of Neu4,5Ac_2, Neu4,5,9Ac_3, and Neu4,5$Ac_2$9Lt the main contribution to the abundancy of *m/e* 143 originates from fragment *G* ($C_6H_9NO_3$; 143.0582). In the mass spectrum of Neu4Ac5Gc the peak at *m/e* 261 is not observed. For this compound fragment *G* shifts to *m/e* 231. By high-resolution mass spectrometry, this fragment ion could not be distinguished from other generally occurring fragment ions in sialic acid, which contribute also to the intensity of the peak at *m/e* 231.

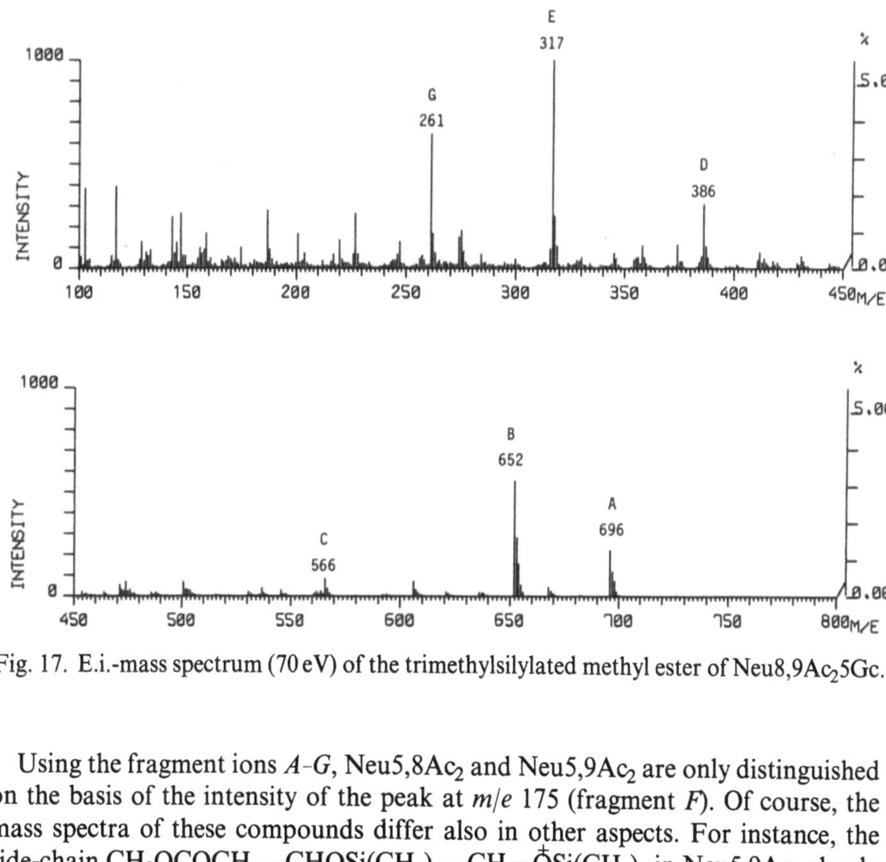

Fig. 17. E.i.-mass spectrum (70 eV) of the trimethylsilylated methyl ester of Neu8,9$Ac_2$5Gc.

Using the fragment ions *A*-*G*, Neu5,8Ac_2 and Neu5,9Ac_2 are only distinguished on the basis of the intensity of the peak at *m/e* 175 (fragment *F*). Of course, the mass spectra of these compounds differ also in other aspects. For instance, the side-chain CH_2OCOCH_3—$CHOSi(CH_3)_3$—$CH{=}\overset{+}{O}Si(CH_3)_3$ in Neu5,9Ac_2 clearly eliminates CH_3COOH, giving rise to the fragment ion *m/e* 217. In Neu5,8Ac_2 the side chain $CH_2OSi(CH_3)_3$—$CHOCOCH_3$—$CH{=}\overset{+}{O}Si(CH_3)_3$ eliminates CH_3COOH (*m/e* 217) as well as $HOSi(CH_3)_3$ (*m/e* 187). See also Kamerling *et al.* (1975c) and Lönngren and Svensson (1974).

It has to be noted that the fragment ion at *m/e* 103 ($CH_2{=}\overset{+}{O}Si(CH_3)_3$) is not characteristic for a primary trimethylsiloxyl group in the sialic acid derivatives, but can also be formed along other routes (Kamerling *et al.* 1978). Furthermore, the fragment ions at *m/e* 186 ($CH_3CO\overset{+}{N}H{=}CH$—$CH{=}CHOSi(CH_3)_3$ and $CH_3CO\overset{+}{N}H{=}CH$—$C(OSi(CH_3)_3){=}CH_2$) in N-acetylneuraminic acids and at *m/e* 274 in N-glycolylneuraminic acids only give information about the type of substitution at C-5 (amino group).

Table 3. *G.l.c. and m.s. data of trimethylsilylated/methylated N,N-acyl,methyl-neuraminic acid methyl ester β-methyl glycosides. The* R_{Neu5Ac}*-values on a packed column (2 m × 4 mm, i.d.) of 3.8% SE-30 on Chromosorb W-AW DMCS HP, 80–100 mesh at 220 °C* (VAN HALBEEK *et al.* 1978) *and on a capillary column (80 m × 0.35 mm, i.d.) wall-coated with OV-101 at 215 °C* (BRUVIER *et al.* 1981) *are given relative to Neu5Ac4,5,7,8,9Me*$_5$ *methyl ester β-methyl glycoside. For the trimethylsilylation procedure, see section II.3*

Sialic acid (as methyl ester β-methyl glycoside)	R_{Neu5Ac} 3.8% SE-30	R_{Neu5Ac} OV-101	*m/e* values								References
			A	*B*	*C*	*D*	*E*	*F*	*G*	*H*	
Neu5Ac4,5,7,8,9Me$_5$	1.00	1.00	392	348	318	254	201	89	129	298	BHATTACHARJEE and JENNINGS 1976, BRUVIER *et al.* 1981, RAUVALA and KÄRKKÄINEN 1977, VAN HALBEEK *et al.* 1978
Neu5Ac4,5,7,8Me$_4$	1.30	1.31	450	406	318	254	201	147	129	298	BHATTACHARJEE and JENNINGS 1976, BRUVIER *et al.* 1981, VAN HALBEEK *et al.* 1978
Neu5Ac4,5,7,9Me$_4$	1.14	1.14	450	406	318	254	201	147	129	356	BHATTACHARJEE and JENNINGS 1976, BRUVIER *et al.* 1981, HAVERKAMP *et al.* 1977a, VAN HALBEEK *et al.* 1978
Neu5Ac4,5,8,9Me$_4$	1.07	1.06	450	406	376	312	201	89	129	298	BRUVIER *et al.* 1981, VAN HALBEEK *et al.* 1978
Neu5Ac5,7,8,9Me$_4$		1.20	450	406	376	254	259	89	187	298	BHATTACHARJEE and JENNINGS 1976, BRUVIER *et al.* 1981
Neu5Ac4,5,7Me$_3$	1.55	1.52	508	464	318	254	201	205	129	356	BRUVIER *et al.* 1981, KAMERLING *et al.* 1978
Neu5Ac4,5,9Me$_3$	1.27	1.27	508	464	376	312	201	147	129	356	BRUVIER *et al.* 1981, VAN HALBEEK *et al.* 1978
Neu5Ac5,7,8Me$_3$		1.60	508	464	376	254	259	147	187	298	BRUVIER *et al.* 1981
Neu5Ac5,7,9Me$_3$		1.39	508	464	376	254	259	147	187	356	BRUVIER *et al.* 1981
Neu5Ac5,8,9Me$_3$		1.23	508	464	434	312	259	89	187	298	BRUVIER *et al.* 1981
Neu5Ac4,5Me$_2$	1.70	1.78	566	522	376	312	201	205	129	356	BRUVIER *et al.* 1981, VAN HALBEEK *et al.* 1978
Neu5Ac5,7Me$_2$		1.91	566	522	376	254	259	205	187	356	BRUVIER *et al.* 1981
Neu5Ac5,9Me$_2$	1.43	1.50	566	522	434	312	259	147	187	356	BRUVIER *et al.* 1981, VAN HALBEEK *et al.* 1978

Neu5Ac5Me	1.89	2.05	624	580	434	312	259	205	187	356	BRUVIER *et al.* 1981, VAN HALBEEK *et al.* 1978
Neu5MeGc4,5,7,8,9Me$_5$ †			422	378	348	284	201	89	159	328	INOUE and MATSUMURA 1979, RAUVALA and KÄRKKÄINEN 1977
Neu5MeGc4,5,7,9Me$_4$ †			480	436	348	284	201	147	159	386	SEKINE *et al.* 1981

† MeGc = methylated glycolyl group.

Table 4. *G.l.c. and m.s. data of acetylated/methylated N,N-acyl,methyl-neuraminic acid methyl ester β-methyl glycosides. The R_{Neu5Ac}-values on a packed column (2 m × 4 mm, i.d.) of 3.8% SE-30 on Chromosorb W-AW DMCS HP, 80–100 mesh at 220 °C* (VAN HALBEEK *et al.* 1978) *are given relative to Neu5Ac4,5,7,8,9Me$_5$ methyl ester β-methyl glycoside. The acetylation is carried out at 100 °C for 30 min with acetic anhydride-pyridine (1 : 1). For an explanation of the — signs, see section III.1*

Sialic acid † (as methyl ester β-methyl glycoside)	R_{Neu5Ac}	*m/e* values							References
		A	*B*	*C*	*D*	*E*	*F*	*G*	
Neu5Ac4,5,7,8Me$_4$	1.47	420	376	318	254	201	117	129	BHATTACHARJEE and JENNINGS 1976, VAN HALBEEK *et al.* 1978
Neu5Ac4,5,7,9Me$_4$	1.25	420	376	318	254	201	—	129	BHATTACHARJEE and JENNINGS 1976, HAVERKAMP *et al.* 1977a, RAUVALA and KÄRKKÄINEN 1977, VAN HALBEEK *et al.* 1978
Neu5Ac4,5,8,9Me$_4$	1.08	420	376	—	—	201	89	129	VAN HALBEEK *et al.* 1978
Neu5Ac5,7,8,9Me$_4$		420	376	346	254	—	89	157	BHATTACHARJEE and JENNINGS 1976, SMIRNOVA and KOCHETKOV 1980
Neu5Ac4,5,7Me$_3$	1.75	448	404	318	254	201	—	129	KAMERLING *et al.* 1978
Neu5Ac4,5,9Me$_3$	1.26	448	404	—	—	201	—	129	VAN HALBEEK *et al.* 1978
Neu5Ac4,5Me$_2$	1.70	476	432	—	—	201	—	129	VAN HALBEEK *et al.* 1978
Neu5Ac5,9Me$_2$	1.63	476	432	—	—	—	—	157	VAN HALBEEK *et al.* 1978
Neu5Ac5Me	2.17	504	460	—	—	—	—	157	VAN HALBEEK *et al.* 1978
Neu5MeGc4,5,7,9Me$_4$ ††		450	406	348	284	201	—	159	INOUE and MATSUMURA 1979

† For data of Neu5Ac4,5,7,8,9Me$_5$ and Neu5MeGc4,5,7,8,9Me$_5$, see Table 3.
†† MeGc = methylated glycolyl group.

3. Sialic Acids and Methylation Analysis

Methylation analysis is generally applied for the determination of the position of glycosidic linkages in glycoconjugates, oligosaccharides, and polysaccharides. After permethylation the biopolymer is solvolyzed (e.g., hydrolyzed or methanolyzed) and the obtained mixture of partially methylated monomers is analyzed (BJÖRNDAL *et al.* 1970, STELLNER *et al.* 1973, LÖNNGREN and SVENSSON 1974, RAUVALA *et al.* 1981). For the linkage analysis of sialic acids in non-reducing and internal positions, methanolysis is the appropriate method of solvolysis. From the analytical data available so far, it can be concluded that after methanolysis the (partially) methylated sialic acids occur predominantly as their methyl ester β-methyl glycosides; only a small percentage of the corresponding α-anomers have been detected (HAVERKAMP *et al.* 1977 a). Partially O-methylated N,N-acyl,methyl-neuraminic acid methyl ester methyl glycosides are analyzed by g.l.c./m.s. after trimethylsilylation or acetylation of free hydroxyl groups.

In literature, the preparation and g.l.c./m.s. data of a large series of partially methylated sialic acid methyl ester methyl glycosides as reference compounds have been reported. For this purpose, specific procedures (VAN HALBEEK *et al.* 1978) as well as methanolysis of permethylated sialobiopolymers (BHATTACHARJEE and JENNINGS 1976, RAUVALA and KÄRKKÄINEN 1977, VAN HALBEEK *et al.* 1978) and non-specific partial methylations (undermethylation) (BRUVIER *et al.* 1981) have been employed. Table 3 summarizes the relative retention times on g.l.c. and the *m/e* values of the fragment ions *A–H* of (partially) O-methylated N,N-acyl,methyl-neuraminic acid methyl ester β-methyl glycosides after trimethylsilylation. Table 4 contains similar information about several sialic acid derivatives after acetylation (fragments *A–G*). In Fig. 3 the mass spectrum of permethylated N,N-acetyl,methyl-neuraminic acid methyl ester β-methyl glycoside is presented. Figs. 18–21 show the mass spectra of the various tri-O-methyl-N,N-acetyl,methyl-neuraminic acid methyl ester β-methyl glycosides after trimethylsilylation. In principle, the latter derivatives are obtained from internal non-branching N-acylneuraminic acids. For additional mass spectra, see BHATTACHARJEE and JENNINGS (1976), RAUVALA and KÄRKKÄINEN (1977), HAVERKAMP *et al.* (1977 a), VAN HALBEEK *et al.* (1978), INOUE and MATSUMURA (1979), and BRUVIER *et al.* (1981).

The choice of the applied methanolysis conditions in relation to the possible release of the N-acyl group is very important for the methylated compounds too. Using 0.5 N methanolic HCl (18 h, 80 °C) the N-acyl (N-acetyl or methylated N-glycolyl) groups of sialic acids in terminal positions of the carbohydrate chain are resistant to methanolic cleavage. However, internal sialic acids are N-deacylated to a large extent. For this reason, after methanolysis (re-)N-acetylation is necessary in the working-up procedure (also perhaps with deuterated acetic anhydride). The use of 0.05 N methanolic HCl (1 h, 80 °C) seems to give no N-deacylation (INOUE and MATSUMURA 1979, 1980). However, these milder conditions do not liberate sialic acid quantitatively from the sialobiopolymer.

In order to verify the general formula of the selected fragment ions, the sialic acid methyl ester β-methyl glycosides of Neu5Ac4,5,7,8,9Me_5 and of Neu5Ac4,5,7,8Me_4, Neu5Ac4,5,7,9Me_4, Neu5Ac4,5,8,9Me_4, Neu5Ac4,5,7Me_3,

Neu5Ac4,5,9Me$_3$, Neu5Ac4,5Me$_2$, Neu5Ac5,9Me$_2$, and Neu5Ac5Me after trimethylsilylation or acetylation, have been studied by high-resolution mass spectrometry (VAN HALBEEK *et al.* 1978). These investigations have indicated that especially the fragment ions *F* and *G* have to be considered in more detail.

Fragment *F*: i. Two different ions contribute to the intensity of the peak at *m/e* 89 in the trimethylsilyl derivative of Neu5Ac4,5,8,9Me$_4$ methyl ester β-methyl glycoside (Table 3), namely CH_2OCH_3—$CH=\overset{+}{O}CH_3$ (*F*; $C_4H_9O_2$) and $\overset{+}{O}Si(CH_3)_3$ (C_3H_9OSi). The fragment C_3H_9OSi can always be detected in the mass spectra of trimethylsilylated carbohydrates. One can assume that for the trimethylsilyl derivatives of Neu5Ac5,7,8,9Me$_4$ and Neu5Ac5,8,9Me$_3$ methyl ester β-methyl glycoside (Table 3) the same reasoning with respect to *m/e* 89 holds. ii. Two different fragment ions contribute to the intensity of the peak at *m/e* 147 in the trimethylsilyl derivatives of Neu5Ac4,5,9Me$_3$ and Neu5Ac5,9Me$_2$ methyl ester β-methyl glycoside (Table 3), namely CH_2OCH_3—$CH=\overset{+}{O}Si(CH_3)_3$ (*F*; $C_6H_{15}O_2Si$) and $(CH_3)_3SiO\overset{+}{S}i(CH_3)_2$ ($C_5H_{15}OSi_2$). The fragment with formula $C_5H_{15}OSi_2$ is generally present in the mass spectra of trimethylsilylated sugars with more than one O-trimethylsilyl group. In this case it can be assumed that for the trimethylsilyl derivatives of Neu5Ac5,7,8Me$_3$ and Neu5Ac5,7,9Me$_3$ methyl ester β-methyl glycoside (Table 3) the peak at *m/e* 147 is also composed of two fragments. iii. In the trimethylsilyl derivatives of Neu5Ac4,5,7Me$_3$, Neu5Ac4,5Me$_2$, and Neu5Ac5Me methyl ester β-methyl glycoside the peak at *m/e* 147 originates only from $C_5H_{15}OSi_2$. The same can be expected for the trimethylsilyl derivatives of Neu5Ac5,8,9Me$_3$ and Neu5Ac5,7Me$_2$ methyl ester β-methyl glycoside (Table 3).

Fragment *G*: i. Mass spectra of trimethylsilylated sugars always contain a peak at *m/e* 129 with low intensity, originating from the fragments $C_5H_9O_2Si$ and $C_6H_{13}OSi$. In the trimethylsilyl derivatives of Neu5Ac5,9Me$_2$ and Neu5Ac5Me methyl ester β-methyl glycoside the peak at *m/e* 129 ($C_5H_9O_2Si$ and $C_6H_{13}OSi$) is of low intensity ($R_4=Si(CH_3)_3$). A similar composition can be expected for the trimethylsilyl derivatives of Neu5Ac5,7,8,9Me$_4$, Neu5Ac5,7,8Me$_3$, Neu5Ac5,7,9Me$_3$, Neu5Ac5,8,9Me$_3$, and Neu5Ac5,7Me$_2$ methyl ester β-methyl glycoside (Table 3). ii. The main contribution to the intense peak at *m/e* 129 in the trimethylsilyl derivatives of Neu5Ac4,5,7,8Me$_4$, Neu5Ac4,5,7,9Me$_4$, Neu5Ac4,5,8,9Me$_4$, Neu5Ac4,5,7Me$_3$, Neu5Ac4,5,9Me$_3$, and Neu5Ac4,5Me$_2$ methyl ester β-methyl glycoside (Table 3) represents fragment *G* ($C_6H_{11}NO_2$; $R_4=CH_3$). iii. In the acetyl derivatives of Neu5Ac4,5,7,8Me$_4$, Neu5Ac4,5,7,9Me$_4$, Neu5Ac4,5,8,9Me$_4$, Neu5Ac4,5,7Me$_3$, and Neu5Ac4,5,9Me$_3$ methyl ester β-methyl glycoside (Table 4) the peak at *m/e* 129 consists mainly of fragment *G*; a small contribution of $C_6H_9O_3$ has been detected.

The foregoing demonstrates that if exact mass measurements are carried out, each fragment as such can provide essential structural information. However, it should be emphasized that, if working under low-resolution conditions only, the whole series of selected fragment ions form a self-consisted system, in which the interpretation of the fragment ions should support each other. It is obvious that to arrive at an unambiguous conclusion about the substitution pattern, the whole mass spectrum should also be considered.

Finally, it has to be mentioned that methylation analysis data have also been

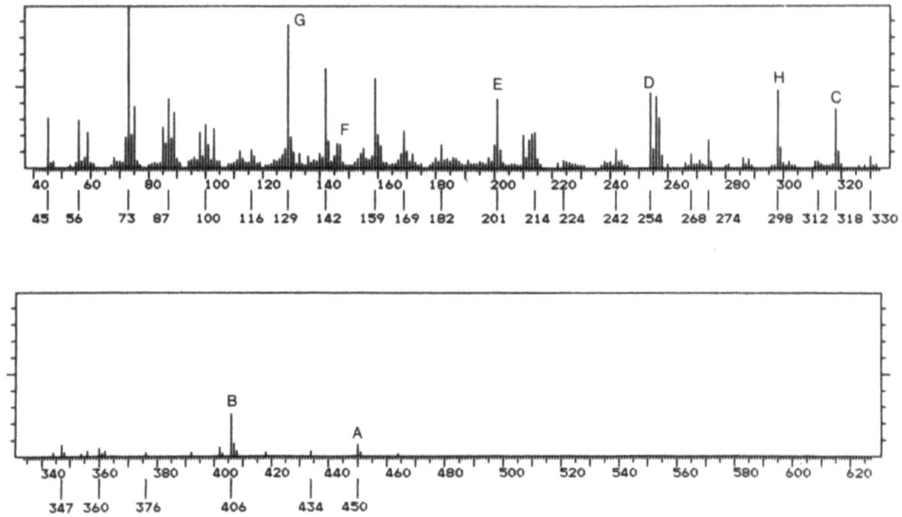

Fig. 18. E.i.-mass spectrum (70 eV) of Neu5Ac4,5,7,8Me_4 methyl ester β-methyl glycoside.

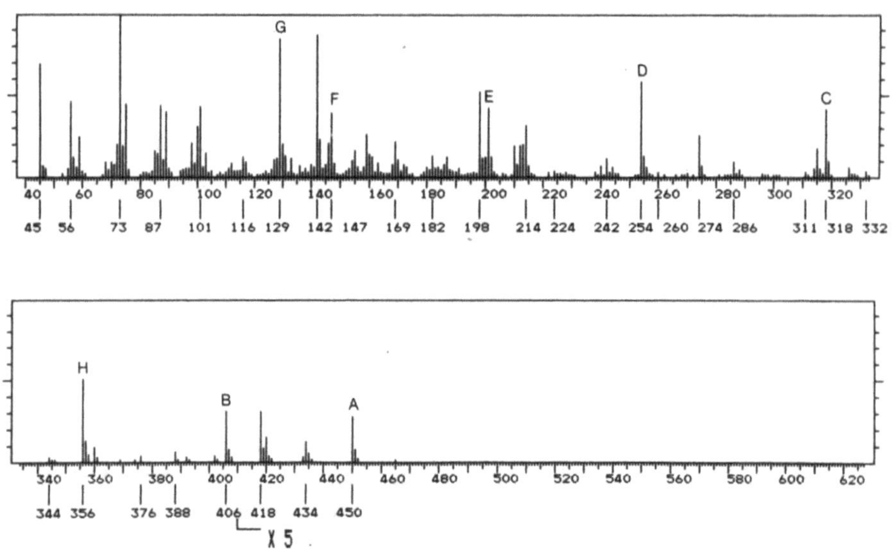

Fig. 19. E.i.-mass spectrum (70 eV) of Neu5Ac4,5,7,9Me_4 methyl ester β-methyl glycoside.

published for neuraminitols (SUGITA 1979 a, b). These derivatives are especially important for oligosaccharides having sialic acid in a reducing position. Before permethylation and methanolysis the oligosaccharides are converted into their corresponding oligosaccharide-alditols. Mass spectra have been reported of two partially methylated N,N-acetyl,methyl-neuraminitol acetate methyl esters, namely, the 2,6,7,8,9-penta-O-methyl and 2,6,7,9-tetra-O-methyl derivatives (SUGITA 1979 a). For mass spectral data of peracetylated N-acetyl- and N-glycolylneuraminitol methyl esters, see SMIRNOVA *et al.* (1977).

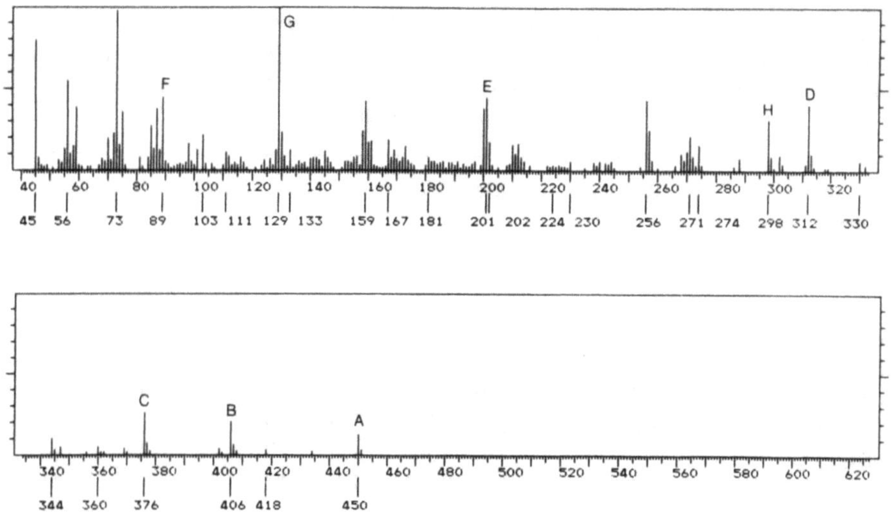

Fig. 20. E.i.-mass spectrum (70 eV) of Neu5Ac4,5,8,9Me$_4$ methyl ester β-methyl glycoside.

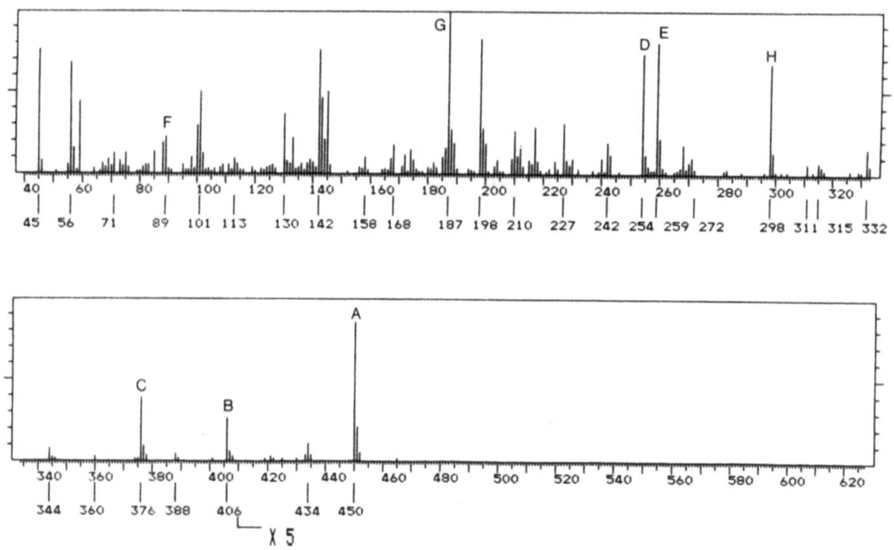

Fig. 21. E.i.-mass spectrum (70 eV) of Neu5Ac5,7,8,9Me$_4$ methyl ester β-methyl glycoside.

4. *Miscellaneous Sialic Acids and Sialic Acid Derivatives*

Partial O-acetylation of Neu5Ac methyl ester β-methyl glycoside led to the formation of several O-acetylated compounds (Haverkamp *et al.* 1975, Van Halbeek *et al.* 1978). In Table 5 the relative retention times on g.l.c. and the m.s. data in terms of the fragment ions *A*–*G* of the various trimethylsilyl derivatives are summarized. Compared with the intensity of the peak at *m/e* 175 (fragment *F*) in the mass spectra of the 9-O-acetylated compounds, this peak is observed only in a

Table 5. *G.l.c. and m.s. data of trimethylsilylated (O-acetylated) Neu5Ac methyl ester β-methyl glycosides. The R_{Neu5Ac}-values on 3.8% SE-30 at 210 °C are given relative to the trimethylsilylated Neu5Ac methyl ester β-methyl glycoside. For the trimethylsilylation procedure and additional analysis conditions, see section II.3. For an explanation of the — signs, see sections III.1 and III.4*

Sialic acid (as methyl ester β-methyl glycoside)	R_{Neu5Ac}	*m/e* values						
		A	*B*	*C*	*D*	*E*	*F*	*G*
Neu5Ac	1.00	610	566	420	298	259	205	173
Neu5,8Ac_2	1.12	580	536	420	298	259	—	173
Neu5,9Ac_2	1.20	580	536	420	298	259	175	173
Neu4,5,8Ac_3	1.28	550	506	390	298	—	—	143
Neu4,5,9Ac_3	1.46	550	506	390	298	—	175	143
Neu4,5,8,9Ac_4	1.42	520	476	390	298	—	—†	143

† The small peak at *m/e* 145 corresponds with $C_6H_{13}O_2Si$.

Table 6. *G.l.c. and m.s. data of trimethylsilylated methyl ester derivatives of degradation products obtained by periodate oxidation of bound Neu5Ac and Neu5Gc* (VEH *et al.* 1977, PFANNSCHMIDT and SCHAUER 1980). *The R_{Neu5Ac}-values on 3.8% SE-30 at 215 °C are given relative to the trimethylsilylated Neu5Ac methyl ester. For the esterification and trimethylsilylation procedure, see section II.3*

Sialic acid derivative	R_{Neu5Ac}	*m/e* values						
		A	*B*	*C*	*D*	*E*	*F*	*G*
C_7-Neu5Ac	0.29	464	420			317		173
C_8-Neu5Ac	0.50	566	522	478	298	317	"103"	173
C_7-Neu5Gc	0.60	552	508			317		261
C_8-Neu5Gc	0.95	654	610	566	386	317	"103"	261

low intensity in the spectra of the 8-O-acetylated ones. For a discussion of fragment *G* (*m/e* 173 or *m/e* 143) and of an additional criterion for the discrimination between 8-O-acetylated and 9-O-acetylated compounds, see section III.2. Some of the mass spectra have been published by HAVERKAMP *et al.* (1975), MONONEN and KÄRKKÄINEN (1975), and SUGITA (1979 b).

Neu5Ac8Me and Neu5Gc8Me have been shown to be constituents of some glycolipids (KOCHETKOV *et al.* 1973, SUGITA 1979 b). Mass spectra of the acetylated and trimethylsilylated methyl ester methyl glycosides of Neu5Ac8Me have been reported by KOCHETKOV *et al.* (1973) and SUGITA (1979 b). In one of these studies (KOCHETKOV *et al.* 1973), the mass spectra of the peracetylated methyl ester methyl

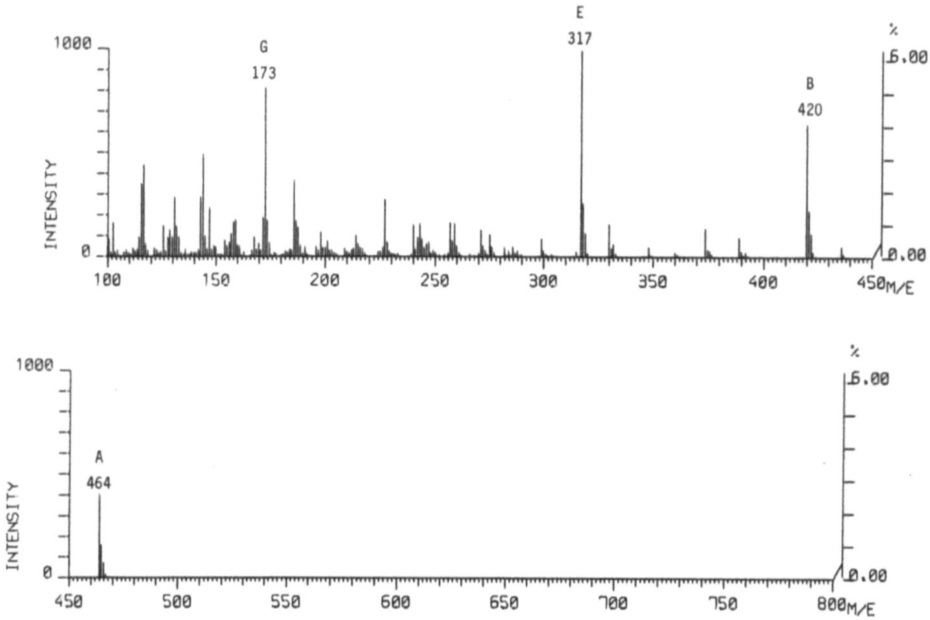

Fig. 22. E.i.-mass spectrum (70 eV) of the trimethylsilylated methyl ester of C_7-Neu5Ac.

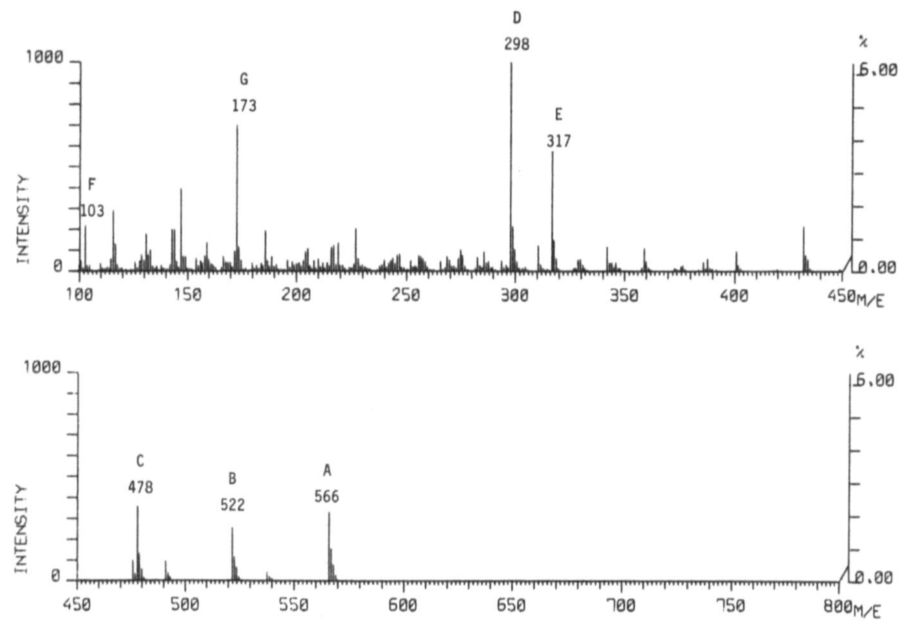

Fig. 23. E.i.-mass spectrum (70 eV) of the trimethylsilylated methyl ester of C_8-Neu5Ac.

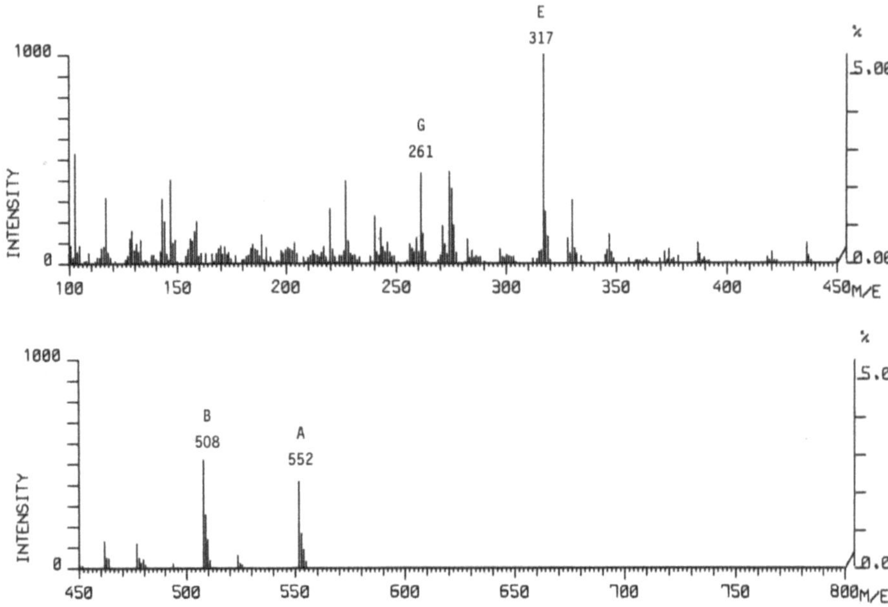

Fig. 24. E.i.-mass spectrum (70 eV) of the trimethylsilylated methyl ester of C_7-Neu5Gc.

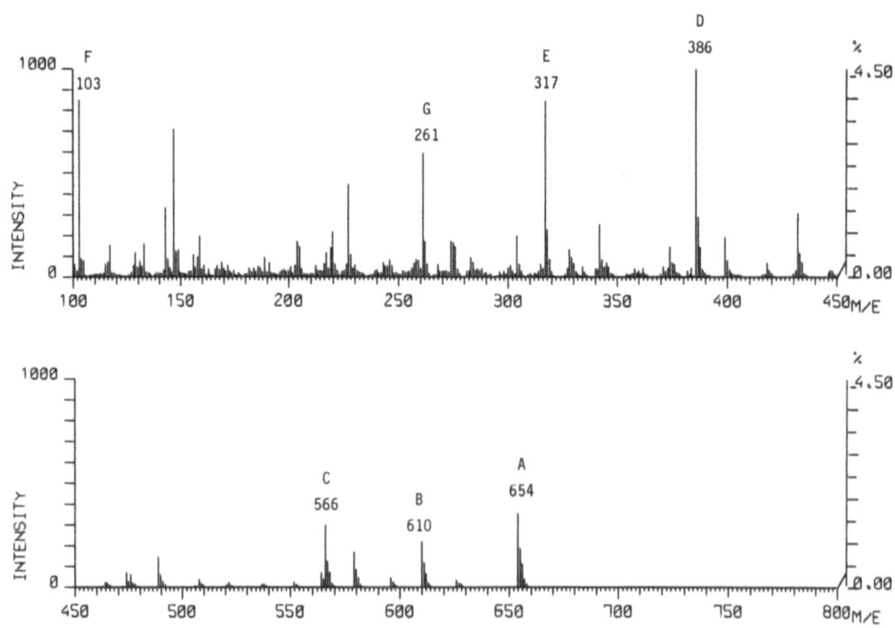

Fig. 25. E.i.-mass spectrum (70 eV) of the trimethylsilylated methyl ester of C_8-Neu5Gc.

glycosides of Neu5Ac and Neu5Gc have also been included. Furthermore, mass spectral data have been published for the trimethylsilylated methyl ester of Neu5Ac4Me (BEAU *et al.* 1978) and for the trimethylsilylated methyl ester methyl glycoside of Neu (SWEELEY and VANCE 1967). For mass spectra of pertri-

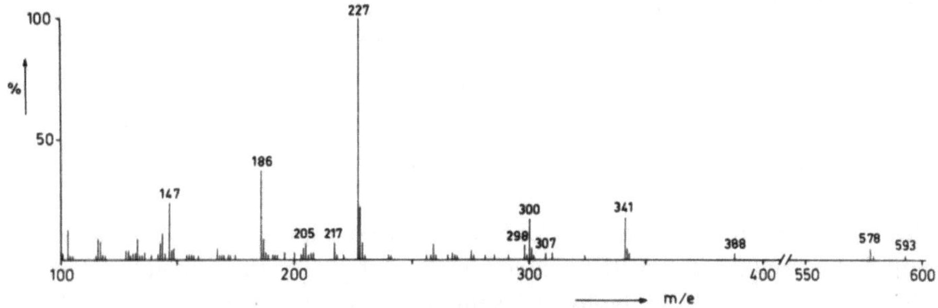

Fig. 26. E.i.-mass spectrum (70 eV) of the trimethylsilylated methyl ester of Neu2en5Ac.

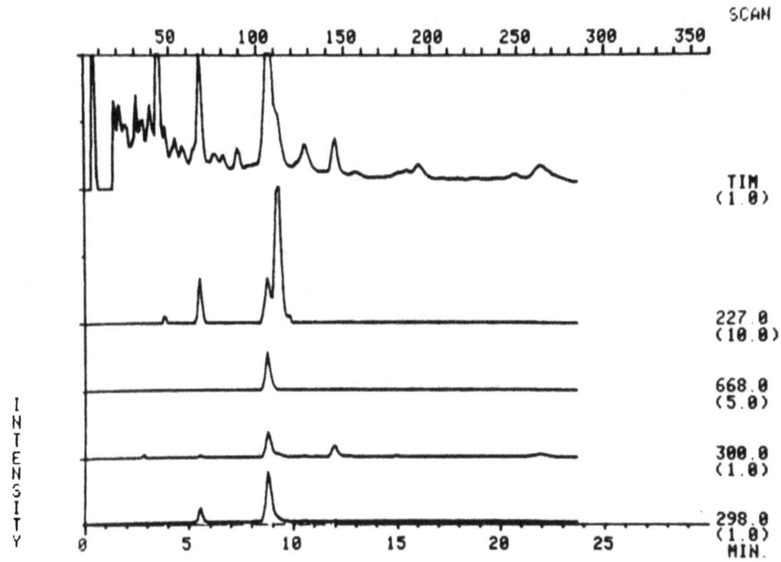

Fig. 27. Mass chromatography of the trimethylsilylated methyl esters of Neu5Ac (*m/e* 668, 300, 298) and Neu2en5Ac (*m/e* 227).

fluoroacetylated Neu and C_7-Neu methyl ester methyl glycosides, see YOHE and YU (1981).

In Table 6 the relative retention times on g.l.c. and the m.s. data of C_7-Neu5Ac, C_8-Neu5Ac, C_7-Neu5Gc, and C_8-Neu5Gc after esterification and trimethylsilylation have been summarized (VEH *et al.* 1977, PFANNSCHMIDT and SCHAUER 1980). In Figs. 22–25 the various mass spectra are presented. G.l.c. data of the trimethylsilyl methyl ester methyl glycosides and of the pertrimethylsilylated derivatives of C_7-Neu5Ac and C_8-Neu5Ac have been reported by MCLEAN *et al.* (1971) and SUTTAJIT and WINZLER (1971), respectively.

Finally, on guidance of the developed fragmentation scheme in Fig. 4, an unsaturated sialic acid, namely Neu2en5Ac, was also found to be present in different biological sources (KAMERLING *et al.* 1975 b, HAVERKAMP *et al.* 1976). In Fig. 26 the mass spectrum of the trimethylsilylated methyl ester of Neu2en5Ac (R_{Neu5Ac}-value 1.09 on 3.8% SE-30 at 215 °C) is depicted. Usually in biological samples, Neu2en5Ac is present in relatively small amounts. To detect Neu2en5Ac as a contaminant in Neu5Ac (R_{Neu5Ac} 1.00) preparations, mass chromatography using the base peak at *m/e* 227 (M-CHOSi$(CH_3)_3$CHOSi$(CH_3)_3$$CH_2OSi(CH_3)_3$-$NH_2COCH_3$) is an excellent method. Fig. 27 shows an example of such an analysis.

5. *Quantitative Analysis of Sialic Acids by g.l.c./m.s.*

MONONEN and KÄRKKÄINEN (1975) have reported a quantitative determination method for sialic acid with g.l.c./e.i.-m.s. using the multiple-ion-detection technique. Sialic acids in biological samples were liberated by methanolysis. Subsequently, the formed methyl ester methyl glycoside of Neu was N-acetylated and trimethylsilylated (see section II.1). Mass spectral data of the formed trimethylsilylated Neu5Ac methyl ester methyl glycoside are included in Table 5. The internal standard Neu5Ac trideuteromethyl ester trideuteromethyl glycoside was added after the N-acetylation. For quantification, the intensities of the *C* fragment ions *m/e* 426 ↔ *m/e* 420 and the *D* fragment ions *m/e* 301 ↔ *m/e* 298 were used. It has been stated by the authors that the accurate determination of 1 ng is possible. The method has been applied for the analysis of the sialic acid content in a crude protein fraction from rat brain.

Another approach has been published by ROBOZ *et al.* (1978). Neu5Ac was released from glycoproteins by sialidase or by acid hydrolysis and subsequently trimethylsilylated with N,O-bis(trimethylsilyl)-trifluoroacetamide plus 1% chlorotrimethylsilane and pyridine (3 : 1) for 1 h at 100 °C, also leading to trimethylsilylation of the N-acetyl function. As the internal standard trimethylsilylated Neu β-methyl glycoside was used. The quantification was carried out with g.l.c./c.i.-m.s. (isobutane) using the intense peak at *m/e* 814 $[(M+H)^+]$ for Neu5Ac and the intense peak at *m/e* 714 $[(M'+H)^+]$ for the internal standard. The detection limit for pure Neu5Ac is 200 pg. The method has been applied in the study of leukemic myeloblasts.

A method to determine sialic acid in erythrocyte ghosts has been developed by ASHRAF *et al.* (1980). For the release of sialic acid, mild methanolysis conditions were chosen (0.1 N methanolic HCl; 1 h, 90 °C). After trimethylsilylation the obtained derivative was analyzed using g.l.c./c.i.-m.s. (methane) with the trimethylsilylated derivative of N-acetylglucosamine α-phenyl glycoside as the internal standard. For quantification, the abundancies of the $[MH\text{-}16]^+$-ions were used: *m/e* 610 for the trimethylsilylated methyl ester methyl glycoside of Neu5Ac and *m/e* 498 for the internal standard. The limit of detection was found to be below 0.4 ng (useful range 10 ng–1 μg).

MIYATAKE *et al.* (1979) reported the use of mass fragmentography in sialidase activity studies. For the analysis of released Neu5Ac, the carboxyl group as well as the various hydroxyl functions were trimethylsilylated. As the internal standard the trimethylsilylated derivative of N-acetylgalactosamine α-phenyl glycoside was employed. The quantification was carried out using fragment *D* of the sialic acid derivative (*m/e* 356) and *m/e* 330 of the internal standard (g.l.c./e.i.-m.s.).

6. *Sialooligosaccharides, Sialoglycolipids, and Sialoglycopeptides*

Mass spectrometric investigations of derivatized sialooligosaccharide-alditols, sialoglycolipids, and sialoglycopeptides have been carried out by several authors. Because of the relatively low volatility of these substances, g.l.c. cannot in general be used as m.s. inlet system. In the framework of this chapter, only some examples will be mentioned.

M.s. data of the pertrimethylsilylated derivatives of the methyl esters of Neu5Ac-α(2→3)- and Neu5Ac-α(2→6)-lactose have been reported by KAMERLING *et al.* (1974). The same type of derivatives were discussed for Neu5Ac-α(2→6)- and β(2→6)-galactose (VAN DER VLEUGEL *et al.* 1982a) and for Neu5Ac-β(2→6)-N-acetylglucosamine (VAN DER VLEUGEL *et al.* 1982b). For Neu4,5Ac_2-α(2→3)-lactose, see KAMERLING *et al.* (1982a). Trimethylsilylated gangliosides as G_{M1}, G_{M2}, G_{M3}, G_{D1a}, G_{D1b}, and G_{D3} have been analyzed by SWEELEY and DAWSON (1969) and DAWSON and SWEELEY (1971).

Data on the mass spectral analysis of permethylated sialooligosaccharide-alditols have been reported, for instance, by VAN HALBEEK *et al.* (1981) (oligosaccharide-alditols from hog-submaxillary-gland mucin glycoproteins), RAUVALA *et al.* (1981), SAITO *et al.* (1981) (oligosaccharide-alditols from bovine colostrum κ-casein), and KAMERLING *et al.* (1982a) (sialyl-lactitols). Comprehensive data on permethylated sialoglycolipids from various biological sources have been published; see for instance KARLSSON (1978). A permethylated sialoglycopeptide (biantennary structure) from human transferrin has been analyzed by KARLSSON *et al.* (1978).

Acknowledgement

The studies from the authors' laboratory were supported by the Netherlands Foundation for Chemical Research (SON) with financial aid from the Netherlands Organization for the Advancement of Pure Research (ZWO).

Bibliography

ASHRAF, J., BUTTERFIELD, D. A., JÄRNEFELT, J., LAINE, R. A., 1980: J. Lip. Res. **21**, 1137—1141.

BEAU, J.-M., SINAŸ, P., KAMERLING, J. P., VLIEGENTHART, J. F. G., 1978: Carbohydr. Res. **67**, 65—77.

BHATTACHARJEE, A. K., JENNINGS, H. J., 1976: Carbohydr. Res. **51**, 253—261.

BISHOP, C. T., 1964: Adv. Carbohydr. Chem. **19**, 95—147.

BJÖRNDAL, H., HELLERQVIST, C.-G., LINDBERG, B., SVENSSON, S., 1970: Angew. Chem. **82**, 643—652.

BRUVIER, C., LEROY, Y., MONTREUIL, J., FOURNET, B., KAMERLING, J. P., 1981: J. Chromatogr. **210**, 487—504.

CASALS-STENZEL, J., BUSCHER, H.-P., SCHAUER, R., 1975: Anal. Biochem. **65**, 507—524.

CHAMBERS, R. E., CLAMP, J. R., 1971: Biochem. J. **125**, 1009—1018.

CLAMP, J. R., BHATTI, T., CHAMBERS, R. E., 1971: Meth. Biochem. Anal. **19**, 229—344.

CRAVEN, D. A., GEHRKE, C. W., 1968: J. Chromatogr. **37**, 414—421.
DAWSON, G., SWEELEY, C. C., 1971: J. Lip. Res. **12**, 56—64.
DUTTON, G. G. S., 1973: Adv. Carbohydr. Chem. Biochem. **28**, 11—160.
— 1974: Adv. Carbohydr. Chem. Biochem. **30**, 9—110.
FINNE, J., KRUSIUS, T., RAUVALA, H., HEMMINKI, K., 1977: Eur. J. Biochem. **77**, 319—323.
GHIDONI, R., SONNINO, S., TETTAMANTI, G., BAUMANN, N., REUTER, G., SCHAUER, R., 1980: J. Biol. Chem. **255**, 6990—6995.
HAVERKAMP, J., KAMERLING, J. P., VLIEGENTHART, J. F. G., VEH, R. W., SCHAUER, R., 1977a: FEBS Lett. **73**, 215—219.
— VAN HALBEEK, H., DORLAND, L., VLIEGENTHART, J. F. G., PFEIL, R., SCHAUER, R., 1982: Eur. J. Biochem. **122**, 305—311.
— VEH, R. W., SANDER, M., SCHAUER, R., KAMERLING, J. P., VLIEGENTHART, J. F. G., 1977b: Hoppe-Seyler's Z. Physiol. Chem. **358**, 1609—1612.
— SCHAUER, R., WEMBER, M., FARRIAUX, J.-P., KAMERLING, J. P., VERSLUIS, C., VLIEGENTHART, J. F. G., 1976: Hoppe-Seyler's Z. Physiol. Chem. **357**, 1699—1705.
— — — KAMERLING, J. P., VLIEGENTHART, J. F. G., 1975: Hoppe-Seyler's Z. Physiol. Chem. **356**, 1575—1583.
HOTTA, K., KUROKAWA, M., ISAKA, S., 1973: J. Biol. Chem. **248**, 629—631.
INOUE, S., MATSUMURA, G., 1979: Carbohydr. Res. **74**, 361—368.
— — 1980: FEBS Lett. **121**, 33—36.
KAMERLING, J. P., DORLAND, L., VAN HALBEEK, H., VLIEGENTHART, J. F. G., MESSER, M., SCHAUER, R., 1982a: Carbohydr. Res. **100**, 331—340.
— GERWIG, G. J., VLIEGENTHART, J. F. G., CLAMP, J. R., 1975a: Biochem. J. **151**, 491—495.
— HAVERKAMP, J., VLIEGENTHART, J. F. G., VERSLUIS, C., SCHAUER, R., 1978: Recent Develop. Mass Spectrom. Biochem. Med. **1**, 503—520.
— MAKOVITZKY, J., SCHAUER, R., VLIEGENTHART, J. F. G., WEMBER, M., 1982b: Biochim. Biophys. Acta **714**, 351—355.
— SCHAUER, R., VLIEGENTHART, J. F. G., HOTTA, K., 1980: Hoppe-Seyler's Z. Physiol. Chem. **361**, 1511—1516.
— VLIEGENTHART, J. F. G., SCHAUER, R., STRECKER, G., MONTREUIL, J., 1975b: Eur. J. Biochem. **56**, 253—258.
— — VERSLUIS, C., SCHAUER, R., 1975c: Carbohydr. Res. **41**, 7—17.
— — VINK, J., 1974: Carbohydr. Res. **33**, 297—306.
KARLSSON, K.-A., 1978: Prog. Chem. Fats other Lipids **16**, 207—230.
— PASCHER, I., SAMUELSSON, B. E., FINNE, J., KRUSIUS, T., RAUVALA, H., 1978: FEBS Lett. **94**, 413—417.
KOCHETKOV, N. K., CHIZHOV, O. S., KADENTSEV, V. I., SMIRNOVA, G. P., ZHUKOVA, I. G., 1973: Carbohydr. Res. **27**, 5—10.
LÖNNGREN, J., SVENSSON, S., 1974: Adv. Carbohydr. Chem. Biochem. **29**, 41—106.
MACHER, B. A., SWEELEY, C. C., 1978: Meth. Enzymol. **50**, 236—251.
MCLEAN, R. L., SUTTAJIT, M., BEIDLER, J., WINZLER, R. J., 1971: J. Biol. Chem. **246**, 803—809.
MIYATAKE, T., SUZUKI, M., YAMADA, T., 1979: Proc. 5th Int. Symp. Glycoconjugates (SCHAUER, R., *et al.*, eds.), pp. 350—351. Stuttgart: G. Thieme.
MONONEN, I., 1981: Carbohydr. Res. **88**, 39—50.
— KÄRKKÄINEN, J., 1975: FEBS Lett. **59**, 190—193.
MONTREUIL, J., 1980: Adv. Carbohydr. Chem. Biochem. **37**, 157—223.
PFANNSCHMIDT, G., SCHAUER, R., 1980: Hoppe-Seyler's Z. Physiol. Chem. **361**, 1683—1695.
RAUVALA, H., FINNE, J., KRUSIUS, T., KÄRKKÄINEN, J., JÄRNEFELT, J., 1981: Adv. Carbohydr. Chem. Biochem. **38**, 389—416.
— KÄRKKÄINEN, J., 1977: Carbohydr. Res. **56**, 1—9.

REUTER, G., PFEIL, R., KAMERLING, J. P., VLIEGENTHART, J. F. G., SCHAUER, R., 1980a: Biochim. Biophys. Acta **630**, 306—310.

— VLIEGENTHART, J. F. G., WEMBER, M., SCHAUER, R., HOWARD, R. J., 1980b: Biochem. Biophys. Res. Commun. **94**, 567—572.

— PFEIL, R., KAMERLING, J. P., VERSLUIS, C., VLIEGENTHART, J. F. G., SCHAUER, R., 1982: in preparation.

ROBOZ, J., SUZUKI, R., BEKESI, J. G., 1978: Anal. Biochem. **87**, 195—205.

SAITO, T., ITOH, T., ADACHI, S., SUZUKI, T., USUI, T., 1981: Biochim. Biophys. Acta **678**, 257—267.

SCHAUER, R., 1978: Meth. Enzymol. **50**, 64—89.

— HAVERKAMP, J., WEMBER, M., VLIEGENTHART, J. F. G., KAMERLING, J. P., 1976: Eur. J. Biochem. **62**, 237—242.

SEKINE, M., ARIGA, T., MIYATAKE, T., 1981: Proc. 6th Int. Symp. Glycoconjugates (YAMAKAWA, T., *et al.*, eds.), pp. 72—73. Tokyo: Japan Scientific Societies Press.

SMIRNOVA, G. P., CHEKAREVA, N. V., CHIZHOV, O. S., ZOLOTAREV, B. M., KOCHETKOV, N. K., 1977: Carbohydr. Res. **59**, 235—239.

— KOCHETKOV, N. K., 1980: Biochim. Biophys. Acta **618**, 486—495.

STELLNER, K., SAITO, H., HAKOMORI, S.-I., 1973: Arch. Biochem. Biophys. **155**, 464—472.

SUGITA, M., 1979a: J. Biochem. **86**, 289—300.

— 1979b: J. Biochem. **86**, 765—772.

SUTTAJIT, M., WINZLER, R. J., 1971: J. Biol. Chem. **246**, 3398—3404.

SWEELEY, C. C., BENTLEY, R., MAKITA, M., WELLS, W. W., 1963: J. Am. Chem. Soc. **85**, 2497—2507.

— DAWSON, G., 1969: Biochem. Biophys. Res. Commun. **37**, 6—14.

— VANCE, D. E., 1967: Lipid Chromatogr. Anal. **1**, 476.

— WALKER, B., 1964: Anal. Chem. **36**, 1461—1466.

VAN DER VLEUGEL, D. J. M., WASSENBURG, F. R., ZWIKKER, J. W., VLIEGENTHART, J. F. G., 1982a: Carbohydr. Res. **104**, 221—233.

— ZWIKKER, J. W., VLIEGENTHART, J. F. G., VAN BOECKEL, S. A. A., VAN BOOM, J. H., 1982b: Carbohydr. Res. **105**, 19—31.

VAN DER MEER, A., KAMERLING, J. P., VLIEGENTHART, J. F. G., 1982: unpublished results.

VAN HALBEEK, H., DORLAND, L., HAVERKAMP, J., VELDINK, G. A., VLIEGENTHART, J. F. G., FOURNET, B., RICART, G., MONTREUIL, J., GATHMANN, W. D., AMINOFF, D., 1981: Eur. J. Biochem. **118**, 487—495.

— HAVERKAMP, J., KAMERLING, J. P., VLIEGENTHART, J. F. G., VERSLUIS, C., SCHAUER, R., 1978: Carbohydr. Res. **60**, 51—62.

VEH, R. W., CORFIELD, A. P., SANDER, M., SCHAUER, R., 1977: Biochim. Biophys. Acta **486**, 145—160.

YU, R. K., LEDEEN, R. W., 1970: J. Lip. Res. **11**, 506—516.

YOHE, H. C., YU, R. K., 1981: Carbohydr. Res. **93**, 1—9.

ZANETTA, J. P., BRECKENRIDGE, W. C., VINCENDON, G., 1972: J. Chromatogr. **69**, 291—304.

G. NMR Spectroscopy of Sialic Acids

JOHANNES F. G. VLIEGENTHART, LAMBERTUS DORLAND, HERMAN VAN HALBEEK, and JOHAN HAVERKAMP[1]

Department of Bio-Organic Chemistry, State University of Utrecht, Utrecht, The Netherlands, and [1]Department of Biomolecular Physics, FOM-Institute for Atomic and Molecular Physics, Amsterdam, The Netherlands

With 18 Figures

Contents

I. Introduction

High-resolution NMR spectroscopy has become an invaluable technique in the study of biopolymers and of their constituents. Using ^{1}H or ^{13}C nuclei as probes information can be obtained about primary structures, conformations and intermolecular interactions of biomolecules in solution. In particular, the possibility to record spectra of underivatized compounds in aqueous solutions allows to afford further insight into the way of action of biomolecules under physiological conditions. For general reviews of high-resolution NMR spectroscopy in the study of biological systems the reader is referred to the recent books of BERLINER and REUBEN (1978, 1980), JARDETZKY and ROBERTS (1981), and SHULMAN (1979).

In 1977 we introduced the application of high-resolution ^{1}H-NMR spectroscopy as a new method for structure elucidation of carbohydrate chains present in glycoproteins. These studies have shown that sialylated carbohydrate chains can completely be characterized with regard to the sialic acid residues. In this chapter relevant ^{1}H- and ^{13}C-NMR parameters of free and glycosidically bound sialic acids will be discussed.

II. ^{1}H-NMR Spectroscopy

1. N-Acetyl- and N-Glycolylneuraminic Acid

As early as 1968 60 MHz ^{1}H-NMR spectra were published of Neu5Ac in ^2H_2O (CHAPMAN *et al.* 1968, KIMURA and TSURUMI 1968, BLIX and JEANLOZ 1969). Such a spectrum is given in Fig. 1. It shows a singlet belonging to the acetamido methyl protons at δ 2.05. Two broad signals are found at δ 3.7 and δ 3.9 comprising most of the skeleton protons. Although the investigators realized that this technique could be very promising for biochemical purposes, hardly any structural information can be obtained from this low-resolution NMR spectrum.

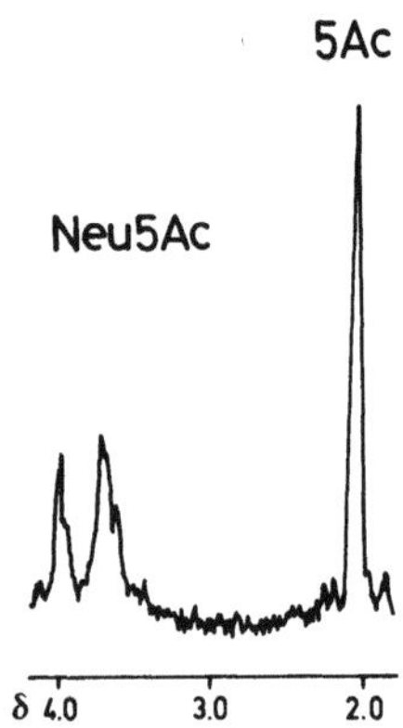

Fig. 1. 60 MHz ^{1}H-NMR spectrum of Neu5Ac dissolved in ^2H_2O, recorded at ambient temperature (taken from KIMURA and TSURUMI 1968).

The recent development of NMR spectrometers operating at higher magnetic fields (up to 11.7 Tesla, equivalent to a frequency of 500 MHz for protons) together with the advances in computer capabilities have led to an enormous improvement in spectral resolution and in sensitivity. This progress is excellently reflected in the 500 MHz ^{1}H-NMR spectrum of Neu5Ac in ^2H_2O, as presented in Fig. 2. The spectrum consists of two subspectra belonging to the α- and β-anomer of Neu5Ac. These anomers occur in a molar ratio of 7 : 93, respectively (JAQUES *et al.* 1977, HAVERKAMP *et al.* 1978, DABROWSKI *et al.* 1979, BEAU *et al.* 1980, FRIEBOLIN *et al.* 1980 a, b, HAVERKAMP *et al.* 1982). The 500 MHz ^{1}H-NMR spectrum given in Fig. 2 allows the assignment of several of the signals for the α-anomer of Neu5Ac. The spectrum of the β-anomer could completely be interpreted (BROWN *et al.* 1975, BEAU *et al.* 1980, HAVERKAMP *et al.* 1982). Chemical shifts and coupling constants are summarized in Table 1. It should be noted that the chemical shifts of Neu5Ac protons are pH-dependent. For the β-anomer the signals shift upfield upon increasing the pH from 2 to 7, e.g. for H3eq from δ 2.313 to δ 2.208 and for H3ax from δ 1.880 to δ 1.827 (see Table 1) (HAVERKAMP *et al.* 1982).

Interestingly, in the pH range 6.5–9.0, H3ax can be exchanged. This could be observed in alkaline ^2H_2O solution of Neu5Ac ($p^2H = 9.0$), wherein a complete replacement of H3ax by ^2H occurred (DORLAND *et al.* 1982). In the ^1H-NMR spectrum this exchange leads to the disappearance of the H3ax signal and of its

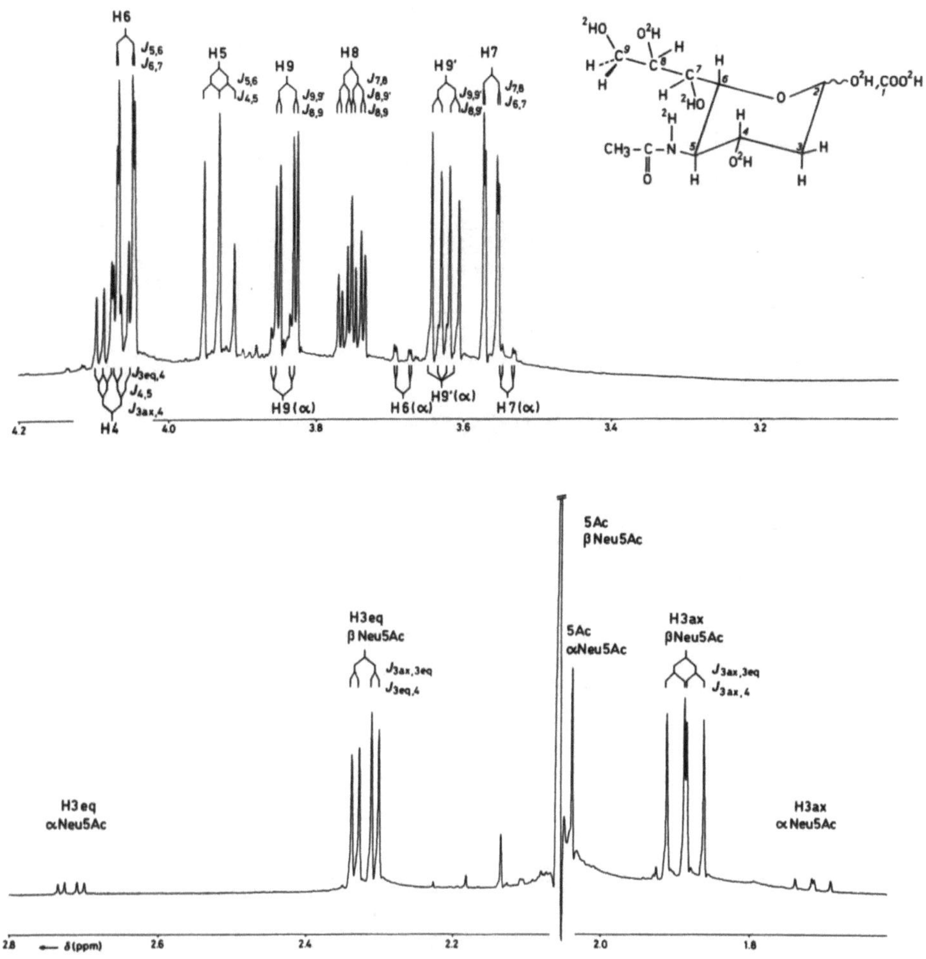

Fig. 2. Resolution-enhanced 500 MHz ^1H-NMR spectrum of Neu5Ac dissolved in ^2H_2O, recorded at p^2H 1.4 and 27 °C. The intensity ratio of the signals from corresponding protons reflects the molar ratio of the α- and β-anomer under the applied measuring conditions to be 7 : 93. Detailed splitting patterns are indicated for all resonances from the β-, and for the greater part of those from the α-anomer.

coupling with H3eq and H4. This phenomenon can excellently be utilized for the preparation of specifically labelled (^2H or ^3H) cytidine-5′-monophospho-N-acetylneuraminic acid (CMP-Neu5Ac), thereby allowing the enzymic introduction of labelled Neu5Ac in glycoconjugates (DORLAND *et al.* 1982). Further increase of the p^2H to 12.4 causes also exchange of H3eq (FRIEBOLIN *et al.* 1981 b). The

Table 1. *^{1}H-NMR data for Neu5Ac and Neu5Gc*

Chemical shift values are given in ppm downfield from DSS for solutions in $^{2}H_2O$ at 25 °C and at the indicated p^2H values. Coupling constants (J) are given in Hz. n.d., value could not be determined

Compound	p^2H	Chemical shift											References
		H3ax	H3eq	H4	H5	H6	H7	H8	H9	H9′	5Ac	5Gc	
βNeu5Ac	1.4	1.880	2.313	4.067	3.931	4.056	3.556	3.750	3.841	3.619	2.053	—	HAVERKAMP *et al.* 1982
αNeu5Ac	1.4	1.705	2.718	n.d.	3.85	3.684	3.53	3.75	3.85	3.62	2.036	—	HAVERKAMP *et al.* 1982
βNeu5Ac	7.0	1.827	2.208	4.024	3.899	3.984	3.514	3.753	3.835	3.608	2.050	—	HAVERKAMP *et al.* 1982
αNeu5Ac	7.0	1.621	2.730	n.d.	n.d.	n.d.	n.d.	n.d.	n.d.	n.d.	2.030	—	HAVERKAMP *et al.* 1982
βNeu5Gc	7.0	1.840	2.243	4.127	4.002	4.106	3.549	3.777	3.821	3.613	—	4.143	HAVERKAMP *et al.* 1982
αNeu5Gc	7.0	1.644	2.749	n.d.	n.d.	n.d.	n.d.	n.d.	n.d.	n.d.	—	4.12	HAVERKAMP *et al.* 1982

Compound	p^2H	Coupling constant										References
		$^{2}J_{3ax,3eq}$	$^{3}J_{3ax,4}$	$^{3}J_{3eq,4}$	$^{3}J_{4,5}$	$^{3}J_{5,6}$	$^{3}J_{6,7}$	$^{3}J_{7,8}$	$^{3}J_{8,9}$	$^{3}J_{8,9'}$	$^{2}J_{9,9'}$	
βNeu5Ac	1.4	—13.2	11.8	5.0	10.4	10.7	1.2	9.4	2.8	6.4	—12.4	BEAU *et al.* 1980
αNeu5Ac	1.4	—13.0	11.5	4.5	n.d.	10.5	1.5	9.0	n.d.	6.5	—12.5	—
βNeu5Gc	7.0	—12.6	11.4	4.6	10.2	10.2	1.0	9.0	2.8	6.2	—11.4	JAQUES *et al.* 1980a

exchange reaction can also be traced in the ^{1}H-NMR spectrum of Neu5Gc in ^2H_2O (p^{2}H ~ 7) as published by JAQUES *et al.* (1980a); the reduced intensity of the H3ax signal has to be ascribed to partial replacement by ^{2}H rather than to saturation effects as proposed by the authors. The ^{1}H-NMR data of Neu5Gc are also compiled in Table 1. It is worthy to note that, except for H9, the chemical shift values for the protons of Neu5Gc are found at lower field than the corresponding protons of Neu5Ac if both measured in neutral solution. These chemical shift differences may result from changes in the microenvironment of the various protons. In this respect it would be interesting to carry out studies on the occurrence of hydrogen bonds of Neu5Gc in a similar way as has been performed

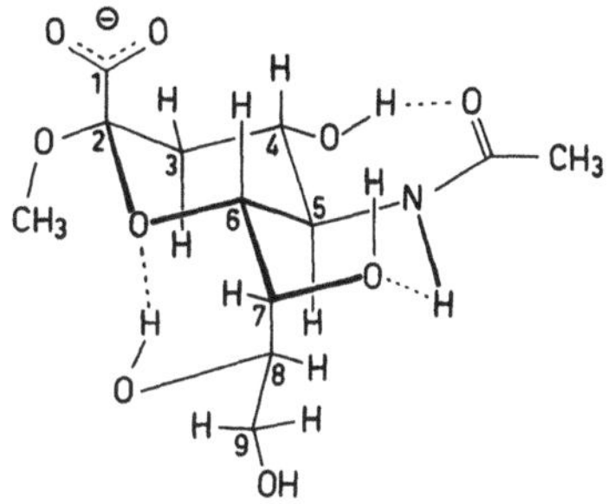

Fig. 3. Conformational model of the anion of αNeu5Ac2Me showing the proposed hydrogen bonds and preferred conformation of the glycerol side chain (taken from CZARNIECKI and THORNTON 1976). Although shown here only for the α-anomer, this conformation is independent of the anomeric configuration.

for Neu5Ac derivatives by CZARNIECKI and THORNTON (1976, 1977a). These authors have studied the spatial structure of the α-methyl glycoside of Neu5Ac in ^2H_2O solution on the basis of the proton-proton coupling constants (see Table 1) in combination with the ^{13}C spin-lattice relaxation times (T_1). The T_1 values provide information about the internal mobility of the molecules in solution. The similarity of the values for C7, C8 and the ring carbon atoms point to an isotropic motion of these atoms. From this observation the authors postulated that the amido-NH is hydrogen-bonded to the oxygen at C7 and that the OH at C8 is hydrogen-bonded to the ring-oxygen. A third hydrogen bond between the acetamido-carbonyl and the OH at C4 was suggested on the basis of model building. The proposed structure is given in Fig. 3. Also REUTER *et al.* (unpublished results) found that the T_1 values for ring and glycerol side chain carbon atoms of derivatives of Neu5Ac are similar. In the latter study the inversion-recovery method was used in combination with T_1 calculation, following the method of SASS and ZIESSOW (1977). This approach gives more accurate T_1 determinations than the fast inversion-recovery method, used by CZARNIECKI and THORNTON (1976, 1977a). Apparently, the anomeric centre is not involved in any hydrogen bond, leading to the same conformation for α- and β-anomers of Neu5Ac.

2. *O-Acetylated Neuraminic Acid Derivatives*

For neuraminic acids bearing O-acetyl groups, it has been demonstrated that high-resolution ^{1}H-NMR spectroscopy can efficiently be employed to determine the number and position of such substituents (HAVERKAMP *et al.* 1982). The number of acetyl groups can be determined on the basis of number and relative intensities of the acetyl signals. O-acetylation causes some specific downfield shifts

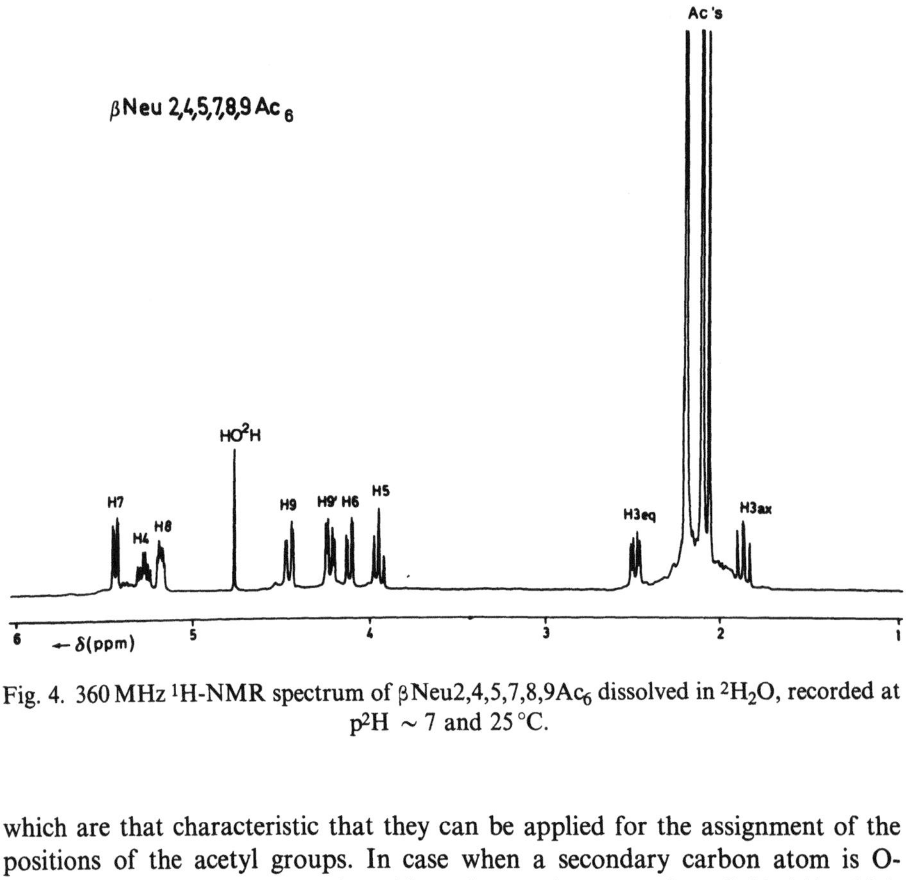

Fig. 4. 360 MHz ^{1}H-NMR spectrum of βNeu2,4,5,7,8,9Ac$_6$ dissolved in ^2H_2O, recorded at p^{2}H ~ 7 and 25 °C.

which are that characteristic that they can be applied for the assignment of the positions of the acetyl groups. In case when a secondary carbon atom is O-acetylated, the proton attached to this carbon undergoes a downfield shift which varies from 1-1.5 ppm. This so-called α-effect amounts to 1.2 ppm for H4 and 1.5 ppm for H7. Upon introduction of an acetyl substituent at the primary hydroxyl group of C9, the effect is shared over the protons H9 and H9′, each being about 0.5–0.6 ppm.

There are also effects on protons at carbon atoms adjacent to the O-acetylated carbon. These β-effects are in the cases of O-acetylation of C4, C7, and C9 about 0.2 ppm. The geminal protons H3eq and H3ax form an exception since they undergo shifts of about 0.04 and 0.14 ppm, respectively. The β-effects on H3eq and H3ax are unequal due to the rigidity of the C3–C4 part of the molecule. The γ- and δ-effects on protons, attached to more remote carbon atoms are small and irregular in direction.

In di-O-acetylated neuraminic acids the effects of O-acetylation on the chemical shifts of skeleton protons are composed of the individual contributions of each O-acetyl group. Apparently, the induced shifts are additive. This is a general phenomenon, which can be illustrated for the peracetylated neuraminic acid. The ^{1}H-NMR spectrum of this compound is shown in Fig. 4. The resonance positions of H5, H6, H7, H8, H9, and H9′ can be estimated on the basis of the presumed α- and β-effects of O-acetylation at position C4, C7, C8, and C9.

In principle, the chemical shifts of the acetyl protons contain information on the position of these substituents. The protons of the acetyl group at C9 are found in the range δ 2.10–2.13, at C4 near δ 2.06 and at C8 near δ 2.09. Furthermore, the occurrence of an acetyl group at C4 or C7 gives rise to a significant upfield shift of the N-acetyl signal. In case when the substituents at C4 and C7 are both present, their effects are additive (HAVERKAMP *et al.* 1982). ^{1}H-NMR data of a series of O-acetylated sialic acids and of some reference compounds are presented in Table 2.

An elegant illustration of the suitability of the above mentioned approach for the determination of the number and position of O-acetyl groups in sialic acids has been given for a trisaccharide from echidna milk (HAVERKAMP *et al.* 1982, KAMERLING *et al.* 1981, 1982). It was found that the trisaccharide is identical to Neu5Acα(2–3)Galβ(1–4)Glc having one O-acetyl substituent at the Neu5Ac residue. The H3, H4, and H5 signals of Neu5Ac have undergone shifts (see Table 2), typical for O-acetylation of C4. The 500 MHz ^{1}H-NMR spectrum of this trisaccharide is presented in Fig. 5.

It may be clear that this NMR method can generally be applied for the determination of the number and positions of O-acetyl groups in glycosidically linked sialic acids, occurring in oligosaccharides or glycopeptides derived from glycoconjugates.

Obviously the isolation procedure for the carbohydrates should be that mild that O-deacetylation is suppressed as far as possible and that migration of O-acetyl groups does not occur. In principle, the method can be extended to other O-acyl substituents, e.g. O-lactyl, provided that adequate reference compounds are available.

On the basis of the shifts which occur upon etherification of sialic acid, e.g. O-methylation, ^{1}H-NMR spectroscopy may also afford valuable information about the positions of these substituents (BEAU *et al.* 1980).

3. *Analysis of Sialic Acid Linkage Types in Glycoconjugates*

High-resolution ^{1}H-NMR spectroscopy has proved to be a suitable method for the determination of the type of linkage of sialic acid residues to carbohydrate units of glycoconjugates (DORLAND *et al.* 1978, SCHUT *et al.* 1978, LEGER *et al.* 1978, DORLAND 1979, VLIEGENTHART 1980, VAN HALBEEK *et al.* 1980, 1981 a, b, NOMOTO *et al.* 1981, INOUE *et al.* 1981, SPIK *et al.* 1982, VLIEGENTHART *et al.* 1981, 1982). First of all, the resonance positions of the sialic acid structural reporter group signals H3eq and H3ax are indicative of the configuration and position of the glycosidic linkage. These chemical shifts are influenced to some extent by the total structure of the carbohydrate chain to which sialic acid is attached. Secondly, several reporter group signals in the carbohydrate chain are shifted in a very

Table 2. ^{1}*H-NMR chemical shift data for O-acetylated sialic*
Chemical shift values are given for solutions in ^2H_2O at 25 °C and at the indicated p^{2}H could not be determined at all, due to complication of the spectra of the corresponding β-

Compound	p^{2}H	Chemical shift					
		H3ax	H3eq	H4	H5	H6	H7
βNeu4,5Ac$_2$	7	1.951	2.249	5.274	4.15	4.15	3.570
βNeu5,7Ac$_2$	4	1.905	2.236	3.950	3.767	4.246	5.045
αNeu5,7Ac$_2$	4	1.649	2.757	n.d.	n.d.	n.d.	n.d.
βNeu5,9Ac$_2$	7	1.833	2.221	4.024	3.913	3.991	3.571
αNeu5,9Ac$_2$	7	1.624	2.720	n.d.	n.d.	n.d.	n.d.
βNeu9Ac5Gc	7	1.842	2.234	4.14	4.006	4.109	3.570
αNeu9Ac5Gc	7	1.649	2.751	n.d.	n.d.	n.d.	n.d.
βNeu5,7,9Ac$_3$	2	1.924	2.303	3.978	3.775	4.293	5.162
αNeu5,7,9Ac$_3$	2	1.686	2.751	n.d.	n.d.	n.d.	n.d.
βNeu5,8,9Ac$_3$	7	1.838	2.189	3.978	3.903	3.780	3.838
βNeu5,8,9Ac$_3$	2	1.862	2.250	4.006	3.912	3.830	3.866
βNeu2,4,5,7,8,9Ac$_6$	7	1.862	2.467	5.276	3.934	4.105	5.438
βNeu5Ac1Me	7	1.913	2.315	4.067	3.916	4.067	3.552
αNeu5Ac1Me	7	1.732	2.723	n.d.	n.d.	n.d.	n.d.
βNeu5Ac2Me	7	1.645	2.337	4.009	3.88	3.785	3.532
αNeu5Ac2Me	7	1.626	2.718	3.675	3.803	3.689	3.586
βNeu5Ac1,2Me$_2$	7	1.784	2.392	4.044	3.917	3.87	3.581
αNeu5Ac1,2Me$_2$	7	1.798	2.676	3.756	3.86	3.83	3.558
βNeu4,5Ac$_2$1,2Me$_2$	7	1.927	2.442	5.260	4.157	4.035	3.623
βNeu5,9Ac$_2$1,2Me$_2$	7	1.788	2.395	4.059	3.938	3.910	3.637
βNeu4,5,9Ac$_3$1,2Me$_2$	7	1.921	2.445	5.299	4.188	4.069	3.686
αNeu4,5,7,8,9Ac$_5$2Me	7	1.701	2.637	4.858	3.834	4.498	5.365
αNeu4,5,7,8,9Ac$_5$1,2Me$_2$	7	1.930	2.718	4.924	3.909	4.273	5.390
Neu5Acα(2-3)Galβ(1-4)Glc	6	1.799	2.757	3.688	3.825	n.d.	n.d.
Neu5Gcα(2-3)Galβ(1-4)Glc	7	1.816	2.777	n.d.	n.d.	n.d.	n.d.
Neu5Acα(2-6)Galβ(1-4)Glc	6	1.739	2.715	3.658	3.836	n.d.	n.d.
Neu4,5Ac$_2$α(2-3)Galβ(1-4)Glc	5	1.926	2.768	4.955	4.088	n.d.	n.d.

[a] Values may be interchanged.
[b] Also including the 2Ac signal.
[c] Individual acetyl signals could not be assigned.

characteristic way upon extension with sialic acid. This information was derived from comparative ^{1}H-NMR studies on a large series of sialo-oligosaccharides and sialo-glycopeptides (DORLAND *et al.* 1978, VLIEGENTHART 1980, VLIEGENTHART *et al.* 1981, 1982, VAN HALBEEK *et al.* 1980, 1981 a, b).

a) The Anomeric Configuration of Sialic Acid

The anomeric configuration of Neu5Ac or Neu5Gc can usually be inferred from the chemical shifts of the H3eq and H4 resonances. Analysis of a series of model

acids and some reference compounds (HAVERKAMP *et al.* 1982)
values. n.d., value could not be determined. Values for αNeu4,5Ac$_2$ and αNeu5,8,9Ac$_3$
anomers as a result of the presence of small non-carbohydrate contaminants in the samples.

Chemical shift										
H8	H9	H9'	5Ac	4Ac	7Ac	8Ac	9Ac	5Gc	1Me	2Me
3.775	3.844	3.619	1.992	2.065	—	—	—	—	—	—
3.911	3.629	3.444	1.976[a]	—	2.144[a]	—	—	—	—	—
n.d.	n.d.	n.d.	1.947	—	2.128	—	—	—	—	—
3.977	4.365	4.187	2.057	—	—	—	2.119	—	—	—
n.d.	n.d.	n.d.	n.d.	—	—	—	n.d.	—	—	—
3.970	4.365	4.183	—	—	—	—	2.115	4.144	—	—
n.d.	n.d.	n.d.	—	—	—	—	n.d.	4.123	—	—
4.140	4.106	4.106	1.981[a]	—	2.134[a]	—	2.106	—	—	—
n.d.	n.d.	n.d.	1.956	—	n.d.	—	n.d.	—	—	—
5.114	4.528	4.287	2.057	—	—	2.089	2.105	—	—	—
5.115	4.545	4.287	2.059	—	—	2.091	2.107	—	—	—
5.182	4.446	4.212	1.940	2.047; 2.081; 2.089; 2.171; 2.179[b,c]				—	—	—
3.731	3.834	3.619	2.051	—	—	—	—	—	3.838	—
n.d.	n.d.	n.d.	2.036	—	—	—	—	—	3.838	—
3.88	3.85	3.662	2.047	—	—	—	—	—	—	3.200
3.886	3.869	3.641	2.033	—	—	—	—	—	—	3.341
3.87	3.841	3.667	2.050	—	—	—	—	—	3.868	3.274
3.84-3.88	3.84-3.88	3.654	2.034	—	—	—	—	—	3.880	3.383
3.87	3.846	3.667	1.990	2.059	—	—	—	—	3.868	3.304
4.087	4.433	4.201	2.057	—	—	—	2.128	—	3.878	3.278
4.109	4.443	4.214	1.999	2.059	—	—	2.125	—	3.885	3.302
5.422	4.382	4.265	1.910	2.028; 2.089; 2.162; 2.162[c]				—	—	3.305
5.428	4.356	4.243	1.922	2.041; 2.093; 2.167; 2.211[c]				—	3.887	3.356
n.d.	n.d.	n.d.	2.030	—	—	—	—	—	—	—
n.d.	n.d.	n.d.	—	—	—	—	—	4.119	—	—
n.d.	n.d.	n.d.	2.030	—	—	—	—	—	—	—
n.d.	n.d.	n.d.	1.963	2.070	—	—	—	—	—	—

substances has shown that for α-anomers the chemical shift of H3eq varies between δ 2.6 and δ 2.8 and that of H4 between δ 3.6 and δ 3.8. For β-anomers these ranges are δ 2.1-δ 2.5 and δ 3.9-δ 4.2, respectively (HAVERKAMP *et al.* 1978, 1982). Furthermore, the signal of the N-acyl group protons furnishes additional evidence for the anomeric configuration (VAN HALBEEK *et al.* 1981 b, HAVERKAMP *et al.* 1982). The signals, which are useful for the determination of the anomeric configuration are compiled in Table 3.

Although O-substitution, e.g. acylation or etherification, may influence the

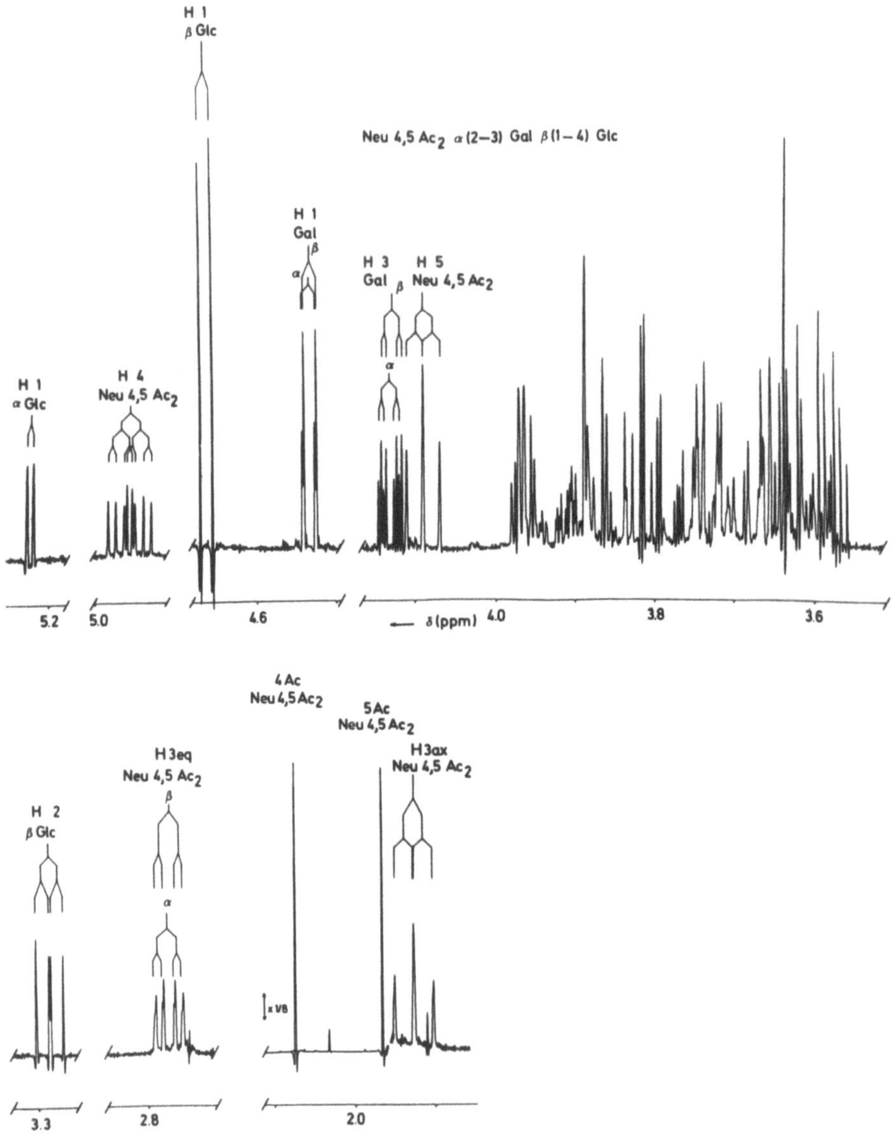

Fig. 5. Resolution-enhanced 500 MHz ^{1}H-NMR spectrum of Neu4,5Ac$_2$α(2-3)Galβ(1-4)Glc dissolved in ^2H_2O, recorded at p^{2}H 7 and 27 °C. Signals of corresponding protons of the two anomers of the trisaccharide (molar ratio α : β = 7 : 10) coincide, unless indicated otherwise. (The relative intensity scale of the acetyl methyl proton signals differs from that of the other parts of the spectrum.)

Table 3. [1]*H-NMR chemical shift data, useful for discrimination between α- and β-glycosidically linked sialic acids* (Haverkamp *et al.* 1978, 1982)
Chemical shift values are given for solutions in ^2H_2O at 25 °C and at p^{2}H ≃ 7. Values have to be modified (see Table 2) if C4 and/or C7 of the sialic acid bear a substituent (e.g., an O-acetyl group)

Configuration of glycosidic linkage of sialic acid	Chemical shift range			
	H3eq	H4	5Ac	5Gc
α	2.6-2.8	3.6-3.8	2.025-2.035	4.11-4.13
β	2.1-2.5	3.9-4.2	2.045-2.055	4.13-4.15

Table 4. [1]*H-NMR chemical shift data discriminative for α(2-3)- or α(2-6)-linkage of sialic acid to galactose* (Dorland *et al.* 1978)
Chemical shift values are given for solutions in ^2H_2O at 25 °C and at p^{2}H ≃ 7

Structural element	Chemical shift	
	H3ax	H3eq
Neu5Acα(2-3)Galβ(1-·)	1.80	2.76
Neu5Acα(2-6)Galβ(1-·)	1.72	2.67

chemical shifts of some of the reporters which are employed for the determination of the anomeric configuration of the glycosidic linkage, they are still useful for this purpose (Beau *et al.* 1980, Haverkamp *et al.* 1982, Kamerling *et al.* 1982).

b) Sialic Acid in Carbohydrate Chains of the N-Acetyllactosamine Type

The diantennary glycan structure

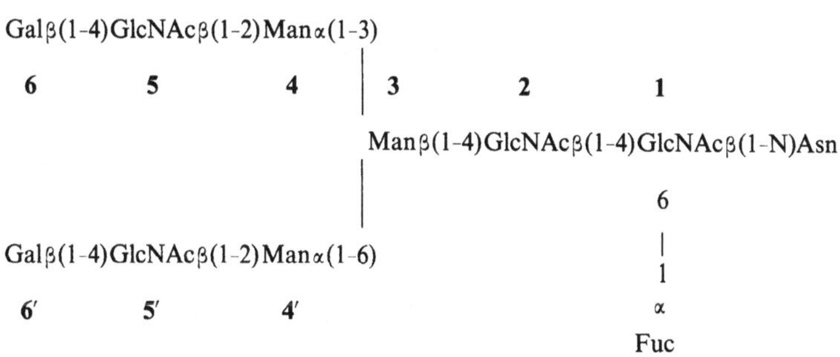

and sialylated extensions thereof which frequently occur in glycoproteins, were extensively studied by ^{1}H-NMR spectroscopy (Dorland *et al.* 1978, Schut *et al.* 1978, Leger *et al.* 1978, Vliegenthart *et al.* 1981, 1982). The N-acetyllactosamine type branches can be terminated by Neu5Ac residues in α(2-6)- or α(2-3)-linkages to the Gal residues. Additional Neu5Ac-N-acetyllactosamine branches can be present at C4 of Man **4** and/or C6 of Man **4'**. In the ^{1}H-NMR spectrum the set of chemical shifts of H3eq and H3ax contains information concerning the type of linkage of the sialic acid. It can generally be stated that the sets of chemical shifts for H3eq and H3ax, given in Table 4, are discriminative for a Neu5Ac residue in α(2-6)- or α(2-3)-linkage to Gal.

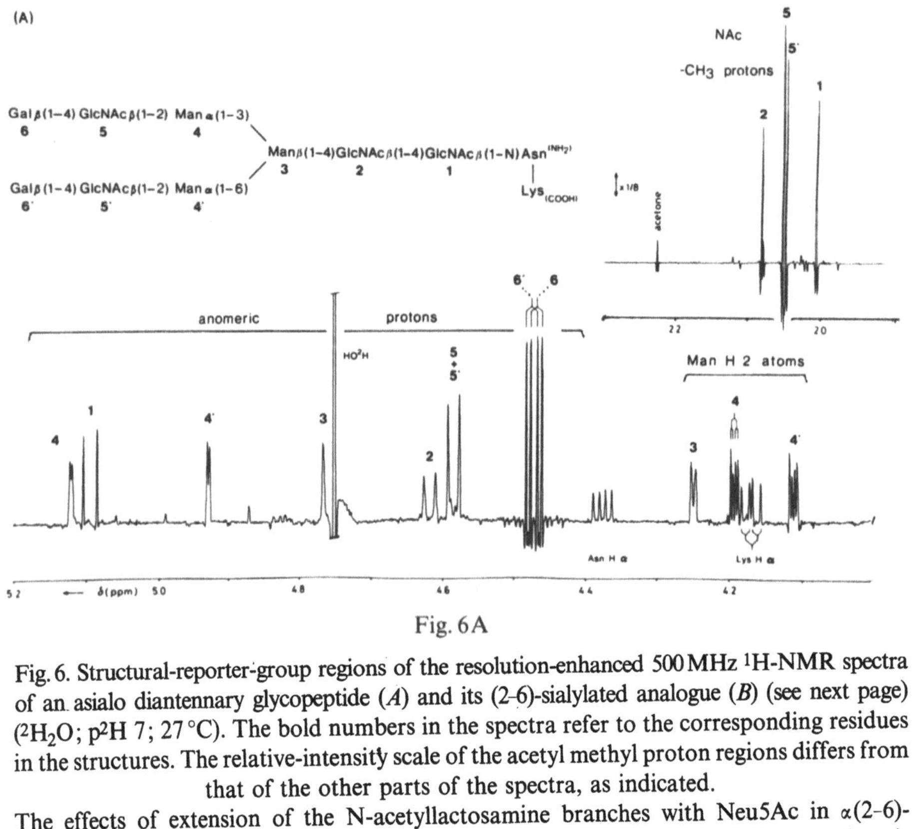

Fig. 6A

Fig. 6. Structural-reporter-group regions of the resolution-enhanced 500 MHz ^{1}H-NMR spectra of an asialo diantennary glycopeptide (*A*) and its (2-6)-sialylated analogue (*B*) (see next page) (^2H_2O; p^{2}H 7; 27 °C). The bold numbers in the spectra refer to the corresponding residues in the structures. The relative-intensity scale of the acetyl methyl proton regions differs from that of the other parts of the spectra, as indicated.
The effects of extension of the N-acetyllactosamine branches with Neu5Ac in α(2-6)-linkage to galactose can be traced from comparison of spectra (*A*) and (*B*). Comparison is facilitated by the fact that sample (*B*) contains a small amount of compound (*A*), as is clearly inferred from the signals marked by asterisks.

To indicate the effects of extension of the N-acetyllactosamine branches with sialic acid, relevant data of the asialo-diantenna are used as a reference for mono- and disialo-compounds. The chemical shift alterations of structural-reporter-group signals in the diantenna due to attachment of two Neu5Ac residues in α(2-6)-linkage to Gal, can be traced from Fig. 6. Downfield shifts are observed for the anomeric proton signals of Man **4** and **4'** and for the anomeric proton signals of GlcNAc **5** and **5'**. The anomeric proton signals of Gal **6** and **6'** shift upfield. For monosialo-diantennae the changes in chemical shift are restricted to the branch which bears the Neu5Ac residue. The sialic acid bearing branch can easily be

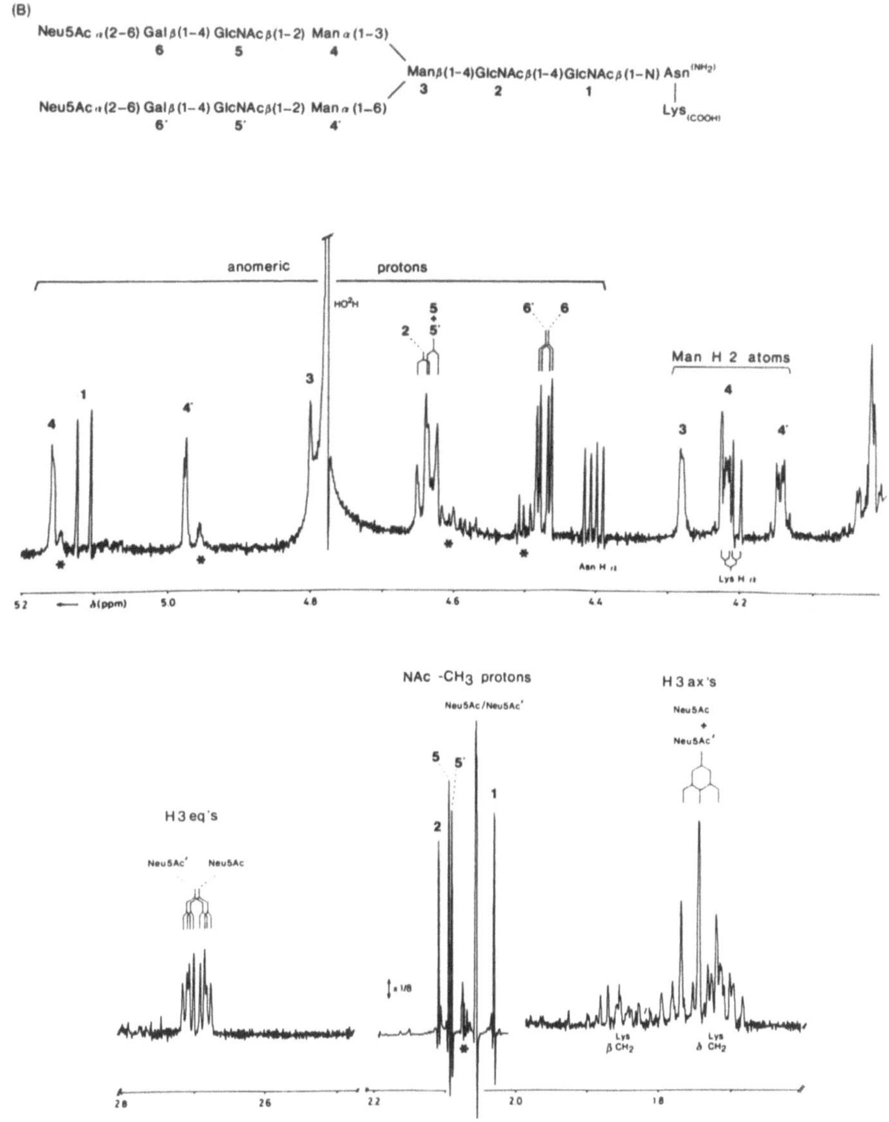

Fig. 6B

identified on the basis of the chemical shifts of H1 of the α-linked mannose residues (Man **4** and **4'**). With regard to monoantennary partial structures it should be noted that the chemical shifts of their structural reporter groups are virtually identical to those of the corresponding branch in the diantennary structure.

In contrast to the α(2-6)-linked sialic acid the α(2-3)-linked analogue influences only pronouncedly the chemical shifts of protons of the sugar residue to which it is attached. In case of a Neu5Ac α(2-3)Gal-linkage, the H1 of Gal shifts from δ 4.47 to δ 4.54 while the H3 of Gal shifts from δ 3.67 to δ 4.11. By consequence,

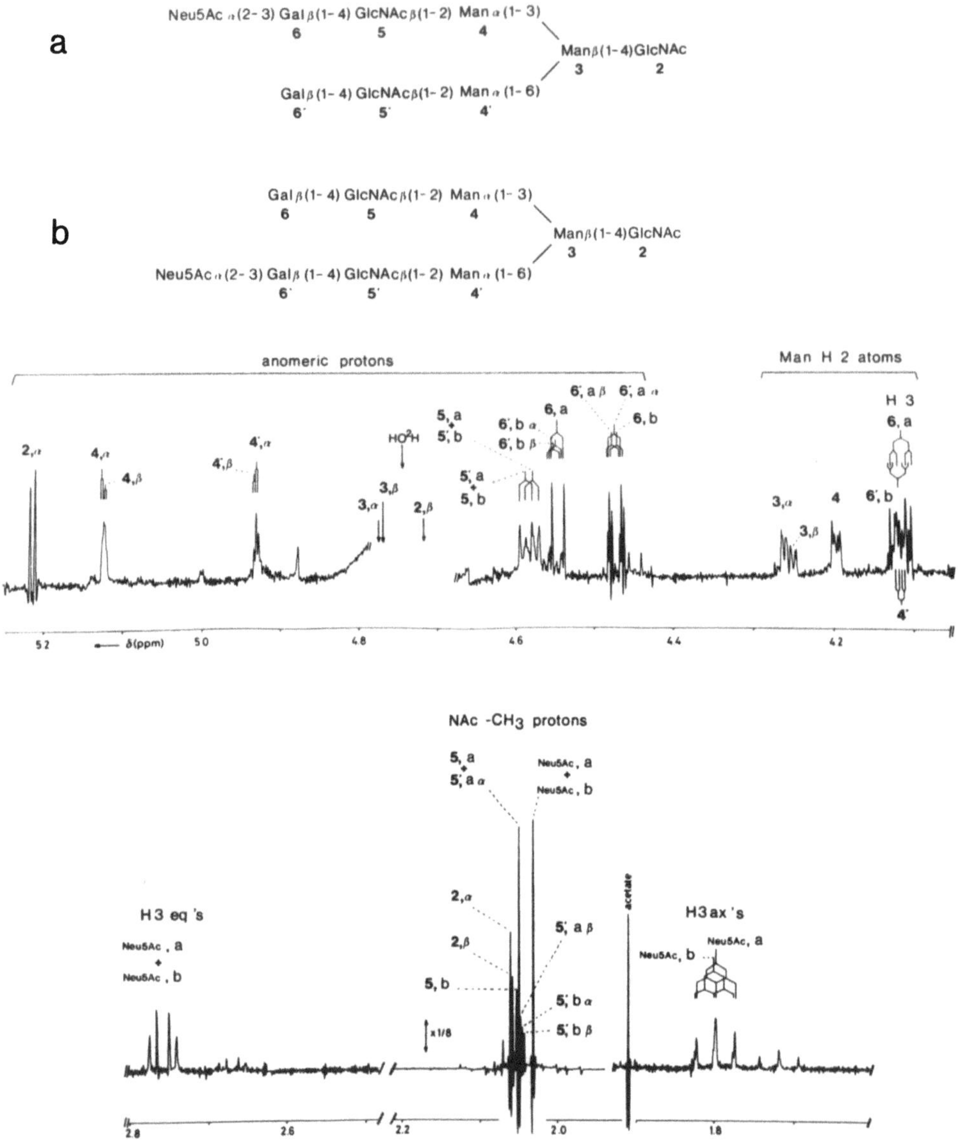

Fig. 7. Structural-reporter-group regions of the resolution-enhanced 500 MHz ^{1}H-NMR spectrum of a mixture containing monosialo diantennary oligosaccharides *a* and *b* in the ratio of 3:1 ($^{2}H_2O$; p^{2}H 7; 27 °C; anomeric ratio for both *a* and *b*, $\alpha:\beta = 2:1$). The bold numbers in the spectrum refer to the corresponding residues in the structures, the letters *a* and *b* to the compounds in the mixture. (E.g., anomeric proton signal designated as **6'**,*a*β means: H1 of Gal **6'** in the β-anomer of compound *a*.) Signals of corresponding protons in *a* and *b* coincide, unless indicated otherwise. The relative-intensity scale of the acetyl methyl proton region differs from that of the other parts of the spectrum. The HO^{2}H-resonance has been omitted; its position is indicated by an arrow.

in monosialo-diantennary compounds it is complicated to determine the branch to which the α(2-3)-linked Neu5Ac residue is attached. This problem can be solved in cases when the signals of the H1's of Gal **6** and **6'** can be assigned unambiguously (see Table 5). This depends on the quality of the sample and the resolution of the ^{1}H-NMR spectrum. Furthermore, as a result of Neu5Ac in α(2-3)-linkage to Gal, a slight effect occurs on the N-acetyl signal of the N-acetyllactosamine unit bearing the Neu5Ac residue. These small effects are useful

Table 5. *^{1}H-NMR chemical shift data, useful for the localization of α(2-6)- and α(2-3)-linked sialic acid in diantennary glycopeptides of the N-glycosidic N-acetyllactosamine type* (VLIEGENTHART *et al.* 1981, 1982)

For numbering of residues, see Figs. 6 and 7. n.d., value could not be determined

Reporter group	Chemical shift in		
	asialo branch	α(2-6)-sialylated branch	α(2-3)-sialylated branch
H1 of Gal **6**	4.467	4.442	4.544
H1 of Gal **6'**	4.473	4.447	4.548
H3 of Gal **6**	3.67	n.d.	4.113
H3 of Gal **6'**	3.67	n.d.	4.115
H1 of GlcNAc **5**/**5'**	4.582	4.603	4.578
NAc of GlcNAc **5**	2.050	2.069	2.048
NAc of GlcNAc **5'**	2.046	2.064	2.043
H1 of Man **4**	5.121	5.135	5.120
H1 of Man **4'**	4.928	4.946	4.926

in case when a mixture of monosialo-compounds has to be analysed. An example is given in Fig. 7, where the spectrum of a mixture of two isomeric monosialo-diantennary oligosaccharides is presented. The reducing character of the compounds implicates that protons in certain positions of the chain i.e. the H1's of GlcNAc **2**, Man **3**, **4**, and **4'**, Gal **6'** and the N-acetyl protons of GlcNAc **2** and **5'**, each show a double set of resonances. Despite of this complication the position of sialic acid can be derived from the H1 signals of the Gal residues and from the N-acetyl signals of the GlcNAc residues **5** and **5'**. The chemical shifts for the key resonances in various asialo-, mono- and disialo-structures are compiled in Table 5.

To illustrate the effect of sialic acid in higher-branched compounds of the N-acetyllactosamine type the asialo-triantenna is used as a reference. In the triantenna bearing three Neu5Ac residues, α(2-6)-linked to Gal, the effects of sialylation of Gal **6** and **6'** are similar to those observed in the corresponding diantenna. Compared with the spectrum of the diantenna the extra branch is

Table 6. *Pertinent* ^{1}*H-NMR chemical shift data for asialo and sialo triantennary structures of the N-glycosidic N-acetyllactosamine type* (VLIEGENTHART *et al.* 1981, 1982)

Reporter group	Chemical shift in				
	asialo glycopeptide **8**-**7** \\ **6**-**5**-**4** \\ **3**-**2**-**1**-Asn / | **6′**-**5′**-**4′** Lys	sialylated glycopeptide N*6-**8**-**7** \\ N6-**6**-**5**-**4** \\ **3**-**2**-**1**-Asn / | N′6-**6′**-**5′**-**4′** Lys	anomer of oligo-saccharide	asialo oligo-saccharide **8**-**7** \\ **6**-**5**-**4** \\ **3**-**2** / **6′**-**5′**-**4′**	sialylated oligo-saccharide N*3-**8**-**7** \\ N6-**6**-**5**-**4** \\ **3**-**2** / N′6-**6′**-**5′**-**4′**
H1 of Gal **6**	4.468	4.440	α, β	4.468	4.443
H1 of Gal **6′**	4.473	4.448	α	4.471	4.443
			β	4.473	4.447
H1 of Gal **8**	4.462	4.439	α, β	4.463	4.545
NAc of GlcNAc **5**	2.048	2.069	α, β	2.050	2.067
NAc of GlcNAc **5′**	2.045	2.065	α	2.048	2.067
			β	2.046	2.065
NAc of GlcNAc **7**	2.075	2.101	α, β	2.078	2.074
H3ax of NeuAc	—	1.717	α, β	—	1.720
H3ax of NeuAc′	—	1.717	α, β	—	1.717
H3ax of NeuAc*	—	1.706	α, β	—	1.801
H3eq of NeuAc	—	2.670	α, β	—	2.670
H3eq of NeuAc′	—	2.674	α, β	—	2.672
H3eq of NeuAc*	—	2.670	α, β	—	2.757

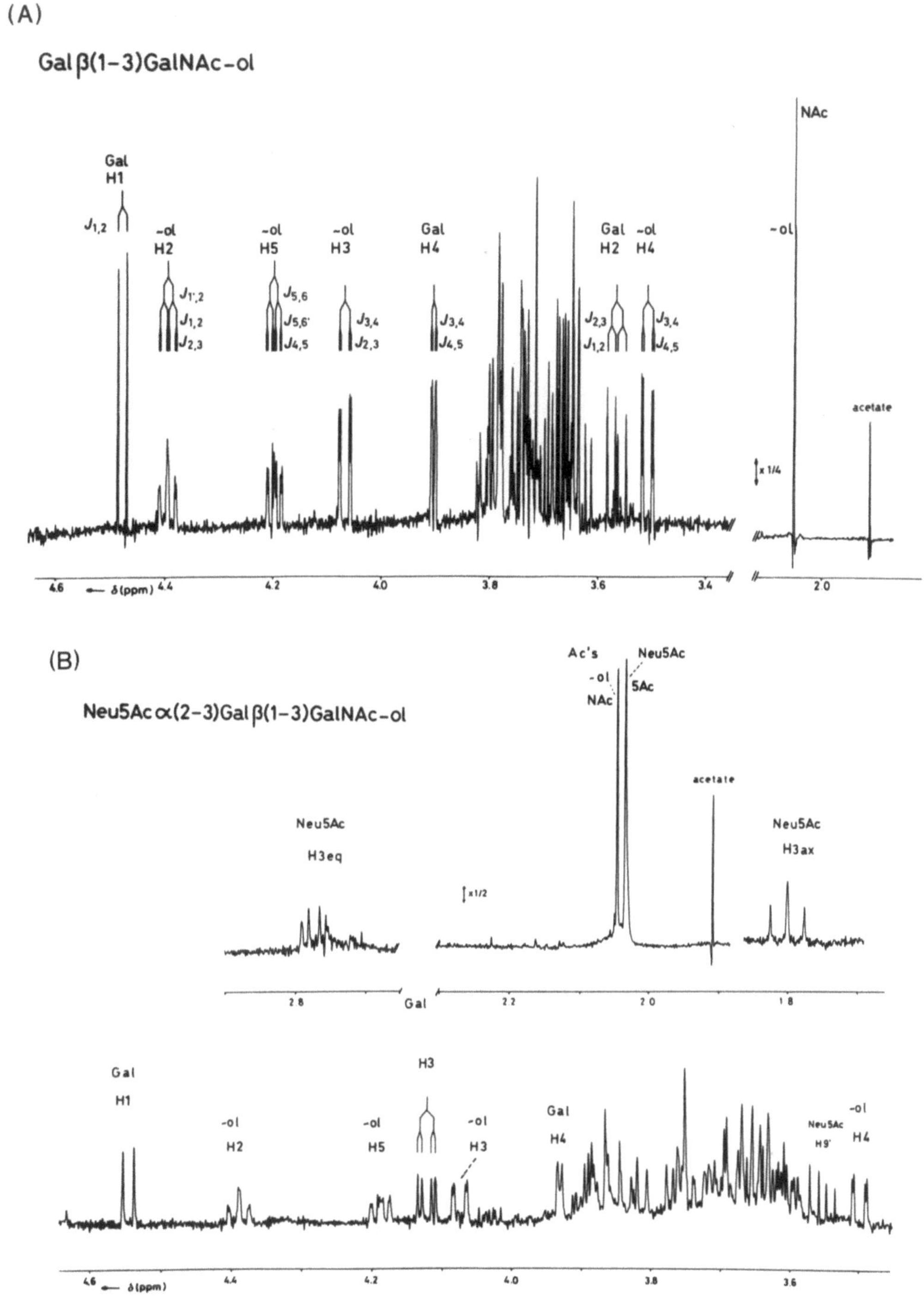

Fig. 8. Resolution-enhanced 500 MHz ^{1}H-NMR spectra of Galβ(1-3)GalNAc-ol (*A*) and Neu5Acα(2-3)Galβ(1-3)GalNAc-ol (*B*), dissolved in ^2H_2O, recorded at 27 °C and p^{2}H 7. Detailed splitting patterns are indicated for structural-reporter-group signals from residues other than Neu5Ac. The relative-intensity scale of the acetyl proton regions differs from that of the other parts of the spectrum, as indicated.

Table 7. [1]*H-NMR chemical shift data for Neu5Ac-containing oligosaccharide-alditols of the* VLIEGENTHART

Reporter group	Residue	Chemical shift in			
		Galβ(1-3)-GalNAc-ol	Neu5Acα(2-3)-Galβ(1-3)-GalNAc-ol*	Neu5Acα(2-6)-GalNAc-ol*	Galβ(1-3)-[Neu5Acα(2-6)]-GalNAc-ol*
H3ax	Neu5Acα(2-3)	—	1.800	—	—
H3eq		—	2.774	—	—
5Ac		—	2.034	—	—
H3ax	Neu5Acα(2-6)	—	—	1.700	1.693
H3eq		—	—	2.728	2.729
5Ac		—	—	2.033	2.034
H1	Gal	4.478	4.547	—	4.475
H3		3.671	4.122	—	3.67
H2	GalNAc-ol	4.395	4.390	4.245	4.378
H3		4.065	4.074	3.842	4.058
H4		3.507	3.498	3.413	3.538
H5		4.196	4.187	4.024	4.245
H6		3.69	3.68	3.837	3.85
H6′		3.628	3.65	3.528	3.492
NAc		2.050	2.046	2.056	2.048

* For the analogues of these compounds containing Neu5Gc instead of Neu5Ac, see Table 8.

characterized by the structural-reporter-group signals of the constituting monosaccharides, in particular the H3 signals of Neu5Ac*, the H1 signal of Gal **8** and the N-acetyl signals of GlcNAc **7** and Neu5Ac*, as indicated in Table 6. For the trisialo-triantennary structure possessing a Neu5Ac residue in α(2-3)-linkage to Gal in the **7**-**8** branch, a downfield shift of H1 of Gal **8** is observed, in comparison with the asialo-analogue (see Table 6). So far, no NMR data have been described for Neu5Ac in other types of glycosidic linkages to N-acetyllactosamine structures.

c) Sialic Acid in O-Glycosidically Linked Carbohydrate Chains

Several types of O-glycosidically linked carbohydrate chains occur in glycoproteins. These carbohydrate chains are usually analysed in the form of oligosaccharide-alditols, prepared by alkaline borohydride reductive cleavage (CARLSON 1966). The obtained oligosaccharide-alditols have frequently in common:

Galβ(1-3)GalNAc-ol.

O-glycosidic mucin type and some reference substances (VAN HALBEEK *et al.* 1980, 1981 a, *et al.* 1981)

Chemical shift in		
Neu5Acα(2-3)-Galβ(1-3)[Neu5Acα(2-6)]-GalNAc-ol	Galβ(1-3)-[Galβ(1-4)GlcNAcβ(1-6)]-GalNAc-ol	Neu5Acα(2-3)Galβ(1-3)-[Galβ(1-4)GlcNAcβ(1-6)]-GalNAc-ol
1.800	—	1.801
2.774	—	2.774
2.032	—	2.033
1.692	—	—
2.723	—	—
2.032	—	—
4.541	4.465	4.534
	4.470	4.470
4.117	3.900	3.922
	3.925	3.931
4.378	4.394	4.390
4.067	4.060	4.072
3.524	3.465	3.456
4.240	4.282	4.272
3.84	3.931	3.927
3.475	3.7	3.7
2.042	2.067	2.066

This structural element can be extended with sialic acid in different ways:

α) in α(2-6)-linkage to GalNAc-ol
β) in α(2-3)-linkage to Gal
γ) in α(2-8)-linkage to the sialic acid residue mentioned under α)
δ) in α(2-3)-linkage to GalNAc, which is β(1-3)-linked to Galβ(1-3)GalNAc-ol.

The NMR parameters of a few of these compounds will be discussed. The 500 MHz ^{1}H-NMR spectrum of Galβ(1-3)GalNAc-ol, as shown in Fig. 8 *A*, could almost completely be interpreted. The H2, H3, H4, and H5 of GalNAc-ol, which resonate outside of the bulk, are suitable structural reporter groups in case of further substitution of this core structure. Extension with Neu5Ac in an α(2-3)-linkage to Gal gives the trisaccharide-alditol, the ^{1}H-NMR spectrum of which is shown in Fig. 8 *B*. Comparison of this spectrum with that of the disaccharide-alditol reveals downfield shifts for H1 and H3 of Gal (VAN HALBEEK *et al.* 1980) and a small but significant upfield shift for the N-acetyl protons of GalNAc-ol. Similar effects have been observed for Neu5Acα(2-3)Galβ(1-3)[Galβ(1-4)GlcNAcβ(1-6)]GalNAc-ol when compared to its asialo-analogue (VAN HALBEEK *et al.* 1981 a). The Neu5Ac residue itself is characterized by the set of chemical shifts of H3ax, H3eq, and 5Ac (see Table 7).

Further extension of the aforementioned trisaccharide with a Neu5Ac residue α(2-6)-linked to GalNAc-ol affords the following tetrasaccharide-alditol:

Neu5Acα(2-3)Galβ(1-3)[Neu5Acα(2-6)]GalNAc-ol.

The additional Neu5Ac residue affects the chemical shifts of H4, H5, and H6′ of GalNAc-ol (see Table 7). The signal of H6′ of GalNAc-ol is now situated apart from the bulk as a result of the substitution at C6 of GalNAc-ol. Furthermore, attachment of Neu5Acα(2-6) to GalNAc-ol introduces a specific shift of the N-acetyl protons of GalNAc-ol, when compared with the trisaccharide-alditol Neu5Acα(2-3)Galβ(1-3)GalNAc-ol. The Neu5Ac residue in α(2-6)-linkage to GalNAc-ol is characterized by the additional set of chemical shifts of its H3 protons (see Table 7).

Fig. 9 shows the ^{1}H-NMR spectrum of a pentasaccharide-alditol containing the bloodgroup A determinant. The NMR parameters of this compound and of its partial structures are presented in Table 8 (VAN HALBEEK *et al.* 1981 b). These compounds were derived from glycoproteins with A^+, H^+ or A^-H^- immunological properties. No structural differences could be found between the oligosaccharide-alditols of the corresponding A^+, H^+, and A^-H^- series, except for the pentasaccharide-alditol which was only present in the A^+ series. The set of chemical shifts for H3eq (δ 2.745) and H3ax (δ 1.721) of Neu5Gc, is indicative of the α(2-6)-linkage of Neu5Gc to GalNAc-ol, bearing no other substituents. This type of attachment of sialic acid to GalNAc-ol is further reflected by oppositely directed shifts for H6 and H6′ of GalNAc-ol and the change in geminal coupling constant $J_{6,6'}$ from —11.7 Hz to —9.8 Hz. Attachment of Gal in β(1-3)-linkage to GalNAc-ol introduces a characteristic upfield shift for H3ax (Δδ—0.008). A similar feature can be observed for the Neu5Ac analogues, as is evident from Table 7 (*cf.* Neu5Acα(2-6)GalNAc-ol, Galβ(1-3)[Neu5Acα(2-6)]GalNAc-ol). The presence of the Fuc and GalNAc residues does not affect the structural-reporter-group signals of Neu5Gc and GalNAc-ol.

A type of sialo-oligosaccharide-alditols, obtained from trout eggs and having in common the following pentasaccharide structure:

GalNAcβ(1-4)Galβ(1-4)GalNAcβ(1-3)Galβ(1-3)GalNAc-ol

were investigated by INOUE *et al.* (1981) and NOMOTO *et al.* (1981) using 270 MHz ^{1}H-NMR spectroscopy. Neu5Gc can be attached to GalNAc-ol in α(2-6)- and/or in α(2-3)-linkage to the internal GalNAc. Furthermore, the former Neu5Gc can be elongated with one or more Neu5Gc residues each in α(2-8)-linkage. The characteristic ^{1}H-NMR data of this class of compounds are compiled in Table 9.

d) CMP-N-acetylneuraminic Acid and Degradation Products

A key intermediate in the biosynthesis of glycoconjugates is the sugar nucleotide cytidine-5′-monophospho-N-acetylneuraminic acid (CMP-Neu5Ac). To check its identity ^{1}H-NMR spectroscopy can effectively be used (HAVERKAMP *et al.* 1979a). The 360 MHz ^{1}H-NMR spectrum of this compound is given in Fig. 10 and the NMR parameters are compiled in Table 10. The chemical shifts for H3eq, H4 and the N-acetyl protons of the Neu5Ac residue point to β-configuration of the glycosidic linkage (see Tables 3 and 10). Another proof for the configuration of the glycosidic bond is found in the heteronuclear long-range coupling constants $^4J_{H3ax,P}$ (6.1 Hz) and $^4J_{H3eq,P}$ (0 Hz) indicating a planar "W"-shape of the bonds in

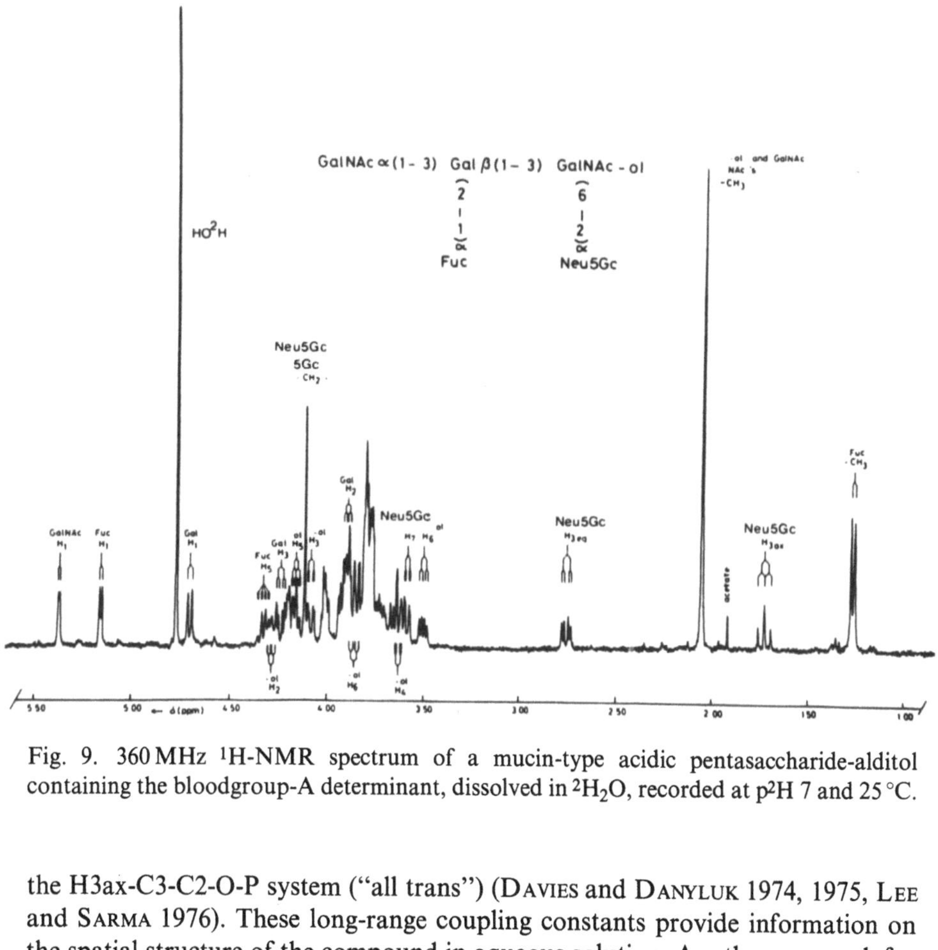

Fig. 9. 360 MHz ^{1}H-NMR spectrum of a mucin-type acidic pentasaccharide-alditol containing the bloodgroup-A determinant, dissolved in ^2H_2O, recorded at p^{2}H 7 and 25 °C.

the H3ax-C3-C2-O-P system ("all trans") (DAVIES and DANYLUK 1974, 1975, LEE and SARMA 1976). These long-range coupling constants provide information on the spatial structure of the compound in aqueous solution. Another approach for the determination of the β-configuration of the glycosidic linkage, based on the heteronuclear vicinal coupling constants $^3J_{C1,H3ax}$ and $^3J_{C1,H3eq}$, will be discussed in section G.III.5.

In studies on the chemical stability of CMP-Neu5Ac in dependence on the pH two interesting decomposition products were observed, *viz.* β-N-acetylneuraminic acid-2-phosphate (Neu5Ac2P) and 2-deoxy-2,3-dehydro-N-acetylneuraminic acid (Neu5Ac2en) (BEAU *et al.*, unpublished results). Their ^{1}H-NMR spectra are given in Fig. 11 and 12, respectively. The presence of a phosphate group in Neu5Ac2P

Table 8. *Relevant* ¹*H-NMR chemical shift data for Neu5Gc-containing oligosaccharide-*

Reporter group	Residue	Chemical shift in			
		GalNAc-ol	Neu5Gcα(2-6)-GalNAc-ol	Galβ(1-3)-GalNAc-ol	Neu5Gcα(2-3)-Galβ(1-3)-GalNAc-ol
H3ax	Neu5Gcα(2-3)	—	—	—	1.817
H3eq		—	—	—	2.787
5Gc		—	—	—	4.122
H3ax	Neu5Gcα(2-6)	—	1.721	—	—
H3eq		—	2.746	—	—
5Gc		—	4.124	—	—
H1	Gal	—	—	4.478	4.547
H3		—	—	3.671	4.132
H2	GalNAc-ol	4.255	4.253	4.395	4.389
H3		3.847	3.842	4.065	4.073
H4		3.385	3.416	3.507	3.495
H5		3.928	4.025	4.196	4.188
H6		3.67	3.842	3.69	3.68
H6′		3.64	3.536	3.628	3.65
NAc		2.055	2.057	2.050	2.045

comes to expression in an additional splitting of the H3ax signal as a result of long-range coupling between P and H3ax. The criteria used for the determination of the β-glycosidic linkage as described for CMP-Neu5Ac (i.e. δ H3eq, H4 and N-acetyl; $^4J_{H3,P}$) also hold for this compound (see Table 10).

The ¹H-NMR data of Neu5Ac2en are compiled in Table 10. The chemical shifts depend strongly on the pH. The double bond formed in the decomposition reaction of CMP-Neu5Ac, is characterized by the doublet near δ 5.7 which is ascribed to H3. The small value observed for $^3J_{3,4}$ is an indication for an almost coplanar orientation of the ring-O-C2-C3-C4 part of the six-membered ring. The other proton-proton coupling constants are similar to those observed in Neu5Ac (Table 1), in CMP-Neu5Ac and in Neu5Ac2P (Table 10). This leads to the conclusion that the conformation of the glycerol side chain, including the hydrogen bridges, is virtually unaltered with respect to the aforementioned compounds. In view of its property to act as an inhibitor of sialidase (MEINDL *et al.* 1974, KAMERLING *et al.* 1975, HAVERKAMP *et al.* 1976; KAMERLING *et al.* 1979) it is relevant to have methods available for the accurate determination of the concentration of Neu5Ac2en in biological samples. High-resolution ¹H-NMR *alditols of the O-glycosidic mucin type and for some reference compounds* (VAN HALBEEK *et al.* 1981 b)

Chemical shift in				
Galβ(1-3)-[Neu5Gcα(2-6)]-GalNAc-ol	Fucα(1-2)-Galβ(1-3)-GalNAc-ol	Fucα(1-2)-Galβ(1-3)-[Neu5Gcα(2-6)]-GalNAc-ol	GalNAcα(1-3)-[Fucα(1-2)]-Galβ(1-3)-GalNAc-ol	GalNAcα(1-3)-[Fucα(1-2)]-Galβ(1-3)-[Neu5Gcα(2-6)]-GalNAc-ol
—	—	—	—	—
—	—	—	—	—
—	—	—	—	—
1.711	—	1.714	—	1.711
2.746	—	2.747	—	2.750
4.123	—	4.122	—	4.123
4.477	4.584	4.586	4.720	4.719
3.669	3.880	3.870	4.254	4.248
4.380	4.398	4.386	4.302	4.291
4.061	4.089	4.087	4.100	4.089
3.541	3.520	3.545	3.606	3.624
4.249	4.162	4.226	4.125	4.185
3.860	3.68	3.84	3.68	3.862
3.497	3.63	3.484	3.64	3.493
2.049	2.046	2.043	2.048	2.046

spectroscopy allows the detection of Neu5Ac2en down to concentrations of about 2% in mixtures of sialic acids. As marker signals for this determination the N-acetyl singlet at δ 2.068 and the doublet of H3 at δ 5.690 are used.

4. *Enzymic and Chemical Conversions*

^{1}H-NMR spectroscopy is suitable for following the course of reactions either directly in the NMR-tube or by analysis of isolated reaction products. Simultaneous measurements of substrate and product concentrations allows under proper conditions the deduction of kinetic reaction parameters (FRIEBOLIN *et al.* 1981 c). Alternatively, the formed products can be isolated and analysed separately by NMR spectroscopy. For example in the saponification of sialic acid methyl esters under mild alkaline conditions both types of approaches are useful (HAVERKAMP *et al.* 1982). The potency of NMR spectroscopy as a real-time analytical probe is in particular evident in the study of enzyme reactions.

^{1}H-NMR evidence for the release of αNeu5Ac as the primary product of sialidase action on various natural and synthetic sialoglycoconjugates was obtained by FRIEBOLIN *et al.* (1980 b, 1981 d). This initial formation of the α-

Table 9. *Some pertinent ^{1}H-NMR chemical shift data for polysialosyl oligosaccharide-alditols of the O-glycosidic mucin type, and for some reference substances* (NOMOTO *et al.* 1981)

Compound	Chemical shift		
	H3ax	H3eq	5Gc
Galβ(1-3)[Neu5Gcα(2-6)]GalNAc-ol*	1.769	2.737	4.125
Galβ(1-3)[Neu5Gcα(2-6)]GalNAc-ol	1.660	2.659	n.d.
8 | 2 α Neu5Gc	1.743	2.779	n.d.
Galβ(1-3)[Neu5Gcα(2-6)]GalNAc-ol	1.668	2.659	n.d.
8 | 2 α Neu5Gc	1.711	2.709	n.d.
8 | 2 α Neu5Gc	1.756	2.779	n.d.
GalNAcβ(1-4)Galβ(1-4)GalNAcβ(1-3)Galβ(1-3)GalNAc-ol 3 | 2 α Neu5Gc	1.855	2.554	4.120
GalNAcβ(1-4)Galβ(1-4)GalNAcβ(1-3)Galβ(1-3)[Neu5Gcα(2-6)]GalNAc-ol	1.669	2.659	n.d.
8 | 2 α Neu5Gc (on the Neu5Gcα(2-6))	1.743	2.780	n.d.
3 (on GalNAcβ(1-3)) | 2 α Neu5Gc	1.865	2.559	n.d.

Table 9 (Continued)

Compound	Chemical shift		
	H3ax	H3eq	5Gc
GalNAcβ(1-4)Galβ(1-4)GalNAcβ(1-3)Galβ(1-3)[Neu5Gcα(2-6)]GalNAc-ol	1.672	2.660	n.d.
3 8–2 α Neu5Gc	1.714	2.709	n.d.
8–2 α Neu5Gc	1.757	2.780	n.d.
2 α Neu5Gc	1.856	2.560	n.d.

* For unclarified reasons, data for this compound are not in accordance with those in Table 8 (Van Halbeek *et al.* 1981b).
n.d. = value not given in quoted reference (Nomoto *et al.* 1981).

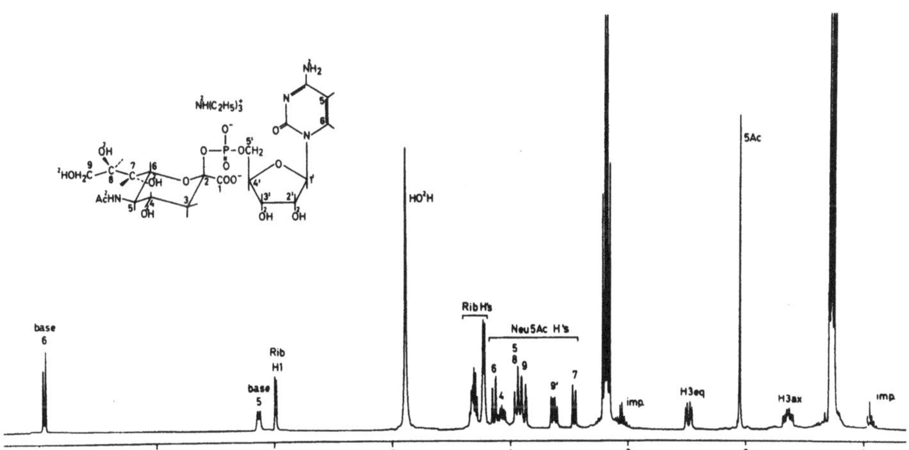

Fig. 10. 360 MHz ¹H-NMR spectrum of CMP-Neu5Ac (as triethylammonium-salt) dissolved in $^{2}H_2O$, recorded at p²H 8 and 25 °C. Multiplets at δ 1.3 and 3.2 represent ethyl groups of triethyl ammonium counter ions.

Table 10. *Relevant ^{1}H-NMR data for CMP-Neu5Ac* (HAVERKAMP *et al.* 1979 a), *Neu5Ac2P and Neu5Ac2en*
Chemical shift values are given in ppm relative to DSS for solutions in ^2H_2O at 25 °C and at the indicated p^{2}H values. Coupling constants (J) are given in Hz

Compound	p^{2}H	Chemical shift									
		H3ax	H3eq	H4	H5	H6	H7	H8	H9	H9′	5Ac
CMP-Neu5Ac	8	1.639	2.484	4.066	3.92	4.141	3.456	3.92	3.90	3.622	2.054
Neu5Ac2P	7.5	1.548	2.403	4.093	3.888	4.239	3.386	4.028	3.883	3.581	2.045
Neu5Ac2en*	6	—	5.690	4.470	4.051	4.213	3.601	3.936	3.885	3.646	2.068

Compound	p^{2}H	Coupling constant											
		$^2J_{3ax,3eq}$	$^4J_{3ax,P}$	$^4J_{3eq,P}$	$^3J_{3ax,4}$	$^3J_{3eq,4}$	$^3J_{4,5}$	$^3J_{5,6}$	$^3J_{6,7}$	$^3J_{7,8}$	$^3J_{8,9}$	$^3J_{8,9'}$	$^2J_{9,9'}$
CMP-Neu5Ac	8	—13.4	6.1	< 1.0	11.6	5.0	10.4	10.5	1.2	9.7	2.8	6.2	—11.7
Neu5Ac2P	7.5	—12.8	4.2	< 1.0	11.2	4.8	9.1	10.5	< 1.0	9.6	2.6	6.9	—11.7
Neu5Ac2en*	6	—	—	—	—	2.3	8.8	10.9	1.0	9.3	2.8	6.3	—11.9

* For comparison with the data for Neu5Ac2en1Me, see KUMAR *et al.* 1981.

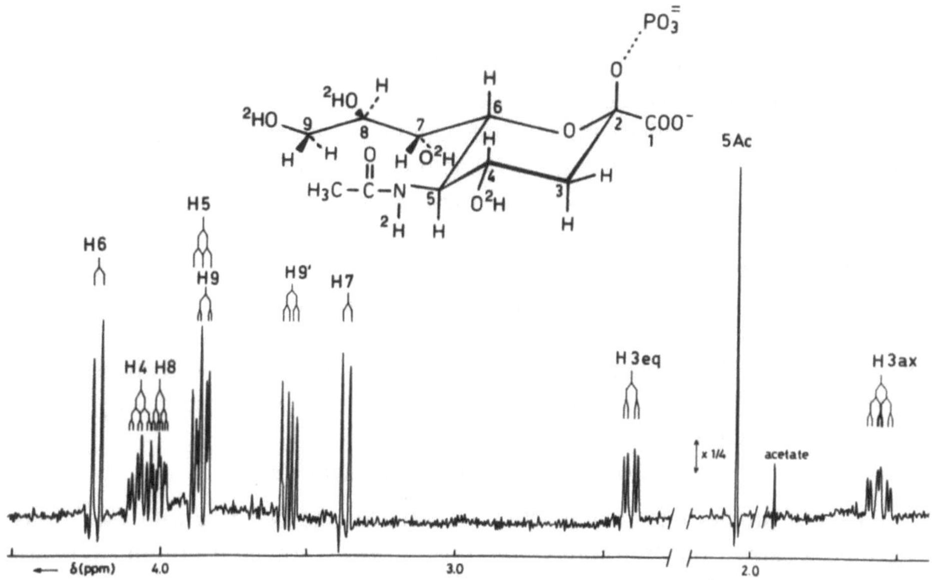

Fig. 11. Resolution-enhanced 360 MHz ^{1}H-NMR spectrum of β-N-acetylneuraminic acid-2-phosphate (βNeu5Ac2P) dissolved in ^2H_2O, recorded at p^{2}H 7.5 and 25 °C. The relative-intensity scale of the 5Ac methyl proton signal differs from that of the other parts of the spectrum, as indicated.

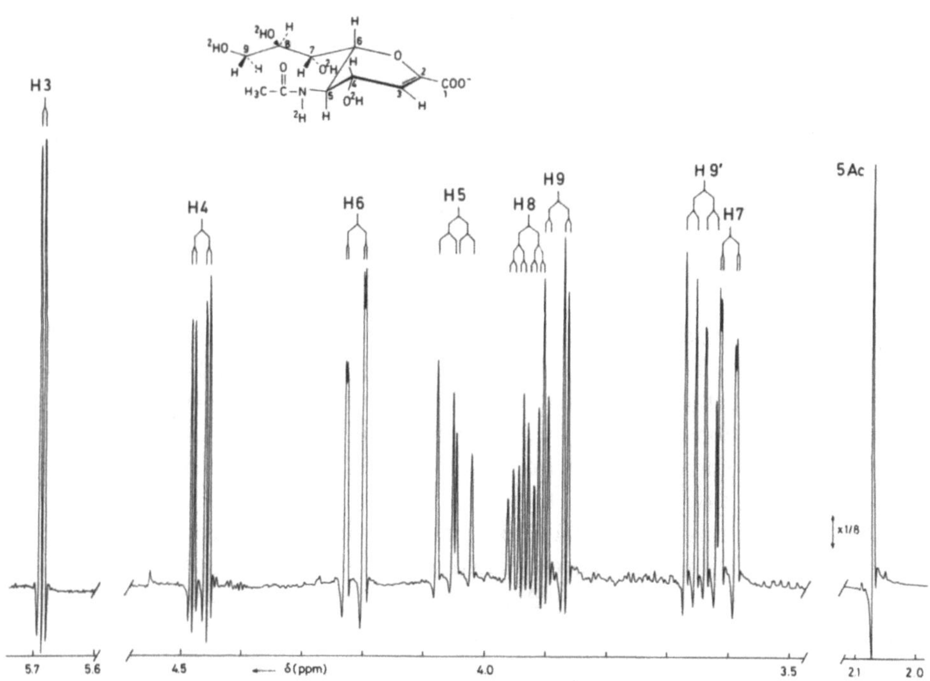

Fig. 12. Resolution-enhanced 360 MHz ^{1}H-NMR spectrum of 2-deoxy-2,3-dehydro-N-acetylneuraminic acid (Neu5Ac2en) dissolved in ^2H_2O, recorded at p^{2}H 6 and 25 °C. The relative-intensity scale of the 5Ac methyl proton signal differs from that of the other parts of the spectrum, as indicated.

anomer offers the possibility to study the kinetics of the mutarotation by means of ^{1}H-NMR spectroscopy. FRIEBOLIN *et al.* (1980 a, 1981 a) also studied the anomerisation in dependence on the pH.

The positional specificity of sialidases can be monitored by the action of the enzymes on mixtures of isomeric sialyllactoses (FRIEBOLIN *et al.* 1981 c). In the same study also kinetic parameters of these enzyme reactions were determined. For sialidase isolated from Newcastle disease virus PAULSON *et al.* (1982) demonstrated by 500 MHz ^{1}H-NMR spectroscopy that this enzyme hydrolyzes preferentially sialic acid residues α(2-3)-linked to α_1-acid glycoprotein.

The specificity of a colostrum sialyltransferase was investigated by VAN DEN EIJNDEN *et al.* (1980) using glycopeptides derived from asialo-α_1-acid glycoprotein as substrates. The positional specificity as well as the branch specificity of the enzyme were studied.

5. *Complexes of Sialic Acids with Ca^{2+}*

^{1}H-NMR studies on the interaction of Neu5Ac and Neu5Gc with Ca^{2+} ions have been carried out by JAQUES *et al.* (1977, 1980 a). Induced proton shifts are observed in the NMR spectra of sialic acids upon increasing the concentration of $CaCl_2$ (the Ca^{2+}/sialic acid ratios varied from 0 to 6.0). The direction of all of the Ca^{2+}-induced chemical shifts is the same for Neu5Ac and Neu5Gc, indicating that the geometries in the two complexes are similar. The signs of the shifts roughly determine the position of the calcium ion relative to the ring and side chain protons. The downfield shifts of H3eq, H4, and H6 and the upfield shift of H3ax indicate that the ion is above the ring. The relatively large downfield Ca^{2+}-induced shifts of H7 and H8 suggest that these protons are closest to the calcium ion but with their C-H bond directions pointing away from it. The major difference in the Ca^{2+}-Neu5Gc complex, when compared to the Ca^{2+}-Neu5Ac complex, is that H7 is somewhat less deshielded. The extra binding of Ca^{2+} to the hydroxyl group of the glycolyl chain as was postulated on the basis of molecular (CPK) model building studies could not clearly be inferred from ^{1}H- and ^{13}C-NMR chemical shift and coupling data. The coupling constants $^3J_{7,8}$, $^3J_{8,9}$, and $^3J_{8,9'}$ have undergone significant changes upon complexation with Ca^{2+}. This indicates a change in the conformation of the glycerol side chain by which the interaction with the ion is enhanced. The coupling constants of the pyranose ring protons remain unaltered, which demonstrates that the $^2C_5$ conformation of the ring does not change upon addition of Ca^{2+}. The approximate position of the Ca^{2+} ion in the Ca^{2+}-sialic acid complex was determined from the Ca^{2+}-induced shifts by using Buckingham's electric-field-shift theory (BUCKINGHAM 1960). For the Ca^{2+}-Neu5Ac complex this theory leads to a model, which is presented in Fig. 18 (section G.III.6). The model is corroborated by ^{13}C-NMR spectroscopy (JAQUES *et al.* 1977, 1980 a). As pointed out by BEHR and LEHN (1973), the glycerol side chain is in a position to effectively "wrap around the Ca^{2+}".

III. ^{13}C-NMR Spectroscopy

1. *N-Acetyl- and N-Glycolylneuraminic Acid*

^{13}C-NMR spectroscopy is another frequently used technique for structure analysis of biomolecules, however, requiring at least 10-100 times more substance.

Carbon spectra provide direct information about the molecular skeleton. Furthermore, ^{13}C resonances are stronger influenced than ^{1}H resonances by neighbouring dipoles and/or electric charges. Hence the ^{13}C signals are observed in a broad range of frequencies and are sensitive to changes in the substitution pattern, molecular conformation and interaction with other compounds. With regard to structural studies of sialic acids, an advantage of ^{13}C-NMR over ^{1}H-NMR is that C1 and C2 which do not bear protons can directly be observed. The

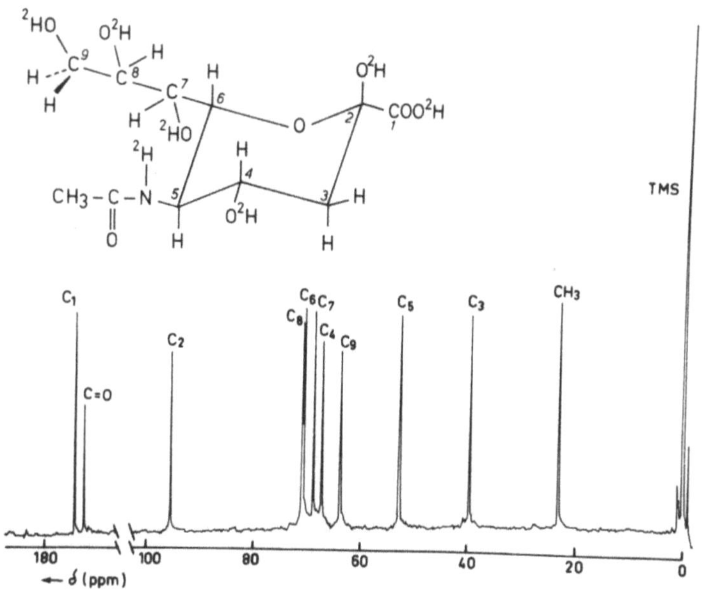

Fig. 13. Proton-noise-decoupled 25 MHz ^{13}C-NMR spectrum of Neu5Ac dissolved in $^{2}H_2O$, recorded at p^{2}H ~ 2 and 25 °C. Only the signals stemming from the β-anomer are clearly observable.

^{13}C-NMR spectrum of Neu5Ac is shown in Fig. 13. This proton noise-decoupled spectrum shows 11 intense singlets. Assignment of the resonances was made by comparison with spectral data of carbon atoms in similar chemical environments of model compounds and by various proton decoupling techniques (Bhattacharjee *et al.* 1975). The relatively low intensities for the carbonyl carbons and for C2 are due to the restricted nuclear Overhauser enhancement as a result of the absence of proton substituents at these carbon atoms. In addition, the spectrum shows several small resonances (see Table 11) belonging to the αNeu5Ac anomer which is present for ~ 9% in the equilibrium mixture (Jaques *et al.* 1977).

It has to be noted that the spectrum of Fig. 13 has been recorded with TMS as an external standard. Addition of an internal standard, e.g. the sodium salt of 2,2,3,3-tetradeutero-3-(trimethyl)propionic acid as applied by Eschenfelder *et al.* (1975) in a study of a series of monomeric αNeu5Ac and βNeu5Ac derivatives (glycosides and esters), can seriously affect several of the carbon resonances if the

Table 11. ^{13}*C-NMR chemical shift data*
Chemical shifts are given in ppm downfield from external TMS for

Compound	p²H	T	Chemical shift					
			C1	C2	C3	C4	C5	C6
βNeu5Ac	1.6	25 °C	176.06	96.46	40.00	67.88	53.26	71.34
βNeu5Ac	7	25 °C	177.87	97.61	40.63	68.51	53.50	71.45
αNeu5Ac	7	25 °C	n.d.	98.42	41.94	69.38	53.07	72.75
βNeu5Ac	7	37 °C	176.0	97.6	40.6	68.5	53.5	71.5*
βNeu5Ac	n.d.	37 °C	176.0	96.4	39.9	67.8	53.2	71.3*
βNeu5Gc	7	25 °C	178.08	97.65	40.61	68.18	53.18	71.20
βNeu5Gc	n.d.	37 °C	176.8	96.4	40.0	67.6	52.9	71.5
Neu5Ac2en	n.d.	28 °C	169.7	148.1	107.8	67.7	50.1	75.5

* Assignments may be interchanged.

addition of internal standard leads to ionization of the carboxyl group. In comparison to the spectrum of the protonated Neu5Ac (pH ~ 2) the signal of C1 shifts downfield (Δδ 1.8; see Table 11). Downfield shifts for carboxyl group carbons are generally observed when going from the free acid to the corresponding carboxylate anion.

Furthermore, significant chemical shift differences between α- and β-anomers are observed for C4 and C6 as a result of 1,3-diaxial interactions between the protons in these positions and the anomeric centre (BHATTACHARJEE *et al.* 1975). The resonance positions of C2, C3, C5, C7, and C9 show only minor differences between α- and β-anomers, whereas C1 and C8 occupy intermediate positions in this respect. The ^{13}C resonances for βNeu5Ac and βNeu5Gc in aqueous solutions are compiled in Table 11. Hydroxylation of the CH_3 group of the N-acetyl substituent results in two alterations in the ^{13}C-NMR spectrum: a strong downfield shift for the resonance of the involved carbon and a small downfield shift for the neighbouring carbonyl carbon.

2. *O-Acetylated Neuraminic Acid Derivatives*

The effect of O-acetylation at various positions of the sialic acid skeleton has been studied for aqueous solutions (BHATTACHARJEE *et al.* 1975). The main effects induced by O-acetylation at C9 of Neu5Ac are the downfield shift of 3.1 ppm for C9 and the upfield shift of -3.1 ppm for C8. In methanol as solvent the directions and magnitudes of these shifts are similar (HAVERKAMP *et al.* 1975); for βNeu4,5,9$Ac_3$1,2Me_2 O-acetylation at C4 causes a downfield shift of 2.9 ppm for C4 and upfield shifts of -3.3 and -3.4 ppm for C3 and C5, respectively. These

for Neu5Ac, Neu5Gc, and Neu5Ac2en
solutions in ^2H_2O at various temperature and p²H values

Chemical shift						References
C7	C8	C9	5Ac (Me)	5Gc (CH_2)	5Ac/Gc (C=O)	
69.43	71.62	64.37	23.27	—	174.49	JAQUES *et al.* 1977
69.82	71.59	64.55	23.34	—	175.98	JAQUES *et al.* 1977, 1980 a
69.38	73.77	64.10	n.d.	—	n.d.	JAQUES *et al.* 1977
69.6	71.7*	64.5	23.3	—	176.0	JENNINGS *et al.* 1977 b
69.4	71.5*	64.3	23.2	—	174.3	BHATTACHARJEE *et al.* 1975
69.63	71.62	64.47	—	62.25	176.85	JAQUES *et al.* 1980 a
69.3	71.5	64.3	—	62.2	174.3	BHATTACHARJEE *et al.* 1975
68.4	70.0	63.3	22.4	—	174.9	CZARNIECKI and THORNTON 1977 a

observations from methanolic solutions may be helpful to predict the magnitude and direction of the acetylation shifts for directly bound and neighbouring carbon atoms in aqueous solutions.

3. *Polysaccharides*

¹³C-NMR spectra are especially useful for studying the structure of sialic acid polymers because ¹H-NMR spectra of such polymers have rather bad resolution (*cf.* EGAN *et al.* 1977). The capsular polysaccharides of *Neisseria meningitidis* serogroups B, C, W-135, Y, BO and the *Escherichia coli* capsular polysaccharides K1 and K92 (strains Bos 12 (O16K92NM), N67 and MT411) were investigated by BHATTACHARJEE *et al.* (1975), EGAN *et al.* (1977), JENNINGS *et al.* (1977 a, b) and LIU *et al.* (1977). The aim of these studies was to establish relations between structure and effectiveness of the polysaccharide antigens as vaccines. The meningococcal serogroups B and C and the *E. coli* K92 polysaccharides are polymers of Neu5Ac with α(2-8)-, α(2-9)- and alternating α(2-8)-, α(2-9)-linkages, respectively. Those of groups W-135 and Y are composed of α(2-6)-linked disaccharide repeating units Galα(1-4)Neu5Ac and Glcα(1-4)Neu5Ac, respectively. The serogroup C, W-135 and Y polysaccharides can partially be O-acetylated to a degree depending on the strain and growth conditions.

Signal assignment in the proton noise-decoupled spectra of the various polysaccharides was made on the basis of comparison with each other and with appropriate sialic acids and hexose derivatives as model compounds. ¹³C-NMR spectral data of the various polymers and relevant model compounds are compiled in Table 12. The meningococcal serogroup B polysaccharide shows a simple spectrum (see Fig. 14 *a*) consisting of 11 singlets (BHATTACHARJEE *et al.* 1975). This

Table 12. ^{13}*C-NMR chemical shift data for*
Chemical shifts are given in ppm downfield from external

Compound	p²H	T	Linkage type	Chemical shift C1	C2	C3	C4
Neisseria meningitidis Serogroup B polysaccharide	7	37 °C	α(2-8)	174.4	102.1	40.9	69.4*
Escherichia coli K1 capsular polysaccharide	7	37 °C	α(2-8)	174.4	102.2	41.0	69.4*
Neisseria meningitidis Serogroup C polysaccharide [a]	<7	37 °C	α(2-9)	172.5	100.1	40.6	68.7
Neisseria meningitidis Serogroup C polysaccharide [a]	7	37 °C	α(2-9)	174.9	101.4	41.2	69.5
Escherichia coli K92 capsular polysaccharide	7	35 °C	α(2-9)	174.5	101.9	40.9	69.8*
	7	35 °C	α(2-8)	174.5	102.0	41.1	69.8
αNeu5Ac2Me	<7	37 °C	—	172.3	100.3	40.0	68.5*
αNeu5Ac2Me	7	37 °C	—	174.6	101.9	41.3	69.5
βNeu5Ac2Me	<7	37 °C	—	174.1	100.6	40.6	67.8
βNeu5Ac2Me	7	37 °C	—	175.9	101.7	41.1	68.4
Neu5Acα(2-9)Neu5Ac	7	37 °C	nr.	174.9	101.6	41.0	69.1*
			r.	175.8*	97.5	40.3	68.3

* Values may be interchanged.
[a] O-deacetylated.
nr. = non reducing unit; r. = reducing unit.

indicates a linear chain with one type of glycosidic linkage. The α(2-8)-linkages introduce characteristic downfield shifts for C2 (1.8 ppm) and C8 (6.8 ppm) with respect to the corresponding resonances in αNeu5Ac2Me. The upfield shift of the C9 resonance (—1.9 ppm) indicates its position vicinal to the glycosidic linkage. Shift effects for the other carbons are less pronounced but fit with the presence of α(2-8)-linkages between the Neu5Ac residues. The α-configuration of the glycosidic linkages was deduced from the chemical shifts of C1, C4, and C6 (see Table 12). The ^{13}C-NMR data for the *E. coli* K1 capsular polysaccharide (colominic acid) match exactly with those described above for the *N. meningitidis* serogroup B capsular polysaccharide indicating identical structures (BHATTACHARJEE *et al.* 1975, JENNINGS *et al.* 1977a, b, LIU *et al.* 1977). However,

Neu5Ac-polymers and for some model compounds
TMS for solutions in ^2H_2O at the indicated p^2H values

Chemical shift								References
C5	C6	C7	C8	C9	5Ac (Me)	5Ac (C=O)	2Me	
53.6	74.3	70.4*	78.8	62.4	23.6	176.1	—	BHATTACHARJEE *et al.* 1975
53.7	74.4	70.6*	78.6	62.5	23.7	176.1	—	BHATTACHARJEE *et al.* 1975
53.1	73.9	69.3	70.9	66.6	23.4	176.1	—	JENNINGS *et al.* 1977b
53.0	73.6	69.5	71.4	66.3	23.3	176.1	—	BHATTACHARJEE *et al.* 1975
52.8	73.5	69.0*	71.5	62.4	23.3	175.9	—	EGAN *et al.* 1977
53.5	74.5	70.4	78.8	66.2	23.5	176.0	—	
52.9	74.0	69.5*	72.0	64.3	23.2	176.1	52.7	BHATTACHARJEE *et al.* 1975
53.2	73.8	69.5	72.9	63.9	23.3	176.3	52.8	JENNINGS *et al.* 1977b
53.1	71.2*	69.4	71.8*	64.7	23.3	176.1	52.1	BHATTACHARJEE *et al.* 1975, JENNINGS *et al.* 1977b
53.3	71.4	69.7	71.4	64.8	23.4	176.0	51.8	JENNINGS *et al.* 1977b
53.0	73.6	69.4*	72.8	63.8	23.1	176.1	—	JENNINGS *et al.* 1977b
53.3	71.3	69.4	70.2	66.7	23.1	176.1*	—	

both polymers may differ in average chain length. The ^{13}C-NMR spectral data of the serogroup C polysaccharide (see Fig. 14*b*, Table 12) clearly demonstrate the α(2-9) type of glycosidic linkage in this polymer. The downfield shift (2.0 ppm) for the resonance of the primary C9 on formation of the glycosidic linkage is only small when compared to the shifts observed for secondary carbon atoms in glycosidic linkage. The chemical shifts for this polysaccharide together with those of the disaccharide Neu5Acα(2-9)Neu5Ac prepared by mild acid hydrolysis of the polymer (JENNINGS *et al.* 1977b) are given in Table 12. Assignment of the anomeric configuration of the monomeric units was made on the basis of the resonance positions of the carboxylate carbon (C1), C4 and C6.

The spectrum of the capsular polysaccharide K92 of *E. coli* is given in Fig. 14*c*

(EGAN *et al.* 1977). The set of 20 resonances forms almost a superposition of the spectra given in Figs. 14*a* and 14*b*. The proposal that this polymer is a linear Neu5Ac chain with alternating α(2-8)- and α(2-9)-linkages was corroborated by ^{13}C-NMR spectroscopy of the periodate-oxidized polymer. The possibility that the

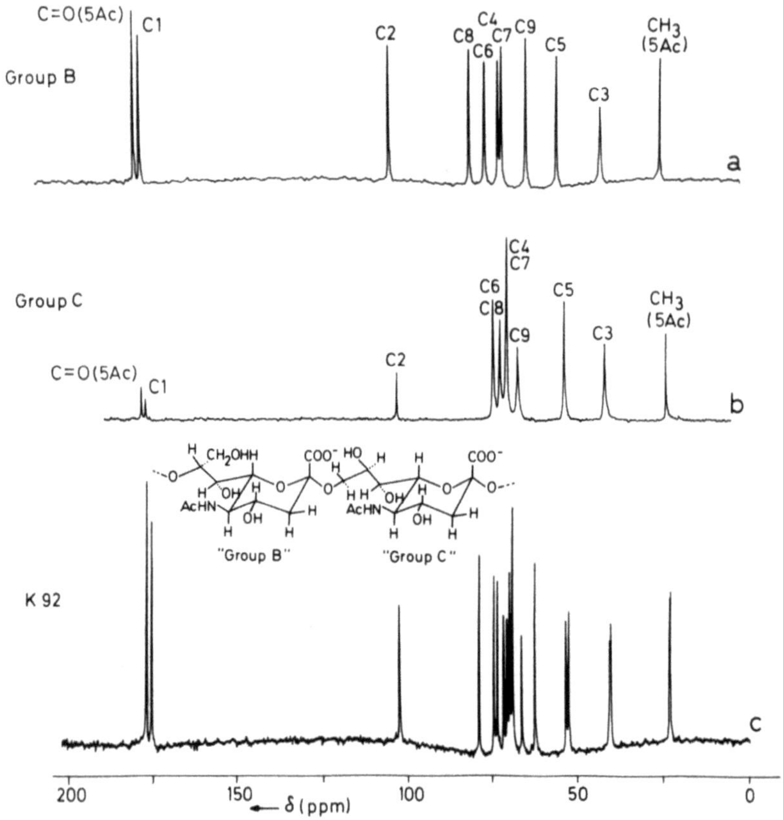

Fig. 14. *a* Proton noise-decoupled 68 MHz ^{13}C-NMR spectrum of meningococcal serogroup B polysaccharide, being an α(2-8)-linked Neu5Ac-homopolymer, dissolved in $^{2}H_2O$, recorded at p²H 7 and 35 °C (taken from EGAN *et al.* 1977). *b* Proton noise-decoupled 25 MHz ^{13}C-NMR spectrum of O-deacetylated meningococcal serogroup C polysaccharide, being an α(2-9)-linked Neu5Ac-homopolymer, dissolved in $^{2}H_2O$, recorded at p²H 7 and 37 °C (taken from BHATTACHARJEE *et al.* 1975). *c* Proton noise-decoupled 68 MHz ^{13}C-NMR spectrum of the capsular polysaccharide K92 from *E. coli*, a Neu5Ac-homopolymer with alternating α(2-8)- and α(2-9)-linkages, dissolved in $^{2}H_2O$, recorded at p²H 7 and 35 °C (taken from EGAN *et al.* 1977).

K92 preparation consists of a mixture of α(2-8)-homopolymer and α(2-9)-homopolymer chains was excluded since no precipitation was observed with anti-group B serum.

The spectra of the capsular polysaccharides of meningococcal serogroups W-135 and Y can easily be interpreted on the basis of sialic acid and hexose reference data (BHATTACHARJEE *et al.* 1976). The spectra are given in Fig. 15 and contain two singlets in the anomeric region, which is consistent with the presence of disaccharide repeating units.

The signal assignments are given in Table 13 and are supported by comparison with the ^{13}C-NMR data of the Hexα(1-4)Neu5Ac disaccharide fragments prepared by mild acid hydrolysis. The glycosylation at position 4 of Neu5Ac is reflected by the downfield shifts for C4 and the upfield shifts for C3 and C5 in

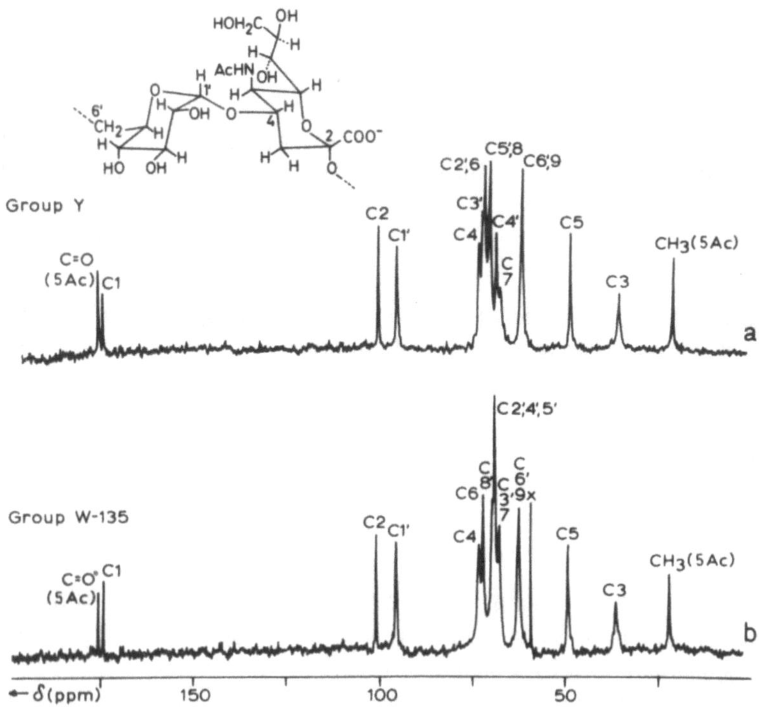

Fig. 15. *a* Proton noise-decoupled 25 MHz ^{13}C-NMR spectrum of O-deacetylated meningococcal serogroup Y polysaccharide, being a {-6)Glcα(1-4)Neu5Acα(2-} polymer, dissolved in ^2H_2O, recorded at p^{2}H 7 and 37 °C (taken from BHATTACHARJEE *et al.* 1976). *b* Proton noise-decoupled 25 MHz ^{13}C-NMR spectrum of meningococcal serogroup W-135 polysaccharide, being a {-6)Galα(1-4)Neu5Acα(2-} polymer, dissolved in ^2H_2O, recorded at p^{2}H 7 and 37 °C (taken from BHATTACHARJEE *et al.* 1976).

disaccharides as well as polysaccharides with respect to αNeu5Ac2Me as a reference (see Table 13). The downfield shifts of the hexose-C6 resonances in the polymers (1.5-2.2 ppm) with respect to the methyl hexoside model compounds are indicative of the involvement of this carbon atom in the glycosidic linkage to Neu5Ac. The α-configuration of the monomeric units is evident from the chemical shifts of hexose C1-C5 and of Neu5Ac C1 and C6.

In a study concerning the immunospecificity of the capsular polysaccharides of *Streptococcus* group B, type III, JENNINGS *et al.* (1981) applied ^{13}C-NMR spectroscopy as a main tool for structure analysis. The native capsular polysaccharide consists of a backbone of -6)GlcNAcβ(1-3)Galβ(1-4)Glcβ(1-

Table 13. ^{13}C-*NMR chemical shift data for Neu5Ac-hexose*
Chemical shifts are given in ppm downfield from external

Compound	Structure	Chemical shift in		
		Neu5Ac		
		C1	C2	C3
Neisseria meningitidis				
Serogroup Y polysaccharide[a]	{-6)Glcα(1-4)Neu5Acα(2-}$_n$	174.2	101.3	38.0
Serogroup Y disaccharide	Glcα(1-4)Neu5Ac	174.4	95.9	36.6
αGlc*p*		—	—	—
αNeu5Ac2Me[b]		174.6	101.9	41.3
Neisseria meningitidis				
Serogroup W-135 polysaccharide	{-6)Galα(1-4)Neu5Acα(2-}$_n$	174.3	101.4	37.8
Serogroup W-135 disaccharide	Galα(1-4)Neu5Ac	174.4	96.0	36.6
αGal*p*		—	—	—

* Values may be interchanged.
[a] O-deacetylated.
[b] JENNINGS *et al.* 1977b ($p^2H = 7$).

repeating units with side chains of Neu5Acα(2-6)Galβ(1- attached to C4 of the core GlcNAc residues. Sialic acid in this (N-acetyllactosamine-type) structural element is characterized by a signal at δ 101.1 for the anomeric C2. The carbon atoms involved in the various glycosidic linkages resonate at δ 78-104. Comparison of these signals in the spectra of the native polysaccharide, the asialo-analogue and the (linear) asialo-agalacto-backbone polysaccharide shows significant shifts in the resonance positions of carbon atoms in corresponding positions in the three polymeric structures. This indicates that distinct interglycose conformational changes, i.e. changes in the torsion angles of the glycosidic bonds occur on specific removal of sialic acid and side chain galactose residues. Serological studies on the native polymer and the derived structures reveal that the sialic acid residues—although not an integral part of the main determinant—influence its specificity by controlling the torsion angles of the glycosidic bond between the penultimate Gal residue and the GlcNAc residue in the polymer backbone. It was postulated that this conformational control could involve specific hydrogen bonds between sialic acid carboxylate groups and the polysaccharide backbone of the native antigen. This suggestion is corroborated by the fact that the carboxylate-reduced polymer behaves both serologically and ^{13}C-NMR spectroscopically like the asialo-polysaccharide. This work clearly demonstrates the power of ^{13}C-NMR spectroscopy as a tool in immunochemical investigations on polysaccharides in general and more in particular for the study of the function of sialic acid residues.

polymers and for some model compounds (BHATTACHARJEE *et al.* 1976)
TMS for solutions in ^2H_2O at 37 °C and unspecified p^{2}H values

Chemical shift in													
Neu5Ac								hexose					
C4	C5	C6	C7	C8	C9	5Ac (Me)	5Ac (C=O)	C1′	C2′	C3′	C4′	C5′	C6′
74.8*	50.8	73.3*	69.4	72.1	63.7	23.4	175.5	96.4	73.3*	74.1	70.2	72.1	63.7
73.3*	50.7	71.3	69.3	71.3	64.3	23.3	175.4	96.3	72.6*	73.9	70.3	72.1*	61.5
—	—	—	—	—	—	—	—	93.1	72.5	73.8	70.7	72.5	61.7
69.5	53.2	73.8	69.5	72.9	63.9	23.3	176.3	—	—	—	—	—	—
74.3*	50.7	73.3*	69.2	70.9	63.8	23.6	175.6	96.1	70.2	69.3	70.2	70.2	63.8
72.5*	50.8	71.5	69.1	71.5	64.4	23.4	175.6	96.6	70.3	69.4	70.7	72.3*	62.0
—	—	—	—	—	—	—	—	93.4	70.3	69.4	70.4	71.6	62.3

4. *Esterified Polysaccharides*

The ^{13}C-NMR spectra of polysaccharides can in principle provide information about the presence of O-acetyl substituents with respect to molar content and localization. The introduction of an O-acetyl group gives rise to characteristic shifts for the O-acetylated carbon atom and for adjacent carbon atoms (see section G.III.2). The degree of O-acetylation can be determined by comparison of the intensities of the characteristic methyl signals of the O-acetyl and N-acetyl groups. For the native meningococcal serogroup C polysaccharide this yields an O-acetyl content of 1.16 mol OAc/Neu5Ac unit (BHATTACHARJEE *et al.* 1975). Comparison of the spectra of the native and O-deacetylated polysaccharide did not allow a precise assignment of the positions of the O-acetyl substituents. This is due to the high complexity of the spectrum of the native polysaccharide as a result of a non-uniform distribution of the O-acetyl groups over the polymer. Using reference chemical shift data of the model compound Neu4,5,7,8,9Ac$_5$ the presence of an O-acetyl group at C4 of the serogroup C polysaccharide could be excluded. Hence, O-acetylation is restricted to C7 and/or C8.

In a similar study of the meningococcal serogroup Y polysaccharide (1.3 mol OAc/Neu5Ac unit) tracing of the positions of the O-acetyl groups was even more difficult since the substituents can be present at C3 or C4 of the Glc units and/or at C7 of the Neu5Ac units (BHATTACHARJEE *et al.* 1976).

An interesting phenomenon observed for the α(2-8)- and α(2-9)-linked sialo-polysaccharides upon storage in slightly acidic solutions (pH 3-6) is the formation

Table 14. ^{13}C-*NMR data for Neu5Ac carbonyl carbons in CMP-Neu5Ac and some reference compounds* (HAVERKAMP *et al.* 1979 b) Chemical shifts (δ) are given in ppm downfield from external TMS, for solutions in ^2H_2O at the indicated temperatures and p^2H values. Coupling constants (J) are given in Hz. n.d., value could not be determined

Compound	p^2H	T	C1				5Ac (C=O)		
			δ	$^3J_{C1,H3ax}$	$^3J_{C1,H3eq}$	$^3J_{C1,H\text{-}ester}$	δ	$^2J_{CAc,HAc}$	$^3J_{CAc,H5}$
βNeu5Ac*	7	25 °C	177.7	≤ 1[a]	≤ 1[a]	—	175.8	6.1	2.8
βNeu5Ac1,2Me$_2$	7	25 °C	171.6	≤ 1[a]	≤ 1[a]	3.7	175.6	5.9	1.8
αNeu5Ac1,2Me$_2$	7	25 °C	171.1	5.9	≤ 1[a]	3.4	175.9	6.2	2.2
αNeu5Ac2Me**	7	25 °C	174.4	5.4	≤ 1[a]	—	176.2	6.2	2.4
CMP-Neu5Ac	8	15 °C	175.3	≤ 1[a]	≤ 1[a]	—	175.7	n.d.	n.d.

[a] Estimated from the line width of the unresolved (double) doublet (see Figs. 16 and 17).
* Compare with Table 11.
** Compare with Tables 12 and 13.

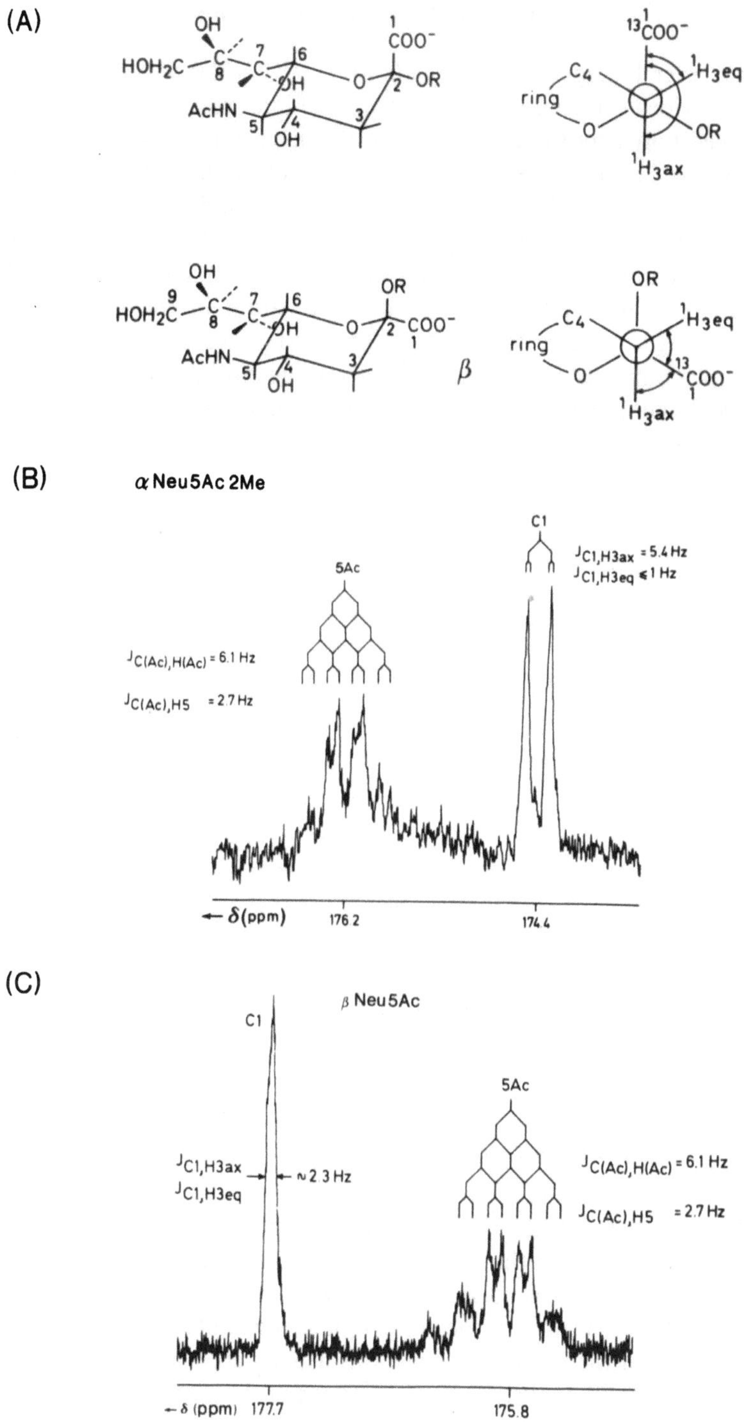

Fig. 16. *A* Structure of Neu5Ac α- and β-anomers and the torsion angles between C1 and the H3 atoms (R=H or aglycon). *B* Carbonyl carbon resonances in the undecoupled 25 MHz ^{13}C-NMR spectrum of αNeu5Ac2Me, dissolved in ^2H_2O, recorded at p^{2}H 7 and 25 °C. *C* Carbonyl carbon resonances in the undecoupled 25 MHz ^{13}C-NMR spectrum of Neu5Ac, dissolved in ^2H_2O, recorded at p^{2}H 7 and 25 °C. Only the signals from the β-anomer are observable.

of internal ester linkages (LIFELY *et al.* 1981). ^{13}C-NMR investigation of these products provided evidence for the occurrence of inter-residue ester bridges. In the α(2-8)-linked polymers these bridges involve the carboxyl group of one residue and the hydroxyl group at C9 of the adjoining residue as becomes evident from the upfield shift (Δδ-7.0) for C1 and the downfield shift (Δδ 5.6) for C9. Although an upfield shift of ~ -3 ppm can be expected for C8 as a result of a neighbouring ester substituent, the much larger upfield shift for C8 (Δδ-7.2) and the upfield shift for C2 (Δδ-4.3) can be explained by assuming a strain in the newly formed 6-membered ring structure between neighbouring Neu5Ac units. ^{13}C-NMR spectroscopy points out that a similar inter-residue ester formation takes place in α(2-9)-linked sialopolymers under mild acidic conditions involving C1 and the hydroxyl group at C8.

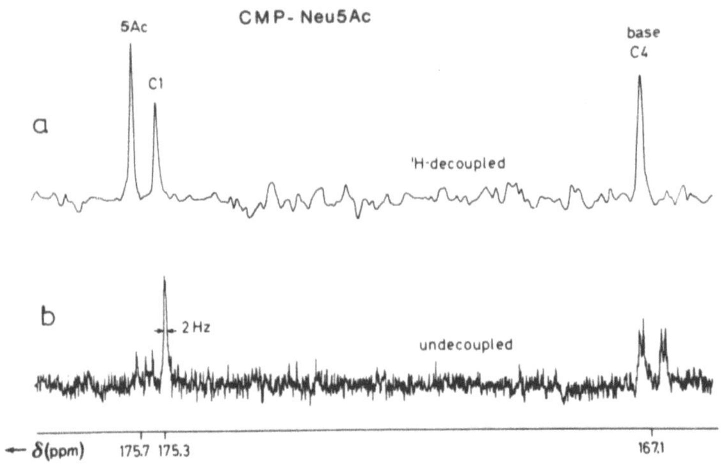

Fig. 17. Partial proton noise-decoupled (*a*) and undecoupled (*b*) 25 MHz ^{13}C-NMR spectrum of CMP-Neu5Ac, dissolved in $^{2}H_2O$, recorded at p^{2}H 8 and 15 °C. The 5Ac carbonyl multiplet in spectrum (*b*) is not well developed. The $^{3}J_{C1,H3ax}$ value (< 1 Hz) indicates the β-configuration of Neu5Ac in this sugar nucleotide.

5. *CMP-N-acetylneuraminic Acid and 2-Deoxy-2,3-dehydro-N-acetylneuraminic Acid*

Another example of structure investigation of a neuraminic acid derivative by ^{13}C-NMR spectroscopy is the elucidation of the anomeric configuration of the Neu5Ac residue in CMP-Neu5Ac (HAVERKAMP *et al.* 1979 b). This study concerns in particular the heteronuclear vicinal coupling constants $^{3}J_{C1,H3ax}$ and $^{3}J_{C1,H3eq}$. The magnitudes of these coupling constants depend according to a Karplus-type relation on the torsion angle between C1 and the protons at position 3. The values observed for CMP-Neu5Ac together with those for some reference compounds are compiled in Table 14. Antiperiplanar orientation of C1 and H3ax, as present in the α-anomers (see Fig. 16) gives rise to a relatively large value for $^{3}J_{C1,H3ax}$ (5.4-5.9 Hz), whereas a synclinal orientation leads to a small value for this vicinal coupling constant (~ 1 Hz). From the measured value for $^{3}J_{C1,H3ax}$, as presented in Fig. 17, it was concluded that the Neu5Ac unit in CMP-Neu5Ac has β-configuration. This result is in agreement with the information obtained via the resonance positions of H6 in the ^{1}H-NMR study of the sugar nucleotide (see section G.II.3.d). By consequence, CMP-Neu5Ac belongs to the class of sugar nucleotides having an axially oriented glycosidic linkage for the activated sugar residue.

The ^{13}C-NMR spectrum of Neu5Ac2en together with its complete interpretation are described by ESCHENFELDER *et al.* (1975) and CZARNIECKI and THORNTON (1977a). The chemical shift values are summarized in Table 11.

6. *Complexes of Sialic Acids with Ca^{2+}*

The property of sialic acids to form complexes with alkaline earth ions has also been investigated by ^{13}C-NMR spectroscopy. This complexation seems to play an important role in the erythrocyte cell membrane where sialic acid units are present as terminal sugar residues at the carbohydrate chains of glycolipids (gangliosides) and glycoproteins. Model studies by ^{13}C-NMR spectroscopy on the interaction of Neu5Ac and Neu5Gc with Ca^{2+} ions have been described in several reports (BEHR and LEHN 1973, CZARNIECKI and THORNTON 1977b, JAQUES *et al.* 1977, JAQUES *et al.* 1980a). The chemical shifts of the sialic acid skeleton carbons are very sensitive to complex formation. The downfield shift in the resonance position of C8 on titration with Ca^{2+} results from a direct involvement of the hydroxyl group at C8 in the Neu5Ac-Ca^{2+} complex formation. However, for the shifts of other carbon resonances CZARNIECKI and THORNTON (1977b) and JAQUES *et al.* (1977) offer different explanations leading to different spatial models for the

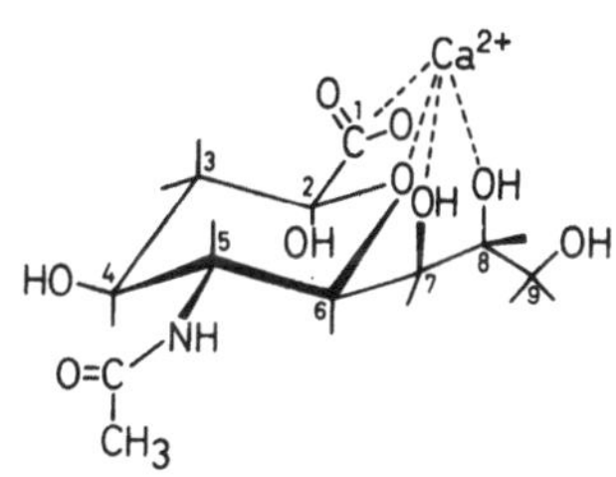

Fig. 18. Conformational model proposed for the complex of αNeu5Ac with Ca^{2+}, based on Ca^{2+}-induced ^{1}H- and ^{13}C-chemical shifts (taken from JAQUES *et al.* 1977).

Neu5Ac-Ca^{2+} complex. The model postulated by the latter authors is given in Fig. 18 and is supported by ^{1}H-NMR studies (see G.II.5). Similar studies with Neu5Gc (JAQUES *et al.* 1980a) reveal that, although the extra hydroxyl function of the glycolyl group could participate in the Ca^{2+} binding, a close resemblance exists between the spatial structures of the Neu5Ac-Ca^{2+} and the Neu5Gc-Ca^{2+} complexes. CZARNIECKI and THORNTON (1977b) propose another model for the Neu5Ac-Ca^{2+} complex in which the anomeric O, the ring-O and the O at C8 are the main binding sites ("oxygen cage") whereas the carboxylate group is not involved in the complexation.

In the complex of αNeu5Ac with Ca^{2+} the set of oxygen atoms involved in the binding differs from that in the βNeu5Ac-Ca^{2+} complex. This leads to a less strong complex for the α-anomer. In view of this observation the rather strong binding of gangliosides with Ca^{2+} cannot exclusively be ascribed to the binding of the αNeu5Ac units of these glycolipids. As also the sequence Neu5Acα(2-3)Galβ(1-4)Glc (and its α(2-6)-analogue) as free oligosaccharides do not show a pronounced complexation behaviour (JAQUES *et al.* 1980 b) this trisaccharide will probably not be the active unit in the complexation of gangliosides with Ca^{2+} provided this structural element will have the same conformation in gangliosides and in the free form. Probably, a more complex interaction of the cation with the carbohydrate chain is involved (BEHR and LEHN 1973, CZARNIECKI and THORNTON 1977 c). The relatively weak complexation of colominic acid with Ca^{2+} (JAQUES *et al.* 1977) may be interpreted in the same way as for monomeric α-anomers (CZARNIECKI and THORNTON 1977 b).

7. ^{13}C-Relaxation Studies on Sialic Acids and Sialyllactoses

Studies on the spatial structure of Neu5Ac derivatives by using ^{13}C spin-lattice relaxation times have already been discussed in conjunction with ^{1}H-NMR parameters in section G.II.1. The measurement of T_1-values has also been applied to obtain information on internal molecular dynamics of sialo-oligosaccharides. CZARNIECKI and THORNTON (1977 c) and JAQUES *et al.* (1980 b) investigated isomers of sialyllactose as model functional units of gangliosides and glycoproteins. Comparison of the relative T_1-values for protonated carbons in the spectra of both Neu5Acα(2-3)Galβ(1-4)Glc and the corresponding α(2-6)-isomer (from bovine colostrum) demonstrates a segmental anisotropic motion of the sugar residues along the main axis of both of the molecules, with the Neu5Ac residue being the least mobile and the Glc residue being the most mobile. Probably the solvated, negatively charged carboxylate group stabilizes the oligosaccharide chain conformation and it is this stabile conformation that is recognized as an immunological determinant.

IV. Analytical Procedures

1. Instrumentation

The most advanced ^{1}H-NMR spectra of carbohydrate chains reported so far, were obtained at 500 MHz on a Bruker WM 500 spectrometer operating in the pulsed Fourier transform mode. These spectra were taken up in 16 k memory with an acquisition time of 3.3 s and a spectral width of 2.5 kHz. Adequate resolution enhancement was achieved by Lorentzian to Gaussian transformation from quadrature phase detection followed by employment of a 32 k point complex Fourier transformation. The probe temperature was stabilized at 300 ± 0.1 K. At this temperature the HO^2H-resonance in neutral aqueous solution is found at δ 4.76. For the detection of proton resonances in this range of the spectrum the HO^2H line can be shifted by slightly changing the temperature (*cf.* VLIEGENTHART *et al.* 1981). Chemical shifts are measured by reference to internal acetone (δ 2.225 ppm downfield from sodium 4,4-dimethyl-4-silapentane-1-sulphonate (DSS)) with an accuracy of 0.001 ppm.

500 MHz ^{1}H-NMR instrumentation as specified above has several advantages over instruments working at lower field strengths. The very strong magnetic field together with the computer resolution enhancement facility allows a precise determination of chemical shifts of structural-reporter-groups and a highly detailed information on splitting patterns of proton multiplets. Furthermore, marked increase in the sensitivity is obtained; samples as small as 25 nmoles of carbohydrate can be adequately analysed with sufficient signal-to-noise ratio. This gain in spectral resolution as well as in sensitivity makes possible the assignment of many structural-reporter-groups and enables the structure elucidation of minor constituents in complex samples or of microheterogeneity in carbohydrate chains.

For instruments operating at lower magnetic fields there is obviously a loss in sensitivity and spectral resolution with respect to the instrument described above. Consequently, the type of problems which can be solved with such less advanced instruments is restricted. Verification of the identity of unknown carbohydrate chains on the basis of comparison with standard reference data sets is possible with 200–360 MHz ^{1}H-NMR instruments.

Usually, ^{13}C-NMR spectra of glycoconjugates and polysaccharides are obtained at 25–50 MHz (corresponding with 100–200 MHz for ^{1}H) instruments operating in the Fourier transform mode. Frequently, spectra are recorded at about 310 K. Chemical shifts are expressed relative to external tetramethyl silane (TMS). Again instruments operating at higher fields will give improvements with regard to spectral resolution and sensitivity. The low natural abundance (1.1%) and the relatively low intrinsic sensitivity of ^{13}C make that sample amounts have to be considerably larger than for ^{1}H-NMR spectroscopy; about 10–100 μmoles of substance is required. For details about special NMR techniques such as specific heteronuclear decoupling, 2D-NMR, relaxation measurements, etc. the reader is referred to the books of Berliner and Reuben (1978, 1980), Jardetzky and Roberts (1981), Shulman (1979), and Breitmaier and Voelter (1974).

2. *Solutions*

Solutions in $^{2}H_2O$ of the compounds to be analysed were in general adjusted to p^{2}H 7 and ^{2}H-exchanged by 5-fold lyophilization of the solution finally using $\geq$ 99.96 atom % ^{2}H. For ^{1}H-NMR spectroscopy in general 0.06–3 mM solutions of the compounds in 0.4 ml $^{2}H_2O$ were used; for ^{13}C-NMR spectroscopy typical sample concentrations lay in the range of 10–100 mM. Amount of material and solubility limit the maximum concentration which can be applied. Too high viscosity of the solution, especially in ^{1}H-NMR spectroscopic measurements, will reduce the spectral resolution. Contamination of the samples with organic or inorganic compounds such as salts and buffer materials has to be avoided, since they affect negatively the spectral quality. Also the presence of paramagnetic materials such as transition metal ions which may be retained or acquired during isolation and purification procedures can disturb spectral resolution. The same holds for contaminants that give rise to sample inhomogeneities like colloidal or particulate matter.

Acknowledgements

The investigations carried out in the authors' laboratory were supported by the Netherlands Foundation for Chemical Research (SON) with financial aid from the Netherlands Organization for the Advancement of Pure Research (ZWO).

Bibliography

BEAU, J.-M., SCHAUER, R., HAVERKAMP, J., DORLAND, L., VLIEGENTHART, J. F. G., SINAŸ, P., 1980: Carbohydr. Res. **82**, 125—129.

BEHR, J.-P., LEHN, J.-M., 1973: FEBS Lett. **31**, 297—300.

BERLINER, L. J., REUBEN, J. (eds.), 1978: Biological Magnetic Resonance, Vol. 1. New York: Plenum Press.

— — (eds.), 1980: Biological Magnetic Resonance, Vol. 2. New York: Plenum Press.

BHATTACHARJEE, A. K., JENNINGS, H. J., KENNY, C. P., MARTIN, A., SMITH, I. C. P., 1975: J. Biol. Chem. **250**, 1926—1932.

— — — — — 1976: Can. J. Biochem. **54**, 1—8.

BLIX, G., JEANLOZ, R. W., 1969: The Amino Sugars, Vol. IA (JEANLOZ, R. W., ed.), pp. 213—265. New York: Academic Press.

BREITMAIER, E., VOELTER, W., 1974: ^{13}C NMR Spectroscopy. Weinheim: Verlag Chemie.

BROWN, E. B., BREY, JR., W. S., WELTNER, JR., W., 1975: Biochim. Biophys. Acta **399**, 124—130.

BUCKINGHAM, A. D., 1960: Can. J. Chem. **38**, 300—307.

CARLSON, D. M., 1966: J. Biol. Chem. **241**, 2984—2986.

CHAPMAN, D., KAMAT, V. B., DE GIER, J., PENKETT, S. A., 1968: J. Mol. Biol. **31**, 101—114.

CZARNIECKI, M. F., THORNTON, E. R., 1976: J. Am. Chem. Soc. **98**, 1023—1025.

— — 1977a: J. Am. Chem. Soc. **99**, 8273—8279.

— — 1977b: Biochem. Biophys. Res. Commun. **74**, 553—558.

— — 1977c: J. Am. Chem. Soc. **99**, 8279—8282.

DABROWSKI, U., FRIEBOLIN, H., BROSSMER, R., SUPP, M., 1979: Tetrahedron Lett. **48**, 4637—4640.

DAVIES, D. B., DANYLUK, S. S., 1974: Biochemistry **13**, 4417—4434.

— — 1975: Biochemistry **14**, 543—554.

DORLAND, L., HAVERKAMP, J., VLIEGENTHART, J. F. G., STRECKER, G., MICHALSKI, J.-C., FOURNET, B., SPIK, G., MONTREUIL, J., 1978: Eur. J. Biochem. **87**, 323—329.

— 1979: Thesis, University of Utrecht, The Netherlands.

— HAVERKAMP, J., SCHAUER, R., VELDINK, G. A., VLIEGENTHART, J. F. G., 1982: Biochem. Biophys. Res. Commun. **104**, 1114—1119.

EGAN, W., LIU, T.-Y., DOROW, D., COHEN, J. S., ROBBINS, J. D., GOTSCHLICH, E. C., ROBBINS, J. B., 1977: Biochemistry **16**, 3687—3692.

ESCHENFELDER, V., BROSSMER, R., FRIEBOLIN, H., 1975: Tetrahedron Lett. **35**, 3069—3072.

FRIEBOLIN, H., SUPP, M., BROSSMER, R., KEILICH, G., ZIEGLER, D., 1980a: Angew. Chem. Int. Ed. Engl. **19**, 208—209.

— BROSSMER, R., KEILICH, G., ZIEGLER, D., SUPP, M., 1980b: Hoppe-Seyler's Z. Physiol. Chem. **361**, 697—702.

— KUNZELMANN, P., SUPP, M., BROSSMER, R., KEILICH, G., ZIEGLER, D., 1981a: Tetrahedron Lett. **22**, 1383—1386.

— SCHMIDT, H., SUPP, M., 1981b: Tetrahedron Lett. **22**, 5171—5174.

— BAUMANN, W., KEILICH, G., ZIEGLER, D., BROSSMER, R., VON NICOLAI, H., 1981c: Hoppe-Seyler's Z. Physiol. Chem. **362**, 1455—1463.

— — BROSSMER, R., KEILICH, G., SUPP, M., ZIEGLER, D., VON NICOLAI, H., 1981d: Biochem. Internat. **3**, 321—326.

Haverkamp, J., Schauer, R., Wember, M., Kamerling, J. P., Vliegenthart, J. F. G., 1975: Hoppe-Seyler's Z. Physiol. Chem. **356**, 1575—1583.
— — — Farriaux, J.-P., Kamerling, J. P., Versluis, C., Vliegenthart, J. F. G., 1976: Hoppe-Seyler's Z. Physiol. Chem. **357**, 1699—1705.
— Dorland, L., Vliegenthart, J. F. G., Montreuil, J., Schauer, R., 1978: Abstr. IXth Int. Symp. Carbohydr. Chem., Chemical Society, London, pp. 281—282.
— Beau, J.-M., Schauer, R., 1979a: Hoppe-Seyler's Z. Physiol. Chem. **360**, 159–166.
— Spoormaker, T., Dorland, L., Vliegenthart, J. F. G., Schauer, R., 1979b: J. Am. Chem. Soc. **101**, 4851—4853.
— Van Halbeek, H., Dorland, L., Vliegenthart, J. F. G., Pfeil, R., Schauer, R., 1982: Eur. J. Biochem. **122**, 305—311.

Inoue, S., Iwasaki, M., Matsumura, G., 1981: Biochem. Biophys. Res. Commun. **102**, 1295—1301.

Jaques, L. W., Brown, E. B., Barrett, J. M., Brey, Jr., W. S., Weltner, Jr., W., 1977: J. Biol. Chem. **252**, 4533—4538.
— Riesco, B. F., Weltner, jr., W., 1980a: Carbohydr. Res. **83**, 21—32.
— Glant, S., Weltner, jr., W., 1980b: Carbohydr. Res. **80**, 207—211.

Jardetzky, O., Roberts, G. C. K., 1981: NMR in Molecular Biology. New York: Academic Press.

Jennings, H. J., Bhattacharjee, A. K., Bundle, D. R., Kenny, C. P., Martin, A., Smith, I. C. P., 1977a: J. Infect. Dis. **136**, S78-S83.
— — 1977b: Carbohydr. Res. **55**, 105—112.
— Lugowski, C., Kasper, D. L., 1981: Biochemistry **20**, 4511—4518.

Kamerling, J. P., Vliegenthart, J. F. G., Schauer, R., Strecker, G., Montreuil, J., 1975: Eur. J. Biochem. **56**, 253—258.
— Strecker, G., Farriaux, J.-P., Dorland, L., Haverkamp, J., Vliegenthart, J. F. G., 1979: Biochim. Biophys. Acta **583**, 403—408.
— Van Halbeek, H., Dorland, L., Vliegenthart, J. F. G., Schauer, R., Messer, M., 1981: Proc. 22th Dutch Fed. Med. Sci. Soc., p. 208.
— Dorland, L., Van Halbeek, H., Vliegenthart, J. F. G., Messer, M., Schauer, R., 1982: Carbohydr. Res. **100**, 331—340.

Kimura, A., Tsurumi, K., 1968: Fukushima J. Med. Sci. **15**, 55—60.

Kumar, V., Kessler, J., Scott, M. E., Patwardhan, B. H., Tanenbaum, S. W., Flashner, M., 1981: Carbohydr. Res. **94**, 123—130.

Lee, C.-H., Sarma, R. H., 1976: Biochemistry **15**, 697—704.

Leger, D., Tordera, V., Spik, G., Dorland, L., Haverkamp, J., Vliegenthart, J. F. G., 1978: FEBS Lett. **93**, 255—260.

Lifely, M. R., Gilbert, A. S., Moreno, C., 1981: Carbohydr. Res. **94**, 193—203.

Liu, T.-Y., Gotschlich, E. C., Egan, W., Robbins, J. B., 1977: J. Infect. Dis. **136**, S71-S77.

Meindl, P., Bodo, G., Palese, P., Schulman, J., Tuppy, H., 1974: Virology **58**, 457—463.

Nomoto, H., Endo, T., Inoue, Y., Inoue, S., Iwasaki, M., 1981: Proc. 6th Int. Symp. Glycoconjugates (Yamakawa, T., *et al.*, eds.), pp. 252—253. Tokyo: Japan Scientific Societies Press.

Paulson, J. C., Weinstein, J., Dorland, L., Van Halbeek, H., Vliegenthart, J. F. G., 1982: J. Biol. Chem. **257** (in press).

Sass, M., Ziessow, D., 1977: J. Magn. Res. **25**, 263—276.

Shulman, R. G. (ed.), 1979: Biological Applications of Magnetic Resonance. New York: Academic Press.

Schut, B. L., Dorland, L., Haverkamp, J., Vliegenthart, J. F. G., Fournet, B., 1978: Biochem. Biophys. Res. Commun. **82**, 1223—1228.

SPIK, G., STRECKER, G., FOURNET, B., BOUQUELET, S., MONTREUIL, J., DORLAND, L., VAN HALBEEK, H., VLIEGENTHART, J. F. G., 1982: Eur. J. Biochem. **121**, 413—419.

VAN DEN EIJNDEN, D. H., JOZIASSE, D. H., DORLAND, L., VAN HALBEEK, H., VLIEGENTHART, J. F. G., SCHMID, K., 1980: Biochem. Biophys. Res. Commun. **92**, 839—845.

VAN HALBEEK, H., DORLAND, L., VLIEGENTHART, J. F. G., FIAT, A.-M., JOLLÈS, P., 1980: Biochim. Biophys. Acta **623**, 295—300.

— — — — — 1981 a: FEBS Lett. **133**, 45—50.

— — HAVERKAMP, J., VELDINK, G. A., VLIEGENTHART, J. F. G., FOURNET, B., RICART, G., MONTREUIL, J., GATHMANN, W. D., AMINOFF, D., 1981 b: Eur. J. Biochem. **118**, 487—495.

VLIEGENTHART, J. F. G., 1980: Adv. Exp. Med. Biol. **125**, 77—91.

— VAN HALBEEK, H., DORLAND, L., 1981: Pure Appl. Chem. **53**, 45—77.

— DORLAND, L., VAN HALBEEK, H., 1982: Adv. Carbohydr. Chem. Biochem. **41** (in press).

H. Histochemistry of Sialic Acids

CHARLES F. A. CULLING† and PHILIP E. REID

Department of Pathology, Faculty of Medicine, University of British Columbia, Vancouver, British Columbia, Canada

With 5 Figures

Contents

I. Introduction

The aim of this chapter will be to escort the reader through the maze of histochemical methodology available for the location and identification of sialic acids and their variants in tissue sections. We shall confine most of our discussion to those forms of sialic acid which can be demonstrated histochemically, the methodology by which they may be differentiated and the degree of specificity which may be achieved when using such methods. We shall endeavour to give both the practical and theoretical reasons for the methodology we advocate, and support such reasoning both by our personal research experience and that published in the literature. Areas which are still undefined will be delineated as such, although suggestions for their elucidation may be advanced.

Sialic acid is apparently ubiquitous in tissues, but is only demonstrable, histochemically, in certain well defined areas. Principally these are the epithelial secretions, although basement membranes (e.g. of the kidney glomeruli), are clearly delineated by those techniques appropriate for the demonstration of sialic acids. Sialic acids occurring in other glycoproteins (e.g. plasma proteins, cell membrane components) may contribute to the diffuse background staining, but due to their low concentration in tissue sections are not readily defined, on a consistent basis, by currently available techniques and therefore are not normally assessed as part of a light microscopic histochemical investigation of epithelial secretions. It appears probable that connective tissues, although containing proteoglycans, do not contain histochemically significant amounts of sialic acid, and conversely that the epithelial mucins do contain sialic acid but not demonstrable proteoglycans and this will be assumed throughout this chapter.

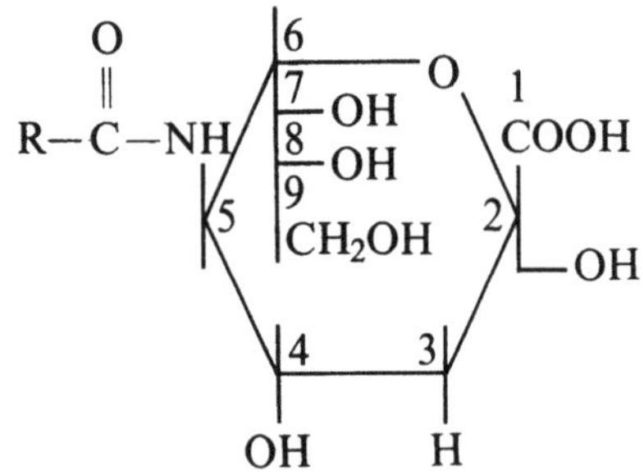

$R = CH_3$ = N-acetylneuraminic acid
$R = CH_2OH$ = N-glycolylneuraminic acid

Fig. 1. Structures of N-acetyl- and N-glycolylneuraminic acids (sialic acids). Sialic acids also occur with ester, usually O-acetyl substituents at position C_4 and/or at various positions (C_7, C_8, C_9) on the glycerol side chain.

Sialic acids occur in biological material as N-acetylneuraminic acid (Fig. 1) and to a lesser extent as N-glycolylneuraminic acid (Fig. 1), but since these acids cannot be differentiated by histochemical procedures when the former is mentioned, it should be assumed to encompass the latter. In addition, sialic acids occur with ester (usually O-acetyl) substituents at position C_4 and at various positions (C_7, C_8, C_9) on the polyhydroxy side chain (Fig. 1) (BUSCHER *et al.* 1974, CASALS-STENZEL *et al.* 1975, CULLING and REID 1980, CULLING *et al.* 1971, 1974a and b, 1975, 1976, 1977a and b, 1979, 1981, DAWSON *et al.* 1978, FENGER and FILIPE 1977, FILIPE 1979, FILIPE and BRANFOOT 1976, FILIPE and COOKE 1974, FILIPE and FENGER 1979, HAVERKAMP *et al.* 1975, KAMERLING *et al.* 1975, NEUBERGER and RATCLIFFE 1972, REID *et al.* 1975a and b, 1977, 1978a and b, 1980, ROGERS *et al.* 1978, VEH *et al.* 1979, WOOD and CULLING 1975). Recent progress in sialic acid histochemistry (CULLING and REID 1980, CULLING *et al.* 1976, 1981, REID *et al.* 1973, 1978, VEH *et al.* 1979) has led to the development of methods for the identification of the position of such substituents but there have been no histochemical procedures developed for the identification of the ester itself.

II. Histochemical Identification of Sialic Acids

1. General

Techniques for the demonstration, investigation and identification of sialic acids, are based primarily upon the presence of two, histochemically reactive structural features in the molecule. These are the carboxyl group located at position C_1 (Fig. 1) which is usually demonstrated by staining with cationic dyes, and the vicinal diols located at positions C_7–C_8, C_8–C_9 (Fig. 1) which are identified by periodate oxidation (or, less commonly, other diol cleavage reagents) followed by visualization of the engendered aldehydes. The interpretation of the results of such techniques is however, complicated by three factors:

i) The methods in current use are not specific for either carboxyl groups or vicinal diols.

ii) The structural features used in the identification of sialic acids are found in other classes of biological macromolecules, e.g. proteoglycans, and,

iii) The presence of substituents which render the sialic acid residues unreactive by one or both of the histochemical techniques. Consequently, the degree of specificity which can be achieved is mainly determined by the enzymic and chemical control methods used, plus, as referred to above, the morphological site of the staining reaction.

An alternative to methods based upon the chemical reactions of carboxyl groups or vicinal diols is the use of lectins labelled with peroxidase or fluorescent tags. Such procedures have been little exploited in the histochemical investigation of sialic acids and indeed have specificity problems peculiar to the mechanism of their reaction but they would appear to be a promising approach to the identification of sialic acids.

Finally it must be remembered that, in common with all histochemical procedures, the absence of a discernible reaction does not necessarily mean absence of the reactant sought, but only that the reactant is not present in sufficient concentration to produce a sufficient density of colour that can be seen by conventional means. It should also be realized that while absolute specificity may not be possible, there are the instances in which the presence of staining together with the morphological site of occurrence, will provide relative specificity.

2. *Methods Based upon Staining with Cationic Dyes*

A great diversity of basic dyes have been, and are being, used to demonstrate anionic sites particularly those present in epithelial mucins. The most commonly employed is unquestionably the copper phthalocyanin dye, Alcian blue which, for reasons not clearly established, seems to have a special affinity for anionic sites in mucins, and results in poor, or the absence of, staining of the phosphate groups in nucleoproteins. Staining with Alcian blue results from the formation of an insoluble salt between the dye cations and tissue anions (Scott *et al.* 1964). The specificity of this staining depends upon two factors, the pKa's of the anions present in the tissues and the pH of the dye bath. In practice the staining of sialic acids with Alcian blue is performed at pH 2.5. At this pH the carboxyl groups of the C-terminal and aspartic and glutamic acid residues of proteins are largely unionized and therefore, staining is considered to be confined to the carboxyl groups of sialic and uronic acids and to half sulphate esters; the latter being located on the carbohydrate portions of some proteoglycans and epithelial glycoproteins. To distinguish between the staining of these classes of reactive groups, three general procedures have been employed.

i) Staining with cationic dyes at different pH's.

ii) The specific removal or chemical blocking of one of the components.

iii) The inhibition of Alcian blue staining by the use of various concentrations of salts in the dye bath.

a) Staining at Different pH's

At one time it was thought that it would be possible to identify most tissue components on the basis of their pKa and therefore assign "pH signatures" to such components. Although this degree of specificity has proved to be impractical, the staining of serial sections with Alcian blue at pH 1.0 and at pH 2.5 will usually serve to distinguish between staining due to half sulphate esters and carboxyl groups respectively, with both groups staining at pH 2.5 while only sulphates stain at pH 1.0 (Lev and Spicer 1964, Spicer 1960). This procedure has the disadvantage that it requires either the use of a comparison microscope or relies upon the visual memory of the observer. Further, although most stainable concentrations of sialic acids are confined to epithelial cells and can therefore be distinguished from connective tissue proteoglycans by morphological criteria, the method does not distinguish between the carboxyl groups of sialic acids and those of the uronic acids of connective tissue proteoglycans. Therefore, it is impossible, without further investigation, to prove or rule out their presence.

An alternative to the use of serial sections stained at different pH's is to stain a single section by applying in sequence and at differing pH's, dyes of contrasting colours. This is the basis of the aldehyde fuchsin-Alcian blue (Spicer and Meyer 1960), and high iron diamine-Alcian blue (Spicer 1965) sequences in which the first named dyes are applied at approximately pH 1.2 while the Alcian blue is used at pH 2.5; such procedures avoid the use of a comparison microscope. However, it may be necessary to show that there is no dye exchange between the dyes in the sequence and potentially a preponderance of one of either carboxyl or sulphate groups will obscure, to some degree, the staining of the other. Thus the staining of the epithelial crypts of the human sigmoid colon with the high iron diamine-Alcian blue sequence implies that sulphomucins are confined to the lower 2/3 of the crypts and sialomucins to the upper 1/3 (Filipe 1979, Filipe and Branfoot 1976) whereas evidence based upon neuraminidase digestion and acid hydrolysis studies implied that sialic acid-containing mucins occur along the entire length of the crypts (Culling *et al.* 1974a).

b) Removal of Specific Components Contributing to the Staining

α) *Use of Enzymatic Procedures*

Digestion with a neuraminidase is the most commonly used histochemical technique for the identification of sialic acids and when positive is the most specific. In this procedure tissue sections are incubated with the neuraminidase and the intensity of their staining with Alcian blue pH 2.5 is compared to that of a serial section incubated under identical conditions, with the buffer but without neuraminidase and also to that of a serial section subjected to no pretreatment. Reduction in staining, in the enzyme treated section only, is evidence for the presence of sialic acids but a negative result (no reduction in staining) cannot be taken as evidence for the absence of such acids until the results of a number of control procedures have been evaluated.

As reviewed by DRZENIEK (1973) and discussed elsewhere in this volume, the removal of sialic acids by digestion with neuraminidases is controlled by a number of factors.

i) The action pattern of the enzyme varies with the source of the neuraminidase and hence the linkage to, and the nature of, the penultimate sugar in the carbohydrate chain must be considered. Since in the great majority of histochemical situations this is unknown, it is usual to use enzymes from *Vibrio cholerae* or *Clostridium perfringens* both of which hydrolyze the largest number of potential linkages.

ii) The presence of an O-acetyl substituent at position C_4 has been shown to block the action of *V. cholerae* neuraminidase, whilst side chain O-acetyl substituents have been shown to decrease the rate at which the sialic acid is removed (DRZENIEK 1973, SCHAUER and FAILLARD 1968). These difficulties may, however, be overcome by removal of the O-acetyl substituents, with alcoholic base, prior to the incubation with the neuraminidase (CULLING *et al.* 1974a), even so it is not always possible to remove all sialic acid residues from gastrointestinal epithelial glycoproteins (CULLING *et al.* 1974a).

iii) The presence of an esterified carboxyl group at C_1 may also inhibit the action of neuraminidase. While this can also be overcome by saponification of the sections prior to digestion, it poses the problem of distinguishing between sialic acids substituted at C_4 and at position C_1. This has been solved by use of the trans-esterification procedure illustrated in Fig. 2 (REID *et al.* 1976). As will be seen, treatment of a C_4-O-acetyl substituted sialic acid (Fig. 2 *A*) with sodium methoxide in anhydrous methanol will remove the substituent thus rendering the sialic acid labile to neuraminidase. In contrast, sodium methoxide/methanol treatment of a C_1-esterified sialic acid (Fig. 2 *B*) will result in trans esterification and the generation of a neuraminidase insensitive methyl ester (Fig. 2 *C*). Saponification of this ester, or the original C_4 and C_1 esters (A and B respectively), will however render all such sialic acids labile to *V. cholerae* neuraminidase (Fig. 2 *D*). Therefore, sialic acids which become neuraminidase labile after treatment with aqueous alcoholic base but not after treatment with sodium methoxide/methanol are considered to be C_1-substituted whereas those that are rendered neuraminidase labile by both pre-treatments are considered to be substituted at position C_4.

iv) Neuraminidases are specific for α-linked sialic acid residues. So far as we are aware, the alternative β-linked acids have not been identified in histochemical studies.

In addition to these problems there are difficulties associated with the nature of histochemical methodology. These include:

i) Observing the slight reduction in staining which occurs following the removal of small quantities of sialic acid when such a reduction must be viewed against a strongly stained background provided for example by large quantities of O-sulphate esters.

ii) The buffer control alluded to above perhaps not being adequate because of the removal of glycoprotein (or other anionic macromolecules) from the sections by non-enzymatic, physical dissolution processes.

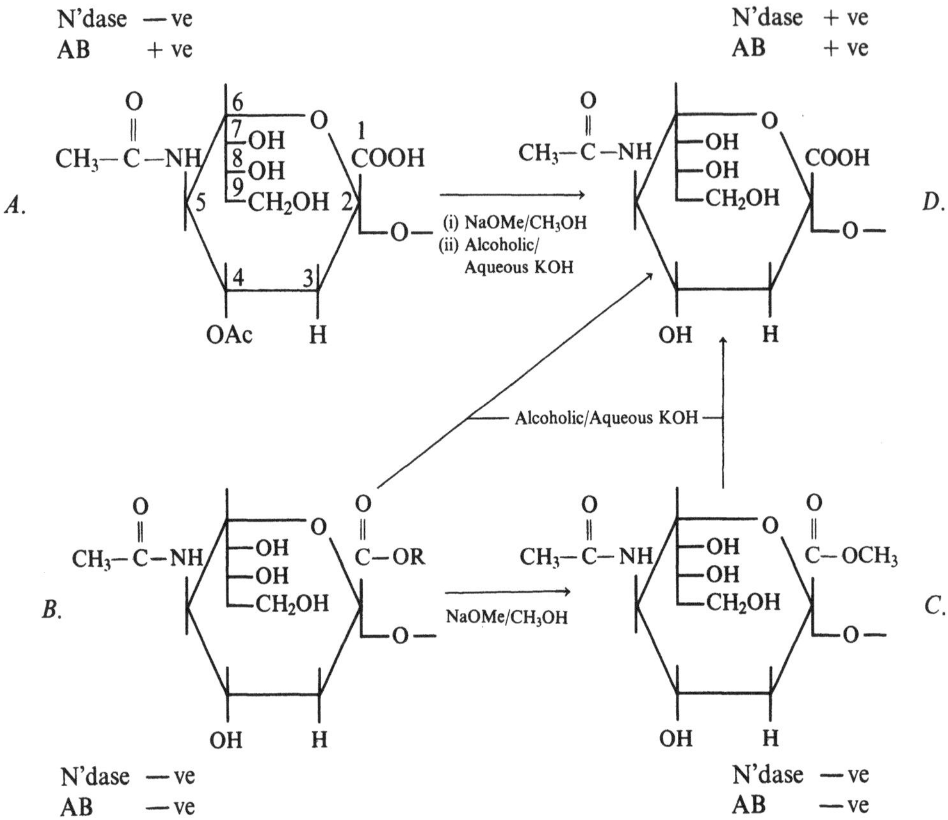

Fig. 2. The mechanism of a transesterification procedure for distinguishing between sialic acids which are neuraminidase resistant because of an ester substituent at either position C_4 or position C_1 (REID *et al.* 1976). *R*, alkyl; *Ac*, CH_3-C=O; *AB*, Alcian blue at pH 2.5; *N'dase*, neuraminidase. Note: Treatment of the ester B with aqueous/alcoholic base to generate sialic acid (*D*) would be expected to increase staining with cationic dyes (e.g. Alcian blue).

iii) Numerous technical factors such as the choice of fixative (SORVARI and LAUREN 1973) and the time and temperature of the incubation influencing the results.

β) *Use of Chemical Procedures*

These consist of the removal of the sialic acid by acid hydrolysis, or the half sulphate ester group by treatment with methanolic hydrogen chloride.

i) *Acid Hydrolysis*. Dilute acid hydrolysis, (e.g. 0.05 *M* sulphuric acid at 80 °C for 1 hour [SPICER and WARREN 1960], or 0.02 *N* sodium acetate-HCl buffer pH 2.5, 75 °C, 2 hours [QUINTARELLI 1961]) is a commonly used procedure for the removal of sialic acids in histochemical investigations. In these procedures a comparison is made between the Alcian blue pH 2.5 staining of serial sections exposed either to acid hydrolysis or to water under the same conditions. Although this procedure is not as theoretically precise as the use of neuraminidase, it avoids many of the problems associated with the use of enzymes and in a modified form we have adopted it for routine use.

Two theoretical problems are associated with this technique. Firstly, studies by NEUBERGER and RATCLIFFE (1972, 1973) have clearly shown that O-acetylated sialic acids are hydrolysed more slowly than their unsubstituted counterparts. This difficulty has been noted in histochemical studies of colonic glycoproteins (CULLING *et al.* 1974a and b). It can be overcome by saponification of the sections prior to hydrolysis (CULLING *et al.* 1974a and b) and this procedure, the saponification/acid hydrolysis/Alcian blue pH 2.5 (KOH/H^+/AB) method is routinely used in our laboratory. Secondly, a potential hazard of methods based upon acid hydrolysis is either the cleavage of glycosidic bonds at points in the oligosaccharide chain which result in the loss of fragments containing O-sulphate ester or loss of O-sulphate ester itself. Although these problems are probably of little practical consequence, they must be considered and, if necessary, controlled by staining sections, treated in parallel, with Alcian blue at pH 1.0.

Two technical problems should also be considered:

1) The potential extraction of glycoproteins under the conditions used.

2) All reagents and glassware must be at 80 °C before commencing the hydrolysis, otherwise hydrolysis may be incomplete (CULLING *et al.* 1974b).

ii) *Methylation.* In the usual procedure sections are treated with 1% methanolic/HCl at 60 °C for 4 hours which, among other reactions (REID *et al.* 1974), esterifies carboxyl groups (CULLING 1974) and brings about the methanolytic cleavage of half sulphate ester (KANTOR and SCHUBERT 1957). Subsequent saponification restores the alcianophilia of the carboxyl groups which can then be seen in the absence of O-sulphate ester. In theory, therefore, sialic acid residues can be visualized. In practice, however, such treatments result in methanolysis of at least some of the sialic acid residues (QUINTARELLI 1963, QUINTARELLI *et al.* 1964, SCHMITZ-MOORMAN 1969) and therefore the procedure is inappropriate for the histochemical demonstration of sialic acids. Other methods of methylation, not in use in routine histochemistry, include thionyl chloride methanol (GEYER 1962, KORHONEN 1967, STOWARD 1967, 1968, SORVARI and STOWARD 1970), diazomethane (SORVARI and STOWARD 1970), methanol-sulphuric acid (LILLIE and JIRGE 1975), methyl iodide (SORVARI and STOWARD 1970, TERNER 1964), anhydrous hydrogen chloride in alcohols and non alcohol solvents (SORVARI and STOWARD 1970 and VILTER 1968) and the mild methylation procedure of SPICER (1960).

c) Inhibition of Staining with Alcian Blue in the Presence of Varying Concentrations of Inorganic Salts (C.E.C. Methods)

The C.E.C. methods developed by SCOTT and his colleagues (KELLY *et al.* 1963, SCOTT and DORLING 1965) whereby a selective staining of various anionic groups is achieved by the use of staining baths containing various concentrations of salts are not in routine use for the histochemical identification of sialic acid.

3. *Methods Based upon the Use of Periodate Oxidation*

The polyhydroxy side chain of sialic acids is rapidly oxidized with periodic acid to the corresponding C_7-aldehyde. This aldehyde can be visualized with a number of Schiff and Schiff type reagents, substituted hydrazines, silver, etc. Of such procedures the most commonly employed is the periodic acid/Schiff technique (PAS) in which the Schiff reagent is prepared from basic fuchsin (CULLING 1974).

a) The PAS Reaction

As generally employed, the PAS reaction is not specific for sialic acid since; (i) periodate reactive vicinal diols on all other carbohydrate residues will react positively in the procedure; (ii) groups unassociated with carbohydrates may be PAS and/or Schiff positive and (iii) sialic acids substituted in their side chain may fail to react and therefore will not stain. In fact, as discussed below, this latter property is used in the identification of such acids. Furthermore, attempts to improve the specificity of the technique by observing the PAS before and after either neuraminidase digestion or acid hydrolysis are complicated not only by those factors affecting these procedures discussed above but also by the position of the linkage between the sialic acid and the adjacent sugar in the carbohydrate chain. Thus as shown in Fig. 3*A*, when the sialic acid is linked to the adjacent residue in such a fashion as to render the residue resistant to periodate oxidation (e.g. the C_3 position of a galactose residue), removal of the sialic acid may have no net effect upon the intensity of PAS staining because removal of the PAS reactive sialic acid residue results in the production of a new PAS reactive vicinal diol. When, however, the linkage between the sialic acid and the adjacent sugar does not interfere with the periodate oxidation of the latter (e.g. C_6 of a galactose residue Fig. 3 *B*) the net effect of the removal of sialic acid should be a reduction in PAS staining. This will only be observable, however, when the sialic acid represents a significant component of the PAS reactivity of the tissue site in question. A further complication is that if the sialic acid residue is itself unoxidizable but is linked to a potential vicinal diol (Fig. 3 *C*), then removal of the sialic acid residue may actually increase PAS staining, whereas if it is linked to an oxidizable residue (Fig. 3 *D*) then no change in staining would be expected. When it is realized that it is possible that one or more or all four of the factors above could be operative in the same section at the same time, it can be seen that interpretation of the results can be difficult if not impossible.

b) The Periodic Acid/Phenylhydrazine/Schiff (PAPS) Procedure

In the periodic acid/phenylhydrazine/Schiff (PAPS) procedure of SPICER (1961) a treatment with phenylhydrazine is interposed between the periodic acid oxidation and Schiff reagent steps of the standard PAS. As phenylhydrazine reacts readily with aldehydes it would be expected to block all subsequent reaction with Schiff reagent. It was found, however, that when this sequence was employed with sections of mouse salivary glands and other tissues rich in sialomucins, PAS staining was reduced but not abolished. When, however, the salivary glands were pre-treated with a neuraminidase the positive Schiff staining in the PAPS

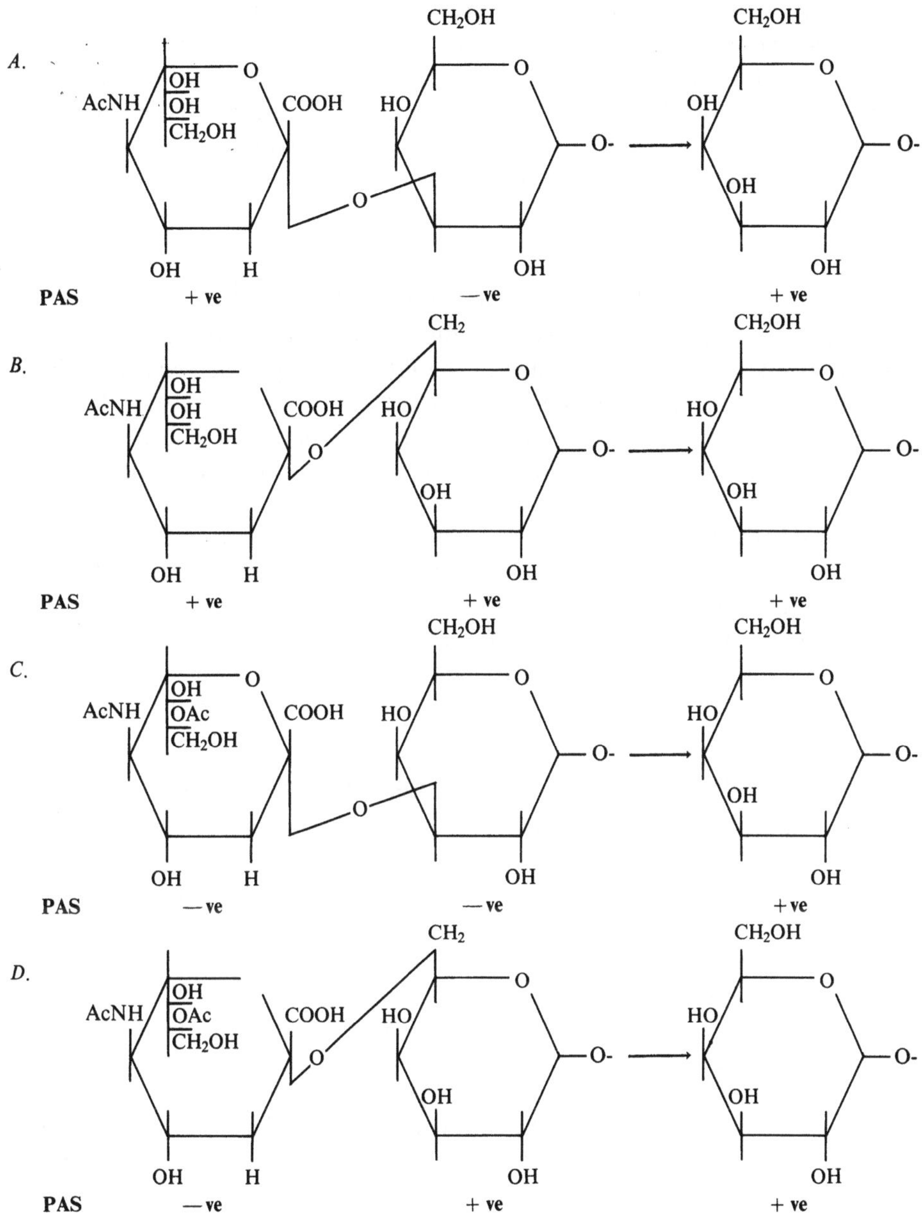

Fig. 3. Predicted effects upon PAS reactivity of the removal of sialic acid and side chain substituted sialic acid linked to either position C_3 or position C_6 of an adjacent galactose residue.

procedure was abolished. This result strongly implied that the PAPS staining of the *untreated* gland sections was due to sialic acid and it was suggested (SPICER 1965) that with such residues the phenylhydrazine formed a salt with the carboxylic acid that hindered any subsequent reaction between the aldehyde group and the phenylhydrazine; consequently, the aldehyde was available for subsequent reaction with the Schiff reagent. This procedure is not necessarily specific for sialic acid however since, as pointed out by SPICER, salt formation with other anionic groups, e.g. sulphate, could prevent blockade of adjacent vicinal diols unassociated with sialic acid residues "per se".

c) The Use of Very Dilute Periodic Acid

An alternative approach is to use very dilute periodic acid and limited oxidation times in the periodate oxidation step of the PAS. This approach which has been exploited by a number of investigators (ROBERTS 1977, VEH *et al.* 1979, WEBER *et al.* 1975) is based upon the observation that upon oxidation of glycoprotein with limited quantities of periodate, the exocyclic diols of sialic acid residues oxidize much more rapidly than the ring diols present in other sugar residues (CULLING and REID 1977, REID and CULLING 1980). It is possible, therefore, to devise conditions under which oxidation is confined entirely or almost entirely to sialic acid residues and hence to confine PAS reactivity to sialic acid residues. It is difficult, however, to know what is a limiting concentration when oxidizing tissue sections and it would seem necessary to include controls which will measure, under the conditions used, the oxidation of vicinal diols of residues other than sialic acid.

d) Identification of Side Chain O-Acetylated Sialic Acid

The identification of these sialic acids in colonic mucins is based upon the observation of an increase in PAS staining after saponification which was shown by us to be due to the removal of an ester group by the saponification step (CULLING *et al.* 1971). We called this the KOH/PAS effect and subsequently showed that in most animals it was largely confined to the lower gastrointestinal tract (CULLING *et al.* 1971, 1974a, 1975, REID *et al.* 1973, 1974). The KOH/PAS effect was isolated by blocking the original PAS reactivity by a periodic acid—borohydride sequence which became the periodic acid/borohydride saponification/periodic acid/Schiff (PBT/KOH/PAS) technique (REID *et al.* 1973). Subsequently we visualized normally PAS reactive material using a thionine Schiff reagent; this was followed by KOH treatment and a routine PAS procedure to visualize the KOH/PAS reactive material (periodic acid/thionine Schiff/saponification/periodic acid/Schiff [PAT/KOH/PAS] [CULLING *et al.* 1976]). These methods have been used in our laboratory for the histochemical identification of side chain O-acetylated sialic acids. The predicted results from the application of these procedures and of the PAS and KOH/PAS methods to various classes of side chain substituted sialic acids are shown in Fig. 4. As will be seen for the purposes of histochemical identification, the sialic acids are divided into four classes:

a) Those without side chain substituents (C_0)
b) Those substituted at C_7 alone (C_7)

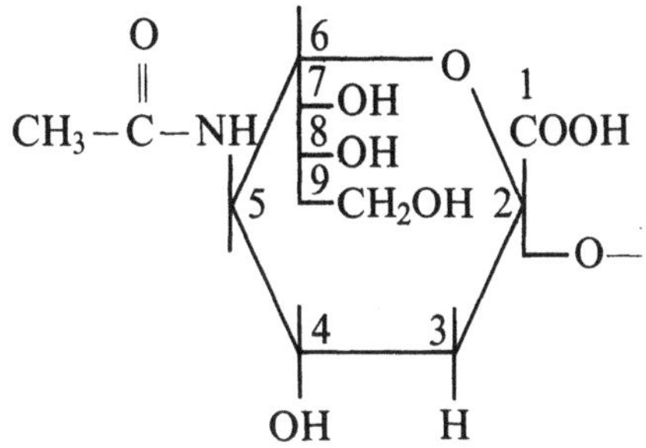

Sialic acid class		PAS	KOH/PAS	PBT/KOH/PAS	PAT/KOH/PAS
C_0	—O —OH —OH CH_2OH	+	+	—ve	blue
C_7	—O —OAc —OH CH_2OH	+	+	+	blue
C_9	—O —OH —OH CH_2OAc	+	+	—ve	blue
C_8*	—O —OH —OAc CH_2OH	—ve	+	+	red

Fig. 4. Predicted reactivity of various classes of side chain substituted sialic acids by the PAS, KOH/PAS, PBT/KOH/PAS, and PAT/KOH/PAS procedures. * C_8 class of sialic acids includes sialic acids which are di(C_7C_8,C_7C_9,C_8C_9)- and tri($C_7C_8C_9$)-substituted sialic acids.

c) Those substituted at C_9 alone (C_9)

d) Those substituted at C_8 or those which are di(C_7C_8,C_7C_9,C_8C_9)- or tri($C_7C_8C_9$)-substituted are collectively assigned to the C_8 class of sialic acids.

α) PBT/KOH/PAS Technique

Fig. 5 shows the sequence of reactions which would occur on the application of the PBT/KOH/PAS procedure to four classes of sialic acids and to tissue diols unassociated with sialic acids. As will be seen, the initial periodate oxidation step

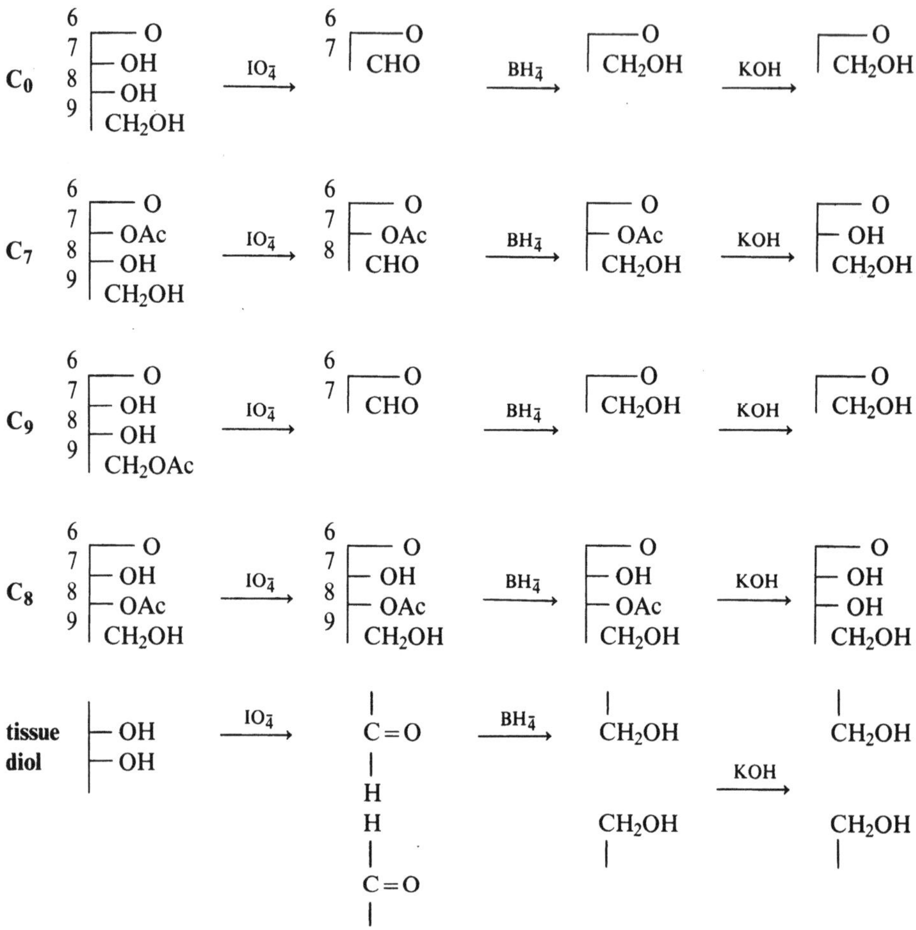

Fig. 5. Mechanism of the PBT/KOH/PAS procedures as applied to various classes of side chain substituted sialic acids and to other PAS reactive tissue vicinal diols.

converts the C_0 and C_9 classes of sialic acids to the corresponding C_7 aldehyde, the C_7 class of sialic acids to the corresponding C_8 aldehyde and the tissue diols to dialdehydes but has no effect on the C_8 class of sialic acids. In the next step the Schiff reactive aldehydes are converted by reduction with sodium borohydride to the corresponding, Schiff negative, primary alcohol. At this stage of the sequence, sections should be (a) Schiff negative indicating complete reduction of the aldehydes and (b) PAS non reactive indicating that the first oxidation step has completely oxidized all the available vicinal diols. Studies in our laboratory indicated that for gastrointestinal mucins, while the oxidation time required to achieve the latter varied from specimen to specimen with the great majority of specimens, 30 minutes was sufficient (REID *et al.* 1973). Therefore this oxidation time was adopted for the initial step of both the PBT/KOH/PAS and the PAT/KOH/PAS. Following periodate oxidation and sodium borohydride reduction, the sections were saponified by treatment with 1% potassium hydroxide in 70% alcohol. As will be seen, this removed the O-acetyl substituents from the C_8 class and the modified C_7 class of sialic acids rendering them PAS positive. Thus staining by the PBT/KOH/PAS technique indicates the presence of the C_7 and/or C_8 classes of sialic acids. It should be noted that potentially it could also indicate the presence of any other tissue diols bearing an alkali labile substituent.

β) *PAT/KOH/PAS Procedure*

The initial step of the PAT/KOH/PAS reaction generates aldehydes from the C_0, C_7, and C_9 classes of sialic acids and from tissue diols unassociated with sialic acid residues. Such aldehydes are stained blue with a Schiff reagent prepared from the phenothiazine dye, thionine. The subsequent saponification step removes acetyl groups from the C_8 class of sialic acids which are then stained red by the standard PAS procedure. Sites containing mixtures of these components stain in various shades of purple. As will be seen in Fig. 4, tissue sites which stain blue in the PAT/KOH/PAS procedure and red (positive) with the PBT/KOH/PAS technique are considered to contain the C_7 class of sialic acids; we have identified a predominance of such acids in a small number of tumours (CULLING *et al.* 1977 a).

Recently COOPER and DURNING (1981) have devised a modification of the PBT/KOH/PAS and the PAT/KOH/PAS procedures. In their modification reaction with 3-hydroxy-2-naphthoic acid hydrazide (ASHBEL and SELIGMAN 1949) which is then visualized with a fresh solution of Fast Black Salt B is substituted for the thionine Schiff reagent. Following saponification the sections are oxidized with dilute periodate and then treated with standard, basic fuchsin, Schiff reagent. The authors consider that this procedure avoids two problems (i) the thionine Schiff reagent was in their hands "both unstable and too weak in its staining performance" and (ii) "anomalous second stage (red) staining of other periodate reactive substances ...". Our preliminary results with their methods to date have not been satisfactory and for the moment we would not recommend their use.

γ) *9-O-Acyl Sialic Acids*

The predicted results outlined in Fig. 4 are based upon the assumption that the C_9 class of sialic acids is completely oxidized under the conditions employed in the PAT/KOH/PAS and PBT/KOH/PAS procedures. HAVERKAMP *et al.* (1975) showed that the oxidation of 9-O-acetyl-N-acetylneuraminic acid and its methylglycoside, with limited quantities of periodate, was very much slower than the oxidation of N-acetylneuraminic acid and its glycoside. Should this phenomenon occur under the very different conditions used in histochemistry, then some, or all, of any C_9 class of sialic acids would be identified as members of the C_8 class of sialic acids.

One solution to this problem (REID *et al.* 1978 a) was to extend the initial periods of oxidation in both the methods. Under these conditions the C_9 class of sialic acids would be expected to oxidize slowly and therefore the red staining of the PBT/KOH/PAS would be expected to gradually diminish while staining with the PAT/KOH/PAS would become progressively bluer. These theoretical results were obtained in practice on sections of bovine submandibular gland, a known site of 9-O-acetyl sialic acids (BUSCHER *et al.* 1974, CASALS-STENZEL *et al.* 1975, HAVERKAMP *et al.* 1975, KAMERLING *et al.* 1975), but were not seen with normal colonic mucins (REID *et al.* 1978 a). One might presume therefore that either 9-O-acyl sialic acids are not present in colonic mucins, or that they are fully oxidized during the initial 30 minutes oxidation step of the PBT/KOH/PAS and PAT/KOH/PAS procedures.

To investigate the latter possibility, sections of bovine submandibular gland and human colon were oxidized for periods of either 10 or 30 minutes and were then reduced with sodium borohydride. The sections were then reoxidized with periodate for various periods and then stained with Schiff reagent. The results obtained showed that when human colonic mucins were oxidized for an initial period of 10 minutes, some PAS reactivity could be regenerated by a second oxidation period of 60 minutes. When however, the initial oxidation period was extended to 30 minutes, no further PAS reactivity could be generated by a second period of oxidation. These results could be interpreted in two ways; either they were the result of the incomplete oxidation of a relatively small number of 9-O-acyl sialic acids or the failure to oxidize other tissue diols. The probability of the latter was lessened by the observation that the oxidation of the crypt mucins of rabbit small intestine which do not contain O-acylated sialic acids was apparently complete in 10 minutes.

In contrast, in bovine submandibular gland mucin, neither of the initial periods of oxidation abolished subsequent PAS reactivity although the PAS reactivity was somewhat weaker in sections exposed to an initial period of 30 minutes. We interpret these results as evidence for the presence of 9-O-acyl sialic acids although not all such acids can be demonstrated by such a method since presumably some oxidation of the 9-O-acyl sialic acids will occur within the initial period of oxidation.

An alternative procedure was adopted by VEH *et al.* (1979). They demonstrated that a mild (lmM) periodate oxidation PAS was specific for the sialic acid residues of bovine submandibular gland. Since these oxidation conditions were not expected to oxidize 9-O-acyl sialic acids, they considered that the procedure demonstrated the C_0 and C_7 classes of sialic acids, whereas using such mild oxidation procedures, the PBT/KOH/PAS demonstrated the C_8 and the C_9 classes of sialic acids. Oxidation with 0.5 *M* periodate for 120 minutes at room temperature, however, resulted in the oxidative cleavage of 9-O-acyl sialic acids between positions C_7 and C_8. Subsequent steps of the PBT/KOH/PAS then demonstrated the C_8 class of sialic acids.

4. Methods Based upon the Use of Lectins

Lectins are mostly obtained from plants, particularly legumes, although they have been shown to occur in some invertebrates (SHARON 1977). They were first called haemagglutinins because of their ability to agglutinate red blood cells. They act like agglutinins with apparently similar specificity. It is now presumed they agglutinate cells by binding to the carbohydrate moieties in/on cell membranes. Their activity can be blocked by pre-treatment with specific carbohydrates in a similar manner to the blocking techniques used in immunology. While there is still some doubt as to their absolute specificity for individual carbohydrates, there is no doubt that they offer a valuable tool in carbohydrate chemistry. By combining a lectin with a peroxidase-anti peroxidase complex (PAPS) which can be visualized, they offer a new tool in histochemical research.

An example of the possibilites of the use of these techniques in the future is the paper by STOWARD, SPICER, and MILLER (1980) in which they use a peanut lectin—PAPS technique to identify galactose as the sub-terminal group, following removal of sialic acid with neuraminidase in, for example, the secretory glycoprotein in mouse duodenal cells. With the number of lectins now available commercially, some of which can be obtained with a PAPS label already attached, it can be anticipated that the histochemical sequencing of carbohydrate chains is well within the realm of possibility in the near future.

III. Staining and Control Procedures

1. Fixation and Processing

Routine formol-calcium fixed, paraplast-embedded, five micron thick, tissue sections were generally used throughout our studies, but cryostat- or paraplast-embedded sections fixed in any manner that does not interfere with vicinal diols may be used equally well. Sections must be floated onto clean, grease-free slides. This should avoid the loss of sections during staining. Chrome alum gelatin (see below) may be used if sections become detached.

Method for Preparing Gelatin—Chrome Alum Adhesive
Chrome-Alum Gelatin

Add 2 g of gelatin to 30 ml of cold water to soften gelatin; then add 170 ml of boiling water, mix and cool to room temperature. Add 0.2 g of chrome alum and shake gently to dissolve.

For Use

Add 10 ml of chrome alum gelatin to the water in a regular sized flotation bath. Sections are then floated on, in the normal manner, and picked up on clean grease-free slides. Allow the slides to drain before placing in the following prepared staining Coplin jar. Line the bottom of the jar with paper (e.g. a towel) moisten with formaldehyde and place lid on. Place the slides, in the Coplin jar, in a 60 °C oven overnight. The following day remove the slides from the formaldehyde container and leave them in the oven for an additional half hour.

2. PAS Technique

Choice of Schiff

Most of the accepted Schiff formulae give equally good results, but we prefer the Schiff reagent of BARGER and DE LAMATER (1948), which gives a brilliant reaction.

BARGER and DE LAMATER's (1948) Schiff Reagent

1. 1 g of basic fuchsin is dissolved in 400 ml of distilled water, using heat if necessary.
2. Add 1 ml of thionyl chloride ($SOCl_2$), stopper the flask and, after shaking, allow to stand for 12 hours.
3. Add 2 g activated charcoal, shake, and immediately filter. This will keep several months in a well-stoppered dark bottle in the refrigerator.

PAS Method

1. Bring sections to water.
2. Oxidize for 10 minutes in freshly prepared 1% aqueous periodic acid.
3. Wash in running water for 5 minutes, and rinse in distilled water.
4. Treat with Schiff reagent for 10–30 minutes.
5. Wash for 10 minutes in running water.
6. Dehydrate, clear and mount in resinous mountant.

Results

PAS positive substances
Bright red

3. PBT/KOH/PAS Technique

This technique, modified from our original method (REID *et al.* 1973), utilizes LILLIE and PIZZOLATO's (1972) borohydride aldehyde blocking method.

1. Sections are brought to water and then treated with 1% periodic acid for 30 min.
2. Washed in running water for 10 min.
3. Treated in 0.1% sodium borohydride in 1% disodium hydrogen phosphate for 30 min.
4. Slides are washed in water.
5. Treated with 0.5% KOH in 70% ethanol for 5 min.
6. Washed in water.
7. Stained by routine periodic acid/Schiff (PAS) method.

Results

Mucin showing any red coloration is interpreted as containing side chain O-acetylated sialic acids; provided a control slide treated by the PBT/PAS (steps 5 and 6 are omitted) is unstained. Optionally, sections may be lightly counterstained with haematoxylin, or tartrazine in cellosolve to visualize the tissue and to aid the orientation of the section; we preferred the former, but do not routinely use any counterstain.

4. PAT/KOH/PAS Technique

Preparation of the Thionin-Schiff Reagent (VAN DUIJN 1956): To 250 ml of distilled water are added 0.5 g thionine. This solution is boiled for 5 minutes and cooled down to room temperature and brought to a volume of 250 ml again by adding distilled water. To the solution are added 250 ml melted tertiary butyl alcohol (melting point approximately 25 °C). This solution is transferred without filtering to a stoppered bottle and 75 ml 1.0 N HCl are added and immediately afterwards 5 g sodium metabisulfite. After shaking for a while the closed bottle is left for 24 hours at room temperature and another 48 hours at 4 °C in the refrigerator. At this stage part of the solution, sufficient for a staining experiment, is filtered. After its use the filtered solution is added again to the unfiltered bulk of the solution. When not in use the solution is kept in the refrigerator at ± 4 °C.

Our original thionin-Schiff reagent. One gram of thionin is dissolved in 100 ml of distilled water, brought to a boil and removed from heat. The solution is then cooled, 0.75 ml thionyl chloride added, and left overnight. It is then filtered through paper and stored in a stoppered dark bottle in the refrigerator (where it will keep for 3–4 weeks). Each batch of dry stain must be tested to ensure that it is satisfactory.

We now use the VAN DUIJN method to prepare the thionin-Schiff reagent.

Staining Method

1. Sections are brought to water.
2. Placed in 1% aqueous periodic acid (freshly prepared) for 30 min.
3. Washed in running tap water for 10 min.
4. Placed in thionin-Schiff reagent for 30 min.
5. Washed in running tap water for 10 min.
6. Rinsed in 70% ethanol, and treated with 0.5% potassium hydroxide in 70% ethanol for 5 min.
7. Washed gently in tap water; taking care to avoid dislodging section.
8. Placed in fresh 1% periodic acid for 10 min.
9. Washed gently in running tap water for 10 min.
10. Placed in standard Schiff reagent for 30 min.
11. Washed gently in running tap water for 10 min.
12. Dehydrated, cleared and mounted in resinous mountant.

Results

Normally PAS positive material will be blue, O-acylated sialic acids will be red, mixtures will be seen as purple.

5. *High Iron Diamine/Alcian Blue Method* (SPICER 1965)

Diamine Solution

N'N-dimethyl-m-phenylenediamine dihydrochloride	120 mg
N'N-dimethyl-p-phenylenediamine hydrochloride	20 mg

Dissolve the diamines in 50 ml of distilled water, then add 1.4 ml of 40 per cent (W/V) ferric chloride. The pH of the prepared solution should be 1.3–1.5. This solution must be freshly prepared and used immediately.

Method

1. Bring sections to water.
2. Stain in diamine solution in Coplin jar for 24 hours.

3. Rinse rapidly in water.
4. Stain in freshly filtered 1 per cent Alcian blue 8GX in 3 per cent acetic acid (pH 2.5) for 30 min.
5. Rinse rapidly in 3% acetic acid.
6. Dehydrate rapidly, clear and mount in a resinous mountant.

Results

Half sulphate esters	Black-brown
Carboxyl groups	Blue

6. Alcian Blue (pH 2.5) (CULLING 1974)

It should be noted there is some diversity in the dyes marketed as Alcian blue and that, therefore new batches of dye should be tested for their performance before use. By purchasing dyes which have been certified (and therefore tested both chemically and biologically) by the Biological Stain Commission one can obviate many problems.

Method

1. Bring sections to water.
2. Stain in freshly filtered 1 per cent Alcian blue 8GX in 3 per cent acetic acid (pH 2.5) for 30 min.
3. Rinse in 3% acetic acid.
4. Dehydrate, clear and mount in resinous mountant.

Results

Acid groups (COOH and OSO_3H)	Blue
(nuclei may stain faint blue)	

7. Alcian Blue (pH 1.0) (LEV and SPICER 1964)

Method

1. Bring sections to water.
2. Stain in 1 per cent Alcian blue 8GX in 0.1 N hydrochloric acid for 30 min. Rinse briefly in 0.1 N HCl.
3. Blot dry with fine filter paper to prevent the non-specific staining which sometimes occurs after dilution with water (which will change the pH) in washing.
4. Dehydrate in alcohol, clear in xylol and mount in resinous mountant.

Results

Sulphated mucosubstances stain blue.

8. Neuraminidase Digestion

1. Bring four (A, B, C, and D) serial sections to water.
2. Treat section "A" with 1% potassium hydroxide in 70% ethanol for 5 min., and wash in water for 5-10 min. (saponification [KOH]).

3. Treat sections "A" and "B" with *V. cholerae neuraminidase* – 100 Behringwerke units per ml in 0.05 M acetate buffer at pH 5.5 (containing approximately 0.1% calcium chloride).
   Treat section "C" with acetate buffer alone.
4. Rinse in water.
5. Stain all four slides by the Alcian blue at pH 2.5 technique (see above).

Results

Loss of staining in "A" and/or "B", by comparison with "C" is due to the removal of sialic acid. Loss of staining in slides A, B, and C by comparison with D indicates non-specific removal of a soluble entity.

9. Acid Hydrolysis

This method, combined with pretreatment in KOH (see above) is used routinely in our laboratory.

Method

1. Bring two sections to water.
2. Treat one section in 0.1 N H_2SO_4 at 80 °C for 1 hour and one in H_2O at 80 °C for 1 hour.
3. Rinse in distilled water.
4. Stain sections by the Alcian blue at pH 2.5 technique (see above).
5. Dehydrate, clear and mount in resinous mountant.

Result

Removal of sialic acid is shown by loss of Alcian blue staining in the acid treated section as compared to that treated in water.

Acknowledgements

It is our pleasure to acknowledge the superior technical assistance we have received during the period covered, and to thank Mrs. Diane Lane for secretarial assistance in the preparation of this paper. We also thank the Canadian Cancer Society and the Medical Research Council of Canada for generous financial support.

Bibliography

Ashbel, R., Seligman, A. M., 1949: Endocrinology **44**, 563–583.
Barger, J. D., De Lamater, E. D., 1948: Science **108**, 121.
Buscher, H.-P., Casals-Stenzel, J., Schauer, R., 1974: Eur. J. Biochem. **50**, 71—82.
Casals-Stenzel, J., Buscher, H.-P., Schauer, R., 1975: Anal. Biochem. **65**, 507—524.
Cooper, J. H., Durning, R. G., 1981: J. Histochem. Cytochem. **29**, 1445–1447.
Culling, C. F. A., 1974: A Handbook of Histopathological and Histochemical Techniques. 3rd ed. London: Butterworth.
— Reid, P. E., 1977: Histochem. J. **9**, 781—785.
— — 1980: J. Microscopy **119**, 415—425.
— — Dunn, W. L., 1971: J. Histochem. Cytochem. **19**, 654—662.
— — — Clay, M. G., 1974a: J. Histochem. Cytochem. **22**, 826–831.

Culling, C. F. A., Reid, P. E., Dunn, W. L., 1974b: Stain Technology **49**, 317—318.
— — Burton, J. D., Dunn, W. L., 1975: J. Clin. Path. **28**, 656—658.
— — Dunn, W. L., 1976: J. Histochem. Cytochem. **24**, 1225—1230.
— — Worth, A. J., Dunn, W. L., 1977a: J. Clin. Path. **30**, 1056—1062.
— — Dunn, W. L., 1977b: J. Clin. Path. **31**, 1063—1067.
— — — 1979: J. Clin. Path. **32**, 1272—1277.
— — Worth, A. J., 1981: In: Meth. Achiev. Exp. Pathol. (Jasmin, G., Cantin, M., eds.), Vol. 10, pp. 73—100. Basel: S. Karger.
Dawson, P. A., Patel, J., Filipe, M. I., 1978: Histochem. J. **10**, 559—572.
Drzeniek, R., 1973: Histochem. J. **5**, 271—290.
Fenger, C., Filipe, M. I., 1977: Acta Path. Microbiol. Scand. **A 85**, 273—285.
Filipe, M. I., 1979: Invest. Cell Path. **2**, 195—216.
— Branfoot, A. C., 1976: In: Current Topics in Pathology (Morson, B. C., ed.), pp. 143—178. Berlin-Heidelberg-New York: Springer.
— Cooke, K. B., 1974: J. Clin. Path. **27**, 315—318.
— Fenger, C., 1979: Histochem. J. **11**, 277—287.
Geyer, G., 1962: Acta Histochem. **14**, 284—296.
Haverkamp, J., Schauer, R., Wember, M., Kamerling, J. P., Vliegenthart, J. F. G., 1975: Z. Physiol. Chem. **356**, 1575—1583.
Kamerling, J. P., Vliegenthart, J. F. G., Versluis, C., Schauer, R., 1975: Carbohyd. Res. **41**, 7—17.
Kantor, T. G., Schubert, M., 1957: J. Amer. Chem. Soc. **79**, 152—153.
Kelly, J. W., Bloom, C. D., Scott, J. E., 1963: J. Histochem. Cytochem. **11**, 791—798.
Korhonen, L. K., 1967: Acta Histochem. **26**, 80—86.
Lev, R., Spicer, S. S., 1964: J. Histochem. Cytochem. **12**, 309.
Lillie, R. D., Pizzolato, P., 1972: Stain Technology **47**, 13—16.
— Jirge, S. K., 1975: Histochem. **41**, 249—256.
Neuberger, A., Ratcliffe, W. A., 1972: Biochem. J. **129**, 683—693.
— — 1973: Biochem. J. **133**, 623—628.
Quintarelli, G., 1961: J. Histochem. Cytochem. **9**, 176—183.
— 1963: Ann. N.Y. Acad. Sci. **106**, 339—363.
— Scott, J. E., Dellovo, M. C., 1964: Histochemie **4**, 99—112.
Reid, P. E., Culling, C. F. A., 1980: J. Histotechnol. **3**, 82—90.
— — Dunn, W. L., 1973: J. Histochem. Cytochem. **21**, 473—482.
— — — 1974: J. Histochem. Cytochem. **22**, 986—991.
— — Tsang, Wai-Chiu, Ramey, C. W., Clay, M. G., 1975a: Can. J. Biochem. **53**, 388—391.
— — Dunn, W. L., Clay, M. G., 1975b: Can. J. Biochem. **53**, 1328—1332.
— — — — 1976: Histochemistry **46**, 203—207.
— — Ramey, C. W., Dunn, W. L., Clay, M. G., 1977: Can. J. Biochem. **55**, 493—503.
— — Dunn, W. L., 1978a: J. Histochem. Cytochem. **26**, 187—192.
— — — Clay, M. G., Ramey, C. W., 1978b: J. Histochem. Cytochem. **26**, 1033—1041.
— — Ramey, C. W., Dunn, W. L., Magil, A. B., Clay, M. G., 1980: J. Histochem. Cytochem. **28**, 217—222.
Roberts, G. P., 1977: Histochem. J. **9**, 97—102.
Rogers, C. M., Cooke, K. B., Filipe, M. I., 1978: Gut **19**, 587—592.
Schauer, R., Faillard, H., 1968: Z. Physiol. Chem. **349**, 961—968.
Schmitz-Moorman, P., 1969: Histochemie **20**, 78—86.
Scott, J. E., Dorling, J., 1965: Histochemie **5**, 221—233.
— Quintarelli, G., Dellovo, M. C., 1964: Histochemie **4**, 73—85.
Sharon, N., 1977: Scient. American **236**, 108—119.

Spicer, S. S., 1960: J. Histochem. Cytochem. **8**, 18—36.
— 1961: Amer. J. Clin. Path. **36**, 393—407.
— 1965: J. Histochem. Cytochem. **13**, 211—234.
— Meyer, D. B., 1960: Amer. J. Clin. Path. **33**, 453—460.
— Warren, L., 1960: J. Histochem. Cytochem. **8**, 135—137.
Sorvari, T. E., Lauren, P. A., 1973: Histochem. J. **5**, 405—412.
— Stoward, P. J., 1970: Histochemie **24**, 106—113.
Stoward, P. J., 1967: J. Roy. Micros. Soc. **87**, 77—103.
— 1968: J. Roy. Micros. Soc. **88**, 119—131.
— Spicer, S. S., Miller, R. L., 1980: J. Histochem. Cytochem. **28**, 979—990.
Terner, J. Y., 1964: J. Histochem. Cytochem. **12**, 504—511.
Van Duijn, P., 1956: J. Histochem. Cytochem. **4**, 55 – 62.
Veh, R. W., Corfield, A. P., Schauer, R., Andres, K. H., 1979: In: Glycoconjugates (Schauer, R., Boer, P., Buddecke, E., Kramer, M. F., Vliegenthart, J. F. G., Wiegandt, H., eds.), pp. 193—194. Stuttgart: G. Thieme.
Vilter, V., 1968: Ann. Histochim. **13**, 205—220.
Weber, P., Harrison, F. W., Hof, L., 1975: Histochemie **45**, 271—277.
Wood, W. S., Culling, C. F. A., 1975: Arch. Path. **99**, 442—445.

I. Metabolism of Sialic Acids

ANTHONY P. CORFIELD and ROLAND SCHAUER

Biochemisches Institut, Christian-Albrechts-Universität, Kiel, Federal Republic of Germany

With 10 Figures

Contents

I. Introduction

The wide occurrence of sialic acids in nature is an indication of their great biological importance. As is described in other chapters in this book, the identification of the sialic acids was pioneered in several laboratories, while the biosynthetic mechanism of sialic acid formation was elucidated essentially by the work of ROSEMAN's and WARREN's groups in the United States (see ROSEMAN 1962, WARREN 1972, SCHACHTER and RODÉN 1973, MCGUIRE 1976, SCHACHTER 1978).

The initial phase was involved with the identification of pathways and the detection of enzymes and intermediates, and included the use of radioactive precursor studies designed to show intermediate pools and end-products of metabolism. It was found that D-glucosamine (GlcN) was a precursor of sialic acid (DELGIACCO and MALEY 1964, MOLNAR *et al.* 1965). Once the metabolic pathway of sialic acid biosynthesis was known and the nucleotide sugar transfer via CMP-N-acetylneuraminic acid (CMP-Neu5Ac) determined, a wave of interest in sialyltransfer reactions led to a rapid expansion in the number of different classes of acceptor molecules identified. This was largely investigated with isolated desialylated glycoconjugates, but experiments with radioactive CMP-Neu5Ac and endogeneous acceptors were also part of the search for sialic acid acceptor molecules. Experiments with radioactive precursors have utilized the knowledge gained in this early period to obtain and observe radio-labelled complex carbohydrates. Further to the anabolic metabolism, interest in sialidases has grown continually and has been boosted by the finding of a deficiency of this enzyme in storage diseases (chapter K, CORFIELD et al. 1981 a). It is primarily the sialidases that have been studied, only occasional interest has been shown in the acylneuraminate pyruvate-lyase (aldolase) and CMP-Neu5Ac hydrolase activities.

II. *De novo* Biosynthesis of Sialic Acids

1. Introduction

The biosynthesis of sialic acids begins with D-glucose (Glc), the common precursor of all monosaccharides in complex carbohydrate metabolism. The complete pathway provides both N-acetylhexosamines and sialic acids for biosynthesis and can be divided up into two halves with UDP-N-acetyl-D-glucosamine (UDP-GlcNAc) as the central metabolite. The individual steps are presented in the schemes in Figs. 1 and 2, and a summary of the regulation on the complete pathways is shown in Fig. 9.

2. From Glucose to UDP-N-Acetylhexosamine

The initiation of sialic acid biosynthesis is the conversion of D-fructose-6-phosphate (Fru-6-P) from the hexose monophosphate pool to the amino sugar phosphate, D-glucosamine-6-phosphate (GlcN-6-P). The conversion of Fru-6-P to GlcN-6-P is the first step in one of several, quantitatively minor, but vital pathways leading to the monosaccharide components found in complex carbohydrates. The initial observations by LELOIR and CARDINI (1953) that hexose phosphate could be aminated, with glutamine serving as the amino donor, were made with *Neurospora crassa*, and the substrates for this enzyme were eventually identified as Fru-6-P and glutamine by GOSH et al. (1960). The enzyme is subject to a feedback inhibition by the product UDP-GlcNAc (KORNFELD et al. 1964, Fig. 1 and 9). This feedback phenomenon is repeated in the second half of the pathway leading to CMP-sialic acid. The importance of the GlcN-6-P synthase is reflected in the existence of this feedback control in all mammalian sources so far assayed (WINTERBURN and PHELPS 1973). Comparison of this enzyme from different tissues reveals a further refinement indicative of its regulatory role. The susceptibility of the enzyme to substrate inhibition by L-glutamine and for Fru-6-P depends on the tissue under study and may in certain cases be independent of the feedback control by UDP-GlcNAc (ELLIS and SOMMAR 1971, 1972, WINTERBURN and PHELPS 1973). The nature of the feedback control by UDP-GlcNAc has been carefully investigated in rat liver (KORNFELD 1967, KIKUCHI et al. 1971 a, MIYAGI and TSUIKI 1971, WINTERBURN and PHELPS 1971 a, b) and the detection of effective modulation of the feedback control of UDP-GlcNAc by glucose-6-phosphate (Glc-6-P), AMP and UTP discussed relative to the physiological role of the enzyme (WINTERBURN and PHELPS 1970, 1971 c). The results of these investigations indicated that the enzyme in rat liver is normally in an inhibited state, only 10% of the maximal activity being expressed, agreeing with the flux rate of 0.1 μmole hexosamine per hour and per g wet weight of rat liver through the pathway, determined with

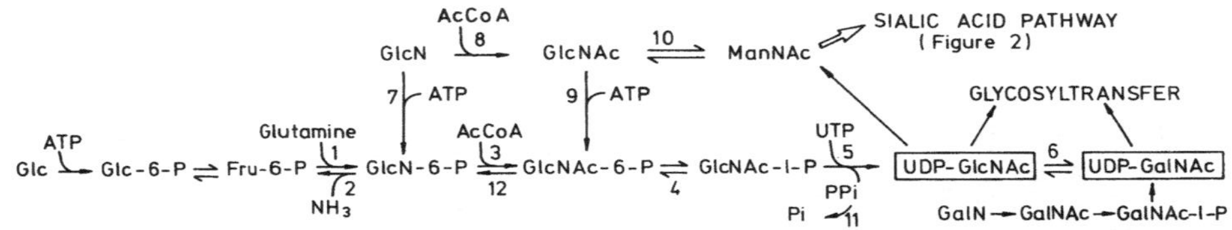

No.	Enzyme	EC number	Substrates	Effectors	Regulatory enzyme	Location	References
1	glucosamine-6-phosphate synthase	5.3.1.19	Fru-6-P, glutamine	Glc-6-P, AMP, UTP UDP-GlcNAc	+	cytosol	Winterburn and Phelps 1971 a
2	glucosamine-6-phosphate deaminase	5.3.1.10	GlcN-6-P	GlcNAc-6-P	?	cytosol, membranes	Comb and Roseman 1958 b
3	glucosamine-6-phosphate N-acetyltransferase	2.3.1.4	GlcN-6-P, AcCoA	—	(+)	cytosol	Corfield 1973
4	N-acetylglucosamine-6-phosphate mutase	2.7.5.2	GlcNAc-6-P	—	—	cytosol	Reissig and Leloir 1966
5	UDP-N-acetylglucosamine pyrophosphorylase	2.7.7.23	GlcNAc-1-P, UTP	?	(+)	cytosol (?)	Strominger and Smith 1959
6	UDP-N-acetylglucosamine 4-epimerase	5.1.3.7	UDP-GlcNAc	—	—	cytosol (?)	Jacobson and Davidson 1963
7	hexokinase/glucokinase	2.7.1.1/2.7.1.2	GlcN, ATP	—	—	cytosol	Spiro 1959
8	glucosamine N-acetyl-transferase	2.3.1.3	GlcN, AcCoA	?	—	cytosol	Chou and Soodak 1952
9	N-acetylglucosamine kinase	2.7.1.59	GlcNAc, ATP	UDP-GlcNAc GlcNAc-6-P	+	cytosol	Datta 1970 a
10	N-acetylglucosamine 2-epimerase	5.1.3.8	GlcNAc	?	?	?	Ghosh and Roseman 1965
11	pyrophosphate	3.6.1.1	PPi	—	(+)	cytosol	Kornfeld *et al.* 1964
12	N-acetylglucosamine-6-phosphate deacetylase	3.5.1.25	GlcNAc-6-P	?	—	?	Matsushita and Takagi 1966

Fig. 1. Mammalian hexosamine metabolism. The *de novo* biosynthesis of UDP-N-acetyl-D-glucosamine (UDP-GlcNAc) from D-glucose (Glc) is shown with known enzymic steps indicated in the table. Note the link with Fig. 2.

radioactive tracer methods by SPIRO (1959). GHOSH *et al.* (1960) showed that the GlcN-6-P-synthase was potently inhibited by 6-diazo-5-oxo-L-norleucine (DON), a glutamine antagonist. The N-acetylated form led to a dramatic depression of the UDP-GlcNAc concentration in rat liver (BATES *et al.* 1966) due to the irreversible inactivation of the synthase. The inactivation could be prevented by addition of glutamine or UDP-GlcNAc and was most efficiently prevented in the presence of both compounds, binding at two enzyme sites being proposed for N-acetyl-DON (BATES and HANDSCHUMACHER 1969).

This initial enzyme step is essentially irreversible. Reversal of the reaction is catalyzed by GlcN-6-P deaminase (Fig. 1) which occurs in bacterial, insect and mammalian tissues, and which leads to the breakdown of GlcN-6-P to Fru-6-P and NH_3. At this point it is important to note that free GlcN enters the pathway via non-specific phosphorylation by hexokinase (McGARRAHAN and MALEY 1962) or after acetylation and phosphorylation by a specific kinase (Fig. 1). Entry via hexokinase phosphorylation leads to GlcN-6-P formation independent of the UDP-GlcNAc feedback control and is widely used in metabolic studies with radioactive GlcN. The absence of a feedback control on this entry into the pathway has been demonstrated on administration of larger amounts of GlcN which leads to dose-dependent depletion of uridine nucleotides and "trapping" as the UDP-N-acetylhexosamine derivatives (BEKESI and WINZLER 1969, DECKER and KEPPLER 1974).

Following the formation of GlcN-6-P is an acetylation reaction which is essentially irreversible (KENT 1970, CORFIELD 1973). GlcN-6-P is acetylated to GlcNAc-6-P, and only a small pool of GlcN-6-P is found e.g. in liver and other tissues (McGARRAHAN and MALEY 1962, MOLNAR *et al.* 1964, TESORIÈRE *et al.* 1971). A deficiency in the production of this enzyme has been detected in mouse fibroblast mutants (NEUFELD and PASTAN 1978), leading to an increase in the GlcN-6-P pool and an absence of N-acetylated hexosamine phosphates after labelling with [^{14}C]GlcN (POUYSSÉGUR and PASTAN 1977). The deficiency could be alleviated by administration of GlcNAc, suggesting that phosphorylation of this intermediate was still functioning, and pinpointing a possible entry point of hexosamine into the pathway, to account for the low but significant levels found in the cells on [^{14}C]GlcN-labelling.

This acetylation reaction is further of importance, as the influence of the free amino group in GlcN-6-P is lost (KENT 1970) and as no evidence for the formation of N-glycolylglucosamine or Neu5Gc from glycolyl-CoA could be found in porcine intestine (KENT and ALLEN 1968). The results of these experiments suggest that the acetylation of GlcN-6-P is the favoured pathway to GlcNAc-6-P formation. However, GlcNAc may enter the pathway via a specific kinase. The GlcNAc kinase has been detected in yeast (BHATTACHARYA *et al.* 1974), bacteria and in porcine spleen (DATTA 1970a), bovine thyroid (TRUJILLO *et al.* 1971), human gastric mucosa (GINDZIENSKI *et al.* 1974) and liver (ALLEN and WALKER 1980a, b). The enzyme has been identified as a regulator in hexosamine metabolism on the basis of its feedback inhibition by UDP-GlcNAc (DATTA 1971). The regulation of the GlcNAc-6-P pool is directly related to this kinase reaction which is inhibited by higher concentrations of GlcNAc-6-P, and to the GlcN-6-P deaminase reaction which requires GlcNAc-6-P in a catalytic function (COMB and

ROSEMAN 1958a). In view of the poor utilization of exogeneous GlcNAc *in vivo* (ROBINSON 1968), probably due to the poor uptake of this monosaccharide into the cell, the regulatory function of the GlcNAc kinase may be related to intracellular recycling of N-acetylhexosamine. This is supported by the absence of free GlcN in the cell.

The pool of GlcNAc-6-P in the cell is in equilibrium with GlcNAc-1-P, due to the presence of a specific mutase (FERNANDEZ-SØRENSEN and CARLSON 1971, Fig. 1). This reaction is in common with other sugar phosphates prior to activation to their respective nucleoside diphosphate derivatives. An equivalent reaction does not occur for sialic acid (Neu5Ac) activation (section II.3).

UDP-GlcNAc is formed as a result of a specific pyrophosphorylase utilizing GlcNAc-1-P and UTP as substrates (STROMINGER and SMITH 1959). The control of this area of the hexosamine pathway has not yet been resolved. The equilibrium of the N-acetylhexosamine phosphate mutase reaction lies in favour of GlcNAc-6-P (REISSIG and LELOIR 1966), and the pyrophosphorylase equilibrium lies in the direction of nucleotide sugar breakdown (PHELPS *et al.* 1975). UDP-GlcNAc formation through the pyrophosphorylase reaction may be influenced by the action of pyrophosphatase removing pyrophosphate, the second product of the pyrophosphorylase reaction, and allowing a flow in a biosynthetic direction. UDP-GlcNAc is a vital metabolite in hexosamine and sialic acid metabolism (see Figs. 2 and 9) influencing GlcN-6-P synthetase, GlcNAc kinase and CMP-Neu5Ac hydrolase activities by feedback inhibition. It serves as a glycosyl donor and as the initial substrate in the biosynthesis of sialic acids.

Epimerization to UDP-GalNAc occurs via a specific 4-epimerase (JACOBSON and DAVIDSON 1963). This enzyme reaction is probably instrumental in the conversion of exogeneous GalNAc into GlcNAc and subsequently sialic acid, as has been found in several studies (MALEY *et al.* 1968, MACNICOLL *et al.* 1978).

3. Formation of Neu5Ac

The second half of the hexosamine pathway leads to the formation of free sialic acid. The initial reaction in this pathway is UDP-GlcNAc 2-epimerase (Fig. 2) catalyzing the reaction: UDP-GlcNAc → [2-acetamidoglucal] → ManNAc. The enzyme, originally discovered in rat liver (CARDINI and LELOIR 1957), catalyzes the release of UDP and epimerization of GlcNAc to N-acetyl-D-mannosamine (ManNAc) (COMB and ROSEMAN 1958b). The product is a monosaccharide and not a nucleotide sugar and is the first of several reactions unique to monosaccharide biosynthesis occurring in the formation of sialic acid. The mechanism of this enzyme reaction has received considerable attention. An involvement of UDP-ManNAc was suggested by SPIVAK and ROSEMAN (1966). However, the work of SALO and FLETCHER (1970a, b) led to an alternative hypothesis involving a 2-acetamidoglucal intermediate. This compound was sought by SOMMAR and ELLIS (1972a, b) who also supported a two-step mechanism via this intermediate. They concluded that hydrolysis of 2-acetamidoglucal to ManNAc was rapid, as none could be detected in incubations with the 2-epimerase. A further line of evidence for this mechanism has come from studies carried out with a sialuria patient (STRECKER and MONTREUIL 1971). This

Figure 1

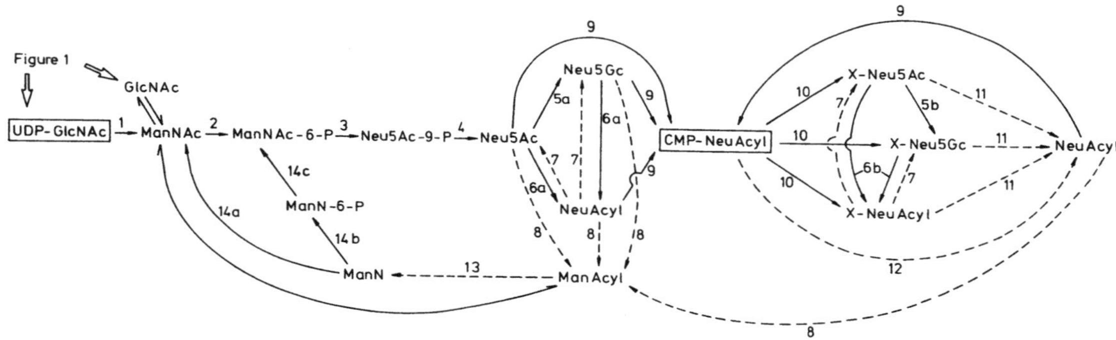

No.	Enzyme	EC number	Substrates	Effectors	Regulatory enzyme	Location	References
1	UDP-N-acetylglucosamine 2-epimerase	5.1.3.14	UDP-GlcNAc	CMP-Neu5Ac	+	cytosol (?)	Spivak and Roseman 1966
2	N-acetylmannosamine kinase	2.7.1.60	ManNAc, ATP	?	(+)	cytosol	Warren and Felsenfeld 1962
3	N-acetylneuraminate-9-phosphate synthase	4.1.3.20	ManNAc-6-P, PEP	?	—	cytosol	Watson *et al.* 1966
4	N-acetylneuraminate-9-phosphate phosphatase	3.1.3.29	Neu5Ac9P	?	—	cytosol	Warren and Felsenfeld 1962
5	N-acetylneuraminate monooxygenase	1.14.99.18	a) free Neu5Ac b) bound Neu5Ac H donor	?	?	a) cytosol b) Golgi	Schauer 1978 b
6	acylneuraminate 4 or 7(9) O-acetyltransferase	2.3.1.44 2.3.1.45	a) free NeuAcyl b) bound NeuAcyl AcCoA	?	?	a) cytosol b) Golgi	Schauer 1978 b

	acylneuraminate esterase	3.1.1.?	free or bound 4,7,8,9-NeuAcyl	?	—	membranes (?)	—
8	acylneuraminate pyruvate-lyase	4.1.3.3	NeuAcyl	—	—	?	Brunetti *et al.* 1962
9	CMP-acylneuraminate synthase	2.7.7.43	NeuAcyl, CTP	CMP, PPi?	+	nucleus	Kean 1970
10	sialyltransferase	2.4.99.1	CMP-NeuAcyl, acceptor	CMP, +?	+	Golgi and plasma membranes, serum	Schachter and Roseman 1980
11	sialidase	3.2.1.18	sialyl-complex carbohydrate	(—)?	+	plasma, microsomal, lysomal membranes	Corfield *et al.* 1981 a
12	CMP-acylneuraminate hydrolase	3.1.4.40	CMP-NeuAcyl	UDP-GlcNAc	+	plasma membrane, serum	Kean and Bighouse 1974
13	N-deacetylase (specific?)	3.5.1.?	ManNAc, ManNGc	?	—	?	—
14 abc	phosphorylation, acetylation (non-specific?)	—	ManN, ManN-6-P AcCoA, ATP	?	—	cytosol	Gan 1975

Fig. 2. Mammalian sialic acid metabolism. The formation and interconversions of the sialic acids continuing Fig. 1 from UDP-GlcNAc and GlcNAc. Acylneuraminic acids (NeuAcyl) include Neu5Ac, Neu5Gc and O-acyl derivatives. CMP-NeuAcyl and the corresponding acylmannosamines (ManAcyl) are indicated in the same way. ——→ Anabolic reactions; — — → catabolic reactions; X-, acceptor complex carbohydrate for sialyltransfer.

patient excretes daily gramme quantities of Neu5Ac, ManNAc, and GlcNAc and in addition 2-deoxy-2,3-dehydro-N-acetylneuraminic acid (Neu5Ac2en) and the 2-acetamidoglucal (KAMERLING *et al.* 1979). A possible defect in metabolism at the UDP-GlcNAc 2-epimerase level has been proposed and discussed (KAMERLING *et al.* 1979, CORFIELD and SCHAUER 1979) and will be considered in section VI.3.d).

The UDP-GlcNAc 2-epimerase is subject to feedback inhibition by CMP-Neu5Ac, the end product in the pathway (KORNFELD *et al.* 1964, KIKUCHI and TSUIKI 1973) (Fig. 9). These authors working with rat liver have used UDP and dithiothreitol to stabilize the labile enzyme and were able to achieve a 500-fold purification. These studies showed that for rat liver the CMP-Neu5Ac inhibition was very sensitive and physiologically significant. A HILL coefficient of 5.7 was measured suggesting the presence of six binding sites for this inhibitor. The binding of UDP-GlcNAc to the enzyme exhibits negative cooperativity such that 2-epimerase activity is augmented at higher UDP-GlcNAc concentrations which lie within physiological levels. Further to its regulation at a feedback level a *de novo* enzyme synthesis control exists dictating the level of the enzyme during development (OKUBO *et al.* 1976), injury and partial hepatectomy (OKUBO *et al.* 1977, OKAMOTO and AKAMATSU 1980). This control appears to mirror that found for glucosamine-phosphate synthase under identical conditions.

The formation of ManNAc is an important junction in hexosamine and sialic acid biosynthesis (Fig. 2), as this metabolite is a substrate for two enzymes which may function in opposing directions. The first of these enzymes is GlcNAc 2-epimerase (Fig. 2) which catalyzes the epimerization of free GlcNAc to ManNAc (GHOSH and ROSEMAN 1965) and has an allosteric site for ATP (DATTA 1970b). The enzyme was found to occur in a number of tissues (DATTA 1970b) and may be responsible for sialic acid formation from GlcNAc via ManNAc-6-P rather than GlcNAc-6-P when the GlcNAc kinase is inhibited by UDP-GlcNAc (ELLIS *et al.* 1972). The enzyme is inducible in bacteria and yeast and functions catabolically (BISWAS *et al.* 1979). The direction of biosynthesis through this epimerase has not been studied in any detail. An equilibrium concentration ratio of 4:1 for GlcNAc : ManNAc has been reported (GHOSH and ROSEMAN 1965, DATTA 1970b).

Evidence exists that the direction of biosynthesis can be altered depending on (i) the tissue itself (ELLIS *et al.* 1972, KIKUCHI and TSUIKI 1979); (ii) the conditions in a tissue (e.g. in liver homogenates at pH 7.4 GlcNAc was converted to sialic acid via UDP-GlcNAc, however, if 10 mM EDTA was included and the pH lowered to 6.3, a direct epimerization to ManNAc could be shown (HULTSCH *et al.* 1972)); or (iii) the metabolic state of the tissue e.g. stimulation of GlcNAc utilization by fructose-1,6-diphosphate and phosphoenol-pyruvate (PEP), the former leading to glycolytic products and the latter stimulating sialic acid production.

The second step in ManNAc utilization is phosphorylation, and a specific ManNAc kinase has been detected in a range of mammals (GHOSH and ROSEMAN 1961, WARREN and FELSENFELD 1962, KUNDIG *et al.* 1966). The rat liver enzyme uses both N-acetyl- and N-glycolylmannosamine as substrate (KUNDIG *et al.* 1966). This phosphorylation step may occupy a regulatory position in the biosynthesis of Neu5Ac. Thus, a reduction of 36% relative to normal activities was found in regenerating livers (OKAMOTO and AKAMATSU 1980), a reduction of 38% occurred in diabetic rat livers (MALEY *et al.* 1971), and very low or no activity was found or suggested for several hepatomas (HARMS *et al.* 1973, KIKUCHI and TSUIKI 1979) and tumours (KIKUCHI *et al.* 1971 b). In some other tissues the enzyme could not be detected, although its activity was suspected due to the occurrence of Neu5Ac formation (KUNDIG *et al.* 1966). In a tracheal epithelial system ELLIS *et al.* (1972) demonstrated that Neu5Ac formation via ManNAc-6-P took place and that the enzyme activity was very labile in this tissue.

An alternative mode of formation of ManNAc-6-P is by epimerization of GlcNAc-6-P. An enzyme catalyzing this reaction has been detected in several bacteria (GHOSH and ROSEMAN 1965), but no evidence for such an epimerization has been reported for animal tissues.

The anabolic route to Neu5Ac proceeds via a N-acetyl-D-neuraminyl-9-phosphate (Neu5Ac9P) synthase which catalyzes Neu5Ac9P formation from ManNAc-6-P and PEP. The enzyme has been found in rat liver and bovine submandibular glands (ROSEMAN *et al.* 1961, WARREN and FELSENFELD 1961 a, b, 1962).

Studies on the influence of hepatic regeneration (OKAMOTO and AKAMATSU 1980), metabolic stimulation by PEP (KIKUCHI and TSUIKI 1979) and malignant transformation in various tumours (KIKUCHI *et al.* 1971 b) on the activity of the synthase revealed significant changes only in the case of PEP stimulation. Here, GlcNAc utilization (relative to normal liver) was channelled in the direction of Neu5Ac formation, possibly through the stimulation of synthase activity due to provision of substrate (KIKUCHI and TSUIKI 1979).

The formation of Neu5Ac is completed by the action of a specific phosphatase acting on Neu5Ac9P. Although Neu5Ac9P can be dephosphorylated by non-specific phosphatases, a highly specific enzyme could be demonstrated in rat liver (WARREN and FELSENFELD 1962) and in erythrocyte lysates (JOURDIAN *et al.* 1964), which acted on the 9-phosphate residues of Neu5Ac and Neu5Gc.

Other modes of Neu5Ac formation include the action of Neu5Ac pyruvate-lyase in animals and bacteria, but the equilibrium of this reaction lies in the catabolic direction, and no activity was found in tissues secreting sialic acid-containing polymers (BRUNETTI *et al.* 1962). The catabolic function of the lyase is discussed below (section IV.4). Furthermore, ManNAc can be condensed with PEP to give Neu5Ac and phosphate. This reaction has been found in *Neisseria meningitidis* (BLACKLOW and WARREN 1962, BROSSMER *et al.* 1980) but never in animal tissue.

4. Activation and Modification of Neu5Ac

After the formation of Neu5Ac a variety of reactions occur leading to modified and activated sialic acids. The occurrence of the different sialic acids is documented in chapter B. The principal modifications are enzymic conversion of the N-acetyl group to an N-glycolyl moiety and the O-acylation at positions 4, 7, 8, and 9 of the neuraminic acid molecule (SCHAUER 1978).

The hydroxylation of Neu5Ac to yield Neu5Gc is catalyzed by a mono-oxygenase specific for Neu5Ac, which requires molecular oxygen and a hydrogen donor ($NADPH_2$ or ascorbate). Fe^{2+} ions are required for maximum activity. The enzyme was found to exist in both soluble and membrane-bound forms, and to act on free and glycosidically bound Neu5Ac (BUSCHER *et al.* 1977, SCHAUER 1978b). The particulate enzyme in porcine submandibular gland tissue was predominantly Golgi membrane-bound activity (BUSCHER *et al.* 1977). Approximately 50% of Neu5Gc found in the completed porcine submandibular gland glycoprotein is hydroxylated in the cytosol, probably by the soluble form of the enzyme, before conversion to CMP-Neu5Gc and transfer to the nascent glycoprotein molecules (SCHAUER and WEMBER 1970, BUSCHER *et al.* 1977, SCHAUER 1978b). The other 50% are formed by hydroxylation of glycosidically bound Neu5Ac, presumably by the Golgi membrane enzyme. Complete cell systems yielded a hydroxylation rate decreasing in the order: free Neu5Ac > nascent glycoprotein-bound Neu5Ac in membranes > soluble cytosolic glycoprotein- or glycopeptide-bound Neu5Ac. The Neu5Gc molecules formed may then be activated to the CMP-glycoside, or alternatively modified further by O-acylation. The enzyme has been implicated in many tissues containing Neu5Gc, but has been assayed only in porcine submandibular glands, liver and serum and in bovine submandibular glands.

The enzymic transfer of O-acetyl groups to positions 4, 7, 8, and 9 has been detected in bovine and equine submandibular glands (see SCHAUER 1978b for a review). The acetyl donor is acetyl coenzyme A (AcCoA), and no other cofactors could be detected. Using radiolabelling techniques it was demonstrated that the spheroplast membranes of *Meningococcus* group C transfer acetyl groups from AcCoA to sialic acids of the polysaccharide synthesized *in vitro* (VANN *et al.* 1978).

As with the monooxygenase, the O-acetyltransferase exists in a soluble, cytosolic and a membrane-bound form. It is envisaged that the soluble enzyme is predominantly responsible for O-acetylation of free Neu5Ac and Neu5Gc, while nascent soluble glycoprotein-bound Neu5Ac or Neu5Gc is O-acetylated by a Golgi membrane-containing fraction from bovine submandibular glands (CORFIELD *et al.* 1976, SCHAUER 1978b, unpublished results). The nature of the O-acetyl transfer is now believed to be due to a single enzyme transfering to position 7 of the sialic acid molecule, as evidence from t.l.c. and n.m.r.-spectroscopy for a non-reversible O-acetyl migration in the sequence 7 > 8 > 9 has been obtained (unpublished results). The formation of 9-mono-O-acetyl and di- or tri-O-acetyl derivatives of sialic acid with ester groups at C-7, C-8, and C-9 is probably due to such O-acetyl migration. The O-acetylation of position 4 depends on a specific enzyme which has been detected in equine submandibular gland (SCHAUER *et al.* 1968, SCHAUER 1971).

Studies with bovine submandibular gland membranes and tissue slices have shown that the order of enzymic activity found for the monooxygenase also applies here. Thus, the sequence: free sialic acid > > > membrane-bound nascent glycoprotein-bound sialic acid > soluble cytosolic glycoprotein- or glycopeptide-bound sialic acid was found showing a very high incorporation rate of O-acetyl groups into the free sialic acid pool (CORFIELD *et al.* 1976, SCHAUER 1978b, unpublished results).

The demonstration of 9-O-lactyl groups in sialic acids from human, bovine and equine tissues (HAVERKAMP *et al.* 1976, SCHAUER *et al.* 1976, REUTER *et al.* 1980) implied the existence of an enzymic transfer or modification. At present no data are available concerning the biosynthesis of the O-lactyl group.

The modifications of sialic acids are important for several reasons: (i) the action of catabolic enzymes, e.g. sialidases and acylneuraminate pyruvate-lyase, may be modified by the presence of N-glycolyl and/or O-acetyl groups. This will be expanded later on in this chapter. (ii) The possibility of modifying sialic acids exists at both monosaccharide and macromolecular levels, thus giving the potential for modification an additional flexibility. (iii) These activities in Golgi membranes suggest that they may be part of a multi-enzyme complex responsible for glycoprotein biosynthesis, being integrated in a membrane system. The presence of O-acetylated sialic acids in the human gastrointestinal tract, saliva, lymphocytes and other tissues (cf. chapter B) implicates similar O-acetylating systems in these tissues, which remain to be studied.

Following modification at the monosaccharide level, activation to the CMP-glycoside takes place. The reaction is catalyzed by CMP-Neu5Ac synthase and the enzyme has been detected in animal and bacterial sources (WARREN 1972). This reaction is unique for the following reasons: (i) The free monosaccharide (Neu5Ac) and not the anomeric phosphate serves as substrate. (ii) The sugar nucleotide is the monophosphate derivative of cytidine and not the usual diphosphate derivatives of uridine (in the case of e.g. Glc, Gal, GlcNAc) or guanosine (Man and Fuc). (iii) The equilibrium of the reaction lies in the direction of synthesis in contrast to other known pyrophosphorylases. (iv) The enzyme has been found in the cell nucleus of spleen, liver, brain, kidney and retina tissues (KEAN 1970, VAN DEN EIJNDEN 1973, GIELEN and HINZEN 1974, COATES *et al.* 1980, SCHAUER *et al.* 1980a).

These data suggest that the enzyme occupies a special role in the regulation of sialic acid metabolism. The high activity associated with the nuclear fraction, readily released on homogenization, has been used as an efficient method of CMP-Neu5Ac preparation with yields of 90% and higher (CORFIELD *et al.* 1979, HAVERKAMP *et al.* 1979a, BEAU and SCHAUER 1980). The high activity of the enzyme is significant in view of the low levels of CMP-Neu5Ac found in different tissues (CORFIELD *et al.* 1976, BUSCHER *et al.* 1977, CAREY and HIRSCHBERG 1979). Thus, a regulation of the activity with regard to the CMP-Neu5Ac pool size appears to exist. A further property of the enzyme from bovine, porcine and equine submandibular glands is the lack of selectivity for the type of sialic acid (i.e. Neu5Ac, Neu5Gc, Neu4,5Ac$_2$, and Neu5,9Ac$_2$; for abbreviations of sialic acids see chapter B) (SCHAUER and WEMBER 1973). Frog liver was active only with Neu5Ac and Neu5Gc, the various O-acetylated sialic acids not being substrates (SCHAUER *et al.* 1980a). The enzymes from rat liver, *Neisseria* and *E. coli* strains utilize 9-azido-9-deoxy-N-acetyl-D-neuraminic acid (Neu5Ac9N$_3$) in addition to Neu5Ac (BROSSMER *et al.* 1979, 1980). CMP-Neu5Ac synthase is inhibited competitively by CMP (e.g. KEAN 1970, SCHAUER *et al.* 1980a), and this may be of physiological significance when considered together with the topographical location (nucleus) of the enzyme. The central position of CMP-sialic acids in sialic acid metabolism is indicated in Fig. 2.

III. Sialyltransfer

1. Introduction

An area of great interest in the field of sialic acid metabolism has been and remains that of sialyltransfer. The literature on this single aspect of sialic acid metabolism is very large (e.g. ROSEMAN 1970, SPIRO 1970, SCHACHTER and RODÉN 1973, SHARON 1975, SCHACHTER 1978, KÖTTGEN *et al.* 1979, SCHACHTER and ROSEMAN 1980, BEYER *et al.* 1981). The sialyltransferases belong to the group of glycosyltransferases and have a number of characteristics in common with the other enzymes in this group: (i) They catalyze the transfer of a monosaccharide from a nucleotide sugar to an acceptor molecule, which may be from any of the groups of complex carbohydrates. (ii) Addition is to the non-reducing end of growing oligosaccharide chains. (iii) A membrane location is suggested for the biosynthesis of complex carbohydrates. (iv) High specificity is shown for the monosaccharide transferred, the nucleotide base, and the sugar (or other) residue at the non-reducing terminal of oligosaccharide chains in acceptor molecules. The sialyltransferases, in addition to these specificities, show characteristics which are not shared by other glycosyltransferases and serve to stress their special role in complex carbohydrate metabolism. These enzymes transfer sialic acids from CMP-β-Neu5Ac to acceptors. Transfer from a nucleoside monophosphate sugar does not occur for other monosaccharides, which involve nucleoside diphosphate sugars (e.g. SCHACHTER and RODÉN 1973, SCHACHTER 1978).

The identification of CMP-NeuAcyl as the sialic acid donor in sialyltransfer reactions was made by COMB *et al.* (1966) in bacterial systems. In mammalian systems isolation of CMP-sialic acid glycosides labelled in the sialic acid moiety has been reported (HARMS *et al.* 1973, CORFIELD *ei al.* 1976, BRILES *et al.* 1977). Conclusive identification of cytosine as the nucleoside in such experiments has been presented by CAREY and HIRSCHBERG (1979) in mouse liver and kidney. The transfer reaction can be formulated as follows:

$$\text{CMP-}\beta\text{-Neu5Ac} + \text{R-OH} \rightarrow \text{Neu5Ac-}\alpha\text{-O-R} + \text{CMP}.$$

The CMP-glycoside contains a β-glycosidic link between the phosphate and the anomeric hydroxyl group at C-2 of the sialic acid. The β-nature of this linkage was originally suggested by COMB *et al.* (1966) and has been confirmed by HAVERKAMP *et al.* (1979 b) using ^{13}C n.m.r.-spectroscopy. This is different to other nucleotide sugars which contain α-linkages (SCHACHTER and RODÉN 1973, SHARON 1975), and is in accordance with the finding of all sialic acid moieties in glycosidic linkages in the α-configuration (chapter B, LEDEEN and YU 1976). This implies that sialyltransfer occurs via a single displacement mechanism with inversion of configuration (SHARON 1975, HAVERKAMP *et al.* 1979 b). A final point of variation with other glycosyltransferases is the apparent lack of requirement of divalent cations (e.g. Mn^{2+}, Mg^{2+}) for activity. Such a requirement appears to be a property of glycosyltransferases with the exception of the sialyltransferases (e.g. SCHACHTER 1978, SCHACHTER and ROSEMAN 1980).

The substrate specificity of sialyltransferases encompasses a wide range of acceptors. According to the one gene—one enzyme hypothesis (SCHACHTER and RODÉN 1973) there should be one sialyltransferase for each type of sialic acid-oligosaccharide linkage. This includes all classes of complex carbohydrates and all types of glycosidic linkage (e.g. α(2-3), (2-4), (2-6), (2-8), and (2-9)). The existence of multiglycosyltransferase complexes was put forward (ROSEMAN 1970) to explain the efficient and accurate biosynthesis of complex carbohydrates. However, direct evidence is still outstanding for such organization. Alternative proposals base on the multiglycosyltransferase complex, where the same glycosyltransferases may be used for the synthesis of similar structures occurring in different classes of complex carbohydrates (RAUVALA and FINNE 1979). The sialyltransferases belong to the glycosyltransferases responsible for the terminal trisaccharide structure Neu5Ac-Gal-GlcNAc in N-glycosidically linked glycoproteins. They do not require a polyprenol intermediate during glycosyltransfer, as is the case for the inner core oligosaccharide in these glycoproteins (STRUCK and LENNARZ 1980). No lipid intermediate involvement could be demonstrated for O-glycosidically linked (serine or threonine to GalNAc) oligosaccharide glycosyltransfer (SCHACHTER 1978, SCHACHTER and ROSEMAN 1980, HANOVER *et al.* 1980). Ganglioside biosynthesis is also believed to proceed without lipid intermediates, although conclusive evidence is still outstanding (DAWSON 1978). In the case of colominic acid biosynthesis a lipid intermediate has been identified, but this remains the only example to date (see section III.8, TROY and MCCLOSKY 1979). The sialic acids occur almost exclusively in terminal positions in the oligosaccharide chains of complex carbohydrates (see chapter B for examples and exceptions). This is therefore one of the terminating signals for biosynthesis (SCHACHTER and RODÉN 1973, CORFIELD and SCHAUER 1979, BEYER *et al.* 1981). The cellular activity of these enzymes has been located predominantly in the smooth Golgi membranes of cells synthesizing complex carbohydrates (SCHACHTER and RODÉN 1973, STURGESS *et al.* 1978), although plasma membrane and soluble activities have been described (see below).

2. Sialyltransfer in Cells and Cell-Free Systems: Cellular Location

a) Intact Cells

Radioactively labelled monosaccharides as precursors of glycosidically bound carbohydrates have been used in tracer experiments in tissues actively secreting glycoproteins (for reviews see SCHACHTER and RODÉN 1973, SCHACHTER 1978, STURGESS *et al.* 1978). In tissues such as liver a larger number of different glycoproteins and other glycoconjugates are synthesized simultaneously, all utilizing the same common precursors as outlined in section II. Furthermore, the choice of precursor is important. As is clear from section II, interconversion of some precursors occurs (SCHACHTER and RODÉN 1973, CORFIELD and SCHAUER 1979), and only identification of the individually labelled monosaccharides from the isolated glycoprotein products will indicate the true rate of incorporation. This is not possible with autoradiographic experiments. An approach to these problems has been the study of individual glycoproteins e.g. thyroglobulin and immunoglobulins (SCHACHTER and RODÉN 1973, SCHACHTER 1978).

α) *N-Glycosidically Linked Oligosaccharide Chains*

The use of [1-^{14}C]glucosamine as a precursor of glycoprotein-bound sialic acid has been demonstrated in liver (e.g. KOHN *et al.* 1962, LAWFORD and SCHACHTER 1966). Radioactivity was incorporated into microsomal glycoprotein (SARCIONE *et al.* 1964), and the sialic acid label could be identified in Golgi fractions (SCHACHTER 1974). However, glucosamine is converted into sialic acid and incorporated into glycoproteins as N-acetylhexosamine and sialic acid. Thus, accurate kinetic analysis required measurement of the individual monosaccharide-specific radioactivities (LAWFORD and SCHACHTER 1966). It was concluded that sequential addition of monosaccharides occurs on subcellular membranes with sialic acid being the final addition in the Golgi membranes (SCHACHTER 1974, 1978). Sialic acid incorporation into thyroid glycoproteins has been studied using N-[^{3}H]acetyl-mannosamine (MONACO *et al.* 1975) and [^{14}C]mannosamine (GAN 1975). The identification of sialyltransfer as the final monosaccharide transfer to nascent thyroglobulin before release from the endoplasmic reticulum could be shown by MONACO *et al.* (1975). The superiority of ManNAc as a sialic acid precursor over GlcNAc could also be demonstrated using this system (MONACO and ROBBINS 1973). Other examples, where similar conclusions could be drawn, were found with lipoprotein (MOOKERJEA and MILLER 1974) and mouse myeloma immunoglobulin light chain (KNOPF *et al.* 1975).

β) *O-Glycosidically Linked Oligosaccharides (Ser/Thr-GalNAc type)*

Much of the pioneering autoradiographic work carried out by LEBLOND and coworkers has established the site of biosynthesis of O-glycosidic oligosaccharide chains in intestinal goblet cells and salivary gland mucus acinar cells (e.g. NEUTRA and LEBLOND 1966). Although sialic acid was not demonstrated autoradiographically, the use of [^{14}C]glucosamine with intact bovine submandibular glands led to the localization of label in smooth endoplasmic reticulum membranes (LAWFORD and SCHACHTER 1967). Porcine and bovine submandibular glands were found to incorporate radioactive acetate into microsomal membrane-bound sialic acids with a peak of specific radioactivity in Golgi-enriched fractions (SCHAUER *et al.* 1974, BUSCHER *et al.* 1977).

Study of N-[^{3}H]acetyl-D-mannosamine incorporation into rat tissues has refined earlier results (BENNETT *et al.* 1981, BENNETT and O'SHAUGHNESSY 1981). The suggestion of Golgi involvement in sialic acid incorporation and a localization detected at the trans-face of the Golgi stack and the adjacent secretory vesicles in hepatocytes was supported. This demonstrated a specific site of sialic acid metabolism within Golgi membranes and led to the suggestion that in hepatocytes incorporation of sialic acids occurs after addition of fucose (BENNETT and O'SHAUGHNESSY 1981). The preferential localization of initial labelling to the trans-face of the Golgi stacks found in hepatocytes was also found in pancreatic acinar and Paneth cells. Other cells such as duodenal villous columnar cells and some kidney proximal tubule cells showed homogenous initial labelling throughout the Golgi stacks (BENNETT *et al.* 1981).

γ) *Glycolipids (Gangliosides)*

Understanding of ganglioside sialic acid biosynthesis has been aided by the use of radioactive ManNAc. This approach has been used to prepare whole rat-brain gangliosides labelled only in the sialic acid moiety (QUARLES and BRADY 1971, RÖSNER *et al.* 1973, CAPUTTO *et al.* 1976, YOHE and ROSENBERG 1977). Studies with radioactive Glc and GlcN suggested that the site of biosynthesis was in the microsomal smooth membranes (ARCE *et al.* 1971, MACCIONE *et al.* 1971), and this was confirmed using [^{3}H]ManNAc (LANDA *et al.* 1977, 1979). These studies showed a distribution of labelled gangliosides in membranes other than the microsomal fraction, e.g. mitochondrial membranes (CAPUTTO *et al.* 1974, 1976).

δ) *Membrane Glycoconjugates*

The brush border of rat small intestine luminal columnar cells has been widely used as a model, since membranes can be readily separated from the mucus glycoproteins of the goblet cells (e.g. FORSTNER 1968, KIM and PERDOMO 1974) and the brush border is rich in carbohydrate and turns over rapidly. Incorporation of radioactive GlcN into rat small intestinal brush border glycoprotein has been investigated extensively by FORSTNER and his group (FORSTNER 1968, 1971, FORSTNER *et al.* 1973). The pattern of labelling was consistent with glycoprotein synthesis in the endoplasmic reticulum and subsequent transfer to the intestinal brush border. Similar results were obtained with sialic acid incorporation occurring in rat liver Golgi apparatus (STURGESS *et al.* 1972, 1978).

Radioactive GlcN has been used in rat liver to detect the biosynthesis of microsomal membrane glycoproteins (AUTUORI *et al.* 1975a). After synthesis of these glycoproteins containing radioactively labelled sialic acid in the Golgi apparatus, the glycoprotein components appear in the cytosol and are selectively incorporated into the endoplasmic reticulum (AUTUORI *et al.* 1975b, ELHAMMER *et al.* 1975, BERGMAN and DALLNER 1977). The addition of sialic acid in the Golgi apparatus was confirmed.

The kinetics of glycosylation and intracellular transport of sialoglycoproteins in mouse liver have been reported (CAREY and HIRSCHBERG 1980). After addition of 9-[^{3}H]Neu5Ac to liver slices or after intravenous injection, the radioactive label appeared first in the Golgi membranes and then decreased, while the plasma membrane fraction increased to a maximum after the decrease in the Golgi membranes. Smooth and rough endoplasmic reticulum showed low labelling. The recovery of radioactive label was very high (~ 100%), and the contribution to secretory glycoproteins which would be lost in the subcellular fractionation calculated at 20% of the incorporated label.

b) Cell-Free Systems

α) *N-Glycosidically Linked Oligosaccharide Chains*

Subcellular fractionation combined with enzymic assay of sialyltransferase and other glycosyltransferase activities provided good evidence for the Golgi apparatus location of the glycosyltransferases involved in the elongation of N-glycosidically linked oligosaccharide chains of the N-acetyllactosamine type (Neu5AcαGalβGlcNAc-). Much of the evidence was gathered in the laboratory of SCHACHTER in Toronto (e.g. SCHACHTER *et al.* 1970, SCHACHTER 1974, 1978, MUNRO *et al.* 1975) with rat liver. Substrates in these experiments, such as asialo-α_1-acid glycoprotein (MUNRO *et al.* 1975) and asialo-transferrin (FLEISCHER 1978) have been employed to demonstrate increases in specific activity of the sialyltransferases in rat liver (SCHACHTER *et al.* 1970, MUNRO *et al.* 1975, FLEISCHER

1978) and in kidney (FLEISCHER 1978) Golgi membranes. Low activities were found in smooth microsomal and plasma membrane fractions. The same conclusions can be drawn from other studies and have been reviewed (SCHACHTER and RODÉN 1973, SCHACHTER 1978, STURGESS *et al.* 1978).

β) *O-Glycosidically Linked Oligosaccharide Chains*

Subcellular fractionation in tissues synthesizing mucus glycoproteins (characterized by the serine/threonine-GalNAc O-glycosidic linkage) has proved difficult due to the presence of the mucus glycoproteins themselves. The preparation of well defined membrane fractions comparable with liver subfractions has not been possible (LAWFORD and SCHACHTER 1967, ROSSIGNOL *et al.* 1969, BUSCHER *et al.* 1977). A membrane location for the sialyltransferase activity in salivary glands has been demonstrated (McGUIRE and ROSEMAN 1967, SCHACHTER *et al.* 1971, CARLSON *et al.* 1973b, SADLER *et al.* 1979a) in ovine and porcine glands. Demonstration of the Golgi membrane nature of sialyltransferase activity has been provided in porcine and bovine submandibular glands (SCHAUER *et al.* 1974, BUSCHER *et al.* 1977, unpublished results).

γ) *Gangliosides*

The identification of the subcellular site of sialyltransferase (and other glycosyltransferases) involved in ganglioside biosynthesis is still uncertain. An approach to the problem by cell fractionation has enabled some progress, but the problems observed remain inherent to the system and are discussed further in section III.5. Sialyltransferase activity has been detected in a number of tissues including kidney, brain, spleen and mammary gland (see DAWSON 1978 for a review). Experiments with rat liver showed a great enrichment of CMP-Neu5Ac: lactosylceramide (LacCer) sialyltransferase in the Golgi membranes, although some activity was also found in the smooth endoplasmic reticulum (KEENAN *et al.* 1974, RICHARDSON *et al.* 1977, PACUSZKA *et al.* 1978). FLEISCHER (1977, 1978) has provided evidence for the localization of glycolipid sialyltransferases in the Golgi membranes of rat liver and kidney. Sucrose gradient fractionation of young rat brain membranes and comparison with marker enzymes led NG and DAIN to propose that the sialyltransferase activity (including glycoprotein-specific sialyltransferase) was located in the smooth endoplasmic reticulum, Golgi complexes and the plasma membranes, but not in mitochondria and synaptosomes (NG and DAIN 1977b, DAIN and NG 1980). In contrast to these findings the localization of glycolipid sialyltransferase activity in calf brain synaptosomal fractions in addition to other microsomal membranes, essentially Golgi and endoplasmic reticulum, was reported (PRETI *et al.* 1980, TETTAMANTI *et al.* 1980).

δ) *Membrane Glycoconjugates*

The demonstration of endogeneous glycoprotein sialyltransferase activity represents a composite of secretory glycoproteins, still membrane-bound during their biosynthesis, and genuine membrane glycoprotein precursors. Thus, no definite evidence for a membrane glycoprotein-specific sialyltransferase in subcellular membranes has been provided. Endogeneous glycoprotein sialyltransferase activities have been reported in microsomal membranes of many tissues including brain (e.g. BRUNNGRABER 1979, WAECHTER and SCHER 1979), liver (SCHACHTER and RODÉN 1973, SCHACHTER 1978), respiratory tissue (BAKER *et al.* 1972), intestinal mucosa (FROT-COUTAZ *et al.* 1973, MARTIN and LOUISOT 1976) and normal and transformed cells (see SCHACHTER 1978).

3. Sialyltransferases with N-Glycosidically Linked Oligosaccharide Substrates

The N-glycosidically linked oligosaccharides of glycoproteins containing sialic acid are the N-acetyllactosamine type (MONTREUIL 1980). The purification of the relevant sialyltransferases (and other glycosyltransferases) to homogeneity has been reported by the group of HILL (e.g. PAULSON *et al.* 1977a, b, SADLER *et al.* 1979b, c, BEYER *et al.* 1981). Consequently, the choice of substrate is very important in allowing conclusions on the specificity to be drawn (SCHACHTER 1978, SCHACHTER and ROSEMAN 1980). Information about the nature of N-glycosidically linked oligosaccharide in glycoproteins has been improved and extended in the last five years (KOBATA 1979, RAUVALA and FINNE 1979, KORNFELD and KORNFELD 1980, MONTREUIL 1980), and with the application of 360 MHz and 500 MHz n.m.r. spectroscopy to purified oligosaccharides (VLIEGENTHART *et al.* 1981) precise data are available. On this basis it is possible to propose a number of sialyltransferase activities. The demonstration of α(2-3), (2-4), (2-6), and (2-8) linkages in N-glycosidically linked oligosaccharides demands at least four sialyltransferases specific for these linkages in accordance with the one gene—one enzyme hypothesis (SCHACHTER and RODÉN 1973). Thus, with the exception of the α(2-8) transfer to sialic acid and the α(2-6) transfer to GlcNAc, respectively, all remaining enzymes catalyze the transfer of sialic acids to β-linked galactose in the non-reducing ends of glycoprotein oligosaccharide chains (Table 2 in chapter B).

Investigations with sialyltransferases have involved the transfer to lactose, to ascertain the glycosidic specificity of the product (e.g. HUDGIN and SCHACHTER 1972, PAULSON *et al.* 1977b). The determination of the glycosidic specificity of purified sialyltransferases has shown that lactose can serve as an acceptor for both N- and O-chain-specific transferases (PAULSON *et al.* 1977b, REARICK *et al.* 1979, SADLER *et al.* 1979b). Thus, results with lactose as acceptor cannot be extended to glycoprotein substrates for prediction of N- or O-chain glycosidic specificity. Sialyltransferase activity has been detected with an ever increasing number of substrates in many tissues (see Table 1 for a selection and some properties). A new methodology in the analysis of *in vivo* sialylation specificity has been introduced by VAN DEN EIJNDEN *et al.* (1977) with rat liver microsomal sialyltransferase. The acceptor molecule (in this case α_1-acid glycoprotein) is desialylated, and the terminal β-galactosyl residues are tritiated with borotritide after generation of the galactosyl-6-aldehyde with galactose oxidase. The acceptor is then used for sialyl- (or other glycosyl-)transfer and the product subjected to permethylation and hydrolysis. Identification of the methylated galactose derivatives allows a conclusion on the site of glycosyltransfer to be drawn.

Table 1. *Sialyltransferase activities from*

Source of enzyme	Substrate	Monosaccharide acceptor
Bacteria		
E. coli K 235	colominic acid Neu5Acα2-8(Neu5Ac)$_n$	α-Neu5Ac
Meningococci C	capsular homopolysaccharide Neu5Acα2-9(Neu5Ac)$_n$	α-Neu5Ac
Amphibia		
Frog liver	asialo-fetuin N-acetyllactosamine lactose	β-Gal/α-GalNAc β-Gal β-Gal
Mammals		
Rat, porcine, bovine, human liver	lactose	β-Gal
Porcine liver	asialo-apolipoprotein (human)	β-Gal GalNAc?
Human liver	asialo-fetuin asialo-ovine submandibular gland glycoprotein (OSM)	β-Gal α-GalNAc
Human liver	asialo-α_1 antitrypsin and asialo-glycopeptides	β-Gal/ α-GalNAc
Rat liver	asialo-α_1 acid glycoprotein	β-Gal
Calf kidney cortex	asialo-α_1 acid glycoprotein asialo-fetuin asialo-OSM Tamm-Horsfall glycoprotein	β-Gal β-Gal/α-GalNAc α-GalNAc ?
Rat intestinal mucosa	asialo-fetuin	β-Gal/α-GalNAc
Human placenta	asialo-fetuin asialo-ceruloplasmin	β-Gal/α-GalNAc β-Gal
Rat brain	asialo-fetuin	β-Gal/α-GalNAc
Mouse sperm	asialo-fetuin	β-Gal/α-GalNAc
Bovine colostrum	asialo-α_1 acid glycoprotein (see Table 2)	β-Gal(1-4)
Rat mammary gland	lactose N-acetyllactosamine	β-Gal(1-4)

* The nature of the glycosidic linkage was not determined in these experiments and is assumed from independent structural analyses. +, Activity was observed, but the type of linkage formed was not determined.

different sources. A selection from the literature

α linkage formed	Location	pH optimum or incubation pH	References
2-8	membranes	8.0	TROY and MCCLOSKEY 1979
2-9	membranes	7.0-8.0	VANN *et al.* 1978
2-3/2-6*	membranes	7.0	KHORLIN *et al.* 1980
2-6	membranes	7.0	KHORLIN *et al.* 1980
2-6	membranes	7.0	KHORLIN *et al.* 1980
2-3 2-6	membranes	6.0-7.0	HUDGIN and SCHACHTER 1972
2-6*	microsomes	7.3	WETMORE *et al.* 1974
2-3 2-6	membranes and supernatant	7.2	ALHADEFF *et al.* 1977
2-3 2-6*	deoxycholate/triton X-100 solubilized homogenate, supernatant	7.0	AGUANNO *et al.* 1978
2-6	microsomes	6.7	VAN DEN EIJNDEN *et al.* 1977
2-6* 2-3/2-6* 2-6* +	microsomes	6.3-6.8	VAN DIJK *et al.* 1977
+	microsomes and soluble	7.4	MARTIN and LOUISOT 1976
2-3/2-6* 2-3*	10,000 g supernatant with triton X-100	6.9	LIU *et al.* 1978
2-3/2-6*	membranes	6.3	NG and DAIN 1977a
2-3/2-6*	cells and supernatant	7.3	DURR *et al.* 1977
2-6	soluble	6.8	PAULSON *et al.* 1977a, b
2-3 (Neu5Ac + Neu5Gc)	membranes	6.9	CARLSON *et al.* 1973a

Table 1 (continued)

Source of enzyme	Substrate	Monosaccharide acceptor
Rat liver	antifreeze glycoprotein	β-Gal/α-GalNAc
Sheep submandibular gland	asialo-OSM	α-GalNAc
Canine respiratory tissue	asialo-OSM	α-GalNAc
	asialo-bovine submandibular glycoprotein (BSM)	α-GalNAc
	asialo-fetuin	β-Gal/α-GalNAc
	asialo-α_1 acid glycoprotein	β-Gal
Bovine mammary gland	asialo-κ casein	β-Gal α-GalNAc
Bovine, porcine, equine submandibular gland	submandibular gland glycoproteins (BSM, PSM, and ESM)	β-Gal α-GalNAc
Bovine and porcine submandibular gland, bovine and frog liver	synthetic glycosylated lysozymes (Gal, GalNAc, GlcNAc)	Gal GalNAc
	asialo-fetuin	β-Gal/α-GalNAc
Rat and calf brain	asialo-α_1 acid glycoprotein	β-Gal
Embryonic chicken brain	asialo-α_1 acid glycoprotein	β-Gal
	asialo-fetuin	β-Gal/α-GalNAc
	lactose	β-Gal
	asialo-α_1 acid glycoprotein	β-Gal
	LacCer	β-Gal
	II^3Neu5GcLacCer	α-Neu5Gc
Rat brain	II^3Neu5AcGgOse$_4$Cer	β-Gal
Mouse brain	Gal-Cer	β-Gal
Mouse neuroblastoma cells,	Lac-Cer	β-Gal
hamster astrocytes, human glioblastoma cells	II^3Neu5AcGgOse$_4$Cer	β-Gal
Mouse neuroblastoma cells	LacCer	β-Gal
Rabbit neurohypophysis	II^3Neu5AcGgOse$_4$Cer	β-Gal
Rat intestinal mucosa	LacCer	β-Gal
Rat kidney	LacCer	β-Gal
Rat liver	II^3Neu5AcLacCer	α-Neu5Ac
Bovine thyroid gland	II^3Neu5AcLacCer	α-Neu5Ac

α linkage formed	Location	pH optimum or incubation pH	References
2-3/2-6*	solubilized (triton X-100)	7.0	Shier and Roloson 1974
2-6	membranes	6.0-6.1	Carlson *et al.* 1973b
2-6*	membranes	6.5	Baker *et al.* 1972
2-6*	membranes	6.5	Baker *et al.* 1972
2-3/2-6*	membranes	6.5	Baker *et al.* 1972
2-6*	membranes	6.5	Baker *et al.* 1972
2-3 2-6	Golgi membranes triton X-100 solubilized	5.5	Keller *et al.* 1979
2-3* 2-6* (Neu5Ac, Neu5Gc, and NeuAcyl)	membranes	6.0	Schauer and Wember 1973
2-3 2-6*	membranes	6.9	Schauer *et al.* 1979b
2-3/2-6*			
2-6/2-3*	membranes (regional distribution in brain found)	6.5-6.9	Van den Eijnden and van Dijk 1974
2-6/2-3*	membranes	6.3	Den *et al.* 1970
2-6/2-3*			
2-3			
2-3/2-6*	synaptosomal membranes	6.3	Den *et al.* 1975
2-3			
2-8			
2-3	smooth microsomal membranes	6.3	Ng and Dain 1977a, b
2-3	microsomal membranes	6.3	Yu and Lee 1976
2-3 2-3	homogenate	6.0 6.5	Duffard *et al.* 1977
2-3	membranes	5.9	Kemp and Stoolmiller 1976
2-3	homogenate	6.5	Clarke and Mulcahey 1976
2-3	membranes	6.3	Glickman and Bouhours 1976
2-3	Golgi membranes	6.5	Fleischer 1977
2-8	Golgi membranes	6.2	Eppler *et al.* 1980a
2-8	membranes	6.4	Pacuszka *et al.* 1978

Table 2. *The specificity and efficiency of some sialyltransferases with various substrates.* transfer to other substrates expressed relative to this; in column 5 substrates within each antifreeze glycoprotein and II³Neu5AcGgOse$_4$Cer as 100 (*) for this group only; in column expressed as % of N-acetyllactosamine or asialo-fetuin; —, not determined. [4], PAULSON *et al.* 1977b; [5], REARICK *et al.* 1979;

Acceptor		[1]Goat colostrum	[2]Rat mammary gland	[3]Porcine liver
Oligosaccharides				
Lactose	Galβ(1-4)Glc	100	100	100
	Galβ(1-3)GalNAc	—	—	—
N-Acetyl-	Galβ(1-4)GlcNAc	770	100	760
lactos-	Galβ(1-3)GlcNAc	238	88	36
amine	Galβ(1-6)GlcNAc	23	91	37
LcOse$_4$		—	80	—
LcnOse$_4$		—	—	—
Glycoproteins				
Asialo-fetuin		1,400	—	332
Asialo-α_1 acid glycoprotein		1,077	5	398
Asialo-ovine submandibular glycoprotein		64	—	116
Asialo-porcine submandibular glycoprotein		387	—	—
Antifreeze glycoprotein		—	—	—
Glycolipids				
II³Neu5AcGgOse$_4$Cer		32	5	—
GgOse$_4$Cer		140	—	—

Details of the specificity of some sialyltransferases are given in Table 2. The best characterized enzyme acting on N-glycosidically linked oligosaccharides has been described by PAULSON *et al.* (1977a). This enzyme is a β-D-galactoside-α(2-6)-sialyltransferase and was isolated from bovine colostrum. Two forms of the enzyme were found which both showed similar kinetic properties. The enzyme was completely specific for α(2-6) transfer. This was demonstrated with Gal-GlcNAc isomers [β(1-3), β(1-4), and β(1-6)] and lactose. The disaccharide Galβ(1-4)GlcNAc was the best acceptor with the other disaccharides showing minimal acceptor activity. The detailed analysis of the purified bovine colostrum enzyme (PAULSON *et al.* 1977b) suggests that earlier experiments (HUDGIN and SCHACHTER 1971, 1972, BARTHOLOMEW *et al.* 1973) contained some α(2-3)-specific transferase, probably the O-glycosidically linked oligosaccharide chain transferase (see Table 2 and sections III.4 and III.6).

Experiments with purified and partially purified sialyltransferase showed greatest transfer to non-reducing, terminal β-galactosyl residues in glycoproteins

In columns 1-4 the activity with lactose or N-acetyllactosamine was taken as 100 and the group were assayed and the activity or % saturation in each case related to Galβ(1-3)GalNAc, 6 figures represent % saturation of available acceptor sites; figures in columns 7-9 are [1], BARTHOLOMEW *et al.* 1973; [2], CARLSON *et al.* 1973a; [3], HUDGIN and SCHACHTER 1971.; [6], SADLER *et al.* 1979b; [7], [8], [9], KHORLIN *et al.* 1980

[4]Bovine colostrum α(2-6)	Porcine submandibular gland α(2-3)[5]	α(2-6)[6]	[7]Frog liver	[8]Normal	[9]Rat liver regenerating
< 1	< 0.3	0	100	50	48
—	100 *	—	—	—	—
100	< 0.3	—	100	100	100
0	0.7	0	85	85	30
0	< 0.3	0	25	15	14
—	1.4	0	—	—	—
—	< 0.3	0	—	—	—
225	27.8	14	100	100	187
218	1.4	2	—	—	—
0	54	70	—	—	—
0	0.5	39	< 10	5	4
0	100*	95	—	—	—
—	100*	—	—	—	—
—	0	—	—	—	—

with N-glycosidically linked oligosaccharide chains (see Table 2, BARTHOLOMEW *et al.* 1973, PAULSON *et al.* 1977b).

Rat liver was found to have only α(2-6)-specific sialyltransferase activity with N-glycosidically linked oligosaccharide substrates (e.g. α_1-acid glycoprotein), as determined by VAN DEN EIJNDEN *et al.* (1977). This result taken together with the studies of HUDGIN and SCHACHTER (1972) and PAULSON *et al.* (1977b) strongly suggests that the α(2-3) transfer to these disaccharides is due to an enzyme specific for O-linked chains. Evidence for an α(2-3) transfer to N-linked oligosaccharide chains could not be found in rat liver (VAN DEN EIJNDEN *et al.* 1977, PAULSON *et al.* 1977b).

Several sialyltransfer activities have not been demonstrated, although structural analysis suggests that they should exist. These include α(2-3)Gal-, (2-4)Gal-, (2-6)GlcNAc- and (2-8)Neu5Ac-specific enzymes (see Table 2 in chapter B). The wide occurrence of α(2-3) linkages (MONTREUIL 1980) indicates that such an enzyme activity should be detectable. The same argument applies to α(2-4) and

(2–8) transfer, and to the (2–6) transfer to GlcNAc found in bovine cold-insoluble globulin (e.g. Kobata 1979, Rauvala and Finne 1979). The purification of sialyltransferases by the techniques developed by Paulson *et al.* (1979) now provides the chance to compare the specificity of these enzymes. This is especially necessary to explain the biosynthesis of glycoconjugates containing different glycosidically linked sialic acids in the same oligosaccharide chains. Examples have been illustrated in urinary oligosaccharides in mucopolysaccharidosis (Strecker and Montreuil 1979), fetuin (Nilsson *et al.* 1979) and horse pancreatic ribonuclease (Fig. 3) (Schut *et al.* 1978). The relation of glycosidic linkage to function has been discussed elsewhere (Montreuil 1975, 1980, Kobata 1979).

Another point which awaits elucidation is the transfer of sialic acids other than Neu5Ac to glycoproteins. The transfer of Neu5Gc from CMP-Neu5Gc could be demonstrated for the colostrum sialyltransferase with N-acetyllactosamine and asialo-α_1-acid glycoprotein as substrates (Bartholomew *et al.* 1973). The transfer of O-acetyl-Neu5Ac to N-glycosidically linked oligosaccharides remains to be demonstrated directly.

Neu5Acα(2→6)Gal β(1→4)GlcNAcβ(1→2)Man α(1→3)\
                                                                                    Manβ(1→4)GlcNAc
Neu5Acα(2→3)Gal β(1→4)GlcNAcβ(1→2)Man α(1→6)/

Fig. 3. The structure of an oligosaccharide isolated from equine pancreatic ribonuclease. The occurrence of α(2–3) and α(2–6) linked sialic acid in a single oligosaccharide chain is shown.

4. *Sialyltransferases with O-Glycosidically Linked Oligosaccharide Substrates*

Initial studies were with ovine and porcine submandibular glands, where the structure of the oligosaccharide chains in the high molecular weight mucus glycoproteins was also determined (Carlson 1977). The properties of some of these sialyltransferases are given in Tables 1 and 2. Two membrane-bound sialyltransferases were purified to homogeneity from porcine submandibular gland and were characterized as β-D-galactoside-α(2–3)-sialyltransferase (Rearick *et al.* 1979, Sadler *et al.* 1979b) and N-acetyl-α-galactosaminide-α(2–6)-sialyltransferase (Sadler *et al.* 1979b, c). The α(2–6)-sialyltransferase was found to show greatest activity with desialylated ovine and porcine submandibular gland mucins and antifreeze glycoprotein. The former two substrates contain terminal α-GalNAc residues which serve as acceptors. Identification of Neu5Acα(2–6)GalNAc-ol after reductive (borohydride) β-elimination of the product of sialyltransfer to asialo-ovine submandibular gland mucin confirmed the specificity of the transfer. Antifreeze glycoprotein contains the disaccharide Galβ(1–3)GalNAcα-polypeptide, which also occurs in the porcine submandibular gland glycoprotein acceptor, and the α(2–6)-sialyltransferase product is

Galβ(1–3)GalNAcα-polypeptide
|α(2–6)
Neu5Ac

(BEYER *et al.* 1979, SADLER *et al.* 1979a). These results together with the substrate specificity studies (see Table 2, column 6) clearly demonstrate the difference to the α(2-6)-specific enzyme isolated from bovine colostrum.

The detection of an α(2-3)-sialyltransferase in porcine submandibular glands was important, as the mucus glycoprotein, the major secretory product in this gland, contains little or no α(2-3)-linked sialic acid. However, the tetrasaccharide

```
Neu5Acα(2-3)Galβ(1-3)GalNAcα-polypeptide
                        |
                        | α(2-6)
                        |
                      Neu5Ac
```

is known in several well characterized glycoproteins including κ-casein, fetuin, and the membrane glycoproteins glycophorin and epiglycanin (see MONTREUIL 1980). The substrate specificity is detailed in Table 2, column 5, transfer to asialo-fetuin was in agreement with the expected results for sialylation of only the O-glycosidically linked chains (similar results were obtained for the α(2-6)-sialyltransferase specific for O-glycosidic chains in fetuin). Several potential glycolipid substrates were assayed and transfer to II^3Neu5AcGgOse$_4$Cer (for structures see chapter B) to yield IV^3Neu5AcII^3Neu5AcGgOse$_4$Cer demonstrated. Mixed gangliosides also gave this product along with three others. The transfer was explained on the basis of the terminal non-reducing disaccharide structure (Galβ(1-3)GalNAc) which occurs in both mucus glycoprotein oligosaccharides and in the glycolipid structure GgOse$_4$Cer (REARICK *et al.* 1979). This raises the question of a dual function of sialyltransferases in glycoprotein and glycolipid metabolism and is discussed further in section VI. Corroboration of O-glycosidic oligosaccharide chain-specific α(2-3)-sialyltransferase activity was obtained in porcine liver, making NeuAcα(2-3)Galβ(1-3)GalNAc- (VAN DEN EIJNDEN *et al.* 1979). The sheep submandibular gland α(2-6)-sialyltransferase was active with both CMP-Neu5Ac and CMP-Neu5Gc (CARLSON *et al.* 1973b), and membrane fractions from bovine, porcine and equine submandibular glands showed similar kinetic constants for the transfer of Neu5Ac, Neu5Gc, Neu5,9Ac$_2$, and Neu4,5Ac$_2$ from their CMP-glycosides (SCHAUER and WEMBER 1973). Using the equine submandibular gland system, transfer of Neu5Ac4Me from its CMP-glycoside to asialo-fetuin could be demonstrated (BEAU and SCHAUER 1980).

The demonstration of α(2-4)GlcNAc-, (2-6)Gal-, and (2-8)Neu5Ac-specific sialyltransferases is lacking at present. The identification of such structural units in mucus glycoproteins (i.e. O-glycosidically linked oligosaccharides) has been documented for rat sublingual gland glycoprotein (α(2-4)GlcNAc- and α(2-6)Gal-linked sialic acid groups; SLOMIANY and SLOMIANY 1978) and in porcine submandibular gland glycoprotein (α(2-8)-linked sialyl residues; SLOMIANY *et al.* 1978), see chapter B. The α(2-4)-sialylation reaction has been discussed as a chain termination mechanism in rat sublingual gland glycoprotein biosynthesis (SLOMIANY and SLOMIANY 1978). The detection of polysialyl chains attached to O-glycosidic oligosaccharides in a trout-egg glycoprotein (INOUE and MATSUMARA 1979) provides another example worthy of substrate specificity studies.

Finally, the relationship of sialyltransferases of different specificities in the biosynthesis of porcine submandibular gland glycoprotein oligosaccharides has been elucidated by BEYER *et al.* (1979, 1981; Fig. 8). These examples of regulation in sialyltransfer will be considered in section VI.

5. *Sialyltransferases with Glycolipid Substrates*

The controversy still existing about the biosynthesis of gangliosides in various tissues has not been resolved by the study of glycosyltransferase enzymes. The

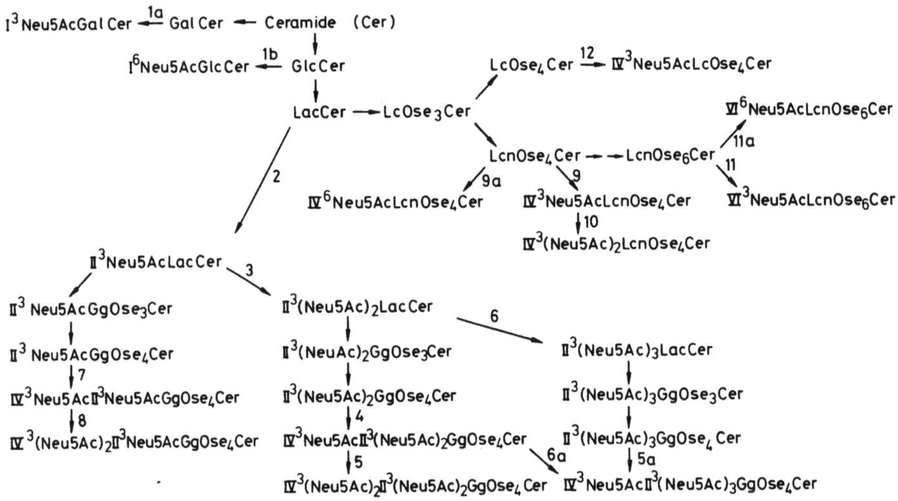

Fig. 4. Ganglioside biosynthetic pathways. An overall view of biosynthetic pathways is presented compiled from data on the vertebrates (YU and ANDO 1980, WIEGANDT 1980, BASU *et al.* 1980). The pathways are discussed in detail in the text. Core structures are abbreviated as follows: *Lac* Galβ(1-4)Glc; *$LcOse_4$* Galβ(1-3)GlcNAcβ(1-3)Galβ(1-4)Glc; *$LcnOse_4$* Galβ(1-4)GlcNAcβ(1-3)Galβ(1-4)Glc; *$GgOse_4$* Galβ(1-3)GalNAcβ(1-4)Galβ(1-4)Glc (see chapter B).

original studies carried out by ROSEMAN's group on embryonic chicken brain are still the accepted biosynthetic routes for ganglioside formation (KAUFMAN *et al.* 1968, DEN *et al.* 1970, 1975). This work and subsequent improvements have been reviewed (FISHMAN 1974, MCGUIRE 1976, DAWSON 1978, 1979, BRUNNGRABER 1979, ROSENBERG 1979).

On the basis of glycosyltransferase studies, the following sequence for ganglioside biosynthesis has been proposed (Fig. 4). The majority of cells contain gangliosides mainly of the LacCer and $GgOse_4$Cer series (LEDEEN 1978, YAMAKAWA and NAGAI 1978). None of the sialyltransferases involved in the biosynthesis of these compounds have been purified to homogeneity. Two approaches to the identification of these enzymes have been utilized: (i) Classical enzyme purification methods, and (ii) use of radioactive nucleotide-sugar substrates as precursors and identification of products in crude extract (homogenate) incubations. Problems have been encountered due to the natural occurrence of gangliosides in membranes, and the exclusive membrane location of such enzymes. In general, the enzymes require detergent for activity and show pH optima in the range of 6–6.5. No complete dependence on divalent cations such as Mg^{2+} and Mn^{2+} has been demonstrated.

As can be seen in Fig. 4, sialyltransfer occurs in at least 17 enzyme-catalyzed reactions in ganglioside biosynthesis. Several of these steps are analogous, e.g. steps 7 and 4, or the transfer reactions to form disialyl groups, e.g. steps 3, 5, 8, and 10. Biosynthesis of the brain-specific ganglioside I^3Neu5AcGalCer in mouse brain and liver microsomes has been described (YU and LEE 1976). The occurrence of I^3Neu5AcGalCer specifically in brain (not in liver) may reflect "sharing" of sialyltransfer specificity, although a completely specific enzyme in brain cannot be ruled out.

Lactosylceramide sialyltransferase (step 2) was originally detected in embryonic chicken brain (KAUFMAN *et al.* 1968), has also been found in calf brain (PRETI *et al.* 1980) and is a key stage in ganglioside biosynthesis (Fig. 4) leading to either II^3Neu5AcGgOse$_3$Cer or II^3(Neu5Ac)$_2$LacCer and their subsequent polysialyl ganglioside derivatives. The enzyme showed an age-dependent activity in rat brain microsomes, and a particulate inhibitor of activity was detected which increased with age (DUFFARD and CAPUTTO 1972). In young rat brain this enzyme could be distinguished from the sialyltransferase converting II^3Neu5AcGgOse$_4$Cer to IV^3Neu5AcII^3Neu5AcGgOse$_4$Cer (step 7) by substrate competition experiments (DAIN and NG 1980), while in chicken brain this was established on the basis of selective heat denaturation (KAUFMAN *et al.* 1968). The enzyme from rat liver was localized in the Golgi apparatus (FLEISCHER 1978) and characterized (RICHARDSON *et al.* 1977). This activity could also be found in retina and isolated neurons (DREYFUS *et al.* 1980).

Sialylation of II^3Neu5AcLacCer to give II^3(Neu5Ac)$_2$LacCer has been ascribed to a specific sialyltransferase (step 3) detected in chicken embryo brain membranes (KAUFMAN *et al.* 1968), bovine thyroid Golgi apparatus (PACUSZKA *et al.* 1978) and rat liver (EPPLER *et al.* 1980a). The latter authors were able to achieve some purification of the enzyme on Concanavalin A (ConA)-Sepharose, illustrating a possible glycoprotein nature of the enzyme (EPPLER *et al.* 1980a). Further sialylation along this branch of the metabolic route leads to the formation of IV^3Neu5AcII^3(Neu5Ac)$_2$Cer from II^3(Neu5Ac)$_2$GgOse$_4$Cer (step 4). This sialyltransferase has been detected in duck embryo brains (ARCE *et al.* 1971, MESTRALLET *et al.* 1977) and is believed to be the same enzyme which converts II^3Neu5AcGgOse$_4$Cer to IV^3Neu5AcII^3Neu5AcGgOse$_4$Cer (step 7). The enzyme was found to differ from the sialyltransferase catalyzing step 2 (MESTRALLET *et al.* 1977).

The sialyltransferase reactions of steps 5, 5a, 6, and 6a have been proposed on the basis of structural studies (ANDO and YU 1979, YU and ANDO 1980, HILBIG *et al.* 1981); enzymic details are not available at present.

The transfer of Neu5Ac leading to IV^3Neu5AcII^3Neu5AcGgOse$_4$Cer (step 7) is distinct from step 2 sialyltransferase (KAUFMAN *et al.* 1968, MESTRALLET *et al.* 1977). This sialyltransferase has also been demonstrated in rat brain by YIP (1973) and in rabbit neurohypophysis by CLARKE and MULCAHEY (1976). Step 8 sialyltransferase is proposed on the basis of structural analysis of natural material (LEDEEN 1978).

The biosynthesis of IV^{3}Neu5AcLcnOse$_4$Cer has been detected in embryonic chicken brain, bovine spleen and human neural cell cultures (BASU *et al.* 1980). On the basis of competition experiments, the enzyme catalyzing step 9 was proposed to be different to step 2 sialyltransferase. Transfer to the IV position of GgOse$_4$Cer could also be demonstrated (BASU *et al.* 1980, STOFFYN and STOFFYN 1980). The transfer of sialic acid to other glycolipids in the LcnOseCer series is hypothetical, thus the natural occurrence of these gangliosides assumes the activity of sialyltransferases catalyzing steps 9 a, 10, 11, and 11 a. This is also the case for the LcOse$_4$Cer series represented by one example (step 12) (RAUVALA *et al.* 1978, SWEELEY and SIDDIQUI 1977, WATANABE *et al.* 1979).

In vivo experiments are of great importance in determining the product-precursor relationships in ganglioside sialyltransfer (ROSENBERG 1979, SCHACHTER and ROSEMAN 1980). The use of cultures of normal and transformed cells has greatly aided confirmation of the proposed pathways. This work is extensive and has been reviewed (FISHMAN 1974, BRADY and FISHMAN 1974, DAWSON 1978, 1979, SCHACHTER and ROSEMAN 1980).

The interest in structures containing Neu5Gc in addition to Neu5Ac even in the same molecule (IWAMORI and NAGAI 1978), and O-acetyl-sialic acids (HAVERKAMP *et al.* 1977, GHIDONI *et al.* 1980, SCHAUER *et al.* 1980 b, and chapter B, Tables 5 and 6) now focusses on the specificity of sialyltransfer. The biosynthesis of such gangliosides has not been studied. The *in vivo* experiments on ganglioside biosynthesis have demonstrated that simple product-precursor relationships do not apply (e.g. MACCIONI *et al.* 1971, CAPUTTO *et al.* 1974). An understanding of this phenomenon requires information about individual enzymes and study of their interactions with each other under simulated membrane conditions.

6. *Sialyltransferases with Oligosaccharide Substrates*

The occurrence of sialyloligosaccharides in the milk and colostrum of mammals has raised the question of their function and biosynthesis (KOBATA 1977). Mammary gland tissue synthesizes sialyloligosaccharides from a wide range of substrates with terminal, non-reducing β-galactose residues (CARLSON *et al.* 1973 a). The best acceptors for sialic acid were lactose and N-acetyllactosamine. Transfer to LcOse$_4$ could also be demonstrated (CARLSON *et al.* 1973 a). Biosynthesis of gangliosides has also been detected in bovine tissue relating to milk fat membrane biosynthesis (BUSHWAY and KEENAN 1978). Good activity with N-acetyllactosamine and lower activity with lactose was found for the purified colostrum α(2-3)-specific enzyme (PAULSON *et al.* 1977 b). The formation of II^{3}Neu5AcLac in colostrum has been described (BARTHOLOMEW *et al.* 1973, PAULSON *et al.* 1977 b). It was suggested that this activity may be due to the presence of an O-glycosidically linked oligosaccharide-specific α(2-3)-sialyltransferase (SCHACHTER and ROSEMAN 1980). Although the role of sialyloligosaccharide formation in milk and colostrum remains unclear, age-dependent sialidase activity correlates well with the concentration of sialyloligosaccharides in milk (KUHN 1972, DICKSON and MESSER 1978). The transfer of sialic acid to $LcOse_4$ yields $III^6Neu5AcLcOse_4$ and other sialyl-lacto-N-tetraoses, as can be concluded from their occurrence in milk (CARLSON *et al.* 1973 a).

In agreement with the occurrence of Neu5Gc in colostrum and milk oligosaccharides from bovine and porcine sources, transfer from CMP-Neu5Gc to lactose could be demonstrated (BARTHOLOMEW *et al.* 1973, CARLSON *et al.* 1973 a).

7. *Sialyltransferases with Membrane Glycoprotein Substrates*

At present little detailed evidence is available for sialyltransferases specifically involved in membrane glycoprotein biosynthesis. It is probable that the membrane-bound sialyltransferase specificities described for glycoproteins (sections III.3 and III.4) and glycolipids (section III.5) will be similar, as the structures of membrane complex carbohydrates show the same features found in soluble molecules (e.g. KOBATA 1979, RAUVALA and FINNE 1979, MONTREUIL 1980). Much indirect evidence has come from cell culture studies, frequently in studies on normal and transformed or malignant cells (HUGHES 1976, HARMON 1978, KÖTTGEN *et al.* 1979).

8. *Sialyltransferases in Bacterial Systems*

The occurrence of sialic acids, mainly as polysaccharides, in bacteria has been described in chapter B. Studies on the sialyltransferases associated with polymer formation have largely been done using *E. coli K 235*, which synthesizes the α(2-8)-linked polymer colominic acid. The system was found to be particle-bound and transferred sialic acid to the non-reducing end of colominic acid. The transfer from CMP-Neu5Gc was only 5% of that with CMP-Neu5Ac (AMINOFF *et al.* 1963, KUNDIG *et al.* 1971). Neu5Gc is not found in colominic acid. Subsequent analysis of this sialyltransferase system has shown that a lipid intermediate is involved in the synthesis of colominic acid (TROY *et al.* 1975, VIJAY and TROY 1975). Transfer of sialic acid from CMP-Neu5Ac into a high molecular weight acceptor and a lipid fraction were found. The lipid could be identified as undecaprenyl-phosphate (TROY *et al.* 1975). The endogeneous membrane-bound sialyl polymers were found to average over 150 Neu5Ac units, and concluse evidence for addition to the non-reducing end was also obtained (ROHR and TROY 1980).

Sialyltransfer in *Meningococci group C* cell-free extracts showed specificity for transfer into group C polysaccharide (only α(2-9) linkages) and *E. coli K 92* polysaccharide (with alternating α(2-9) and α(2-8) linkages), but not into *Meningococci group B* polymer (only α(2-8) linkages). No involvement of a lipid intermediate could be detected here (VANN *et al.* 1978). The sialyltransfer reactions in bacterial systems thus show differences in specificity, synthesizing some structures (e.g. α(2-9) linkages) which have not yet been found in animal systems.

9. *Ectosialyltransferases*

The presence of glycosyltransferase activities at the surface of cells, i.e. located in the plasma membrane, remains controversial. Very low or no activity of sialyltransferases could be detected in rat kidney and liver plasma membrane

(MUNRO *et al.* 1975, FLEISCHER 1977, 1978). However, after the suggestion that such enzymes could play a role in the regulation of complex intercellular reactions such as cell-cell-recognition and adhesion (ROSEMAN 1970), a great effort has been directed towards the identification of such activities (see SCHUR and ROTH 1975 for a review). Evidence in favour of sialyltransferase activity at the surface of cells has been presented for mouse leukaemic cells (L-1200) in an ultrastructural study (PORTER and BERNACKI 1975), rat lymphocytes (CACAN *et al.* 1977, VERBERT *et al.* 1977, HOFLACK *et al.* 1979), mouse thymocytes (PAINTER and WHITE 1976) and Ehrlich ascites cells (CÉRVEN 1977, 1978). These are only a few examples from a large literature. Arguments against such activities point out that due to the low specific activity of sialyl(glycosyl)transferase and high specific activity of nucleotide-sugar hydrolases in the plasma membranes, a functional glycosyltransfer at this site is unlikely (SCHACHTER 1978). In accordance with this, such activities have been measured for CMP-Neu5Ac hydrolysis (KEAN and BIGHOUSE 1974, HIRSCHBERG *et al.* 1976, SPIK *et al.* 1979). Apparent sialyltransferase activity has been ascribed to this nucleotide-sugar hydrolysis and subsequent uptake of the sugar into the cell where incorporation occurs (DEPPERT *et al.* 1974, DEPPERT and WALTER 1978). The transport of sialyl(glycosyl)transferase out of the cell also may account for some activities measured in the plasma membrane (KEENAN and MORRÉ 1975). The lack of identification of the membrane product of the sialyltransferase reaction in most cases has precluded confirmation of the endogeneous substrate and does not eliminate the possibility of soluble glycoprotein adsorption. Further, technical difficulties have been encountered due to the leakage of CMP-Neu5Ac out of broken cells in cultures, thus invalidating quantitative results (HOFLACK *et al.* 1979). Further characterization and comparison is necessary in this field to clarify this situation.

10. Soluble Serum Sialyltransferases

Soluble sialyltransferase activity has often been measured in the serum of tumour patients and significant elevation detected (BERNACKI and KIM 1977, GANZINGER 1977, HENDERSON and KESSEL 1977, COOMBES *et al.* 1978, IP and DAO 1978, FOX *et al.* 1979). Liver, platelets and erythrocytes have been proposed as sources of the enzyme under normal conditions (KIM *et al.* 1971, BOSMANN 1972, MOOKERJEA *et al.* 1972). Neoplastic cells have been implicated as a source of the transferase in disease (BOSMANN and HALL 1974, BOSMANN and HILF 1974, BERNACKI and KIM 1977). At present the use of such measurements in diagnosis is not practicable, as too little characterization has been carried out.

IV. Catabolism of Sialic Acids

1. Introduction

The catabolism of the sialic acids is inseparable from the biosynthesis of the monosaccharides themselves and the various sialyltransfer reactions described in the previous subsections. Most information is available in the field of sialidase research, as will be evident below. However, the complete catabolism of the sialic acids entails several components, shown in Fig. 2, which exhibit links to all parts of the metabolic routes involving sialic acids.

2. *CMP-Sialic Acid Hydrolase*

The occurrence of an enzymic activity cleaving CMP-Neu5Ac to CMP and Neu5Ac has been described in sheep brain (SHOYAB and BACHHAWAT 1967), rat liver and isolated hepatocytes (KEAN and BIGHOUSE 1974, VAN DIJK *et al.* 1977), calf kidney cortex (VAN DIJK *et al.* 1976) and human ovary and serum (CHATTERJEE *et al.* 1978). The existence in serum has allowed monitoring of this activity in patients with malignant disease, where an elevation of activity was observed in cases of ovarian cancer (CHATTERJEE *et al.* 1978).

Careful analysis of the subcellular location has identified the plasma membrane as the predominant cellular site (KEAN and BIGHOUSE 1974). VAN DIJK *et al.* (1977) have further demonstrated that the enzyme activity is regional and with its functional centre directed towards the outside of the cell.

The enzyme is quite specific for CMP-β-Neu5Ac and is distinct from non-specific phosphodiesterase of nucleotide pyrophosphatase activities which do not cleave CMP-Neu5Ac. It may thus be considered as a β-sialidase, CMP-Neu5Ac being the only known natural sialic acid glycoside in β-linkage (HAVERKAMP *et al.* 1979 b).

The importance of CMP-Neu5Ac hydrolase in sialic acid metabolism has not yet been assessed. The occurrence in the plasma membranes, separated from the site of CMP-Neu5Ac synthesis (nucleus) or consumption (Golgi membranes), and the observation that the enzyme is feedback-inhibited by UDP-GlcNAc, are indications that a close integration of this enzyme in sialic acid metabolism exists.

3. *Sialidases*

The group of enzymes designated N-acetylneuraminosyl-glycohydrolase (EC 3.2.1.18), also termed sialidase or neuraminidase, comprise a large number of enzymes whose occurrence and range of substrate specificity bear witness to their vital role in sialic acid metabolism (DRZENIEK 1972, GOTTSCHALK and DRZENIEK 1972, ROSENBERG and SCHENGRUND 1976, CORFIELD *et al.* 1981 a).

The sialidases have been reported in a variety of microorganisms: the ortho- and paramyxoviruses; the *Pseudomonales*, *Eubacteriales*, *Mycoplasmatales*, *Rickettsiales* and the genus *Arthrobacter* under the bacteria; in *Protozoa*; *Mastigophora* and *Sporozoa* (ROMANOVSKA and WATKINS 1963, CRAMPEN *et al.* 1979), in fungi under the *Actinomycetales* (UCHIDA *et al.* 1974, MÜLLER 1975) and in the *Mycoplasma* (MÜLLER 1974). They are also widespread in the vertebrates. Many of the enzyme activities discovered have not been studied in detail, and the majority of the data available are concerned with a few well defined, mostly bacterial examples. Mammalian sialidases in general have not been widely studied due to their predominantly membrane-bound nature and the complex problems of tissue fractionation. Recent trends, however, have shown that this line of research is particularly valuable and important for a general understanding of complex carbohydrate and sialic acid metabolism.

A special feature of the sialidases is the wide range of potential substrates available (GOTTSCHALK and DRZENIEK 1972, DRZENIEK 1973, CORFIELD *et al.* 1981 a). Thus, substrate specificities have been observed ranging from very broad, relatively non-selective enzymes, e.g. *C. perfringens* sialidase, to those cleaving particular glycosidic linkages in a limited number of complex carbohydrates, e.g. lysosomal, ganglioside-specific sialidase from human liver. Examples of these complex carbohydrate substrates and the different sialic acids occurring in them are given by STRECKER and MONTREUIL (1979), CORFIELD *et al.* (1981 a) and in chapter B. Furthermore, these may be in cytosolic, soluble form or part of macromolecular complexes and membranes. With this in mind it is not surprising to find the mammalian sialidases showing a tissue-determined specificity much narrower than that observed for most microorganism sialidases.

The discovery of a number of metabolic diseases involving defective sialic acid metabolism, partially due to sialidase activity deletion or reduction, has been invaluable in pinpointing areas of importance for future study. These disorders called sialidoses are discussed in chapter K.

A further aspect of the wide range of substrates for sialidases has been the analysis of these substrates. The determination of oligosaccharide chain structures has allowed precise study of sialidase specificity. A need for a spectrum of well defined substrates has developed as a prerequisite for the elucidation of sialidase specificity. The large body of data which has accumulated on the sialidases points to a number of characteristics general to sialidase action, and forms the basis of current knowledge about sialidases.

a) Anomeric Specificity

The sialidases cleave α-glycosidically linked sialic acid residues (GOTTSCHALK 1958, KUHN and BROSSMER 1958). The sialic acids occurring in nature are found in the α-glycosidic conformation (see chapter B) with the exception of CMP-Neu5Ac which is a β-glycoside (HAVERKAMP *et al.* 1979 b) and which is not a substrate for sialidases.

The initial demonstration of the α-anomeric specificity was made with II^{3}Neu5AcLac (KUHN and BROSSMER 1958) with influenza and mumps virus sialidases. The preparation of synthetic β-glycosides of sialic acid allowed this specificity to be tested and verified for several viral and bacterial sialidases (MEINDL and TUPPY 1965, FAILLARD *et al.* 1966, KUHN *et al.* 1966). A further analysis was carried out on synthetic α-N- and α-S-glycosides of Neu5Ac, which proved not to be substrates for viral and bacterial sialidases (KHORLIN *et al.* 1970). Subsequent studies with *V. cholerae* sialidase and 4-nitrophenyl-α-Neu5Ac, however, showed low but significant release by the enzyme (ESCHENFELDER 1979).

The nature of the cleavage mechanism has been studied using n.m.r.-spectroscopy leading to the proposal that the α-glycoside is cleaved to yield α-Neu5Ac which then leaves the active site of the sialidase molecule and mutarotates in water to give over 90% of the β-anomer (HOLMQUIST and ÖSTMAN 1975, FRIEBOLIN *et al.* 1980).

b) The Carboxyl Group

Sialidases require a negatively charged carboxyl function on the α-linked sialic acid-containing substrate for catalytic activity. *V. cholerae* sialidase did not cleave the sialic acid methyl esters of bovine submandibular gland mucus glycoproteins (GOTTSCHALK 1962) or II^{3}Neu5AcLac (KUHN *et al.* 1966). In colominic acid, esters of sialic acid carboxyl groups with the C-7 or C-9 hydroxyl group of the adjacent sialic acid moiety blocked sialidase action (AMINOFF 1961), which could be alleviated by saponification (McGUIRE and BINKLEY 1963). The reduction of the carboxyl function to a primary alcohol and preparation of the benzyl-α-N-acetyl-D-nonulosamine or the corresponding II³Neu5AcLac derivative prevented the action of bacterial, viral and mammalian sialidases (BROSSMER and HOLMQUIST 1971, and unpublished).

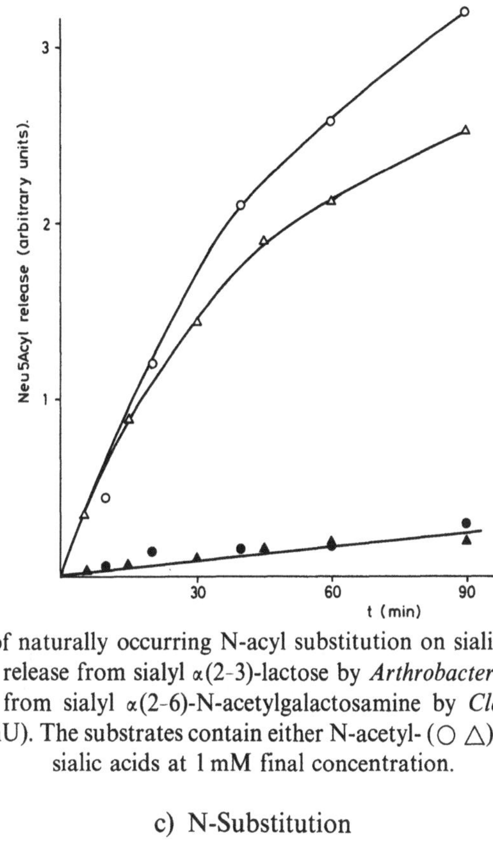

Fig. 5. The effect of naturally occurring N-acyl substitution on sialidase activity. A time curve of Neu5Acyl release from sialyl α(2-3)-lactose by *Arthrobacter ureafaciens* sialidase (△ ▲, 1 mU) and from sialyl α(2-6)-N-acetylgalactosamine by *Clostridium perfringens* enzyme (○ ●, 3.2 mU). The substrates contain either N-acetyl- (○ △) or N-glycolyl- (●, ▲) sialic acids at 1 mM final concentration.

c) N-Substitution

The nature of the N-substituent in sialic acid glycosides has a marked influence on the rate of cleavage by all classes of sialidases. The synthetic modification of the N-substituent to give a smaller residue as in N-formylneuraminic acid (BROSSMER and NEBELIN 1969), or larger residues as in N-propionylneuraminic acid (MEINDL and TUPPY 1966 b) and N-carbobenzoxyneuraminic acid (FAILLARD *et al.* 1969) resulted in reduced cleavage of the glycosides of N-formyl- and N-propionyl-neuraminic acid and loss of cleavage in the case of the N-carbobenzoxy derivative. Comparison of the cleavage rates of synthetic N-acetyl- and N-glycolylneuraminic acid-α-methyl glycosides revealed that the Neu5Ac derivative was released more rapidly by *V. cholerae* and influenza A virus sialidases (MEINDL and TUPPY 1966 b). The naturally occurring Neu5Ac and Neu5Gc derivatives are hydrolyzed by sialidases of viral, bacterial and mammalian origin with a marked preference for Neu5Ac. This was demonstrated using II³Neu5Ac(Gc)Lac, Neu5Ac(Gc)α(2-6)GalNAc and II³Neu5Ac(Gc)LacCer (CORFIELD *et al.* 1981 b) and is shown in Fig. 5. Viral sialidases showed largely reduced V_{max} values for the Neu5Gc-containing substrates, while bacterial sialidases showed increases in K_m and in some cases additional decreases in V_{max} values for the Neu5Gc substrates (CORFIELD *et al.* 1981 b).

d) O-Substitution

The discovery of a reduced cleavage rate of O-acetylated sialic acid by sialidase (SCHAUER and FAILLARD 1968) has now been extended to all classes of sialidase. The occurrence of O-acetyl-sialic acids is described in chapter B. In the initial investigation it was observed that O-acetylation in the sialic acid side chain led to a reduction of about 50% in cleavage rate from bovine submandibular gland mucin by *C. perfringens* sialidase (SCHAUER and FAILLARD 1968). A similar reduction in activity was observed with *V. cholerae* and *C. perfringens* sialidases using rabbit Tamm-Horsfall glycoprotein (NEUBERGER and RATCLIFFE 1972) or with *V. cholerae* and viral sialidases using 9-O-acetylated II^3Neu5AcLac isolated from rat urine (unpublished) as substrates. A particularly interesting ganglioside, II^3Neu5AcIV^3(Neu5,9Ac_2Neu5Ac)GgOse_4Cer, has been used to demonstrate the influence of this 9-O-acetyl substitution on the O-acetylated sialic acid moiety and also on another non-O-acetylated sialic acid in the same molecule. The IV^3Neu5,9Ac_2 was released more slowly than the corresponding non-O-acetylated residue, while the II^3Neu5Ac moiety was also released more slowly; all rates were compared to the non-O-acetylated ganglioside II^3Neu5Ac$IV^3$$(NeuAc)_2GgOse_4$Cer (GHIDONI *et al.* 1980).

The acetylation of the hydroxyl group at C-4 of the sialic acid molecule was shown to block sialidase action. Thus, Neu4,5Ac_2 found in equine submandibular gland mucin was resistant to the action of *C. perfringens* and *V. cholerae* sialidases (SCHAUER and FAILLARD 1968), as was the sialic acid moiety in II^3Neu4,5Ac_2Lac (MESSER 1974). The isolation of II^3Neu4Ac5GcLacCer from horse erythrocyte membranes and a horse serum glycoprotein containing Neu4,5Ac_2 provided two further substrates which were also resistant towards the action of *V. cholerae, C. perfringens* and *Arthrobacter ureafaciens* sialidases. No release was found with human brain, heart and liver and horse liver sialidases (SCHAUER *et al.* 1979 a, 1980 b, VEH *et al.* 1979, unpublished). In contrast, viral sialidases (Newcastle disease virus (NDV), fowl plague virus (FPV) and influenza A_2 virus (IA_2V)) were able to release low but significant amounts of 4-O-acetylated sialic acids.

Additional information on the effect of the substitution at C-4 has been obtained using asialo-fetuin resialylated with Neu5Ac4Me using equine sialyltransferase. FPV sialidase could cleave Neu5Ac4Me at a rate greater than for asialo-fetuin resialylated with Neu5Ac, while *V. cholerae* sialidase showed no activity with the Neu5Ac4Me substrate (BEAU and SCHAUER 1980).

e) Length of the Side Chain

The influence of the C-7 to C-9 side chain of the sialic acid molecule on sialidase action has been demonstrated using complex carbohydrates subjected to mild periodate oxidation and subsequent borohydride reduction to yield C-7 and C-8 analogues of sialic acid. The progressive shortening of the chain led to a progressive decrease in bacterial and viral sialidase action (SUTTAJIT and WINZLER

1971, Veh *et al.* 1977). Although a reduction in cleavage rate is observed for the C-8 and especially the C-7 derivatives, the introduction of a tritium label into the sialic acid side chain by reduction of the periodate-oxidized product with borotritide has provided a sensitive sialidase assay with several substrates including glycoproteins (Suttajit and Winzler 1971, Bernacki and Bosmann 1973) and gangliosides (Tallman and Brady 1973, Veh and Schauer 1978). Closer study of the action of mammalian sialidases on C-7 and C-8 sialic acid-containing substrates has revealed different specificities. In horse liver, two activities could be detected on the basis of their activity towards C-7 and C-8 sialic acid-containing gangliosides and glycoproteins (Sander *et al.* 1979 a, b).

f) Side Chain or Branch Sialic Acids

Sialic acid has been detected in oligosaccharide chains as a branch of an internal monosaccharide. Resistance of such sialyl groups to sialidase cleavage has been observed in some cases. Three examples illustrate this effect.

(i) The milk oligosaccharide IV^{3}Neu5AcIII^{6}Neu5AcLcOse$_4$,

Neu5Acα(2-3)Galβ(1-3)GlcNAcβ(1-3)Galβ(1-4)Glc
|α(2-6)
Neu5Ac

is a substrate for bacterial, viral and mammalian sialidases. The IV^{3}Neu5Ac residue is readily cleaved, while the side-positioned, α(2-6)-linked sialic acid is released much slower on incubation with NDV and FPV sialidases, partly due to the α(2-6) nature of the glycosidic linkage (see below), leading to III^{6}Neu5AcLcOse$_4$ as a product (von Nicolai *et al.* 1971). In contrast, bacterial sialidase released both sialic acids at similar rates, and the cleavage of the IV^{3}Neu5Ac residue did not appear to be influenced by the presence of the III^{6}NeuAc moiety (von Nicolai *et al.* 1971).

(ii) The tetrasaccharide O-glycosidically linked to serine and threonine found in e.g. fetuin, glycophorin A and κ-casein,

Neu5Acα(2-3)Galβ(1-3)GalNAc-O-Ser/Thr,
|α(2-6)
Neu5Ac

is cleaved by bacterial and viral sialidases at rates similar to those observed with the first example, the major difference in cleavage being due to the glycosidic linkages (α(2-3) or (2-6); unpublished). The significance of this structure for mammalian sialidases must await studies with suitable substrates.

(iii) The ganglioside II^{3}Neu5AcGgOse$_4$Cer and its oligosaccharide II^{3}Neu5AcGgOse$_4$,

Galβ(1-3)GalNAcβ(1-4)Galβ(1-4)Glc
|α(2-3)
Neu5Ac

are not substrates for *V. cholerae* and several viral sialidases including NDV, FPV, and IA$_2$V (SCHAUER *et al.* 1980 b, SUZUKI *et al.* 1980). The resistance to cleavage may be due to the substitution of the internal galactose C-4 atom by GalNAc, constituting a steric hindrance (KUHN and WIEGANDT 1963). Furthermore, n.m.r. analysis of II^{3}Neu5AcGgOse$_4$Cer led to the proposal of an oxygen cage involving the carbonyl oxygen of the N-acetyl group of GalNAc and the carbonyl moiety of sialic acid (HARRIS and THORNTON 1978). 500 MHz n.m.r. studies of II^{3}Neu5AcGgOse$_4$ and IV^{3}Neu5AcII^{3}Neu5AcGgOse$_4$, isolated from II^{3}Neu5AcGgOse$_4$Cer and IV^{3}Neu5AcII^{3}Neu5AcGgOse$_4$Cer, respectively, have shown that the glycosidic linkage of the side-chain sialic acid (II^{3}Neu5Ac) has much less conformational freedom than that of the terminal residue (IV^{3}Neu5Ac) in IV^{3}Neu5AcII^{3}Neu5AcGgOse$_4$ (unpublished).

In contrast to the non-cleavage of II^{3}Neu5AcGgOse$_4$Cer and its oligosaccharide by the sialidases mentioned above, *C. perfringens* sialidase was found to act on the ganglioside substrate (but not on the oligosaccharide), albeit at greatly reduced rates relative to substrates such as II^{3}Neu5AcLac or fetuin. This activity could be enhanced by bile salts (WENGER and WARDELL 1972), and a change in micellar structure proposed as a mechanism (RAUVALA 1976). The course of action of *C. perfringens* sialidase on II^{3}Neu5AcGgOse$_4$Cer without bile salt showed a transition below $10^{<4}$M, where a dramatic increase in cleavage rate took place (RAUVALA 1976, 1979, CORFIELD *et al.* 1980).

A number of sialidases have been discovered which will cleave II^{3}Neu5AcGgOse$_4$Cer without bile salt at rates significantly greater than *C. perfringens* sialidase. These include the sialidases from *A. ureafaciens* (SAITO *et al.* 1979) and Sendai virus (SUZUKI *et al.* 1980). Mammalian sialidases have also been described which act on this ganglioside in human liver (MICHALSKI *et al.* 1981, unpublished), and on II^{3}Neu5AcGgOse$_3$Cer from intestine (KOLODNY *et al.* 1971), rat and human heart (TALLMAN and BRADY 1973, PARKER *et al.* 1979) and human brain (PARKER *et al.* 1979). Other mammalian sialidases do not hydrolyze these internal sialic acids, and such compounds inhibit II^{3}Neu5AcLacCer and II^{3}Neu5AcIV^{3}Neu5AcGgOse$_4$Cer cleavage (e.g. SANDHOFF and PALLMANN 1978).

g) Influence of Aglycone and Natural Glycosidic Binding Partners

The aglycone of synthetic sialidase substrates showed no great differences between *V. cholerae* and influenza virus sialidase activities (MEINDL and TUPPY 1966 a, 1967). The K_m values measured for α-ketosides with methyl, pentyl, decyl, benzyl and phenyl groups amongst other aglycones were all within the range of 10^{-3}-7×10^{-3}M. The introduction of a positive charge into the aglycone led to increased K_m values (HOLMQUIST and BROSSMER 1972) with *V. cholerae* sialidase.

Recent studies have implicated the existence of a hydrophobic centre distinct from the catalytic centre in *V. cholerae* (KEILICH *et al.* 1979) and *C. perfringens* (CORFIELD *et al.* 1980) sialidases. The detection of multiple sialidase activities in bacterial, protozoan and mammalian sialidase preparations using immobilized N-(4-nitrophenyl)-oxamic acid columns has supported these observations (BROSSMER *et al.* 1977 b, CRAMPEN *et al.* 1979, ZIEGLER *et al.* 1980). Those sialidases showing affinity for the immobilized ligand also showed low activity with synthetic substrates containing hydrophobic aglycones (e.g. the benzyl-α-ketoside of Neu5Ac), while the enzyme not binding to the column cleaved such substrates readily and at similar rates to other non-aromatic aglycones (KEILICH *et al.* 1979, ZIEGLER *et al.* 1980).

The extension of such studies to natural substrates is difficult. Evidence is now accumulating that a specificity for the size and structure of the oligosaccharide chain and the nature of the complex carbohydrate, i.e. glycoprotein or glycolipid etc., is important. These phenomena can be illustrated for all classes of sialidases. A study of virus sialidase specificity has shown that, compared to II^6Neu5AcLac (as 100), the rates of cleavage for Neu5Acα(2-6)Galβ(1-4)GlcNAc, Neu5Acα(2-6)Galβ(1-4)GlcNAc-asparagine, Neu5Acα(2-6)Galβ(1-4)GlcNAcβ(1-2)Manα(1-3)Manβ(1-4)GlcNAc and Neu5Acα(2-6)GalNAc were 1,000, 700, 300, and 350, respectively, for NDV sialidase, while IA_2V sialidase showed rates of 425, 1.2, 225, and 225, respectively (CORFIELD *et al.* 1981 a, and unpublished).

Bacterial sialidases behave similarly with bovine submandibular gland mucus glycoprotein. Using the saponified intact mucin (containing only Neu5Ac or Neu5Gc), mucin glycopeptides and the disaccharides Neu5Acα(2-6)GalNAc and Neu5Gcα(2-6)GalNAc from the same mucin, large differences in cleavage rate are observed in the order glycopeptides > > glycoprotein > Neu5Acα(2-6)GalNAc > Neu5Acα(2-6)GalNAc (CORFIELD *et al.* 1981 a, and unpublished).

Mammalian sialidases show a narrower substrate specificity. For instance, the sialidase isolated from rat heart by TALLMAN and BRADY (1973) cleaved fetuin at twice the rate of II^3Neu5AcIV^3Neu5AcGgOse$_4$Cer, but did not act on II^3Neu5AcLac. Rat liver Golgi membrane sialidase was found to be active only with sialyllactose isomers but not with fetuin, ovine submandibular gland mucin and a ganglioside mixture (KISHORE *et al.* 1975). A sialidase isolated from human liver lysosomes by affinity chromatography was active with gangliosides and bovine submandibular mucin, but did not hydrolyze II^3Neu5AcLac or fetuin (MICHALSKI *et al.* 1981).

h) Glycosidic Linkage

The occurrence of α(2-3), (2-6), and (2-8) linkages in different complex carbohydrates (see chapter B) has raised the question of glycosidic linkage specificity of the sialidases. The viral sialidases in general cleave the α(2-3) linkage at a much greater rate than the α(2-6) bonds, while hydrolysis of α(2-8) linkages ranged between moderate to poor with different viruses (DRZENIEK 1972, 1973, CORFIELD *et al.* 1981 a). Bacterial sialidases show a glycosidic specificity, too, but this is less marked than for the viral enzymes and only becomes significant at low enzyme concentrations (DRZENIEK 1972, 1973, ROSENBERG and SCHENGRUND 1976, CORFIELD *et al.* 1981 a). Mammalian sialidases have not been widely studied in this respect, but some recent results suggest that the same type of specificity exists (MICHALSKI *et al.* 1981).

It is important to realize that α(2-3), (2-6), and (2-8) linkages occur in all complex carbohydrate classes, and may be combined within one molecule in some cases (see chapter B, STRECKER and MONTREUIL 1979, CORFIELD *et al.* 1981 a), and that the specificity outlined above also relates to the different classes of complex

Table 3. *The glycosidic linkage specificity of some sialidases.* The rates of reaction for are expressed relative to II³Neu5AcLac as 100%; —, not

Sialidase	α(2-3) II³Neu5Ac-Lac	α(2-3) II³Neu5Gc-Lac	α(2-6) II⁶Neu5Ac-Lac	α(2-8)/(2-3) II³(Neu5Ac)₂-Lac
Clostridium perfringens	100	20	44	44
Vibrio cholerae	100	25	63	31
Arthrobacter ureafaciens	100	7	166	54
Newcastle disease virus	100	11	0.2	78
Influenza A_2 virus	100	15	0.4	3
Fowl plague virus	100	37	2	14
Rat liver Golgi	100	—	58	—
Human liver lysosomal	100	85	27	13
Human fibroblast	100	—	25	10

carbohydrates. A comparison of glycosidic specificities is given in Table 3 for several sialidases.

New studies have expanded the work of DRZENIEK (1972, 1973) with regard to the glycosidic linkage specificity of viral and bacterial sialidases (CORFIELD *et al.* 1981 a, b), allowing discrimination between three different viral sialidase activities on the basis of α(2-3) and (2-6) linkages and the nature of the oligosaccharide chain in substrates as described under IV.3.g).

A general division of sialidases into two groups, being either oligosaccharide- or glycoprotein-specific has been made (VON NICOLAI 1976). A third group can be added, represented by *A. ureafaciens* sialidase, showing glycolipid preference. A further subdivision of the bacterial sialidases has been proposed by VON NICOLAI (1976) based on their behaviour with oligosaccharide substrates, and the relationship of pathogenicity to this specificity discussed (MÜLLER 1974, VON NICOLAI 1976).

The cleavage rate for mammalian sialidases can be ordered in the following sequence: α(2-3) > (2-6) > (2-8) bonds (see CORFIELD *et al.* 1981 a for a review). Differences in the rates of cleavage of these three linkages in sialyl- and disialyllactose have been documented for human fibroblast sialidase (CAIMI *et al.* 1979), of α(2-3) and (2-6) linkages in Neu5AcαGalβ(1-4)GlcNAcβ(1-2)Manα(1-3)Man(1-4)GlcNAc for human fibroblast sialidase (STRECKER and MICHALSKI 1978, CANTZ and MESSER 1979, POTIER *et al.* 1979) and of α(2-3) and (2-8) linkages in gangliosides (CAIMI *et al.* 1979, CANTZ and MESSER 1979) for the same enzyme. Although these differences in α(2-3), (2-6), and (2-8) cleavage rates have been demonstrated, the existence of α(2-3)- or (2-6)-specific enzymes remains a controversy. These studies have been carried out largely with cell homogenates, and fractionation and solubilization of the different activities is still outstanding.

i) Molecular Parameters

The bacterial sialidases generally fall in the molecular weight range of 50,000 to 100,000 daltons. No definitive evidence for subunit structure is available. Multiple viral, bacterial and mammalian sialidases are compared with different substrates. Results determined; *, in the presence of detergent

α(2-3) II³Neu5Ac-LacCer	α(2-3) II³Neu5Gc-LacCer	α(2-3) > (2-6) Fetuin	References
300	17	272	CORFIELD *et al.* 1981 a, b
220	32	340	SCHAUER 1982
31 (99)*	9	41	SAITO *et al.* 1979
32	30	6	CORFIELD *et al.* 1981 a, b
23	18	14	
42	35	5	SCHAUER 1982
—	—	0	KISHORE *et al.* 1975
480	120	310	CORFIELD *et al.* 1981 a and unpublished
25	—	26	CAIMI *et al.* 1979

forms of sialidase from pneumococcal cultures proved to have different charges but the same molecular weight (TANENBAUM and SUN 1971). In contrast, multiple forms of the enzyme from other bacteria, e.g. *V. cholerae*, *C. perfringens* and *Pasteurella multocida* (SCHRAMM and MOHR 1959, ROSENBERG *et al.* 1960, MÜLLER 1971, NEES *et al.* 1975) could be related to the age of the culture and may represent the products of proteolytic digestion.

Viral sialidases are membrane components, and as such show greater thermal stability relative to bacterial enzymes. They require purification after release from the virus. Complexes in the range of $1\text{-}2 \times 10^5$ daltons, or monomers in the range of $5\text{-}6 \times 10^4$ daltons have been reported for sialidase preparations without proteolytic degradation (see DRZENIEK 1972, ROSENBERG and SCHENGRUND 1976).

Electron microscopic studies with influenza B/Lee sialidase have demonstrated a tetrameric structure consisting of four coplanar glycoprotein units (WRIGLEY *et al.* 1973). The influenza N2 sialidase was found to contain 46% carbohydrate and to consist of four subunits giving a total molecular weight of approximately 150,000. The subunits show a molecular weight of 33,500 (GROOME *et al.* 1977). Other examples of viral sialidases with similar structural features have been reviewed by DRZENIEK (1972) and ROSENBERG and SCHENGRUND (1976).

j) Kinetic Parameters and pH Optima

The measurement of K_m values for sialidases from viral, bacterial and mammalian sources has yielded values of approximately 10^{-3} M with sialyllactose as substrate (DRZENIEK 1972, 1973, ROSENBERG and SCHENGRUND 1976, CORFIELD *et al.* 1981 a). Serum glycoproteins have given values in the range of 10^{-5}-10^{-6} M for viral and bacterial enzymes, while mammalian enzymes exhibit K_m values at about 10^{-4} M (DRZENIEK 1972, 1973, CORFIELD *et al.* 1981 a). Gangliosides present a special case, as early reports of K_m values in the region of 10^{-5} M without detergent were found to be increased on addition of such compounds with bacterial enzymes (e.g. *A. ureafaciens* sialidase, SAITO *et al.* 1979). Mammalian sialidase preparations also showed K_m values in the range of 10^{-5}M, but these values have subsequently been revised to approximately 10^{-3}M to allow for adsorbance of the glycolipid substrate to the enzyme membrane preparation (TETTAMANTI *et al.* 1976, VEH and SCHAUER 1978). Smaller differences in K_m and V_{max} values can be observed between substrates in the different groups of complex carbohydrates, e.g. the acetyl and glycolyl derivatives of neuraminic acid in the substrates discussed under IV.3c).

The pH optima of viral and bacterial sialidases generally lie in the range of pH 5-6 (DRZENIEK 1972, 1973, CORFIELD *et al.* 1981a). Variations depend on the nature of the substrate, e.g. II^{3}Neu5AcLac at pH 5.0-5.5 and colominic acid at pH 4.3-4.5 for *A. ureafaciens* sialidase (UCHIDA *et al.* 1979), the nature of the buffer system (DRZENIEK 1972) and the ionic strength with glycolipid substrates (LIPOVAC *et al.* 1973, SAITO *et al.* 1979). Mammalian sialidases show lower pH optima, around pH 4.0, with a wide range of substrates for lysosomal or other membrane-bound enzymes. Soluble sialidase activity has been reported with pH optima in the range of 5-6 (see CORFIELD *et al.* 1981a).

k) Activators and Inhibitors

Initial studies with sialidases demonstrated that some enzymes need divalent cation cofactors, Ca^{2+} being the most effective (DRZENIEK 1972, 1973). For instance, *V. cholerae*, *Diplococcus pneumoniae* and some influenza virus sialidases require Ca^{2+} ions for activity and are consequently inhibited by the addition of EDTA. In the case of *V. cholerae* sialidase it was concluded that the Ca^{2+} ions are required for stabilization of the enzyme rather than for direct enzyme-substrate binding. Mammalian enzymes are more complex and some examples of activation due to mono- and divalent cations were found to be concentration-dependent and related to the nature of the substrate (gangliosides), thus suggesting a substrate-enzyme interaction (SCHENGRUND and NELSON 1975).

Sialidase inhibitors can be divided up into four groups: (i) Naturally occurring high molecular weight compounds, (ii) naturally occurring low molecular weight compounds, (iii) synthetic compounds which may or may not resemble sialic acids and (iv) inorganic ions and EDTA. Some examples are given in Table 4. The different specificities of sialidase inhibitors with respect to the substrate for viral and bacterial enzymes is of great value in the analysis of mixtures of sialidase activities, especially in the case of mammalian systems where several sialidase activities of different specificity may occur (CORFIELD *et al.* 1981a, MICHALSKI *et al.* 1981, SCHAUER and CORFIELD 1981). In addition, the immobilization of inhibitors has proved invaluable in the purification of sialidase activities (e.g. ZIEGLER *et al.* 1980, SCHAUER and CORFIELD 1981).

4. Acylneuraminate Pyruvate-Lyase

a) Bacteria

The enzymic conversion of Neu5Ac to ManNAc and pyruvate was initially identified in *V. cholerae* (HEIMER and MEYER 1956) and *C. perfringens* (COMB and ROSEMAN 1960). The enzyme responsible for the conversion, acylneuraminate

Table 4. *The influence of inhibitors on viral, bacterial and mammalian sialidases*
The strength of inhibition is indicated as weak (+) to very strong (+ + +) or no inhibition (—). High (H) and low (L) molecular weight inhibitors are indicated and inhibition as competitive (C) or non-competitive (NC). ND, not determined

Inhibitor	Sialidase			Molecular weight	Type of inhibition	References
	viral	bacterial	mammalian			
Natural Compounds						
DNA, RNA, heparin	+ + +	+	ND	H	NC	DRZENIEK 1972, 1973
PSM	+ + +	+ +	ND	H	C	DRZENIEK 1972, 1973
ESM	+ +	+ +	ND	H	C	SANDER *et al.* 1979 a, b
Neuraminin	+ +	—	ND	H	C	LIN *et al.* 1977
Concanavalin A	+ +	—	ND	H	NC	ZALAN *et al.* 1975
II³Neu5AcGgOse$_4$Cer micelles	ND	+ +	+ +	H	ND	VEH and SCHAUER 1978
Neu5Ac	+	+	+	L	C	DRZENIEK 1972, 1973, VEH and SCHAUER 1978
Neu5Ac2en	+ + +	+ + +	+ +/—	L	C	VEH and SCHAUER 1978, MEINDL *et al.* 1974, MILLER *et al.* 1978
II⁶Neu5AcLac	+	—	ND	L	C	DRZENIEK 1972
Siastatin A, B	—	+ +/—	+ +/—	L	C	AOYAGI *et al.* 1975
Panosialin	+ +	ND	ND	L	C	AOYAGI *et al.* 1971
Synthetic Compounds						
Dextran sulphate	+ + +	+	ND	H	NC	DRZENIEK 1972, 1973
Neu5Acyl2en	+ + +	+ + +	+ +	L	C	MEINDL *et al.* 1974, VEH and SCHAUER 1978
S and N-ketosides of Neu5Ac	+ + +	+ + +	ND	L	C	KHORLIN *et al.* 1970
3-Aza-2,3,4-tri-deoxy-4-oxo-D-arabino-octonic acid δ lactone	+ + +	+ + +	ND	L	C	KHORLIN *et al.* 1970
N-substituted oxamic acids	+	+ +/—	+ +/—	L	C	CORFIELD *et al.* 1981 a, BROSSMER *et al.* 1977 a, VEH and SCHAUER 1978
Ions						
Cu^{2+}	+ +	+ +	+ +	L	C	ROSENBERG and SCHENGRUND 1976
Hg^{2+}	+ +	+ +	+ +	L	C	DRZENIEK 1972, 1973
Fe^{3+}	ND	ND	+ +	L	C	CORFIELD *et al.* 1981 a, CABEZAS 1978

pyruvate-lyase (EC 4.1.3.3), has been detected in a number of bacteria (FAILLARD *et al.* 1969, BARNETT *et al.* 1971, DEVRIES and BINKLEY 1972), often together with sialidase (MÜLLER 1974), to enable a conversion of Neu5Ac to glucose for energy metabolism via the enzymic conversions discussed under II.

The enzyme from *C. perfringens* has been prepared to homogeneity with 770-fold purification, yielding 12 mg enzyme protein/kg wet cell weight and a specific activity of 167 nkat/mg protein (NEES *et al.* 1976). The enzyme belongs to the group of class I aldolases, and the formation of a Schiff's base between substrate and an enzyme lysine moiety was confirmed (BARNETT *et al.* 1971, SCHAUER and WEMBER 1971, NEES *et al.* 1976). Irreversible inhibition of the enzyme occurs after borohydride reduction in the presence of the substrates Neu5Ac or pyruvate

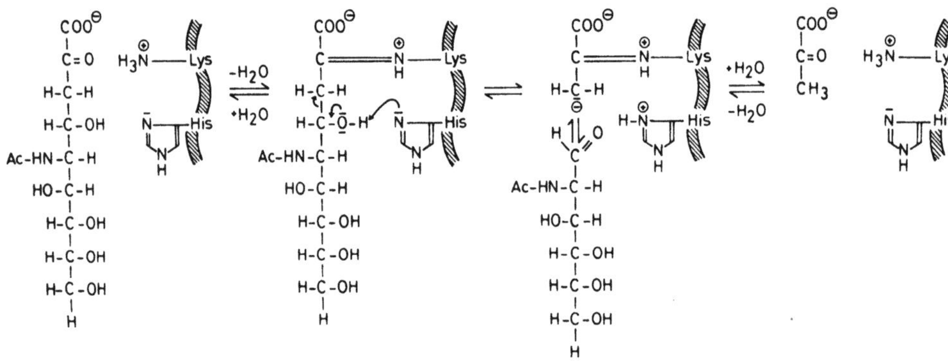

Fig. 6. Proposed mechanism of reversible cleavage of Neu5Ac by acylneuraminate pyruvate-lyase (from NEES *et al.* 1976).

(SCHAUER and WEMBER 1971). The nature of the nucleophilic amino acid accepting a proton from the C-4 hydroxyl group of Neu5Ac during the cleavage reaction was proposed to be histidine (SCHAUER and WEMBER 1971) or cysteine, the latter based on inhibitory experiments with bromopyruvate (BARNETT 1967, BARNETT *et al.* 1971) and chloropyruvate (BARNETT and KOLISIS 1974). Evidence in favour of histidine and supporting the earlier photooxidation experiments (SCHAUER and WEMBER 1971) came from inhibition studies with 5-diazonium-1H-tetrazole at different pH values and after alkylation with iodoacetamide to allow discrimination between tyrosine and histidine (NEES *et al.* 1976). In addition, the pH dependency of K_m showed a catalytic centre ionizing group with a pK of 6.4, compatible with the histidine imidazole, but not with cysteine. These experiments led to the following reaction scheme (Fig. 6).

A molecular weight in the range of 90,000–100,000 has been reported (DEVRIES and BINKLEY 1972, NEES *et al.* 1976). The purified enzyme could be dissociated into two subunits each with a molecular weight of 50,000. Such subunits were also demonstrated directly by electron microscopy (NEES *et al.* 1976) and the binding of two molecules of chloropyruvate per enzyme molecule (BARNETT and KOLISIS 1974) suggested the occurrence of two active centres per molecule.

The enzyme catalyzes an equilibrium reaction which is in favour of the breakdown of Neu5Ac, and is therefore a degradative enzyme in bacteria. The acylmannosamine produced can be metabolized to produce energy (as shown e.g. in the yeast *Candida albicans* by BISWAS *et al.* (1979)).

The K_m for Neu5Ac has been reported in the range of 2.8–3.9 mM (COMB and ROSEMAN 1960, BARNETT *et al.* 1971, NEES *et al.* 1976) for the *C. perfringens* lyase, the lower value being with the purified enzyme. Pyruvate is a competitive inhibitor with a K_i of approximately 3 mM (BARNETT *et al.* 1971, NEES *et al.* 1976), and 10 mM pyruvate was sufficient to completely inhibit the purified lyase. This observation has been correlated with the increased stability of the enzyme to heat denaturation in the presence of pyruvate. The purified enzyme was found to have a pH optimum at pH 7.2 (BRUNETTI *et al.* 1962, NEES *et al.* 1976).

The enzyme was shown to be active only with the free monosaccharide; glycosidically linked sialic acids as in α_1-acid glycoprotein or colominic acid and also the β-methyl glycoside were not cleaved (COMB and ROSEMAN 1960, COMB *et al.* 1966). O-Acetylated sialic acids were cleaved at slower rates than Neu5Ac and Neu5Gc, which showed similar rates of cleavage (100 and 93, respectively). Thus, Neu5,9Ac_2 exhibited 97, Neu5,7Ac_2 56 and Neu4,5Ac_2 22 relative cleavage rates, and the O-acetylated derivatives of Neu5Gc reacted in a similar way (SCHAUER *et al.* 1971, CORFIELD *et al.* 1976). The significant reduction in cleavage ($\sim$ 80%) observed for Neu4,5Ac_2 when compared with Neu5Ac is in good agreement with the requirement of a proton from the C-4 hydroxyl group in the reaction mechanism (Fig. 6). There is further support for this proposal using Neu5Ac4Me which was not cleaved at all by the *C. perfringens* lyase (BEAU and SCHAUER 1980). Neu5Ac9P (BRUNETTI *et al.* 1962) and Neu5Ac2en (SCHAUER *et al.* 1971) were also not cleaved.

A role for the glycerol side chain in the enzymic mechanism has been demonstrated using the C-7 and C-8 analogues of Neu5Ac. The C-8 derivative showed only 20% of the activity with Neu5Ac and the K_m was two times larger, while the C-7 derivative was not cleaved (SUTTAJIT *et al.* 1971). A requirement for an ionizable carboxyl group was demonstrated by the non-cleavage of the methyl ester of Neu5Ac (SCHAUER *et al.* 1971). The N-substituent appears to be less important for binding and catalysis of the lyase reaction. No great influence on the rate of catalysis was seen for a variety of N-substituted derivatives of neuraminic acid with the exception of N-fluoroacetylneuraminic acid, which showed only 28% of the rate relative to Neu5Ac (FAILLARD *et al.* 1969, SCHAUER *et al.* 1971). Two other synthetic derivatives of sialic acid, which are not split by the bacterial lyase, are 3-fluoro-Neu5Ac and 3-hydroxy-Neu5Ac (GANTT *et al.* 1964, DEVRIES and BINKLEY 1972).

The 4-O-acetyl derivative of Neu5Ac is a competitive inhibitor and substrate along with N-monofluoroacetylneuraminic acid (SCHAUER *et al.* 1971). The C-8 analogue of Neu5Ac inhibited competitively (SUTTAJIT *et al.* 1971), while 3-fluoro-Neu5Ac (GANTT *et al.* 1964) and 3-hydroxy-Neu5Ac (DEVRIES and BINKLEY 1972) also inhibited but were not substrates.

Further non-specific inhibitors include the heavy metal ions Cu^{2+}, Fe^{2+}, and Hg^{2+} (NEES *et al.* 1976) as well as p-chloromercuribenzoate, diisopropylfluorophosphate and o-phenanthroline (SCHAUER and WEMBER 1971).

b) Mammals

Acylneuraminate pyruvate-lyase has been demonstrated in hog kidney (BRUNETTI *et al.* 1962), beef kidney (SIRBASKU and BINKLEY 1970), mouse (NÖHLE and SCHAUER 1981), rat intestine (WITT-KRAUSE *et al.* 1979) and human liver (unpublished). Studies on the uptake of Neu5Ac and Neu5Gc in the mouse and rat have provided evidence that an acylneuraminate pyruvate-lyase is active in most tissues utilizing free sialic acid (NÖHLE and SCHAUER 1981).

The enzyme from hog kidney was purified approximately 1700-fold and showed similar K_m, pH optima and substrate specificities to the *C. perfringens* lyase (BRUNETTI *et al.* 1962). These authors concluded that kidney was the major site of this enzyme. Experiments in the authors' laboratory have shown that the lyase from hog kidney can be stabilized by pyruvate, as suggested by KOLISIS *et al.* (1981) for the *C. perfringens* enzyme, and affinity chromatography on columns of Neu-β-Me linked to a polymethylacrylate hydrazide-Sepharose 4B co-polymer via the sialic acid amino group has been used to improve the purification of the enzyme (unpublished). Studies on the mechanism of enzyme catalysis have shown that this lyase also belongs to the class I aldolases.

The lyase from beef kidney was partially purified and showed similar properties to the bacterial enzyme with respect to K_m and pH optimum; it had a molecular weight of 55,000 on Sephadex G-200 and was competitively inhibited by Neu5Ac (SIRBASKU and BINKLEY 1970).

The cellular localization of the acylneuraminate pyruvate-lyase has not been identified. Human liver, fractionated by classical ultracentrifugation methods, showed activity only in the 100,000 g supernatant (unpublished).

A catabolic role for the lyase in mammalian tissues is favoured, as several mucin-synthesizing tissues, e.g. submandibular glands, were found to be lacking lyase activity (BRUNETTI *et al.* 1962). Furthermore, rapid decomposition of Neu5Ac and Neu5Gc to acylmannosamines in various rat and mouse tissue homogenates was observed (NÖHLE and SCHAUER 1981, NÖHLE *et al.* 1982).

V. Uptake and Recycling of Sialic Acids in Cells

Very little work has been concerned with the transport of sialic acids into cells, its utilization within the cell and the possibilities of recycling. As outlined in section II, free Neu5Ac occurs as an intermediate in complex carbohydrate biosynthesis. In spite of this fact several lines of evidence support the slow or poor uptake of free Neu5Ac into intact tissues (HARMS *et al.* 1973, HIRSCHBERG and YEH 1977), the most significant being the comparison of the rate of uptake with GlcN or ManNAc (KOHN *et al.* 1962, HARMS *et al.* 1973). While GlcN is converted to Neu5Ac and other monosaccharides, ManNAc is more specific (although some conversion to GlcNAc occurs in e.g. liver (HARMS *et al.* 1973) as a Neu5Ac precursor (e.g. QUARLES and BRADY 1971, YOHE *et al.* 1980).

The significance of the poor uptake of Neu5Ac has still to be evaluated. Evidence for a Neu5Ac permease activity has only been presented in *C. perfringens* (NEES and SCHAUER 1974). The administration of Neu5Ac orally, intravenously or intraperitoneally leads to partial uptake and incorporation into glycoproteins and glycolipids (CAREY and HIRSCHBERG 1979, MORGAN and WINICK 1979, WITT *et al.* 1979, FAN and DATTA 1980, NÖHLE and SCHAUER 1981). The use of radioactively labelled Neu5Ac has been assumed to lead directly to radioactive Neu5Ac in complex carbohydrates and in some cases has been identified (DATTA 1974, HIRSCHBERG *et al.* 1976, CAREY and HIRSCHBERG 1979, FAN and DATTA 1980, NÖHLE and SCHAUER 1981). However, a single radioactive label in C-1 to C-3 or C-4 to C-9 will not allow any conclusion to be drawn whether the extracellular

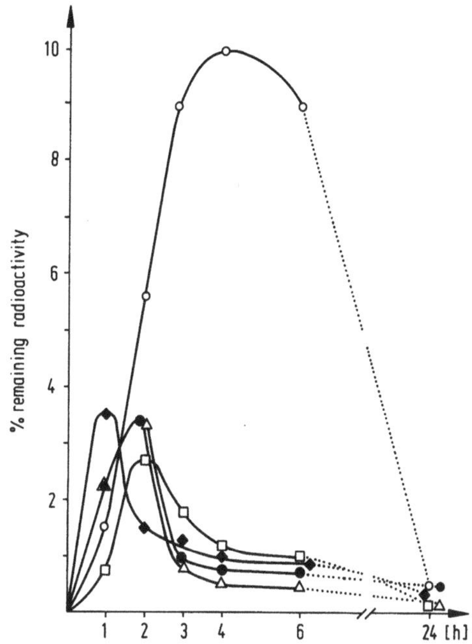

Fig. 7. Oral administration of sialic acids to mice. The sum of remaining radioactivity in blood, liver, spleen, kidney, and brain after oral application each of 2 mg [2-^{14}C,9-^{3}H]Neu5Ac, 2 mg [2-^{14}C,9-^{3}H]Neu5Gc and 0.087 mg [^{3}H]Neu5Ac2en to fasted mice (strain C 57) is shown (NÖHLE and SCHAUER 1981, NÖHLE *et al.* 1982). Symbols: ○ ^{3}H; □ ^{14}C from Neu5Ac; ● ^{3}H; △ ^{14}C from Neu5Gc; ◆ ^{3}H from Neu5Ac2en.

Neu5Ac was cleaved by Neu5Ac pyruvate-lyase and entered the cell as ManNAc, followed by resynthesis and incorporation, or whether the same sequence occurs within the cell after transport of Neu5Ac across the cell membrane. A final possibility is that the Neu5Ac is not cleaved at all by the lyase and is taken up as Neu5Ac and activated.

Experiments with cultured cells (DATTA 1974, HIRSCHBERG *et al.* 1976, HIRSCHBERG and YEH 1977, FAN and DATTA 1980) and intact animals (CAREY and HIRSCHBERG 1979, NÖHLE and SCHAUER 1981) have identified Neu5Ac as the radioactive product taken up by the cell. CAREY and HIRSCHBERG (1979) used dual labelled Neu5Ac to detect cleavage of the Neu5Ac skeleton on or after uptake. They found isotope ratio differences between free (corresponding to the administered Neu5Ac dose) and covalently bound (CMP-Neu5Ac and glycoconjugates) sialic acid to be less than 12% and proposed that no significant cleavage of Neu5Ac occurred during uptake and activation. Experiments performed in the authors' laboratory *in vivo* with mice and rats confirmed these findings. However, on study of the $^{14}C/^{3}H$ ratio for longer times (up to 6 h at least), the experiments demonstrated a much greater difference in this ratio (NÖHLE and SCHAUER 1981). This variation in the ratio was especially evident when Neu5Ac was orally administered, first in the intestine and then being mirrored in whole blood, brain, liver, kidney and spleen (Fig. 7). Excretion of 60–90% of the Neu5Ac in the urine within 6 h occurred without alteration of the isotopic ratio. Intravenous injection in rats led to a more rapid excretion in unchanged form in the urine, and the change in isotopic ratio in label incorporated into liver, spleen, kidney and brain. Neu5Ac pyruvate-lyase activity was present in liver, spleen, kidney and brain, and a cleavage of Neu5Ac is proposed. Whether the lyase activity is at the cell surface or inside the cell is still not clear, and a transport system for Neu5Ac into the cell remains to be demonstrated. When Neu5Gc was used in such experiments similar results were obtained except that the percentage uptake was less than that observed for Neu5Ac.

Further studies using Neu5Ac have been related to the metabolism of extracellular CMP-Neu5Ac and the detection of ectosialyltransferase activity. On the basis of a lag period observed for Neu5Ac uptake and incorporation into glycoconjugates, and the absence of such a lag for CMP-Neu5Ac incorporation, two routes have been proposed. The uptake of Neu5Ac and metabolism as described above has been studied in hamster and mouse fibroblasts, and cell surface labelling of glycoprotein and glycolipid demonstrated (DATTA 1974). The breakdown of CMP-Neu5Ac was shown, and incorporation due to Neu5Ac uptake rather than direct CMP-Neu5Ac transfer proposed (HIRSCHBERG *et al.* 1976). The uptake of CMP-Neu5Ac into the cells (NIL, BHK and 3T3 fibroblasts) could be ruled out, and the K_m for Neu5Ac uptake was estimated to be 10 mM. Other experiments with CMP-Neu5Ac and intact cell cultures (PAINTER and WHITE 1976, CERVÉN 1977, see section III.9) pointed to surface sialyltransferase. Further studies by FAN and DATTA (1980) provided evidence that both transfer and transport occur, by localization of acceptors within the cell and on the cell surface (plasma membrane), and direct demonstration of the presence of a plasma membrane sialyltransferase. The sialylation due to Neu5Ac uptake occurs (at least initially) with different acceptors in comparison with CMP-Neu5Ac plasma membrane sialylation.

A further important observation has been that of a microsomal intraluminal pool of CMP-Neu5Ac (CAREY *et al.* 1980). Penetration of CMP-Neu5Ac into microsomal vesicles occurs via a carrier-mediated transport system, and is in accordance with the site of sialylation in these membranes (CAREY *et al.* 1980, CREEK and MORRÉ 1981).

A study in newborn rats compared the uptake of [4,5,6,7,8,9-^{14}C]Neu5Ac and II^{3}-[^{14}C]Neu5Ac-Lac administered orally to the animals (WITT *et al.* 1979). The uptake and distribution of Neu5Ac was similar to that reported above, but differences were observed for II^{3}-[^{14}C]Neu5Ac-Lac. The absorption of radioactivity from the trisaccharide was delayed by 30 min relative to Neu5Ac, and slower but higher incorporation in the same tissues found for Neu5Ac.

Presumably the II³Neu5AcLac is cleaved to Neu5Ac and lactose by intestinal sialidase (DICKSON and MESSER 1978, WITT *et al.* 1979).

Administration of Neu5Ac *in vivo* leads to incorporation into glycoproteins and glycolipids (DATTA 1974, HARMS *et al.* 1973, HIRSCHBERG *et al.* 1976, CAREY and HIRSCHBERG 1979, MORGAN and WINICK 1979, 1980b). The Neu5Ac content of these substances early in development was related to learning ability (MORGAN and WINICK 1979) and environmental stimulation (MORGAN and WINICK 1980b). Intraperitoneal injection of Neu5Ac in malnourished rats was found to prevent the expected behavioural abnormalities, and increased glycoprotein and ganglioside Neu5Ac content significantly. It remains unclear whether Neu5Ac was decomposed in the peritoneal cavity before uptake.

The final area of discussion in this section is the turnover of sialic acid in glycoproteins and glycolipids. Evidence for an individual turnover of sialic acids in glycoconjugates relative to the protein or lipid and other monosaccharide components has been presented for brain tissue (MARGOLIS and MARGOLIS 1973, 1977, LEDEEN *et al.* 1976), in rat liver and hepatoma plasma membrane (HARMS and REUTTER 1974, GRÜNHOLZ *et al.* 1977) and in regenerating rat liver (TAUBER and REUTTER 1978). In brain, evidence for two pools of sialic acid turning over with a half-life of 4.6 and 15 days respectively, was found (MARGOLIS and MARGOLIS 1973). Only sulphate and fucose were found to turnover more rapidly. Protein turnover was significantly slower than all oligosaccharide components and sulphate. Rat liver plasma membrane in normal and regenerating liver showed a more rapid turnover of sialic acid relative to the cell homogenate (GURD and EVANS 1973, HARMS and REUTTER 1974, TAUBER and REUTTER 1978). In normal liver the turnover of Neu5Ac was at a constant rate, while that of fucose was found to be biphasic. This linearity for Neu5Ac was also observed in regenerating liver, but the half-life of protein-bound Neu5Ac was doubled (TAUBER and REUTTER 1978). The protein and carbohydrate moieties in a single plasma membrane glycoprotein from rat liver turned over at different rates, with protein showing the greater half-life (70–78 h). Fucose (12.5 h), galactose (20 h) and Neu5Ac (33 h) turned over 2–6 times during the normal life-time of the glycoprotein (KREISEL *et al.* 1980). Turnover of the oligosaccharide moiety in this glycoprotein appears to be regulated with regard to terminal monosaccharides.

VI. Regulation of Sialic Acid Metabolism

1. Introduction

The complexity of sialic acid metabolism is immense and persists throughout metabolism, from monosaccharide to complex carbohydrate. The study of such regulation must attempt to account for the formation and activation of Neu5Ac, its transfer into many complex carbohydrates at different rates and in different compartments within one cell. It must allow for changes during development and short term fluctuations due to dietary or pathological changes. The metabolism of the sialic acids cannot be separated from the general metabolism of complex carbohydrates, as an integration of all components exists. However, it may be concluded that due to the special roles fulfilled by the sialic acids, a unique regulation may exist. The different levels of regulation possible have been discussed by RODÉN (1970), and general regulation of complex carbohydrate metabolism reviewed (e.g. SCHACHTER and RODÉN 1973, DAWSON 1978, SCHACHTER 1978, CORFIELD and SCHAUER 1979).

2. *Genetic Regulation*

Genetic regulation may occur in two ways. The information for the specific "invariant" polypeptide portion of the glycoprotein and proteoglycan core is coded in the genes. The glycosyl(sialyl)transfer reaction is a post-translational, non-template event. The enzymes responsible for glycosyl transfer are translated from gene templates in the classical manner. This has been shown for a sialyltransferase in mouse liver, sialylating lysosomal α-mannosidase (DIZIK and ELLIOT 1978). Thus, the coding of genes for protein in both structural and enzymic molecules involved in complex carbohydrate metabolism represents an accurate and essentially error-free mechanism.

3. *Regulation of Enzyme Activities*

This area of control results largely from the study of single enzymic reactions.

a) Specificity

Control through specificity is one way of producing defined products. For example, the phosphorylation of ManNAc to ManNAc-6-P is catalyzed by an enzyme which has little activity with GlcNAc (section II.3). The transfer of Neu5Ac from CMP-Neu5Ac to the oligosaccharide chains of glycoproteins is regulated by the structure of the oligosaccharide (PAULSON *et al.* 1978, BEYER *et al.* 1979, 1981; see section III and Fig. 8). Thus, the presence of fucose, GalNAc or Neu5Ac in specific linkages may lead to a block of sialyltransfer.

The sialidases provide another example of a group of enzymes which show specificity for the type of complex carbohydrate, the nature of the glycosidic linkage (see section IV.3.h) and the type of sialic acid (sections IV.3.e) and f).

Due to such mechanisms the generation, removal or protection of particular populations of sialic acids in defined molecules can be envisaged and a relationship to the specific function of sialic acids may be proposed.

b) Kinetic Effects

α) *Substrate and Enzyme Concentration*

Availability of substrate is an obvious regulatory factor in any enzyme-catalyzed reaction. In bisubstrate reaction it is important to consider the availability of each substrate, especially when they are synthesized in different compartments (e.g. sialyltransfer, requiring asialoacceptor and CMP-Neu5Ac). Coenzymes are particularly important in this respect, as they will also be substrates in other metabolic reactions. Thus, acetyl-CoA, ATP and CTP levels are important in sialic acid metabolism. Acetyl-CoA is required for the N- and O-acetylation reactions in Neu5Ac biosynthesis. In tissues synthesizing the highly O-acetylated sialic acids of mucus glycoproteins (e.g. bovine submandibular gland, human colon) the requirement for an O-acetylating potential to match the secretion rate of the mucus glycoprotein demands a high acetyl-CoA : CoA ratio.

In view of the concentration of substrates in the cells under study it is valuable to assess the requirement of enzymes in the pathway for these components. Measurement of K_m and V_{max} values for individual enzymes will give an indication

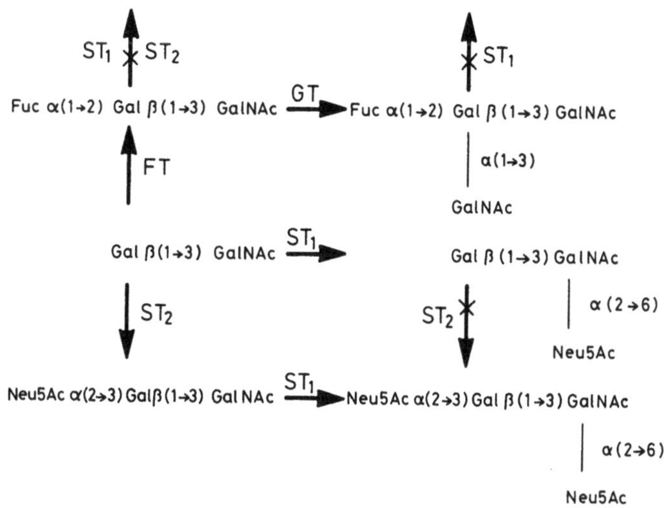

Fig. 8. Specificity of sialyltransfer to Galβ(1-3)GalNAc by purified sialyltransferases. Glycosyltransfer to the disaccharide using purified glycosyltransferase enzymes is shown. The absence of transfer or very slow rates are indicated by ⟶ (crossed arrow). *ST*$_1$, N-acetyl-α-galactosaminide-α-(2-6)-sialyltransferase; *ST*$_2$, β-galactoside-α-(2-3)-sialyltransferase; *FT*, β-galactoside-α-(1-2)-fucosyltransferase; *GT*, α-(1-2)-fucosyl-β-galactoside-α-(1-3)-N-acetyl-galactosaminyl transferase. (From Beyer *et al.* 1979.)

of the maximum rate possible through the complete pathway and help identify control points and limiting substrates. Comparison of kinetic constants for Neu5Ac pyruvate-lyase and Neu5Ac9P synthase allowed a catabolic role to be ascribed to the lyase, while the synthase was believed to be part of the *de novo* biosynthetic pathway of Neu5Ac (Warren and Felsenfeld 1962, section II.3). Although the lyase probably functions catabolically in most tissues, it is possible that under certain circumstances (e.g. high ManNAc concentrations) the reaction can be reversed. This is possible as the reaction is reversible, and an example is given below (section VI.3.c). This type of approach is more difficult to assess where interactions between cytosolic and membrane compartments exist, as is the case in sialic acid metabolism. A continuation of this aspect is the inhibition by substrate and products in sialic acid metabolism. This will be considered in section VI.3.b)β).

Regulation of synthesis via enzyme concentration may be related to hormonal stimulation of *de novo* enzyme synthesis, as has been demonstrated in regenerating rat liver for UDP-GlcNAc 2-epimerase (Okubo *et al.* 1976, Okamoto and Akamatsu 1980), or may simply function by absence or low activity of the enzyme. This is believed to be the case for Neu5Ac pyruvate-lyase in submandibular glands (Brunetti *et al.* 1962), and precludes breakdown of Neu5Ac to ManNAc by such a mechanism in this tissue.

A consequence of the non-template-mediated nature of glycosyltransfer is a loss of accuracy in transfer relative to protein synthesis. This is believed to be one of the reasons for the microheterogeneity observed in many glycoconjugates. An incomplete glycosylation or lack of specificity of glycosyltransferases have been suggested as possible explanations (Gottschalk 1969, Huang *et al.* 1970, Schachter and Rodén 1973). The problem remains open, as examples of "unfinished" oligosaccharides have been associated with biological function (e.g. immunoglobulin IgA oligosaccharide chains; Baenziger and Kornfeld 1974). Further to this, evidence for a decreased polydispersity in glycoprotein structure at increased biosynthetic rates has been reported (Gallagher and Corfield 1978), and such evidence supports an intentional molecular variability in nature.

Termination of oligosaccharide structure has been predicted from structural studies, and this approach is indispensible in predicting possible regulatory mechanism. Sialic acid has been associated with terminating mechanisms, and evidence at the sialyltransferase level has now been obtained (Paulson *et al.* 1978, Beyer *et al.* 1979, 1981). The interaction of fucosyl-, N-acetylgalactosaminyl- and sialyltransferase activities impose structural limitations on further glycosylation (Beyer *et al.* 1979, 1981, Fig. 8).

A final consideration in this respect can be accommodated with the proposal of Rauvala and Finne (1979) for the sharing of basic "core" structure glycosyltransferases in multiglycosyltransferase systems. The omission of particular glycosyltransferases in such a complex would lead to "incomplete" (terminated) chains, and such complexes would then be programmed by tissue- or cell-specific "complex forming" factors.

β) *Inhibition and Activation*

Inhibition of enzyme activity may be due to substrate, product or other effector molecules, and the concentrations of these components in the cell may lead to regulation of the pathway. Competitive product inhibition has been ascribed a significant role in sialyltransferase reactions (Bernacki 1975, Klohs *et al.* 1979, Eppler *et al.* 1980 b) with CMP yielding apparent K_i values of 50 μM (Klohs *et al.* 1979). This inhibition may be a self-regulatory effect *in situ* in view of the high nucleotide phosphatase activity of Golgi apparatus. A non-competitive inhibition of sialyltransferase activity was also observed with other nucleotides (Bernacki *et al.* 1978, Klohs *et al.* 1979).

Inhibition of sialidase by Neu5Ac2en may be important in specialized tissue sites, e.g. saliva, where sialidase-producing bacteria may influence tooth decay and mouth hygiene (Haverkamp *et al.* 1976).

Enhancement of enzyme activity may be demonstrated in relation to regulation. Glucosamine-6-phosphate deaminase is activated by GlcNAc-6-P, an intermediate in the hexosamine pathway leading to UDP-GlcNAc. The deaminase is

No.	Enzyme	Reaction	Feedback	Effectors	Location
1	GlcN-6-P synthase	Fru-6-P + glutamine → GlcN-6-P + glutamate	UDP-GlcNAc	AMP, Glc-6-P, UTP	cytosol
2	GlcNAc kinase	GlcNAc + ATP → GlcNAc-6-P	UDP-GlcNAc	GlcNAc-6-P	cytosol
3	UDP-GlcNAc 2 epimerase	UDP-GlcNAc → ManNAc + UDP	CMP-Neu5Ac	?	cytosol
4	CMP-Neu5Ac hydrolase	CMP-Neu5Ac → Neu5Ac + CMP	UDP-GlcNAc	?	plasma membrane

Fig. 9. Feedback inhibition in sialic acid metabolism.

believed to be primarily a catabolic enzyme, and thus a regulation of the GlcNAc-6-P pool may be achieved. Here it is also interesting to note that GlcNAc-6-P inhibits its own formation via GlcNAc kinase, suggesting a carefully regulated level of this intermediate in the cell (see section II.3, Fig. 1).

c) Feedback Inhibition

As outlined in sections II.2–4, several enzymes are subject to negative feedback control during sialic acid formation. These are listed in Fig. 9. The central intermediates are UDP-GlcNAc and CMP-Neu5Ac. As has been described for other monosaccharide activation pathways, the negative feedback influences the first enzymic steps in the pathway (Schachter and Rodén 1973, Schachter 1978).

Feedback control on the GlcN-6-P synthase has been studied in different tissues and animals (see section II.2). The enzyme in rat liver therefore normally operates at only 10% of its possible maximal rate, while the effectors Glc-6-P, AMP, and UTP can modulate this inhibition and provide a link with the energetic state of the cell (Winterburn and Phelps 1970, 1971 c). This regulation is very flexible and can easily accomodate large and rapid increases in activity. GlcNAc kinase is also feedback-inhibited by UDP-GlcNAc and represents the first step in the salvage pathway of free GlcNAc in the cell. Thus, a regulation of entry into the GlcNAc-6-P pool is effected by the end product UDP-GlcNAc. UDP-GlcNAc is also the substrate for the initial reaction in the formation of Neu5Ac. The feedback control on the enzyme catalyzing the corresponding reaction, the UDP-GlcNAc 2-epimerase, is effected by the natural end product of the pathway, CMP-Neu5Ac. As described in section II.3, the feedback mechanism has been characterized (Kikuchi and Tsuiki 1973), and is a central reaction in sialic acid metabolism. The formation of the end product CMP-Neu5Ac is linked to the substrate, UDP-GlcNAc, as CMP-Neu5Ac hydrolase is feedback-inhibited by this intermediate. Unless other factors play a role (e.g. topographical effects, membrane integration), it would appear that the intracellular pool of CMP-Neu5Ac is protected by high UDP-GlcNAc concentrations (inhibition of CMP-Neu5Ac breakdown) and may be linked to the capacity for sialyltransfer. At low concentrations of UDP-GlcNAc the CMP-Neu5Ac pool would be more susceptible to depletion.

d) Sialuria, a Defect in Sialic Acid Metabolism

The discovery of a young patient excreting daily gramme quantities of Neu5Ac in the urine has been documented by Strecker and Montreuil (1971). Measurements of the excess excretion of different intermediates has allowed several predictions to be made. A scheme of sialic acid metabolism in this patient is put forward in Fig. 10. As indicated by Kamerling *et al.* (1979) the most likely defect is one involving the UDP-GlcNAc 2-epimerase (step 2a–2b). If the feedback control on this enzyme step does not function, the rate through the pathway may increase dramatically, such that phosphorylation of ManNAc or GlcNAc, or the initial formation of GlcN-6-P from Fru-6-P becomes rate-limiting. The latter two steps are regulated by UDP-GlcNAc (feedback inhibition), which is also the substrate for the UDP-GlcNAc 2-epimerase (step 2a–2b). If the phosphorylation of ManNAc becomes rate-limiting, the concentration of

ManNAc will increase and may lead to conditions where the Neu5Ac pyruvate-lyase (step 4) reverses its catalytic function and synthesizes Neu5Ac (CORFIELD and SCHAUER 1979). The absence of increased serum glycoprotein sialylation indicates that sialyltransfer is probably not affected. The elevated Neu5Ac2en concentration may arise from CMP-Neu5Ac, if this intermediate is increased in the cell. A mechanism for Neu5Ac2en formation from CMP-Neu5Ac has been proposed (unpublished). Increase in the concentration of 2-acetamidoglucal to detectable levels suggests involvement of the UDP-GlcNAc 2-epimerase

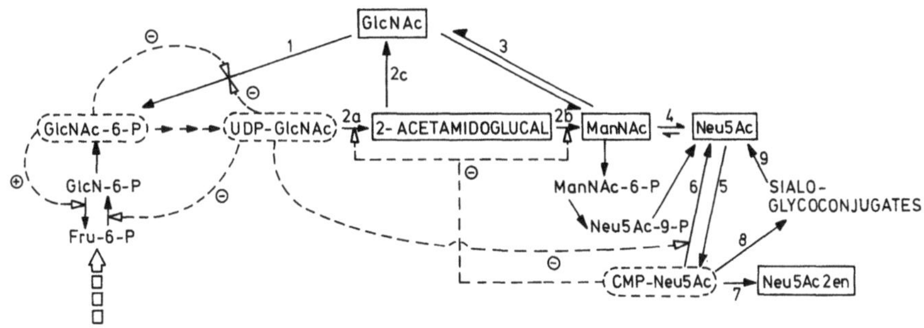

Fig. 10. Hexosamine and sialic acid metabolism in sialuria. The possible relationship of intermediates in sialuria is presented in the light of known data. *1* GlcNAc kinase; *2a, b* UDP-GlcNAc 2-epimerase; *2c* spontaneous breakdown of 2-acetamidoglucal in aqueous solution; *3* GlcNAc 2-epimerase; *4* Neu5Ac pyruvate-lyase; *5* CMP-Neu5Ac synthase; *6* CMP-Neu5Ac hydrolase; *7* elimination reaction leading to Neu5Ac2en; *8* sialyltransferase; *9* sialidase. Intermediates in boxes ▭ are elevated in sialuria; intermediates in broken circles ( ) have regulatory functions in the pathway; ⊖ – – – → ⊕ negative or positive feedback control.

(KAMERLING *et al.* 1979), although this has not been demonstrated. An alternative to the lack of feedback inhibition by CMP-Neu5Ac is a block in the second step (2b) of the 2-epimerase reaction (CORFIELD and SCHAUER 1979) leading to large amounts of 2-acetamidoglucal, which may be converted to GlcNAc in aqueous media (SALO and FLETCHER 1969, SOMMAR and ELLIS 1972b). ManNAc can only arise due to the enzymic catalysis of step 2b (SOMMAR and ELLIS 1972b). The formation of GlcNAc may be followed by phosphorylation by the specific kinase (step 1) or epimerization to ManNAc by a second 2-epimerase (step 3). The equilibrium concentration ratio for this ManNAc 2-epimerase (step 3) has been found to be GlcNAc : ManNAc 4 : 1 (GHOSH and ROSEMAN 1965), but a ratio of 1 : 4 was measured in the urine of the sialuria patient (KAMERLING *et al.* 1979).

Increased Neu5Ac excretion in other diseases (e.g. Salla disease) has been reported (RENLUND *et al.* 1979), but the levels of Neu5Ac were not as high as in the sialuria patient.

4. Nutritional Factors and Energy

Although many essential elements and vitamins are involved in the *de novo* synthesis of Neu5Ac and its subsequent metabolism, no agent has been described having an influence specifically on sialic acid metabolism. A discussion in general is given elsewhere (RODÉN 1970, SCHACHTER 1978). The influence of "ethanol poisoning" has been studied in adult and young rats. Changes in the levels of brain glycoprotein- and glycolipid-bound sialic acid were observed (KLEMM and ENGEN 1978, 1979). The mechanism of "ethanol poisoning" has been correlated with the spatial orientation of brain cell membrane-bound sialic acid and altered sialic acid biosynthesis and sialidase action (NOBLE *et al.* 1976, ROSS *et al.* 1977, KLEMM and ENGEN 1979).

Energy requirements in sialic acid metabolism have not been directly assessed, although general studies of glycoconjugate biosynthesis have given relevant information. Studies of hexosamine and 2-deoxy-glucose "poisoning" have demonstrated a reduction of the uridine nucleotide pool, due to metabolic "trapping" of these nucleotides as their nucleotide sugar derivatives (DECKER and KEPPLER 1974). Monitoring of intracellular cytosine nucleotide levels will give an indication of sialic acid energy requirement. Furthermore, close integration of secretion and energy formation could be demonstrated in cat submandibular gland, with only small reductions in the adenylate charge ratio after 2 h of continual gland stimulation (PHELPS and YOUNG 1977).

5. *Spatial Organization*

Throughout this chapter reference has been made to the topographical separation of enzymes involved in sialic acid metabolism. Regulation through physical separation, i.e. different locations in the cell, is a well documented phenomenon (see RODÉN 1970). The *de novo* formation of Neu5Ac from Fru-6-P is essentially a cytosolic event. No evidence has been presented for membrane location of these enzymes. The activation of Neu5Ac to the sugar nucleotide takes place in the nucleus (KEAN 1970, COATES *et al.* 1980, section II.3). This suggests an accumulation of sialic acid within the nucleus and subsequent transport of CMP-Neu5Ac to the Golgi apparatus, where sialyltransfer takes place. Furthermore, the plasma membrane location of CMP-Neu5Ac hydrolase introduces another subcellular site into the metabolic scheme within the cell.

The isolation of pure membrane sialyltransferases enables a study of these enzymes with other membrane components including other glycosyltransferases to be made. It is interesting to speculate on the suggestion by RAUVALA and FINNE (1979) that "core structures" may be synthesized by basic units of multiglycosyltransferases and that initiation and completion specified by incorporation of the appropriate enzyme into the complex. However, this area remains largely hypothetical through technical problems and sparse attention.

6. *Hormonal and Nervous Regulation*

Hormonal regulation is on a short or long term basis. The short term influence of hormones causes changes in intermediate levels which in turn can modify the substrate level control of sialic acid metabolism, either as substrates or effectors. Cellular ratios of e.g. ATP:(AMP + ADP) or AcCoA:CoA and hexosemonophosphate levels due to insulin, glucagon and glucocorticosteroid action may be altered. These effects will be reflected in the K_m and V_{max} values of the individual enzymes, and their interaction with metabolic inhibitors and effectors as discussed above (section VI.3. b)α).

Long term hormonal control functions in the expression of enzyme concentrations and *de novo* acceptor molecule synthesis in the case of sialyltransfer reactions. Examples in development have been cited (sections II.2, II.3, and III.10), and the regenerating liver has proved valuable in studies of this kind, e.g. UDP-GlcNAc 2-epimerase (OKUBU *et al.* 1976, OKAMOTO and AKAMATSU 1980). Elevation of enzyme concentration could be shown to be a function of *de novo* protein synthesis. Study of hormone effects (oestrogen, progesterone) on sialyltransferase activity and glycosidically linked sialic acid levels in the cervical cyclic phenomena have been reported (MOGHISSI and SYNER 1976, CHANTLER and DEBRUYN 1977, HATCHER *et al.* 1977, NASIR-UD-DIN *et al.* 1979, WOLF *et al.* 1980). Similar effects have been demonstrated in rat endometrium (NELSON *et al.* 1975).

The specific induction of *de novo* synthesis of sialyltransferase for glycolipid substrates by fatty acids has been studied in detail, and a hormonal regulation discussed by FISHMAN *et al.* (1976). Morphological changes in HeLa cells could be associated with the specific induction of LacCer sialyltransferase (II^3Neu5AcLacCer formation), which was blocked by protein synthesis inhibitors.

Regulation at the nervous level falls into two groups. First, a direct involvement with serotonin binding to glycosidically linked sialic acid at the synaptosomal membrane has been demonstrated (e.g. KRISHNAN and BALARAM 1976, DETTE and WESEMANN 1978), and a role for synaptosomal membrane surface glycoconjugates together with the interaction of sialyltransferases and sialidases proposed (SCHENGRUND and NELSON 1975, RAHMANN *et al.* 1976, SVENNERHOLM 1980, TETTAMANTI *et al.* 1980, VEH and SANDER 1981).

The second area of interest involves stimulated secretion in salivary glands. DISCHE *et al.* (1962) observed that electrical stimulation of canine salivary gland leads to an increase in sialic acid concentration and a decrease in fucose concentration in the secreted glycoprotein in the course of stimulation. Pilocarpine had the reverse effect. These results were extended by LOMBART and WINZLER (1974) to include sulphate. The stimulation of isolated cat submandibular gland led to similar changes of the glycoprotein sialic acid content with time (PHELPS and YOUNG 1977). Identification of individual glycoproteins or sites of synthesis have not been correlated with these results.

A role for glycosidically bound sialic acid in brain complex carbohydrates in development, learning ability and environmental stimulation has been described (MORGAN and WINICK 1979, 1980a, b). The regulation in these events probably exists at hormonal and nervous levels and may involve mediation of Neu5Ac uptake into the cell as discussed in section V.

Bibliography

AGUANNO, J. J., ROLL, D. E., GLEW, R. H., 1978: J. Biol. Chem. **253**, 6997—7004.

ALHADEFF, J. A., CIMINO, G., JANOWSKY, A., O'BRIEN, J. S., 1977: Biochim. Biophys. Acta **484**, 307—321.

ALLEN, M. B., WALKER, D. G., 1980a: Biochem. J. **185**, 565—575.

Allen, M. B., Walker, D. G., 1980b: Biochem. J. **185**, 577—582.
Aminoff, D., 1961: Biochem. J. **81**, 384—392.
— Dodyk, F., Roseman, S., 1963: J. Biol. Chem. **238**, PC 1177—PC 1178.
Ando, S., and Yu, R. K., 1979: J. Biol. Chem. **254**, 12224—12229.
Aoyagi, T., Komiyama, T., Nerome, K., Takeuchi, T., Umezawa, H., 1975: Experientia **31**, 896—897.
— Yagisawa, M., Kumagai, M., Hamada, M., Okami, Y., Takeuchi, T., Umezawa, H., 1971: J. Antibiotics **24**, 860—869.
Arce, A., Maccioni, H. J. F., Caputto, R., 1971: Biochem. J. **121**, 483—493.
Autuori, F., Svensson, H., Dallner, G., 1975a: J. Cell Biol. **67**, 687—699.
— — — 1975b: J. Cell Biol. **67**, 700—714.
Baenziger, J., Kornfeld, S., 1974: J. Biol. Chem. **249**, 7270—7281.
Baker, A. P., Sawyer, J. L., Munro, J. R., Weiner, G. P., Hillegass, L. M., 1972: J. Biol. Chem. **247**, 5173—5179.
Barnett, J. E. G., 1967: Biochem. J. 105, 42P.
— Corina, D. L., Rasool, G., 1971: Biochem. J. **125**, 234—242.
— Kolisis, F., 1974: Biochem. J. **143**, 487—490.
Bartholomew, B. A., Jourdian, G. W., Roseman, S., 1973: J. Biol. Chem. **248**, 5751—5762.
Basu, S., Basu, M., Chien, J.-L., Presper, K. A., 1980: In: Structure and Function of Gangliosides (Svennerholm, L., Mandel, P., Dreyfus, H., Urban, P. F., eds.), pp. 213—226. New York-London: Plenum Press.
Bates, C. J., Handschumacher, R. E., 1969: In: Advances in Enzyme Regulation (Weber, G., ed.), Vol. 7, pp. 183—204. Oxford-New York: Pergamon Press.
— Adams, W. R., Handschumacher, R. E., 1966: J. Biol. Chem. **241**, 1705—1712.
Beau, J.-M., Schauer, R., 1980: Eur. J. Biochem. **106**, 531—540.
Bekesi, J. G., Winzler, R. J., 1969: J. Biol. Chem. **244**, 5663—5668.
Bennett, G., O'Shaughnessy, D., 1981: J. Cell Biol. **88**, 1—15.
— Kan, F. W. K., O'Shaughnessy, D., 1981: J. Cell Biol. **88**, 16—28.
Bergman, A., Dallner, G., 1977: FEBS Lett. **82**, 359—362.
Bernacki, R. J., 1975: Eur. J. Biochem. **58**, 477—481.
— Bosmann, H. B., 1973: Eur. J. Biochem. **34**, 425—433.
— Kim, U., 1977: Science **195**, 577—580.
— Porter, C., Korytnyk, W., Mihich, E., 1978: Adv. Enz. Regul. **16**, 217—237.
Beyer, T. A., Rearick, J. I., Paulson, J. C., Prieels, J.-P., Sadler, J. E., Hill, R. L., 1979: J. Biol. Chem. **254**, 12531—12541.
— Sadler, J. E., Rearick, J. I., Paulson, J. C., Hill, R. L., 1981: In: Advances in Enzymology (Meister, A., ed.), Vol. 52, pp. 23—175. New York: Wiley Interscience.
Bhattacharya, A., Banerjee, S., Datta, A., 1974: Biochim. Biophys. Acta **374**, 384—391.
Biswas, M., Singh, B., Datta, A., 1979: Biochim. Biophys. Acta **585**, 535—542.
Blacklow, R. S., Warren, L., 1962: J. Biol. Chem. **237**, 3520—3526.
Bosmann, H. B., 1972: Biochem. Biophys. Res. Commun. **49**, 1256—1262.
— Hall, T. C., 1974: Proc. Natl. Acad. Sci. U.S.A. **71**, 1833—1837.
— Hilf, R., 1974: FEBS Lett. **44**, 313—316.
Brady, R. O., Fishman, P. H., 1974: Biochim. Biophys. Acta **355**, 121—148.
Briles, E. B., Li, E., Kornfeld, S., 1977: J. Biol. Chem. **252**, 1107—1116.
Brossmer, R., Holmquist, L., 1971: Hoppe-Seyler's Z. Physiol. Chem. **352**, 1715—1719.
— Nebelin, E., 1969: FEBS Lett. **4**, 335—336.
— Keilich, G., Ziegler, D., 1977a: Hoppe-Seyler's Z. Physiol. Chem. **358**, 391—396.
— Ziegler, D., Keilich, G., 1977b: Hoppe-Seyler's Z. Physiol. Chem. **358**, 397—400.
— Rose, U., Unger, F. M., Grasmuk, H., 1979: In: Glycoconjugates (Schauer, R., Boer,

P., BUDDECKE, E., KRAMER, M. F., VLIEGENTHART, J. F. G., WIEGANDT, H., eds.), pp. 242—243. Stuttgart: G. Thieme.
BROSSMER, R., ROSE, U., KASPER, D., SMITH, T. L., GRASSMUK, H., UNGER, F. M., 1980: Biochem. Biophys. Res. Commun. **96**, 1282—1289.
BRUNETTI, P., JOURDIAN, G. W., ROSEMAN, S., 1962: J. Biol. Chem. **237**, 2447—2453.
BRUNNGRABER, E. G., 1979: Neurochemistry of Aminosugars. Springfield, Ill.: Ch. C Thomas.
BUSCHER, H.-P., CASALS-STENZEL, J., SCHAUER, R., MESTRES-VENTURA, P., 1977: Eur. J. Biochem. **77**, 297—310.
BUSHWAY, A. A., KEENAN, T. W., 1978: Lipids **13**, 59—65.
CABEZAS, M., 1978: Int. J. Biochem. **9**, 47—49.
CACAN, R., VERBERT, A., HOFLACK, B., MONTREUIL, J., 1977: FEBS Lett. **81**, 53—56.
CAIMI, L., LOMBARDO, A., PRETI, A., WIESMANN, U., TETTAMANTI, G., 1979: Biochim. Biophys. Acta **571**, 137—146.
CANTZ, M., MESSER, H., 1979: Eur. J. Biochem. **97**, 113—118.
CAPUTTO, R., MACCIONI, H. J., ARCE, A., 1974: Mol. Cell. Biochem. **4**, 97—106.
— — ARCE, A., CUMAR, F. A., 1976: In: Ganglioside Function (PORCELLATI, G., CECCARELLI, B., TETTAMANTI, G., eds.), pp. 27—44. New York-London: Plenum Press.
CARDINI, C. E., LELOIR, L. F., 1957: J. Biol. Chem. **225**, 317—324.
CAREY, D. J., HIRSCHBERG, C. B., 1979: Biochemistry **18**, 2086—2092.
— — 1980: J. Biol. Chem. **255**, 4348—4354.
— SOMMERS, L. W., HIRSCHBERG, C. B., 1980: Cell **19**, 597—605.
CARLSON, D. M., 1977: In: Mucus in Health and Disease (ELSTEIN, M., PARKE, D. V., eds.), pp. 597—610. New York-London: Plenum Press.
— JOURDIAN, G. W., ROSEMAN, S., 1973a: J. Biol. Chem. **248**, 5742—5750.
— McGUIRE, E. J., JOURDIAN, G. W., ROSEMAN, S., 1973b: J. Biol. Chem. **248**, 5763—5773.
CERVÉN, E., 1977: Biochim. Biophys. Acta **467**, 72—85.
— 1978: Life Sciences **23**, 2769—2778.
CHANTLER, E., DEBRYNE, E., 1977: In: Mucus in Health and Disease (ELSTEIN, M., PARKE, D. V., eds.), pp. 131—141. New York-London: Plenum Press.
CHATTERJEE, S. K., BHATTACHARYA, M., BARLOW, J. J., 1978: Biochem. Biophys. Res. Commun. **80**, 826—832.
CHOU, T. C., SOODAK, M., 1952: J. Biol. Chem. **196**, 105—109.
CLARKE, J. T. R., MULCAHEY, M. R., 1976: Biochim. Biophys. Acta **441**, 146—154.
COATES, S. W., GURNEY, T., jr., SOMMERS, L. W., YEH, M., HIRSCHBERG, C. B., 1980: J. Biol. Chem. **255**, 9225—9229.
COMB, D. G., ROSEMAN, S., 1958a: J. Biol. Chem. **232**, 807—827.
— — 1958b: Biochim. Biophys. Acta **29**, 653—654.
— — 1960: J. Biol. Chem. **235**, 2529—2537.
— WATSON, D. R., ROSEMAN, S., 1966: J. Biol. Chem. **241**, 5637—5642.
COOMBES, R. C., ELLISON, M. L., NEVILLE, A. M., 1978: Br. J. Dis. Chest **72**, 263—287.
CORFIELD, A. P., 1973: Ph.D. Thesis, University of Wales, Cardiff, U.K.
— FERREIRA DO AMARAL, C., WEMBER, M., SCHAUER, R., 1976: Eur. J. Biochem. **68**, 597—610.
— SCHAUER, R., 1979: Biol. Cellulaire **36**, 213—226.
— — WEMBER, M., 1979: Biochem. J. **177**, 1—7.
— — SCHWARZMANN, G., WIEGANDT, H., 1980: Hoppe-Seyler's Z. Physiol. Chem. **361**, 231.
— MICHALSKI, J.-C., SCHAUER, R., 1981a: In: Sialidases and Sialidoses, Perspectives in Inherited Metabolic Diseases (TETTAMANTI, G., DURAND, P., DI DONATO, S., eds.), Vol. 4, pp. 3—70. Milano: Edi Ermes.
— VEH, R. W., WEMBER, M., MICHALSKI, J.-C., SCHAUER, R., 1981b: Biochem. J. **197**, 293—299.

CRAMPEN, M., VON NICOLAI, H., ZILLIKEN, F., 1979: Hoppe-Seyler's Z. Physiol. Chem. **360**, 1703—1712.

CREEK, K. E., MORRÉ, D. J., 1981: Biochim. Biophys. Acta **643**, 292—305.

DAIN, J. A., NG, S.-S., 1980: In: Structure and Function of Gangliosides (SVENNERHOLM, L., MANDEL, P., DREYFUS, H., URBAN, P. F., eds.), pp. 239—245. New York-London: Plenum Press.

DATTA, A., 1970a: Biochim. Biophys. Acta **220**, 51—60.

— 1970b: Biochemistry **9**, 3363—3370.

— 1971: Arch. Biochem. Biophys. **142**, 645—650.

DATTA, P., 1974: Biochemistry **13**, 3987—3991.

DAWSON, G., 1978: In: The Glycoconjugates (HOROWITZ, M. I., PIGMAN, W., eds.), Vol. 2, pp. 255—284. New York: Academic Press.

— 1979: In: Complex Carbohydrates of Nervous Tissue (MARGOLIS, R. U., MARGOLIS, R. K., eds.), pp. 291—325. New York-London: Plenum Press.

DECKER, K., KEPPLER, D., 1974: Rev. Physiol. Biochem. Pharmac. **71**, 77—106.

DELGIACCO, R., MALEY, F., 1964: J. Biol. Chem. **239**, PC 2400—PC 2402.

DEN, H., KAUFMAN, B., ROSEMAN, S., 1970: J. Biol. Chem. **245**, 6607—6615.

— — MCGUIRE, E. J., ROSEMAN, S., 1975: J. Biol. Chem. **250**, 739—746.

DEPPERT, W., WALTER, G., 1978: J. Supramol. Struct. **8**, 19—37.

— WERCHAU, H., WALTER, G., 1974: Proc. Natl. Acad. Sci. U.S.A. **71**, 3068—3072.

DETTE, G. A., WESEMANN, W., 1978: Hoppe-Seyler's Z. Physiol. Chem. **359**, 399—406.

DEVRIES, G. H., BINKLEY, S. B., 1972: Arch. Biochem. Biophys. **151**, 234—242.

DICKSON, J. J., MESSER, M., 1978: Biochem. J. **170**, 407—413.

DISCHE, Z., PALLAVICINI, C., KAVASAKI, H., SMIRNOV, N., CIZEK, L. J., CHIEN, S., 1962: Arch. Biochem. Biophys. **97**, 459—469.

DIZIK, M., ELLIOTT, R. W., 1978: Biochem. Genet. **16**, 247—260.

DREYFUS, H., HARTH, S., YUSUFI, A. N. K., URBAN, P. F., MANDEL, P., 1980: In: Structure and Function of Gangliosides (SVENNERHOLM, L., MANDEL, P., DREYFUS, H., URBAN, P. F., eds.), pp. 227—237. New York-London: Plenum Press.

DRZENIEK, R., 1972: Curr. Top. Microb. Immunol. **59**, 35—74.

— 1973: Histochem. J. **5**, 271—290.

DUFFARD, R. O., CAPUTTO, R., 1972: Biochemistry **11**, 1396—1400.

— FISHMAN, P. H., BRADLEY, R. M., LAUTER, C. J., BRADY, R. O., TRAMS, E. G., 1977: J. Neurochem. **28**, 1161—1166.

DURR, R., SHUR, B., ROTH, S., 1977: Nature **265**, 547—548.

ELHAMMER, Å., SVENSSON, H., AUTUORI, F., DALLNER, G., 1975: J. Cell Biol. **67**, 715—724.

ELLIS, D. B., SOMMAR, K. M., 1971: Biochim. Biophys. Acta **230**, 531—534.

— — 1972: Biochim. Biophys. Acta **267**, 105—112.

— MUNRO, J. R., STAHL, G. H., 1972: Biochim. Biophys. Acta **289**, 108—116.

EPPLER, C. M., MORRÉ, D. J., KEENAN, T. W., 1980a: Biochim. Biophys. Acta **619**, 318—331.

— — — 1980b: Biochim. Biophys. Acta **619**, 332—343.

ESCHENFELDER, V., 1979: Habilitationsschrift, Ruprecht-Karl-University, Heidelberg, Federal Republic of Germany.

FAILLARD, H., KIRCHNER, G., BLOHM, M., 1966: Hoppe-Seyler's Z. Physiol. Chem. **347**, 87.

— FERREIRA DO AMARAL, C., BLOHM, M., 1969: Hoppe-Seyler's Z. Physiol. Chem. **350**, 798—802.

FAN, P., DATTA, P., 1980: Biochemistry **19**, 1893—1900.

FERNANDEZ-SØRENSEN, A., CARLSON, D. M., 1971: J. Biol. Chem. **246**, 3485—3493.

FISHMAN, P. H., BRADLEY, R. M., HENNEBERRY, R. C., 1976: Arch. Biochem. Biophys. **172**, 618—626.

Fishman, W. H., 1974: Chem. Phys. Lipids **13**, 305—326.
Fleischer, B., 1977: J. Supramol. Struct. **7**, 79—89.
— 1978: In: Cell Surface Carbohydrate Chemistry (Harmon, R. E., ed.), pp. 27—47. New York-London: Academic Press.
Forstner, G. G., 1968: Biochim. Biophys. Acta **150**, 736—738.
— 1971: Biochem. J. **121**, 781—789.
Forstner, J. F., Jabbal, I., Forstner, G. G., 1973: Can. J. Biochem. **51**, 1154—1166.
Fox, O. F., Kishore, G. S., Carubelli, R., 1979: Cancer Letts. **7**, 251—257.
Friebolin, H., Brossmer, R., Keilich, G., Ziegler, D., Supp, M., 1980: Hoppe-Seyler's Z. Physiol. Chem. **361**, 697—702.
Frot-Coutaz, J., Dubois, P., Berthillier, G., Got, R., 1973: Exp. Cell. Res. **77**, 223—231.
Gallagher, J. T., Corfield, A. P., 1978: Trends Biochem. Sci. **3**, 38—41.
Gan, J. C., 1975: Biochim. Biophys. Acta **385**, 412—420.
Gantt, S., Millner, S., Binkley, S. B., 1964: Biochemistry **3**, 1952—1960.
Ganzinger, U., 1977: Wien. klin. Wschr. **89**, 594—597.
Ghidoni, R., Sonnino, S., Tettamanti, G., Baumann, N., Reuter, G., Schauer, R., 1980: J. Biol. Chem. **255**, 6990—6995.
Ghosh, S., Roseman, S., 1961: Proc. Natl. Acad. Sci. U.S.A. **47**, 955—958.
— — 1965: J. Biol. Chem. **240**, 1531—1536.
— Blumenthal, H. J., Davidson, E., Roseman, S., 1960: J. Biol. Chem. **235**, 1265—1273.
Gielen, W., Hinzen, D. H., 1974: Hoppe-Seyler's Z. Physiol. Chem. **355**, 895—901.
Gindzienski, A., Glowacka, D., Zwierz, K., 1974: Eur. J. Biochem. **43**, 155—160.
Glickman, R. M., Bouhours, J. F., 1976: Biochim. Biophys. Acta **424**, 17—25.
Gottschalk, A., 1958: Adv. Enzymol. **20**, 135—146.
— 1962: Perspectives Biol. Med. **5**, 327.
— 1969: Nature **222**, 452—454.
— Drzeniek, R., 1972: In: Glycoproteins, Their Composition, Structure and Function (Gottschalk, A., ed.), 2nd ed., pp. 381—402. Amsterdam: Elsevier.
Groome, N. P., Belyavin, G., Lansdell, A., Ashford, D., 1977: Biochim. Biophys. Acta **495**, 58—70.
Grünholz, H.-J., Vischer, P., Reutter, W., 1977: Hoppe-Seyler's Z. Physiol. Chem. **358**, 1209—1210.
Gurd, F. W., Evans, W. H., 1973: Eur. J. Biochem. **36**, 273—279.
Hanover, J. A., Lennarz, W. J., Young, J. D., 1980: J. Biol. Chem. **255**, 6713—6716.
Harmon, R. E. (ed.), 1978: Cell Surface Carbohydrate Chemistry. New York: Academic Press.
Harms, E., Reutter, W., 1974: Cancer Res. **34**, 3165—3172.
— Kreisel, W., Morris, H. P., Reutter, W., 1973: Eur. J. Biochem. **32**, 254—262.
Harris, P. L., Thornton, E. R., 1978: J. Am. Chem. Soc. **100**, 6738—6745.
Hatcher, V. B., Schwarzmann, G. O. H., Jeanloz, R. W., McArthur, J. W., 1977: Fertil. Steril. **28**, 682—688.
Haverkamp, J., Schauer, R., Wember, M., Farriaux, J.-P., Kamerling, J. P., Versluis, C., Vliegenthart, J. F. G., 1976: Hoppe-Seyler's Z. Physiol. Chem. **357**, 1699—1705.
— Veh, R. W., Sander, M., Schauer, R., Kamerling, J. P., Vliegenthart, J. F. G., 1977: Hoppe-Seyler's Z. Physiol. Chem. **358**, 1609—1612.
— Beau, J.-M., Schauer, R., 1979a: Hoppe-Seyler's Z. Physiol. Chem. **360**, 159—166.
— Spoormaker, T., Dorland, L., Vliegenthart, J. F. G., Schauer, R., 1979b: J. Am. Chem. Soc. **101**, 4851—4853.
Heimer, R., Meyer, K., 1956: Proc. Natl. Acad. Sci. U.S.A. **42**, 728—734.
Henderson, M., Kessel, D., 1977: Cancer **39**, 1129—1134.
Hilbig, R., Rösner, H., Rahmann, H., 1981: Comp. Biochem. Physiol. **68B**, 301—305.

HIRSCHBERG, C. B., YEH, M., 1977: J. Supramol. Struct. **6**, 571—577.
— GOODMAN, S. R., GREEN, C., 1976: Biochemistry **15**, 3591—3599.
HOFLACK, B., CACAN, R., MONTREUIL, J., VERBERT, A., 1979: Biochim. Biophys. Acta **568**, 348—356.
HOLMQUIST, L., ÖSTMAN, B., 1975: FEBS Lett. **60**, 327—330.
— BROSSMER, R., 1972: FEBS Lett. **22**, 46—48.
HUANG, C.-C., MAYER, H. E., jr., MONTGOMERY, R., 1970: Carbohydr. Res. **13**, 127—137.
HUDGIN, R. L., SCHACHTER, H., 1971: Can. J. Biochem. **49**, 829—837.
— — 1972: Can. J. Biochem. **50**, 1024—1028.
HUGHES, R. C., 1976: Membrane Glycoproteins. A Review of Structure and Function. London: Butterworths.
HULTSCH, E., REUTTER, W., DECKER, K., 1972: Biochim. Biophys. Acta **237**, 132—140.
INOUE, S., MATSUMURA, G., 1979: Carbohydr. Res. **74**, 361—368.
IP, C., DAO, T., 1978: Cancer Res. **38**, 723—728.
IWAMORI, M., NAGAI, Y., 1978: J. Biol. Chem. **253**, 8328—8331.
JACOBSON, B., DAVIDSON, E. A., 1963: Biochim. Biophys. Acta **73**, 145—151.
JOURDIAN, G. W., SWANSON, A. L., WATSON, D., ROSEMAN, S., 1964: J. Biol. Chem. **239**, PC2714—PC2715.
KAMERLING, J. P., STRECKER, G., FARRIAUX, J.-P., DORLAND, L., HAVERKAMP, J., VLIEGENTHART, J. F. G., 1979: Biochim. Biophys. Acta **583**, 403—408.
KAUFMAN, B., BASU, S., ROSEMAN, S., 1968: J. Biol. Chem. **234**, 5804—5807.
KEAN, E. L., 1970: J. Biol. Chem. **245**, 2301—2308.
— BIGHOUSE, K. J., 1974: J. Biol. Chem. **249**, 7813—7823.
KEENAN, T. W., MORRÉ, D. J., 1975: FEBS Lett. **55**, 8—13.
— — BASU, S., 1974: J. Biol. Chem. **249**, 310—315.
KEILICH, G., ZIEGLER, D., BROSSMER, R., 1979: In: Glycoconjugates (SCHAUER, R., BOER, P., BUDDECKE, E., KRAMER, M. F., VLIEGENTHART, J. F. G., WIEGANDT, H., eds.), pp. 346—347. Stuttgart: G. Thieme.
KELLER, S. J., KEENAN, T. W., EIGEL, W. N., 1979: Biochim. Biophys. Acta **566**, 266—273.
KEMP, S. F., STOOLMILLER, A. C., 1976: J. Neurochem. **27**, 723—732.
KENT, P. W., 1970: Expos. Ann. Biochim. Méd. **32**, 97—120.
— ALLEN, A., 1968: Biochem. J. **106**, 645—658.
KHORLIN, A. YA., PRIVALOVA, I. M., ZAKSTELSKAYA, L. YA., MOLIBOG, E. V., EVSTIGNEEVA, N. A., 1970: FEBS Lett. **8**, 17—19.
— GABRIELYAN, N. D., ZURABYAN, S. E., SHULMAN, M. L., 1980: In: Frontiers of Bioorganic Chemistry and Molecular Biology (ANANCHENKO, S. N., ed.), pp. 73—79. Oxford: Pergamon Press.
KIKUCHI, H., KOBAYASHI, Y., TSUIKI, S., 1971 a: Biochim. Biophys. Acta **237**, 412—421.
KIKUCHI, K., KIKUCHI, H., TSUIKI, S., 1971 b: Biochim. Biophys. Acta **252**, 357—368.
— TSUIKI, S., 1973: Biochim. Biophys. Acta **327**, 193—206.
— — 1979: Biochim. Biophys. Acta **584**, 246—253.
KIM, Y. S., PERDOMO, J. M., 1974: Biochim. Biophys. Acta **342**, 111—124.
— — BELLA, A., NORDBERG, J., 1971: Biochim. Biophys. Acta **244**, 505—512.
KISHORE, G. S., TULSIANI, D. R. P., BHAVANANDAN, V. P., CARUBELLI, R., 1975: J. Biol. Chem. **250**, 2655—2659.
KLEMM, W. R., ENGEN, R. L., 1978: J. Neurosci. Res. **3**, 341—352.
— — 1979: J. Stud. Alcohol. **40**, 554—561.
KLOHS, W. D., BERNACKI, R. J., KORYTNYK, W., 1979: Cancer Res. **39**, 1231—1238.
KNOPF, P. M., SASSO, E., DESTREE, A., MELCHERS, F., 1975: Biochemistry **14**, 4136—4143.
KOBATA, A., 1977: In: Glycoconjugates (HOROWITZ, M. I., PIGMAN, W., eds.), Vol. 1, pp. 423—440. New York: Academic Press.

KOBATA, A., 1979: Cell Struct. Funct. **4**, 169—181.
KOHN, P., WINZLER, R. J., HOFFMAN, R. C., 1962: J. Biol. Chem. **237**, 304—308.
KOLISIS, F. N., SOTIROUDIS, T. G., EVANGELOPOULOS, A. E., 1981: FEBS Lett. **121**, 280—282.
KOLODNY, E. H., KANFER, J., QUIRK, J. M., BRADY, R. O., 1971: J. Biol. Chem. **246**, 1426—1431.
KORNFELD, R., 1967: J. Biol. Chem. **242**, 3135—3141.
— KORNFELD, S., 1980: In: The Biochemistry of Glycoproteins and Proteoglycans (LENNARZ, W. J., ed.), pp. 1—34. New York-London: Plenum Press.
KORNFELD, S., KORNFELD, R., NEUFELD, E., O'BRIEN, P. J., 1964: Proc. Natl. Acad. Sci. U.S.A. **52**, 371—379.
KÖTTGEN, E., BAUER, CH., REUTTER, W., GEROK, W., 1979: Klin. Wschr. **57**, 151—159, 199—214.
KREISEL, W., VOLK, B. A., BÜCHSEL, R., REUTTER, W., 1980: Proc. Natl. Acad. Sci. U.S.A. **77**, 1828—1831.
KRISHNAN, K. S., BALARAM, P., 1976: FEBS Lett. **63**, 313—315.
KUHN, N. J., 1972: Biochem. J. **130**, 177—180.
KUHN, R., BROSSMER, R., 1958: Angew. Chem. **70**, 25—26.
— WIEGANDT, H., 1963: Chem. Ber. **96**, 866—880.
— LUTZ, P., MACDONALD, D. L., 1966: Chem. Ber. **99**, 611—617.
KUNDIG, F. D., AMINOFF, D., ROSEMAN, S., 1971: J. Biol. Chem. **246**, 2543—2550.
KUNDIG, W., GHOSH, S., ROSEMAN, S., 1966: J. Biol. Chem. **241**, 5619—5626.
LANDA, C. A., MACCIONI, H. J. F., ARCE, A., CAPUTTO, R., 1977: Biochem. J. **168**, 325—332.
— — CAPUTTO, R., 1979: J. Neurochem. **33**, 825—838.
LAWFORD, G. R., SCHACHTER, H., 1966: J. Biol. Chem. **241**, 5408—5418.
— — 1967: Can. J. Biochem. **45**, 507—522.
LEDEEN, R. W., 1978: J. Supramol. Struct. **8**, 1—17.
— YU, R. K., 1976: In: Biological Roles of Sialic Acid (ROSENBERG, A., SCHENGRUND, C.-L., eds.), pp. 1—57. New York-London: Plenum Press.
— SKRIVANEK, J. A., TIRRI, L. J., MARGOLIS, R. K., MARGOLIS, R. U., 1976: In: Ganglioside Function (PORCELLATI, G., CECCARELLI, B., TETTAMANTI, G., eds.), pp. 83—103. New York-London: Plenum Press.
LELOIR, L. F., CARDINI, C. E., 1953: Biochim. Biophys. Acta **12**, 15—22.
LIN, W., OISHI, K., AIDA, K., 1977: Virology **78**, 108—114.
LIPOVAC, V., BARTON, N., ROSENBERG, A., 1973: Biochemistry **12**, 1858—1861.
LIU, C.-K., SCHMIED, R., GREENSPAN, E. M., WAXMAN, S., 1978: Biochim. Biophys. Acta **522**, 375—384.
LOMBART, C. G., WINZLER, R. J., 1974: Eur. J. Biochem. **49**, 77—86.
MACCIONI, H. J. F., ARCE, A., CAPUTTO, R., 1971: Biochem. J. **125**, 1131—1137.
MACNICOLL, A. D., WUSTEMAN, F. S., POWELL, G. M., CURTIS, C. G., 1978: Biochem. J. **174**, 421—426.
MALEY, F., TARENTINO, A. L., MCGARRAHAN, J. F., DELGIACCO, R., 1968: Biochem. J. **107**, 637—644.
— GHAMBEER, R., DELGIACCO, R., 1971: Biochem. J. **124**, 661—663.
MARGOLIS, R. K., MARGOLIS, R. U., 1973: Biochim. Biophys. Acta **304**, 413—420.
MARGOLIS, R. U., MARGOLIS, R. K., 1977: Int. J. Biochem. **8**, 85—91.
MARTIN, A., LOUISOT, P., 1976: Int. J. Biochem. **7**, 501—505.
MATSUSHITA, Y., TAKAGI, Y., 1966: Biochim. Biophys. Acta **124**, 204—207.
MCGARRAHAN, J. F., MALEY, F., 1962: J. Biol. Chem. **237**, 2458—2465.
MCGUIRE, E. J., 1976: In: Biological Roles of Sialic Acid (ROSENBERG, A., SCHENGRUND, C.-L., eds.), pp. 123—158. New York: Plenum Press.
— BINKLEY, S. E., 1963: Biochemistry **3**, 247—251.

McGUIRE, E. J., ROSEMAN, S., 1967: J. Biol. Chem. **242**, 3745—3755.
MEINDL, P., TUPPY, H., 1965: Monatsh. Chem. **96**, 802—815.
— — 1966a: Monatsh. Chem. **97**, 990—999.
— — 1966b: Monatsh. Chem. **97**, 1628—1647.
— — 1967: Monatsh. Chem. **98**, 53—60.
— BODO, G., PALESE, P., SCHULMAN, J., TUPPY, H., 1974: Virology **58**, 457—463.
MESSER, M., 1974: Biochem. J. **139**, 415—420.
MESTRALLET, M. G., CUMAR, F. A., CAPUTTO, R., 1977: Mol. Cell. Biochem. **16**, 63—70.
MICHALSKI, J.-C., CORFIELD, A. P., SCHAUER, R., 1981: Hoppe-Seyler's Z. Physiol. Chem. **362**, 222—223.
MILLER, C. A., WANG, P., FLASHNER, M., 1978: Biochem. Biophys. Res. Commun. **83**, 1479—1487.
MIYAGI, T., TSUIKI, S., 1971: Biochim. Biophys. Acta **250**, 51—62.
MOGHISSI, K. S., SYNER, F. N., 1976: Int. J. Fertil. **21**, 246—250.
MOLNAR, J., ROBINSON, G. B., WINZLER, R. J., 1964: J. Biol. Chem. **239**, 3157—3162.
— LUTES, R. A., WINZLER, R. J., 1965: Cancer Res. **25**, 1438—1445.
MONACO, F., ROBBINS, J., 1973: J. Biol. Chem. **248**, 2072—2077.
— SALVATORE, G., ROBBINS, J., 1975: J. Biol. Chem. **250**, 1595—1599.
MONTREUIL, J., 1975: Pure Appl. Chem. **42**, 431—477.
— 1980: Adv. Carbohyd. Chem. Biochem. **37**, 157—223.
MOOKERJEA, S., MILLER, C., 1974: Can. J. Biochem. **52**, 767—773.
— MICHAELS, M. A., HUDGIN, R. L., MOSCARELLO, M. A., CHOW, A., SCHACHTER, H., 1972: Can. J. Biochem. **50**, 738—740.
MORGAN, B. L. G., WINICK, M., 1979: Proc. Soc. Exp. Biol. Med. **161**, 534—537.
— — 1980a: J. Nutr. **110**, 416—424.
— — 1980b: J. Nutr. **110**, 425—432.
MUNRO, J. R., NARASIMHAN, S., WETMORE, S., RIORDAN, J. R., SCHACHTER, H., 1975: Arch. Biochem. Biophys. **169**, 269—277.
MÜLLER, H. E., 1971: Zbl. Bakt. Hyg. I. Abtl. Orig. A **217**, 326—344.
— 1974: Behring Inst. Mitt. **55**, 34—56.
— 1975: Zbl. Bakt. Hyg. I. Abtl. Orig. A **232**, 365—372.
NASIR-UD-DIN, JEANLOZ, R. W., REINHOLD, V. N., McARTHUR, J. W., 1979: Carbohydr. Res. **75**, 349—356.
NEES, S., SCHAUER, R., 1974: Behring Inst. Mitt. **55**, 68—78.
— VEH, R. W., SCHAUER, R., 1975: Hoppe-Seyler's Z. Physiol. Chem. **356**, 1027—1042.
— SCHAUER, R., MAYER, F., EHRLICH, K., 1976: Hoppe-Seyler's Z. Physiol. Chem. **357**, 839—853.
NELSON, J. D., JATO-RODRIGUEZ, J. J., MOOKERJEA, S., 1975: Arch. Biochem. Biophys. **169**, 181—191.
NEUBERGER, A., RATCLIFFE, W. A., 1972: Biochem. J. **129**, 683—693.
NEUFELD, E. J., PASTAN, I., 1978: Arch. Biochem. Biophys. **188**, 323—327.
NEUTRA, M., LEBLOND, C. P., 1966: J. Cell Biol. **30**, 137—150.
NG, S.-S., DAIN, J. A., 1977a: J. Neurochem. **29**, 1075—1083.
— — 1977b: J. Neurochem. **29**, 1085—1093.
NILSSON, B., NORDÉN, N. E., SVENSSON, S., 1979: J. Biol. Chem. **254**, 4545—4553.
NOBLE, E. P., SYAPIN, P. J., VIGRAN, R., ROSENBERG, A., 1976: J. Neurochem. **27**, 217—221.
NÖHLE, U., SCHAUER, R., 1981: Hoppe-Seyler's Z. Physiol. Chem. **362**, 1495—1506.
— BEAU, J.-M., SCHAUER, R., 1982: Eur. J. Biochem. **126**, 543 – 548.
OKAMOTO, Y., AKAMATSU, N., 1980: Biochem. J. **188**, 905—911.
OKUBO, H., SHIBATA, K., ISHIBASHI, H., YANASE, T., 1976: Proc. Soc. Exp. Biol. Med. **152**, 626—630.

Okubo, H., Shibata, K., Ishibashi, H., Yanase, T., 1977: Proc. Soc. Exp. Biol. Med. **155**, 152–156.

Pacuszka, T., Duffard, R. O., Nishimura, R. N., Brady, R. O., Fishman, P. H., 1978: J. Biol. Chem. **253**, 5839—5846.

Painter, R. G., White, A., 1976: Proc. Natl. Acad. Sci. U.S.A. **73**, 837—841.

Parker, T. L., Veh, R. W., Schauer, R., 1979: In: Glycoconjugate Research (Gregory, J. D., Jeanloz, R. W., eds.), Vol. 2, pp. 917—921. New York: Academic Press.

Paulson, J. C., Beranek, W. E., Hill, R. L., 1977a: J. Biol. Chem. **252**, 2356—2362.

— Rearick, J. I., Hill, R. L., 1977b: J. Biol. Chem. **252**, 2363—2371.

— Prieels, J.-P., Glasgow, L. R., Hill, R. L., 1978: J. Biol. Chem. **258**, 5617—5624.

— Sadler, J. E., Hill, R. L., 1979: J. Biol. Chem. **254**, 2120—2124.

Phelps, C. F., Young, A. M., 1977: In: Mucus in Health and Disease (Elstein, M., Parke, D. V., eds.), pp. 143—154. New York-London: Plenum Press.

— Stevens, R. A. J., Young, A. M., Luscombe, M., 1975: In: Protides of the Biological Fluids (Peeters, H., ed.), pp. 205—210. Oxford-New York: Pergamon Press.

Porter, C. W., Bernacki, R. J., 1975: Nature **256**, 648—650.

Potier, M., Beauregard, G., Bélisle, M., Mameli, L., Nguyen Hong, V., Melançon, S. B., Dallaire, L., 1979: Clin. Chim. Acta **99**, 97—105.

Pouysségur, J., Pastan, I., 1977: J. Biol. Chem. **252**, 1639—1646.

Preti, A., Fiorilli, A., Lombardo, A., Caimi, L., Tettamanti, G., 1980: J. Neurochem. **35**, 281—296.

Quarles, R. H., Brady, R. O., 1971: J. Neurochem. **18**, 1809—1820.

Rahmann, H., Rösner, H., Breer, H., 1976: J. Theor. Biol. **57**, 231—237.

Rauvala, H., 1976: FEBS Lett. **65**, 229—233.

— 1979: Eur. J. Biochem. **97**, 555—564.

— Finne, J., 1979: FEBS Lett. **97**, 1—8.

— Krusius, T., Finne, J., 1978: Biochim. Biophys. Acta **531**, 266—274.

Rearick, J. I., Sadler, J. E., Paulson, J. C., Hill, R. L., 1979: J. Biol. Chem. **254**, 4444—4451.

Reissig, J. L., Leloir, L. F., 1966: Methods Enzymol. **8**, 175—178.

Renlund, M., Chester, M. A., Lundblad, A., Aula, P., Raivo, K. O., Autio, S., Koskela, S.-L., 1979: Eur. J. Biochem. **101**, 245—250.

Reuter, G., Pfeil, R., Kamerling, J. P., Vliegenthart, J. F. G., Schauer, R., 1980: Biochim. Biophys. Acta **630**, 306—310.

Richardson, C. L., Keenan, T. W., Morré, D. J., 1977: Biochim. Biophys. Acta **488**, 88—96.

Robinson, G. B., 1968: Biochem. J. **108**, 275—280.

Rodén, L., 1970: In: Metabolic Conjugation and Metabolic Hydrolysis (Fishman, W. H., ed.), Vol. 2, pp. 345—442. New York: Academic Press.

Rohr, T. E., Troy, F. A., 1980: J. Biol. Chem. **255**, 2332—2342.

Romanovska, E., Watkins, W. M., 1963: Biochem. J. **87**, 37P.

Roseman, S., 1962: Fed. Proc. **21**, 1075—1083.

— 1970: Chem. Phys. Lipids **5**, 270—297.

— Jourdian, G. W., Watson, D., Rood, R., 1961: Proc. Natl. Acad. Sci. U.S.A. **47**, 958—961.

Rosenberg, A., 1979: In: Complex Carbohydrates of Nervous Tissue (Margolis, R. U., Margolis, R. K., eds.), pp. 25—43. New York-London: Plenum Press.

— Schengrund, C.-L., 1976: In: Biological Roles of Sialic Acid (Rosenberg, A., Schengrund, C.-L., eds.), pp. 295—359. New York-London: Plenum Press.

— Binnie, B., Chargaff, E., 1960: J. Am. Chem. Soc. **82**, 4113—4114.

Ross, D. H., Kibler, B. C., Cardenas, H. L., 1977: Drug Alc. Depend. **2**, 305—315.

ROSSIGNOL, B., HERMAN, G., CLAUSER, H., 1969: Biochem. Biophys. Res. Commun. **34**, 111—119.
RÖSNER, H., WIEGANDT, H., RAHMANN, H., 1973: J. Neurochem. **21**, 655—665.
SADLER, J. E., PAULSON, J. C., HILL, R. L., 1979a: J. Biol. Chem. **254**, 2112—2119.
— REARICK, J. I., PAULSON, J. C., HILL, R. L.: 1979b: J. Biol. Chem. **254**, 4434—4443.
— — HILL, R. L., 1979c: J. Biol. Chem. **254**, 5934—5941.
SAITO, M., SUGANO, K., NAGAI, Y., 1979: J. Biol. Chem. **254**, 7845—7854.
SALO, W. L., FLETCHER, H. G., jr., 1969: J. Org. Chem. **34**, 3189—3191.
— — 1970a: Biochemistry **9**, 878—881.
— — 1970b: Biochemistry **9**, 882—885.
SANDER, M., VEH, R. W., SCHAUER, R., 1979a: In: Glycoconjugate Research (GREGORY, J. D., JEANLOZ, R. W., eds.), Vol. 2, pp. 927—931. New York: Academic Press.
— — — 1979b: In: Glycoconjugates (SCHAUER, R., BOER, P., BUDDECKE, E., KRAMER, M. F., VLIEGENTHART, J. F. G., WIEGANDT, H., eds.), pp. 358—359. Stuttgart: G. Thieme.
SANDHOFF, K., PALLMANN, B., 1978: Proc. Natl. Acad. Sci. U.S.A. **75**, 122—126.
SARCIONE, E. J., BOHNE, M., LEAHY, M., 1964: Biochemistry **3**, 1973—1976.
SCHACHTER, H., 1974: Biochem. Soc. Symp. **40**, 57—71.
— 1978: In: Glycoconjugates (HOROWITZ, M. I., PIGMAN, W., eds.), Vol. 2, pp. 87—181. New York: Academic Press.
— RODÉN, L., 1973: In: Metabolic Conjugation and Metabolic Hydrolysis (FISHMAN, W. H., ed.), Vol. 3, pp. 1—149. New York: Academic Press.
— ROSEMAN, S., 1980: In: The Biochemistry of Glycoproteins and Proteoglycans (LENNARZ, W. J., ed.), pp. 85—160. New York-London: Plenum Press.
— JABBAL, I., HUDGIN, R. L., PINTERIC, L., MCGUIRE, E. J., ROSEMAN, S., 1970: J. Biol. Chem. **245**, 1090—1100.
— MICHAELS, M. A., CROOKSTON, M. C., TILLEY, C. A., CROOKSTON, J. H., 1971: Biochem. Biophys. Res. Commun. **45**, 1011—1018.
SCHAUER, R., 1971: Hoppe-Seyler's Z. Physiol. Chem. **352**, 1282—1290.
— 1978a: Methods Enzymol. **50 C**, 64—89.
— 1978b: Methods Enzymol. **50 C**, 374—386.
— 1982: Adv. Carbohydr. Chem. Biochem. **40**, 131—234.
— CORFIELD, A. P., 1981: In: Medicinal Chemistry Advances (DE LAS HERAS, F. G., VEGA, S., eds.), pp. 423—434. Oxford: Pergamon Press.
— FAILLARD, H., 1968: Hoppe-Seyler's Z. Physiol. Chem. **349**, 961—968.
— WEMBER, M., 1970: Hoppe-Seyler's Z. Physiol. Chem. **351**, 1353—1358.
— — 1971: Hoppe-Seyler's Z. Physiol. Chem. **352**, 1517—1523.
— — 1973: Hoppe-Seyler's Z. Physiol. Chem. **354**, 1405—1414.
— SCHOOP, H. J., FAILLARD, H., 1968: Hoppe-Seyler's Z. Physiol. Chem. **349**, 645—652.
— WEMBER, M., WIRTZ-PEITZ, F., FERREIRA DO AMARAL, 1971: Hoppe-Seyler's Z. Physiol. Chem. **352**, 1073—1080.
— BUSCHER, H.-P., CASALS-STENZEL, J., 1974: Biochem. Soc. Symp. **40**, 87—116.
— HAVERKAMP, J., WEMBER, M., VLIEGENTHART, J. F. G., KAMERLING, J. P., 1976: Eur. J. Biochem. **62**, 237—242.
— SANDER, M., VEH, R. W., WEMBER, M., 1979a: In: Glycoconjugates (SCHAUER, R., BOER, P., BUDDECKE, E., KRAMER, M. F., VLIEGENTHART, J. F. G., WIEGANDT, H., eds.), pp. 360—361. Stuttgart: G. Thieme.
— MOCZAR, E., WEMBER, M., 1979b: Hoppe-Seyler's Z. Physiol. Chem. **360**, 1587—1593.
— HAVERKAMP, J., EHRLICH, K., 1980a: Hoppe-Seyler's Z. Physiol. Chem. **361**, 641—648.
— VEH, R. W., SANDER, M., CORFIELD, A. P., WIEGANDT, H., 1980b: In: Structure and Function of Gangliosides (SVENNERHOLM, L., MANDEL, P., DREYFUS, H., URBAN, P. F., eds.), pp. 283—294. New York-London: Plenum Press.

SCHENGRUND, C.-L., NELSON, J. T., 1975: Biochem. Biophys. Res. Commun. **63**, 217—223.
SCHRAMM, G., MOHR, E., 1959: Nature **183**, 1677—1678.
SCHUT, B. L., DORLAND, L., HAVERKAMP, J., VLIEGENTHART, J. F. G., FOURNET, B., 1978: Biochem. Biophys. Res. Commun. **82**, 1223—1228.
SHARON, N., 1975: Complex Carbohydrates. Their Chemistry, Biosynthesis and Functions (SHARON, N., ed.), pp. 155–176. Reading, Mass.: Addison-Wesley.
SHIER, W. T., ROLOSON, G., 1974: Biochem. Biophys. Res. Commun. **59**, 51—56.
SHOYAB, M., BACHHAWAT, B. K., 1967: Biochem. J. **102**, 13C—14C.
SHUR, B. D., ROTH, S., 1975: Biochim. Biophys. Acta **455**, 473—512.
SIRBASKU, D. A., BINKLEY, S. B., 1970: Biochim. Biophys. Acta **198**, 479—482.
SLOMIANY, A., SLOMIANY, B. L., 1978: J. Biol. Chem. **253**, 7301—7306.
SLOMIANY, B. L., SLOMIANY, A., HERP, A., 1978: Eur. J. Biochem. **90**, 255—260.
SOMMAR, K. M., ELLIS, D. B., 1972a: Biochim. Biophys. Acta **268**, 581—589.
— — 1972b: Biochim. Biophys. Acta **268**, 590—595.
SPIK, G., SIX, P., MONTREUIL, J., 1979: Biochim. Biophys. Acta **584**, 203—215.
SPIRO, R. G., 1959: J. Biol. Chem. **234**, 742—748.
— 1970: Glycoproteins. Ann. Rev. Biochem. **39**, 599—638.
SPIVAK, C. T., ROSEMAN, S., 1966: Methods Enzymol. **9**, 612—615.
STOFFYN, P., STOFFYN, A., 1980: Carbohydr. Res. **78**, 327—340.
STRECKER, G., MICHALSKI, J.-C., 1978: FEBS Lett. **85**, 20—24.
— MONTREUIL, J., 1971: Clin. Chim. Acta **33**, 253—255.
— — 1979: Biochimie **61**, 1199—1246.
STROMINGER, J. L., SMITH, M. S., 1959: J. Biol. Chem. **234**, 1822—1827.
STRUCK, D. K., LENNARZ, W. J., 1980: In: The Biochemistry of Glycoproteins and Proteoglycans (LENNARZ, W. J., ed.), pp. 35—83. New York-London: Plenum Press.
STURGESS, J. M., MITRANIC, M. M., MOSCARELLO, M. A., 1972: Biochem. Biophys. Res. Commun. **46**, 1270—1277.
STURGESS, J., MOSCARELLO, M., SCHACHTER, H., 1978: Curr. Top. Memb. Transp. **11**, 15—105.
SUTTAJIT, M., WINZLER, R., 1971: J. Biol. Chem. **246**, 3398—3404.
— URBAN, C., MCLEAN, R. L., 1971: J. Biol. Chem. **246**, 810—814.
SUZUKI, Y., MONOKA, T., MATSUMOTO, M., 1980: Biochim. Biophys. Acta **619**, 632—639.
SVENNERHOLM, L., 1980: In: Structure and Function of Gangliosides (SVENNERHOLM, L., MANDEL, P., DREYFUS, H., URBAN, P. F., eds.), pp. 533—544. New York-London: Plenum Press.
SWEELEY, C. C., SIDDIQUI, B., 1977: In: The Glycoconjugates (HOROWITZ, M. I., PIGMAN, W., eds.), Vol. 1, pp. 459—540. New York: Academic Press.
TALLMAN, J. F., BRADY, R. O., 1973: Biochim. Biophys. Acta **293**, 434—443.
TANENBAUM, S. W., SUN, S.-C., 1971: Biochim. Biophys. Acta **229**, 824—828.
TAUBER, R., REUTTER, W., 1978: FEBS Lett. **87**, 135—138.
TESORIÈRE, G., VENTO, R., MAGISTRO, D., DONES, F., 1971: Eur. J. Biochem. **19**, 289—293.
TETTAMANTI, G., VENERANDO, B., CESTARO, B., PRETI, A., 1976: In: Ganglioside Function (PORCELLATI, G., CECARELLI, B., TETTAMANTI, G., eds.), pp. 65—79. New York-London: Plenum Press.
— PRETI, A., CESTARO, B., VENERANDO, B., LOMBARDO, A., GHIDONI, R., SONNINO, S., 1980: In: Structure and Function of Gangliosides (SVENNERHOLM, L., MANDEL, P., DREYFUS, H., URBAN, P. F., eds.), pp. 263—281. New York-London: Plenum Press.
TROY, F. A., MCCLOSKEY, M. A., 1979: J. Biol. Chem. **254**, 7377—7387.
— VIJAY, I. K., TESCHE, N., 1975: J. Biol. Chem. **250**, 156—163.
TRUJILLO, J. L., HORNG, W. J., GAN, J. C., 1971: Biochim. Biophys. Acta **252**, 443—452.
UCHIDA, Y., TSUKADA, Y., SUGIMORI, T., 1974: Biochim. Biophys. Acta **350**, 425—431.

UCHIDA, Y., TSUKADA, Y., SUGIMORI, T., 1979: J. Biochem. **86**, 1573—1585.
VAN DEN EIJNDEN, D. H., 1973: J. Neurochem. **21**, 949—958.
— VAN DIJK, W., 1974: Biochim. Biophys. Acta **362**, 136—149.
— STOFFYN, P., STOFFYN, A., SCHIPHORST, W. E. C. M., 1977: Eur. J. Biochem. **81**, 1—7.
— BARNEVELD, R. A., SCHIPHORST, W. E. C. M., 1979: Eur. J. Biochem. **95**, 629—637.
VAN DIJK, W., MAIER, H., VAN DEN EIJNDEN, D. H., 1976: Biochim. Biophys. Acta **444**, 816—834.
— — — 1977: Biochim. Biophys. Acta **466**, 187—197.
VANN, W. F., LIU, T.-Y., ROBBINS, J. B., 1978: J. Bacteriol. **133**, 1300—1306.
VEH, R. W., SANDER, M., 1981: In: Sialidases and Sialidoses, Perspectives in Inherited Metabolic Diseases (TETTAMANTI, G., DURAND, P., DI DONATO, S., eds.), Vol. 4, pp. 71—109. Milano: Edi Ermes.
— SCHAUER, R., 1978: In: Enzymes of Lipid Metabolism (GATT, S., FREYSZ, L., MANDEL, P., eds.), pp. 447—462. New York-London: Plenum Press.
— CORFIELD, A. P., SANDER, M., SCHAUER, R., 1977: Biochim. Biophys. Acta **486**, 145—160.
— SANDER, M., HAVERKAMP, J., SCHAUER, R., 1979: In: Glycoconjugate Research (GREGORY, J. D., JEANLOZ, R. W., eds.), Vol. 1, pp. 557—559. New York: Academic Press.
VERBERT, A., CACAN, R., DEBEIRE, P., MONTREUIL, J., 1977: FEBS Lett. **74**, 234—238.
VIJAY, I. K., TROY, F. A., 1975: J. Biol. Chem. **250**, 164—170.
VLIEGENTHART, J. F. G., VAN HALBEEK, H., DORLAND, L., 1981: Pure Appl. Chem. **53**, 45—77.
VON NICOLAI, H., 1976: Habilitationsschrift, University of Bonn, Federal Republic of Germany.
— DRZENIEK, R., ZILLIKEN, F., 1971: Z. Naturforschung **26**, 1049—1051.
WAECHTER, C. J., SCHER, M. G., 1979: In: Complex Carbohydrates of Nervous Tissue (MARGOLIS, R. U., MARGOLIS, R. K., eds.), pp. 75—102. New York-London: Plenum Press.
WARREN, L., 1972: In: Glycoproteins, Their Composition, Structure and Function (GOTTSCHALK, A., ed.), part B, 2nd ed., pp. 1097—1126. Amsterdam: Elsevier.
— FELSENFELD, H., 1961 a: Biochem. Biophys. Res. Commun. **4**, 232—235.
— — 1961 b: Biochem. Biophys. Res. Commun. **5**, 185—190.
— — 1962: J. Biol. Chem. **237**, 1421—1431.
WATSON, D. R., JOURDIAN, G. W., ROSEMAN, S., 1966: J. Biol. Chem. **241**, 5627—5636.
WATANABE, K., POWELL, M. E., HAKOMORI, S. I., 1979: J. Biol. Chem. **254**, 8223—8229.
WENGER, D. A., WARDELL, S., 1972: Physiol. Chem. Physics **4**, 224—230.
WETMORE, S., MAHLEY, R. W., BROWN, W. V., SCHACHTER, H., 1974: Can. J. Biochem. **52**, 655—664.
WIEGANDT, H., 1980: In: Ganglioside Structure and Function (SVENNERHOLM, L., MANDEL, P., DREYFUS, H., URBAN, P. F., eds.), pp. 3—10. New York: Plenum Press.
WINTERBURN, P. J., PHELPS, C. F., 1970: Nature **228**, 1311—1313.
— — 1971 a: Biochem. J. **121**, 701—709.
— — 1971 b: Biochem. J. **121**, 711—720.
— — 1971 c: Biochem. J. **121**, 721—730.
— — 1973: In: The Enzymes of Glutamine Metabolism (PRUSINIER, S., STADTMAN, E. R., eds.), pp. 343—363. New York: Academic Press.
WITT, W., VON NICOLAI, H., ZILLIKEN, F., 1979: Nutr. Metab. **23**, 51—61.
WITT-KRAUSE, W., VON NICOLAI, H., ZILLIKEN, F., 1979: Hoppe-Seyler's Z. Physiol. Chem. **360**, 404.
WOLF, D., SOKOLOSKI, J. E., LITT, M., 1980: Biochim. Biophys. Acta **630**, 545—558.
WRIGLEY, N. G., SKEHEL, J. J., CHARLWOOD, P. A., 1973: Virology **51**, 525—529.
YAMAKAWA, T., NAGAI, Y., 1978: Trends Biochem. Sci. **3**, 128—131.

Yip, M. C. M., 1973: Biochim. Biophys. Acta **306**, 298—306.
Yohe, H. C., Rosenberg, A., 1977: J. Biol. Chem. **252**, 2412—2418.
— Ueno, K., Chang, N.-C., Glaser, G. H., Yu, R. K., 1980: J. Neurochem. **34**, 560—568.
Yu, R. K., Ando, S., 1980: In: Ganglioside Structure and Function (Svennerholm, L., Mandel, P., Dreyfus, H., Urban, P. F., eds.), pp. 33—45. New York: Plenum Press.
— Lee, S. H., 1976: J. Biol. Chem. **251**, 198—203.
Zalan, E., Wilson, C., Freitag, R., 1975: Arch. Virol. **47**, 177—179.
Ziegler, D., Keilich, G., Brossmer, R., 1980: Fresenius Z. Anal. Chem. **301**, 99—100.

J. Biological Significance of Sialic Acids

WERNER REUTTER, ECKART KÖTTGEN*, CHRISTIAN BAUER, and WOLFGANG GEROK*

Institut für Molekularbiologie und Biochemie, Freie Universität Berlin (Dahlem),
* Medizinische Klinik der Universität, Freiburg i. Br., Federal Republic of Germany

With 2 Figures

Contents

I. Introduction

Sialic acids are essential constituents of many glycoproteins, glycopeptides and glycolipids. This widespread occurrence in glycoconjugates of different origin indicates that a variety of biological functions should be associated with this sugar. There are valid indications that sialic acids influence or even determine the recognition of low and high molecular-weight compounds, the action of certain hormones, physicochemical and catalytic properties of enzymes, hemostasis, cellular adhesiveness, antigenicity, transport processes and synaptic transmission. Despite this apparently heterogenous spectrum of biological actions, four main functions—according to the concept of SCHAUER (1982)—can be attributed to sialic acid: (1) Endowment of glycoconjugates and cellular membranes with a negative charge, (2) Influence on the macromolecular structure of certain glycoproteins, (3) Information transfer, (4) Protection of glycoconjugates and cells from recognition and degradation. This rather physicochemical classification could be of use when starting to elucidate the biological significance of sialo-glycoconjugates. Reviews on sialic acids have recently been published by

ROSENBERG and SCHENGRUND 1976, NEUFELD and ASHWELL 1980, and SCHAUER 1982. The aim of this contribution, however, is to emphazise the important role of sialic acids in the occurrence of pathobiochemical alterations found in animals and human beings.

II. Role in Cellular Adhesiveness

1. Adhesiveness of Benign Cells

A presupposition for the coordinate development of any multicellular system or tissue is the exchange of information (i) between neighbouring cells (cellular contact) and (ii) between the cell surface and the environmental fluid phase (humoral contact). Fluid phase contact mainly comprises the recognition and uptake of informatory molecules by receptors and the secretion of cellular components. Contact formation and information transfer are mediated by the plasma membrane and numerous studies indicate that membrane-bound glycoconjugates are responsible for this interrelation between cell and environment.

A glycoconjugate is characterized by at least two functional regions, the hydrophobic domain is inserted more or less tightly into the protein-lipid bilayer of the membrane, while the hydrophilic oligosaccharide "tail" emerges into the surrounding fluid. This external orientation of plasma membrane-bound sugars could be verified for different cells by applying different techniques (CARRAWAY 1975, GAHMBERG and HAKOMORI 1975, HYNES 1976, PETERS and RICHARDS 1977). It is widely accepted that sialic acids, due to their predominantly terminal position within the carbohydrate chain, are involved in cellular adhesion. Moreover, it is unquestionable that this sugar contributes a great deal to the negative surface charge of cells (COOK *et al.* 1961) with the consequence of a possible electrostatic repulsion as described for platelets, erythrocytes and some carcinoma cells (for review see JEANLOZ and CODINGTON 1976, SCHAUER 1982). Yet, this negative charge is also responsible for the formation of Ca^{2+}-bridges between adjacent chick embryonic muscle cells (KEMP 1970). Similarly, the attachment of both endothelium and epithelium to the glomerular basement membrane is facilitated by sialic acid (KANWAR and FARQHUAR 1980) and it is certain that the surface charge can markedly influence the contact formation between cells (CURTIS 1967, WEISS 1973, LLOYD 1975, JEANLOZ and CODINGTON 1976). It is necessary, however, to emphasize that these findings cannot be generalized. Treatment of CHO cells or 3T3 cells with neuraminidase had no effect upon their subsequent attachment and spreading on a glass surface (KRAEMER 1966, KOLODNY 1972), while in the case of many other cell lines the removal of sialic acid increased cellular adhesion (for review see LLOYD 1975, ROSENBERG and SCHENGRUND 1976, SCHAUER 1982).

Carbohydrates exposed on the cell surface apparently have an additional function, namely the binding of adhesion proteins, for example fibronectin. Several ricin-resistant mutants of BHK cells, which are deficient in some surface sugars (MEAGER *et al.* 1976), because the equivalent glycosyltransferases are absent or possess only minimal activity, showed an abnormal response to fibronectin (PENA and HUGHES 1978, VISCHER and HUGHES 1979). These altered cells adhered quantitatively less to a collagen film in the presence of fibronectin compared to normal BHK cells (HUGHES *et al.* 1979a). Furthermore, surface carbohydrate-deficient mutant cells were unable to maintain a stable spread out configuration on a fibronectin lattice (HUGHES *et al.* 1979a) and tunicamycin, which interferes with protein glycosylation, affected the cellular response to fibronectin (BUTTERS *et al.* 1980).

It is known that desialylation induces conformational changes (GOTTSCHALK 1960a, BEZKOROVAINY 1965, DUC DODON and QUASH 1981, KÖTTGEN *et al.* 1981) and there is good evidence that this phenomenon is not restricted to soluble glycoconjugates. Rat dermal fibroblasts, which had been exposed to neuraminidase, showed an increased degree of deformability and it has been discussed that sialic acid controls the membrane "tone", thus enabling the cell to protrude at deformable regions (WEISS 1967, LLOYD 1975). Treatment of isolated liver plasma membranes with *Vibrio cholerae* neuraminidase apparently causes alterations in membrane structure as demonstrated by the determination of the circular dicroism (positive cotton effect after desialylation) (REUTTER *et al.* 1978). Hence it is reasonable to conclude that sialic acids have a stabilizing effect on glycoprotein conformation and one might speculate that in the case of multiantenneric glycoconjugates one oligosaccharide chain does not project into the aqueous environment, but forms a bridge to a positively charged group in close vicinity. This behaviour would further strengthen membrane stability independently from microfilaments and microtubules, while the occurrence of neuraminidase would loosen these bonds. The presence of membrane-bound neuraminidase has been demonstrated in the membrane of platelets (BOSMANN 1972), erythrocytes (BOSMANN 1974) and liver (PRICER and ASHWELL 1971, SCHENGRUND *et al.* 1972, VISSER and EMMELOT 1973).

The content of sialic acid in the plasma membrane can also be subject to a possible hormonal regulation, a point neglected so far. HeLa S_3 cells grown in the presence of corticosteroids contain markedly more sialic acid than their corresponding untreated controls (CARUBELLI and GRIFFIN 1967). Therefore it is conceivable that changes of hormonal levels could influence cellular interaction by altering the amount of membrane-bound sialic acid.

2. *Adhesiveness of Transformed and Malignant Cells*

The important role of sialic acid in the adsorption of polyoma virus and the subsequent hemagglutination of erythrocytes was first indicated after treatment with neuraminidase which leads to loss of agglutinability (STEWART *et al.* 1957, EDDY *et al.* 1958, SACHS *et al.* 1959, HARTLEY *et al.* 1959). In a recent publication evidence has been presented that the virus is bound by specific sialyloligosaccharide receptors exposed on the erythrocyte's surface (CAHAN and PAULSON 1980). In order to ascertain the structure and the biological significance of this receptor, neuraminidases of different origin and substrate specificity have been used. Red blood cells treated with neuraminidase of Newcastle disease virus, which hydrolyses sialic acids arranged in the sequence Neu5Acα(2-3)Gal and Neu5Acα(2-8)Neu5Ac, lost about 40% of their surface-bound sialic acid, but were not agglutinable by polyoma virus. Similarly, exposure to *Vibrio cholerae* neuraminidase, which removes virtually all sialic acid residues, prevents agglutination by polyoma and influenza virus. Selective replacement of sialic acid with purified galactoside-α(2-3)-sialyltransferase completely restored hemagglutination by polyoma virus, while sialyltransferases which add Neu5Ac to the C-atom 6 of either galactose or N-acetylgalactosamine were ineffective (CAHAN and PAULSON 1980). Hence it is reasonable to assume that the sialyloligosaccharide receptor has the sequence Neu5Acα(2-3)Galβ(1-3)GalNAc-Thr/Ser. This conclusion is further supported by studies on 3T3 cells that can only be infected by polyoma virus, if this specific receptor is present (FRIED *et al.* 1981) and it is worth mentioning that the binding of Sendai virus to the host cell is mediated by a similar carbohydrate sequence [Neu5Acα(2-8)Neu5Acα(2-3)Galβ(1-3)GalNAc], though this oligosaccharide chain is part of a ganglioside (HOLMGREN *et al.* 1980).

Evidence indicating that Neu5Ac is associated with a variety of alterations observed in the malignant cell is accumulating, though the results are in part contradictory. There seems to exist a positive correlation between the increased content of plasma membrane-bound sialic acid and the metastatic potential of different cell lines derived from renal sarcoma (YOGEESWARAN *et al.* 1978), virus transformed BALB/C3T3 cells (PEARLSTEIN *et al.* 1980) or murine melanoma variants (DOBROSSY *et al.* 1981), while the loss of metastatic properties is accompanied by a decline of the sialic acid content on the cell surface (BURGER *et al.* 1979). In a highly metastatic melanoma cell line the increase of Neu5Ac was paralleled by a twofold elevation of ectosialyltransferase activity (DOBROSSY *et al.* 1981). Therefore it may be tempting to speculate that this increased ecto-enzyme activity is responsible for the increased transfer, or alternatively, for an accelerated reinsertion of sialic acid molecules. In contrast to these findings some other melanoma variants exhibit an inverse relationship between the degree of sialylation of a major membrane glycoprotein and the tendency to form metastases (RAZ *et al.* 1980). Therefore it is not possible to fit the different results into a scheme of general validity.

Adhesive properties of normal fibroblasts are markedly reduced upon transformation by oncogenic viruses or other agents *in vitro* and these alterations have been implicated with the tendency of cells to segregate and to form metastases *in vivo* (VASILIEV and GELFAND 1977). Though different findings indicate that fibroblasts show a variety of cell surface changes following transformation (HYNES 1976, SHIU *et al.* 1977, VAHERI and MOSHER 1978, YAMADA and OLDEN 1978, POUYSSÉGUR and YAMADA 1978, BRAMWELL and HARRIS 1978, VAHERI *et al.* 1980, CHEN and SINGER 1980, ATKINSON and HAKIMI 1980, SIDEBOTTOM 1980), no crucial role, at least at the moment, can be attributed to sialic acid. Things are further complicated by the fact that most studies have been done with single cells in culture and hence it cannot be excluded that the regulation of the cell contact in solid tumors differs considerably from the situation found in the petri dish. Moreover, it is indispensable to perform appropriate control experiments, because many of the changes found after malignant transformation could be due to rapid proliferation, instead of being closely related to malignancy. When comparing the findings summarized in Tables 1 and 2, there are substantial differences in the metabolism of Neu5Ac, but the possibly underlying significance is still obscure. The capacity to transfer sialic acid to glycoprotein or ganglioside acceptors in transformed or malignant cells can be found increased, decreased or is unchanged.

Table 1. *Alterations of the N-acetylneuraminic acid metabolism in transformed and neoplastic cells*

Parameter	Mode of alteration	Tissue, cell (-organelle)	References
1. Protein-bound Neu5Ac	decrease	whole cells (Py-3T3-cells)	Ohta *et al.* 1968
		membrane fraction (SV-40-3T3)	Wu *et al.* 1969, Grimes 1970, 1973
		plasma membranes (hepatoma cells)	Shimizu and Funakoshi 1973, Leblond-Larouche *et al.* 1975
		thyroglobulin (rat)	Monaco and Robbins 1973
		malignant lung cells (mouse)	Vilarem *et al.* 1981
	increase	Morris hepatoma	Harms *et al.* 1973
		Py-BHK, SV-40-3T3, malignant lymphoblasts	van Beek *et al.* 1973
		serum (rat, dog)	Bernacki and Kim 1977, Kloppel *et al.* 1978
		serum (human)	Winzler 1955, Watkins 1974, Bradley *et al.* 1977, Silver *et al.* 1979, 1980, Hogan-Ryan *et al.* 1980, Hogan-Ryan and Fennelly 1981 Harvey *et al.* 1981
2. CMP-Neu5Ac	moderate increase	Morris hepatoma 7777, 3924 A	Harms *et al.* 1973, Reutter and Bauer 1978, Grünholz 1978
3. UDP-N-Acetylglucosamine-2′-epimerase	strong decrease	Morris hepatomas	Reutter *et al.* 1970, Kikuchi *et al.* 1971, Harms *et al.* 1973, Grünholz 1978
		Yoshida sarcoma, Yoshida ascites tumor	Kikuchi *et al.* 1971
4. N-Acetylmannosamine-kinase	decrease	Morris hepatomas, Yoshida sarcoma, Yoshida ascites hepatoma	Kikuchi *et al.* 1971, Grünholz 1978
5. CMP-Neu5Ac synthase	moderate decrease	Morris hepatoma 7777, 3924 A	Grünholz 1978
6. CMP-Neu5Ac hydrolase	decrease	Morris hepatoma 7777, 3924 A	Grünholz 1978
		ovarian tumors (human)	Chatterjee *et al.* 1978
	increase	serum (ovarian cancer patients)	Chatterjee *et al.* 1978

Table 2. *Alteration of sialyltransferase activities in transformed and neoplastic cells*

Mode of alteration	Tissue, cell (-organelle)	References
Decrease	SV-40-3T3 cells	GRIMES 1970, 1973
	Py-BHK cells	DEN *et al.* 1971
	Morris hepatoma 7777, 7800	HUDGIN *et al.* 1971
	thyroid tumor (rat)	MONACO and ROBBINS 1973
	ascites hepatoma cells	SAITO *et al.* 1974
	Morris hepatoma 9618 A_2, 3924 A	BAUER *et al.* 1977, REUTTER and BAUER 1978
	mammary tumors (rat)	KEENAN and MORRÉ 1973, CHATTERJEE 1979
Increase	Py-3T3-, MSV-3T3-, RSV-3T3 cells	BOSMANN 1972
	RSV-BHK cells	WARREN *et al.* 1972
	mammary and colon carcinomas (human)	BOSMANN and HALL 1974
	serum (carcinoma-bearing rats)	BOSMANN *et al.* 1975, IP and DAO 1977
	serum (mammary tumor-bearing rats)	BOSMANN and HILF 1974, BERNACKI and KIM 1977, CHATTERJEE 1979
	serum (melanoma-bearing mice)	KONDO *et al.* 1981
	liver (Walker sarcoma-bearing rats)	IP and DAO 1977
	serum (patients with different malignant, and benign, diseases)	KESSEL and ALLEN 1975, BAYER *et al.* 1977, GANZINGER 1977, HENDERSON and KESSEL 1977, IP and DAO 1978, DAO *et al.* 1980, RONQUIST *et al.* 1980
Unchanged	Morris hepatoma 5123 D	HUDGIN *et al.* 1971
	colon carcinoma (human)	KIM *et al.* 1974

Similarly, the content of membrane-bound sialic acid is subject to profound variations. In most Morris hepatoms the content of protein-bound Neu5Ac is increased from 4.5 ± 0.5 nmol (normal liver) to 5.9–8.4 nmol/mg protein, while the concentration of CMP-sialic acid remains unchanged (HARMS *et al.* 1973, REUTTER and BAUER 1978). The level of free sialic acid, however, as well as the activity of UDP-N-acetylglucosamine-2′-epimerase, the key enzyme of Neu5Ac synthesis, are markedly decreased (REUTTER *et al.* 1970, KIKUCHI *et al.* 1971, HARMS *et al.* 1973). This finding seems to be specific for malignant growth, because in regenerating liver after partial hepatectomy the activity of 2′-epimerase remains unchanged. It should be emphasized that all these measurements have been performed with "crude" protein fractions and not with defined glycoproteins occurring in both normal and transformed tissue. Attention should also be focused on the localization of sialic acid-containing glycoconjugates within the domains of the plasma membrane, because differences could reflect specific functions of certain membrane areas. Moreover, even mitosis of normal cells can modify the content of protein-bound sialic acid (GLICK and BUCK 1973).

Yet, this finding is of special interest, because there exists a positive correlation between several surface characteristics of proliferating cells, as inability to produce stable lateral cell contact, increased exposition of receptor sites or "leakness" for intracellular proteases and the surface density of sialic acid (for review see VAN BEEK *et al.* 1973). These results are compatible with data that neuraminidase treatment of density-inhibited cell cultures stimulates cell division (VAHERI *et al.* 1972).

III. Influence on the Life-Span of Blood Components

1. Erythrocytes

Glycophorin A, a major glycoprotein of the erythrocyte's membrane, is characterized by a high content of sialic acid and consequently has a decisive influence on the negative charge of red blood cells (EYLAR *et al.* 1962). Glycophorin carries the antigenic determinants of AB and MN blood groups, the receptor-like function for influenza viruses (FURTHMAYR 1978), exhibits a lectin-like behaviour (BOWLES and HANKE 1977), probably prevents agglutination and phagocytosis of erythrocytes and is able to activate the lysosomal proteinase cathepsin L (BÖHMER *et al.* 1979). Different studies have shown that the content of surface-bound sialic acid may depend on the age of the erythrocytes (DANON and MARIKOVSKY 1961, YAARI 1969, GREENWALT and STEANE 1973); immature or young cells possess more sialic acid residues (YACHNIN and GARDNER 1961), while aging reduces its number (COHEN *et al.* 1976, LUNER *et al.* 1977, GATTEGNO *et al.* 1981). Despite some contradictory results, showing either decreased or normal surface charge of immature erythrocytes (VAN GASTEL and DEVIT 1970, ACKERMANN 1972), and normal endowment of reticulocytes with sialic acid (TISHKOFF 1966), Neu5Ac is still the most likely candidate that determines the life-span of erythrocytes (for review see SCHAUER 1982).

Treatment of erythrocytes with neuraminidase (STEWART *et al.* 1955, 1957, GARDNER *et al.* 1961, PERONA *et al.* 1964) or intravenous injection of this enzyme (LANDAW *et al.* 1973) leads to a drastic decrease of their half-life in the circulation, in man it is reduced from about 120 days to a few hours (HALBHUBER *et al.* 1972, DUROCHER *et al.* 1975, JANCIK *et al.* 1978). In vitro experiments revealed that the recognition mechanism of asialo-erythrocytes has very much in common with the uptake of soluble asialo-glycoproteins, though the blood cells are degraded exclusively in Kupffer cells (HALBHUBER *et al.* 1972), while asialo-glycoproteins are taken up by hepatocytes (ASHWELL and MORELL 1974).

The adhesion of asialo-erythrocytes (AMINOFF *et al.* 1977) is mediated by lectin-like receptors on Kupffer cells which recognize the unmasked galactosyl residue (JANCIK *et al.* 1978, KOLB and KOLB-BACHOFEN 1978, KOLB *et al.* 1979, 1981, SCHLEPPER-SCHÄFER *et al.* 1980). This process, however, can be inhibited by oligosaccharides or glycoproteins that possess galactose in terminal position (lactose, asialo-fetuin, asialo-glycophorin, galactosyl-albumin, galactosyl-lysozym) (SCHLEPPER-SCHÄFER *et al.* 1980, KOLB *et al.* 1980, KÜSTER and SCHAUER 1981). By using different experimental conditions it could be ruled out that neuraminidase, which had been used to produce asialo-erythrocytes, acts as a linker between the red blood cells and the macrophages (KÜSTER and SCHAUER 1981), and there is good evidence that antibodies are not prerequisite for this cellular interaction (KOLB *et al.* 1981).

It is necessary to stress, however, that adhesion of asialo-erythrocytes does not necessarily entail their uptake and degradation (KÜSTER and SCHAUER 1981). Furthermore, the complement system is apparently involved in the endocytosis of attached cells, because heat inactivation of serum reduces considerably the rate of phagocytosis.

There exists still some controversity about the question, whether most of the findings obtained under experimental conditions really reflect the situation *in vivo* and caution is necessary in the interpretation of experiments in which neuraminidase has been used to prepare aged erythrocytes (NORDT *et al.* 1981). Senescent erythrocytes possess 10 to 15% less sialic acid when compared to young cells (COHEN *et al.* 1976, LUNER *et al.* 1977), while glycophorins extracted from young and aged erythrocytes contain nearly identical amounts of sialic acid (LUTZ and FEHR 1979). In order to solve this discrepancy it has been suggested that erythrocytes lose intact glycophorin molecules together with membrane fractions during cellular senescence (LUTZ and FEHR 1979). There is hardly any doubt that binding of asialo-erythrocytes to Kupffer cells follows the same basic mechanism *in vivo* and *in vitro* (KOLB *et al.* 1981, MÜLLER *et al.* 1981). Though autoantibodies may ultimately induce phagocytosis of aged erythrocytes (KAY 1975), sialic acid probably protects red blood cells from premature recognition and degradation.

It may be of clinical interest to refer to an autoimmune disease induced by the hydrolytic action of bacterial neuraminidases. The occurrence of these enzymes in the blood stream leads to a demasking of the Thomsen-Friedenreich cryptantigen on the erythrocyte's surface. Since in the serum of most human beings equivalent antibodies are present, any sudden or uncontrolled exposure of the Thomsen-Friedenreich antigen could cause anemia (FISCHER and POSCHMANN 1976). There are also certain indications that sickle cell anemia (RIGGS and INGRAM 1977), β-thalassemia (KAHANE *et al.* 1978, 1980) and paroxysmal nocturnal hemoglobinuria (BALDUINI *et al.* 1977) are related to a reduced metabolism of sialic acid. This fact may explain the shortened half-life of erythrocytes, a typical feature associated with these diseases.

2. *Thrombocytes*

Comparable with red blood cells, desialylation of platelets leads to a rapid removal from the circulation (CHOI *et al.* 1972, GREENBERG *et al.* 1979). Despite contradictory results (NURDEN and CAEN 1976, GARTNER *et al.* 1977, PATSCHKE *et al.* 1977), there are valid indications that membrane-bound sialic acid does play an important role in the adhesion and aggregation of thrombocytes. (For further details see section IV, Blood Clotting.)

3. *Lymphocytes*

Desialylated lymphocytes lose their orientation in the organism, show an increased agglutinability and are reversibly trapped in the liver (WOODRUFF and GESNER 1969). After resialylation during the following two days the lymphocytes leave the liver again and show normal behaviour in the circulation (WOODRUFF

1974). It could be shown that asialo-lymphocytes adhere *in vitro* to both Kupffer cells and hepatocytes. Adhesion is mediated by stereospecific interaction between unmasked galactose residues after neuraminidase treatment and lectin-like receptors on the liver cell surface (KOLB *et al.* 1978). A subpopulation of asialo-lymphocytes remains in the liver (WOODRUFF and WOODRUFF 1976) which can lead to an autoimmune reaction (KOLB-BACHOFEN and KOLB 1979). Repeated weekly injections of asialo-lymphocytes apparently cause liver damage. It may be assumed that desialylation of lymphocytes by viral neuraminidases also occurs *in vivo*, suggesting the following pathobiochemical sequence: virus infection—enzymatic liberation of sialic acid residues—fixation of lymphocytes to the liver cell surface—mitogenic transformation of lymphocytes—release of cytotoxic compounds—liver damage. Further support for this concept comes from investigations of NOVOGRODSKY and ASHWELL 1977, which show that mitogenesis of lymphocytes is induced by a mammalian liver protein that specifically binds desialylated glycoproteins. Hence it is intelligible that viral infections are often accompanied by a slight or moderate liver damage. Furthermore, the question arises if not the preponderance of the liver for attachment of metastatic cells may be related to lectin—like proteins for these cells which are probably altered in their surface carbohydrate content.

Recent studies have shown that T-lymphocytes of peritoneal origin possess both Neu5Ac and Neu5Gc on their surface (KAUFMANN *et al.* 1981). Since this is the first demonstration of Neu5Gc on murine lymphocytes, it would be of special interest to learn, whether this derivative has the same properties in respect to migration behaviour and life span of cell.

For further information see section V, Immunology and Infection.

4. Serum Glycoconjugates

Liver is not only the site of synthesis of the bulk of plasma proteins, but also the main target of their catabolism. After an average half-life of about 100 h, serum glycoproteins are eliminated from the blood stream predominantly via uptake by hepatocytes, a process that will be enhanced dramatically, if sialic acid residues are split off by enzymatic or acid hydrolysis. This characteristic feature has been established for orosomucoid, fetuin, coeruloplasmin, haptoglobin, α_2-macroglobulin, thyroglobulin, lactoferrin, erythropoietin, human chorionic gonadotropin, follicle stimulating hormone. More recently the same mechanism has been proven for plasma esterase, plasmin, kallikrein, activated Hagemann factor and transcortin (for review see ASHWELL and MORELL 1974, NEUFELD and ASHWELL 1980), while high doses of asialo-fetuin inhibit the uptake, mainly due to a competition at the receptor (HOSSNER and BILLIAR 1979). This basic mechanism is also valid for the accelerated uptake of desialylated low-ρ lipoproteins by smooth muscle cells (FILIPOVIC *et al.* 1979).

Removal of the penultimate sugar, mostly galactose, leads to a normalization of both the survival time in the circulation and the distribution among organs. On the first sight the extraordinary behaviour of asialo-transferrin did not fit in this general scheme, because desialylation apparently did not reduce the half-life. More detailed studies finally revealed that small amounts of asialo-transferrin are rapidly taken up by the liver, possibly resialylized intracellularly and secreted afterwards again (DEBANNE and REGOECZI 1981). This is a mass-dependent process in which individual asialo-transferrin molecules act synergistically in promoting endocytosis of each other (REGOECZI et al. 1978). This endo- and exocytotic process has been termed diacytosis. Human asialotransferrin is diacytosed with a half-life of approximately 20 min (TOLLESHAUG et al. 1981). This implies that a target signal is necessary for intralysomal degradation of "aged" asialo-transferrin, possibly achieved by differences in the degree of glycosylation or by changes of glycoprotein conformation.

The receptor responsible for the binding of asialo-glycoconjugates was first purified from rabbit liver (HUDGIN et al. 1974) and a few years later characterized by KAWASAKI and ASHWELL (1976). In rats two variants exist with a molecular weight of 55,000 and 65,000 (SCHWARTZ et al. 1981 a, b) and by applying immunological techniques a transmembrane nature of the receptor has been demonstrated (HARFORD and ASHWELL 1981). The receptor is not exclusively localized in the plasma membrane, but is also found in the lysosomes, the Golgi apparatus and the smooth endoplasmic reticulum (PRICER and ASHWELL 1976). This subcellular localization led to the hypothesis of a receptor recycling during ligand translocation, and in the meantime evidence has been gained that the receptor is indeed internalized and later reinserted into the plasma membrane (TOLLESHAUG et al. 1977, TANABE et al. 1979). Rat hepatocytes can bind 500,000 molecules of asialo-orosomucoid per cell (SCHWARTZ et al. 1980) with a rate constant of internalization of 0.040 min^{-1} at 20 °C, 0.18 min^{-1} at 30 °C and 0.28 min^{-1} at 40 °C (TOLLESHAUG et al. 1980). Above 10 °C the rate of endocytosis is clearly temperature-dependent (WEIGEL and OKA 1981). Preincubation of hepatocytes in the presence of a high concentration of asialo-orosomucoid did not influence the uptake of the subsequently added asialo-fetuin (TOLLESHAUG et al. 1980), but an intact microfilamentous and microtubular system is necessary for normal function of this system (KOLSET et al. 1979). Furthermore, exposure to neuraminidase leads to a loss of receptor activity and chloroquine reduces the receptor number on the hepatocytic surface (TOLLESHAUG and BERG 1979).

On freshly isolated hepatocytes the receptors are randomly distributed (WEIGEL 1980), but binding to a galactoside-coated surface is mediated by large clusters of receptor complexes and it has been discussed that the threshold of binding response is essential for optimal interaction (WEIGEL et al. 1979). Concerning the distribution in different plasma membrane domains, the studies indicate that the receptor is preferentially localized in the sinusoidal and lateral, but not in the canalicular area (WALL et al. 1980). Interestingly, the receptor associates spontaneously with small unilamellar lipid vesicles, whereby the protein undergoes conformational changes, reflected by an increase in the β-pleated sheet structure (KLAUSNER et al. 1980). This result is in contrast to the association of free apolipoproteins with lipids which is followed by a marked increase in α-helicity (ASSMANN and BREWER 1974).

During receptor-mediated endocytosis (for rev. see GOLDSTEIN et al. 1979) the ligand remains associated with the "transport vehicle"; within the cell the complex dissociates and the ligand is finally degraded (BRIDGES et al. 1982). The constant of internalization is $3.1 \times 10^{-5} \cdot sec^{-1}$, while the dissociation constant has a value of

$6 \times 10^{-6} \cdot sec^{-1}$ for both asialo-orosomucoid and asialo-fetuin, a finding that can be reconciled with the concept of receptor recycling (STEER and ASHWELL 1980, TOLLESHAUG 1981). The underlying mechanism why the receptor molecules are protected from degradation is not yet clear. Taking into account that terminal carbohydrates, especially L-fucose, galactose and Neu5Ac (KREISEL *et al.* 1980, TAUBER *et al.* 1982, VOLK *et al.* 1982), show a markedly higher rate of degradation, it is conceivable that during receptor-mediated endocytosis terminal and subterminal sugars are split off, thus protecting the receptor "remnant" from recognition by lysosomes. Simultaneously, the exposure of signals for the recognition of intermediate vesicles, that should be endowed with the necessary equipment for reglycosylation would restore the biological activity of the receptor. Ultrastructural studies of liver sections after the uptake of lactosaminated ferritin molecules indicate, that the ligand ferritin is taken up by intermediate vesicles soon after endocytosis (WALL *et al.* 1980).

In the course of galactosamine-induced liver injury (REUTTER *et al.* 1968, KEPPLER *et al.* 1968), a decrease of the asialo-glycoprotein receptor has been described (SAWAMURA *et al.* 1981). This finding might explain the decreased uptake of asialo-fetuin by hepatoma cells (KOLB *et al.* 1979, unpublished) and the accumulation of asialo-glycoproteins in the serum of patients with liver cirrhosis or hepatoma (MARSHALL *et al.* 1974, MARSHALL and WILLIAMS 1978, STEER and CLARENBURG 1979, SOBUE and KOSAKA 1980). It should be added that in galactosamine-damaged liver as well as in experimentally induced hepatomas glycosylation is disturbed (BAUER *et al.* 1974, 1980, RUPPRECHT *et al.* 1976, VISCHER and REUTTER 1978, REUTTER and BAUER 1978, KISS and KATTERMANN 1979, KISS *et al.* 1981).

IV. Blood Clotting

Blood coagulation and fibrinolysis are controlled by various plasma glycoproteins and cellular elements. The carbohydrate moiety of the soluble coagulation factors, fibrinolytic enzymes and glycoproteins of the thrombocyte plasma membrane protects the protein from degradation, serves to stabilize its structure and plays a part in signal mediation. The terminal sialic acid of the carbohydrate often makes an important contribution to these various functions.

The carbohydrate sequence of some coagulation factors has been determined (prothrombin or F II, thromboplastin or F X, fibrinogen or F I, and the fibrinolytic serine protease, plasmin), but only a few investigations have been reported on the functional significance of the carbohydrate residues (MIZUOCHI *et al.* 1979, MIZUOCHI *et al.* 1980, GATI and STRAUB 1978, HAYES and CASTELLINO 1979, LIJNEN *et al.* 1981).

Desialylation destroys the protease activity of plasminogen. Functional prothrombin arises from its precursor by two post-translational modifications: a vitamin K-dependent carboxylation of glutamic acid residues, followed by glycosylation in the carboxylated region (prothrombin fragment 1: AS 1-156). Deglycosylation of prothrombin in the presence of Ca^{2+} leads to greatly increased self-association; this occurs at certain contact points on the molecule, which are probably concealed by carbohydrate chains in the intact molecule (PLETCHER *et al.*

1980). Fibrinogen is the precursor of fibrin, which inter alia in the presence of the protease, thrombin, forms a fibrin network. Partial functions of fibrinogen related to this process are markedly decreased by desialylation. The so-called thrombin and reptilase times are shortened, and one can observe an increased tendency for the aggregation of monomers; on the other hand, the release of fibrinopeptides and cross-linking are unchanged (PALASCAK and MARTINEZ 1977, MARTINEZ *et al.* 1978, SCHWICK *et al.* 1977). As already reported for transferrin, native and asialo-fibrinogen have about the same *in vivo* half-life. The pathobiochemical significance of an increased sialylation of fibrinogen is discussed later.

F VIII/von Willebrand factor is a dissociable complex with functionally different subunits. F VIII activity, whose absence causes haemophilia, is unaffected by desialylation. The von Willebrand factor, a glycoprotein polymer (mol. wt. 1,000,000) is very probably responsible for the binding of thrombocytes to the endothelium of the blood vessel wall. Desialylation of this complex decreases its ability to bind to the thrombocyte surface, and at the same time shortens the *in vivo* half-life of both F VIII and von Willebrand factor (SODETZ *et al.* 1977, 1978). This observation is relevant to the pathobiochemical aspects, which will be discussed later. Thrombocytes, the most important cellular elements in blood coagulation, possess various groups of glycoproteins on their surface. These are divided according to molecular weight into groups GP I (MW 155,000), GP II (MW 135,000), and GP III (MW 105,000). GP I, the probable precursor of the extremely protease-labile glycocalicin of the thrombocyte plasma membrane, contains 60% carbohydrate, 22% being due to Neu5Ac. GP II and GP III possess essentially fewer carbohydrate residues (CLEMETSON *et al.* 1981, OKUMURA *et al.* 1978, LAWLER *et al.* 1978). The involvement of thrombocytes in thrombus formation can be subdivided into three different processes; in each case, the carbohydrate moiety of the participating glycoproteins or glycosyltransferases is important. Firstly, thrombocytes adhere to the blood vessel wall. The ligands for this adhesion are probably glycoconjugates (collagen, elastin, fibronectin, glycosaminoglycans) of subendothelial structures, which are exposed by damage to the endothelial lining (SOLUM and HAGEN 1978, SAKARIASSEN *et al.* 1979, KOTELIANSKY *et al.* 1981). It has been proposed that this binding is the result of interaction between glycosyltransferase of the thrombocyte membrane and the above mentioned glycoconjugates (WU and KU 1978, CARTRON and NURDEN 1979). In the adhesion process, the thrombocyte changes its shape from discoid into spheroid (SOLUM *et al.* 1977, BLAJCHMAN *et al.* 1981, LÜSCHER 1978, BUNTING *et al.* 1978, MOSHER *et al.* 1979, NURDEN and CAEN 1976). During adhesion, the thrombocytes also develop their ability to aggregate, possibly by secretion of a lectin, which in combination with fibrinogen acts as a bridging agent between the thrombocytes (GARTNER *et al.* 1981, 1977, 1978, 1980, MARGUERIE *et al.* 1979, BOWLES and ROTMAN 1978). In addition, it is also proposed that thrombocyte aggregation is mediated by the reaction of an ecto-sialyltransferase of the thrombocyte plasma membrane with GP IIb (BAUVOIS *et al.* 1981). Finally, thrombocyte aggregation results in the secretion of mediators (serotonin, platelet-derived growth factor) and other groups of substances (SHATILL and BENNETT 1981). Some studies claim to show that secretion is accompanied by a pronounced release of Neu5Ac, but these results are doubted by other authors. It is certain that free Neu5Ac blocks thrombocyte aggregation, so that Neu5Ac secretion would represent an important regulation mechanism (PACKHAM *et al.* 1980, KOVÁCS and GÖRÖG 1979, PATSCHEKE 1981, COSTELLO *et al.* 1979). Neu5Ac is also important in the pathology of various disorders of blood coagulation:

(i) Thrombocytopenia, i.e. a decreased level of thrombocytes in the blood, can be caused by desialylation of the thrombocyte plasma membrane; after treatment with neuraminidase, thrombocytes show a much shorter half-life (PODOLSAK and BRUNSWIG 1976).

(ii) Adhesion disorders. Von Willebrand disease is characterized by a defect in the ability of thrombocytes to adhere to subendothelial glycoconjugates of the blood vessel wall. In these patients, glycosylation (sialylation) of the F VIII-associated von Willebrand factor is absent or deficient, thus resulting in an impairment of function (GRALNICK *et al.* 1976, NURDEN *et al.* 1981, CAEN *et al.* 1976, KUNICKI *et al.* 1978).

(iii) Aggregation disorders. Glanzmann's thrombastenia is an example of a congenital thrombocyte aggregation disorder. Nowadays, it is thought to be caused by defective glycoproteins GP IIb and GP IIIa of the thrombocyte membrane. The glycoproteins show decreased glycosylation, and especially GP IIb has a reduced content of Neu5Ac. As a result, the ability of the thrombocytes to bind fibrinogen is decreased, even after activation with the most varied types of aggregation stimulators (ADP, collagen, epinephrine, thrombin). As already mentioned, an ecto-sialyltransferase of the thrombocyte membrane is involved in aggregation. This is confirmed by the fact that aggregation-promoting substances markedly increase the activity of the enzyme, whereas aspirin (acetylsalicylic acid) inhibits both aggregation and sialyltransferase activity (BAUVOIS *et al.* 1981, NURDEN and CAEN 1978, COHEN *et al.* 1981, MCGREGOR *et al.* 1981, PHILLIPS and AGIN 1977). Increased sialylation of glycoproteins of the thrombocyte plasma membrane likewise causes a decreased capacity for thrombocyte aggregation. In this case, sialyltransferase is probably denied access to its substrate on the neighbouring cell, due to blockage by the excess carbohydrate residues. A similar mechanism may explain the above-mentioned inhibition of aggregation by free Neu5Ac.

Thrombocyte aggregation is also inhibited by asialo-orosomucoid, which inter alia is elevated in the serum during chronic inflammation (COSTELLO *et al.* 1979). It has not yet been investigated whether this "acute phase protein" can take over the role of substrate for thrombocyte sialyltransferase. Thrombocytes from tumor patients may show up twice the normal level of sialyltransferase activity. This explains the increased tendency to thrombosis sometimes observed in these patients (SCIALLA *et al.* 1979). At the same time, it is known that thrombocytes preferentially aggregate at or on tumor cells. In various sarcoma cell lines a direct relationship has been shown between the degree of sialylation of the surface, the aggregation-capacity of the thrombocytes and the tendency of these tumors to metastase (PEARLSTEIN *et al.* 1980). It is possible that immunological defence reactions could be prevented by layering of thrombocyte aggregates on transformed cells. In addition, invasive growth is favoured by the secretion of vasoactive amines, which accompanies thrombocyte aggregation.

(iv) Fibrin polymerization can also be disturbed by increased sialylation of the fibrin molecule. Thus, a faulty aggregation of fibrin monomers has been described in patients with hepatocellular carcinoma (GRALNICK *et al.* 1978). Finally, patients with alcoholic liver cirrhosis show a decreased fibrin polymerization capacity (ORDINAS *et al.* 1978, KLINGEMANN *et al.* 1980). In this same disorder other proteins, such as gamma-glutamyl transferase, also show an increase in the degree of sialylation (KÖTTGEN 1980).

V. Immunology and Infection

In immunological processes, sialic acid as well as neuraminidase and sialyltransferase have important regulator properties. Here again, the terminal sialic acid can have a masking or receptor function. In the many examples discussed here, however, it is not certain whether sialic acid functions directly, or whether sialylation/desialylation of proteins leads to a conformational change of the whole molecule, which in turn affects the receptor-ligand interaction. An important question that still awaits final resolution is whether Neu5Ac itself can act as an antigen, or whether it is only active as the specific determinant of a macromolecule. This is a long standing question with respect to the MN blood group antigens. These MN determinants are localized on the very highly sialylated glycophorin A of the erythrocyte membrane. In the N-terminal octapeptide, two tetra- and one trisaccharide are linked O-glycosidically to the serine or threonine residues at positions 2, 3, and 4. According to recent work, the carbohydrate moieties of M and N antigens are the same, but their peptide sequences differ in positions 1 and 5, which are occupied by serine and glycine in M antigen and leucine and glutamic acid in N antigen. Nevertheless, desialylation of the above-mentioned octapeptide leads to a loss of M and N antigenicity. Furthermore, limited treatment of MN-phenotype erythrocytes with neuraminidase leads to a transitory increase in N-antigenicity, followed by the loss of both M and N behaviour upon further desialylation. It now seems improbable that the terminal Neu5Ac residue of the N-terminal glycopeptide is antigenically active; rather, the carbohydrate moiety is a codeterminant in the antigenic character of M or N. Some authors propose that the amino acid sequence has no importance for the antigenic character of M or N blood groups. According to their investigations, M antigen is synthesized by a specific sialyltransferase or a "sialyltransferase modifier protein". On the other hand, there is also good evidence that M and N antigens are determined primarily by the above mentioned dissimilar peptide sequences, which can cause secondary differences in sialylation. Further investigations on the possible antigenic nature of Neu5Ac in glycoproteins and glycolipids are reviewed elsewhere (YOSHIDA *et al.* 1979, DESAI and SPRINGER 1980, SADLER *et al.* 1979, ANSTEE 1981). The antigenic determinants in the human erythrocyte Pr system and Sa cold agglutinins have been shown to include sialic acid (ROELCKE *et al.* 1980). Modifications of the sialic acid moieties lead to a loss of antigenicity (LICHTHARDT *et al.* 1981). The Sa antigen includes the structure Neu5Acα(2-3)Gal (ROELCKE *et al.* 1980). A study of the occurrence of Neu5Ac and Neu5Gc in II^3NeuAcylLacCer in dogs revealed that II^3Neu5GcLacCer is found only in east asian, especially japanese dogs and was absent in european dogs. This antigen is autosomally dominant and seems to be a dog blood group substance (YASUE *et al.* 1978). The antigenicity of several meningococcal serotypes has been attributed to specific carbohydrate structures including sialic acid polymers. These include the serotypes B, C, W-135, and Y (LIU *et al.* 1977, GLODE *et al.* 1979).

Of special interest is the finding that structurally related polymers of sialic acid containing O-acetyl groups are weaker immunogens than their de-O-acetylated counterparts (GLODE *et al.* 1979, ØRSKOV *et al.* 1979). Antibodies to gangliosides containing Neu5Gc have been identified as serum-sickness type heterophile antibodies (HIGASHI *et al.* 1977). Recently, antibodies against sialo-oligosaccharides coupled to bovine serum albumin as phenethylamine derivatives could be obtained (SMITH and GINSBURG 1980).

The masking action of sialic acid residues is very important in biological regulation. The protective role of sialic acid may be direct or indirect. In direct masking, an antigenic carbohydrate residue of a glycoconjugate is covered by the terminal sialic acid. In indirect masking, the sialic acid residues of neighbouring glycoconjugates within the plasma membrane cover the actual antigen, which itself does not necessarily possess a carbohydrate residue. In both cases sialic acid is an effective masking agent, because it is surrounded by a hydration shell as a result of its molecular charge. In general, there is an inverse relationship between the H_2O binding capacity of polysaccharides, proteoglycans and glycoproteins and their antigenicity. On the other hand, neutral and especially fucose-rich glycoconjugates (with hydrophobic methyl groups) have higher antigenic activity (APFFEL and PETERS 1970, BEUKERS *et al.* 1980, HERBERMAN 1977, KNOP *et al.* 1978). Typical examples of non-fortuitous antigen masking by acidic glycoproteins are found in tumor cells, trophoblasts, glycoproteins of the embryo, spermatozoa and glycoproteins of seminal plasma: Some of these will be referred to again later. The term "immunologically privileged" is applied to certain parts of the organism. These parts, e.g. anterior chamber and lens of the eye, testes, brain, thyroid and connective tissue, characteristically have a high content of sialo-glycoconjugates. It is precisely in these tissues and organs that auto-immune diseases are most frequently manifested, possibly by the unmasking of antigenic material by the action of bacterial or viral neuraminidases (APFFEL and PETERS 1970). A further example is the Thomsen-Friedenreich antigen of the erythrocyte membrane, which will be discussed later. Other Neu5Ac-masked cryptic antigens of the erythrocyte membrane are discussed elsewhere.

Carbohydrate residues are not only important in the antigen, but possibly also in the immunoglobulin. About one third of the carbohydrate chains of the IgG molecules are localized on the Fab fragment. It can therefore be assumed that here also they contribute to the formation of the tertiary and quaternary structures of the antigen-binding region of the immunoglobulin. It is also supposed that terminal Neu5Ac residues of IgG exhibit virus-neutralizing properties, since IgG prevents virus adhesion to the sialo-glycoconjugates of the cell membrane (WINKELHAKE *et al.* 1980, MATSUUCHI *et al.* 1981, BERGMAN *et al.* 1981). In patients with pathological IgG complex formation, an auto-immune disease, Neu5Ac residues of the Fab fragment participate in the formation of the complex (HYMES *et al.* 1979). Furthermore, deglycosylation of IgG leads to a decreased capacity for binding complement (C1q), which must be seen as an essential prerequisite for the immunologically directed cytolysis of foreign cells (WINKELHAKE *et al.* 1980). Sialic acid residues also have a regulatory function in unspecific complement activation, which does not involve immunoglobulins. The relative affinity of factor "B" for the particle-bound complement C3b is codetermined by the presence of sialic acid residues on the cell surface (KAZATCHKINE *et al.* 1979). Only Neu5Ac-deficient erythrocyte membranes can activate the "human alternative complement pathway" (NYDEGGER *et al.* 1978). Finally, as already described for other plasma glycoproteins, the half-life of C1 esterase inhibitor is determined by its content of terminal Neu5Ac. It is so far unknown whether the hereditary deficiency of this activator is caused by a decreased synthesis, or by a shorter half-life of Neu5Ac (HIRSCH *et al.* 1981, MINTA 1981). In addition to T-lymphocytes and macrophages, the cellular immune response also involves mediator proteins. It is established that lymphocytes possess on their surface ectosialyltransferases, which are strongly activated by ConA or PHA, particularly after lymphocyte stimulation. At the same time, the content of protein-bound Neu5Ac increases in the plasma membrane of these cells (PAINTER and WHITE 1976, VERBERT *et al.* 1977, LUI and MUN 1980, PARISH *et al.* 1981). The functional role of this process is still unexplained. The mechanism whereby various neuraminidases can stimulate lymphocytes mitogenically is also still unknown (SEMENZATO *et al.* 1977, KUPPERS and HENNEY 1979, NOVOGRODSKY *et al.* 1980). On the other hand, the intact Neu5Ac residues of the glycoconjugates of the lymphocyte plasma membrane are necessary for colony formation (TONELLI and MEINTS 1977) and mediate the delayed type hypersensitivity (KAUFMANN *et al.* 1981).

The mediator proteins required for transformation and further proliferation are also functionally regulated by their degree of sialylation. Macrophage inhibitor factor is inactivated by desialylation. The same is true for a plasma protein, which in conjunction with interleukin-1, stimulated the synthesis and secretion of interleukin-2 by T-helper cells (KÖTTGEN *et al.* 1981). Desialylation does not inactivate interferon, but renders it much more labile to proteases (MIZRAHI *et al.* 1978). The interferon receptor of the plasma membrane, monosialo-ganglioside GM_1, is destroyed by neuraminidase treatment. The receptor for the activation of natural killer cells appears to be an asialo-GM_1 of the lymphocyte membrane (DYATLOVITSKAYA *et al.* 1980, HOESSLI *et al.* 1980, KASAI *et al.* 1980, AKAGAWA *et al.* 1981, SPIEGEL and WILCHEK 1981). From several investigations, it now seems probable that it is precisely the increased sialylation of α_1-fetoprotein that gives the ability to induce T-suppressor cells (LESTER *et al.* 1976, 1978, ZIMMERMAN *et al.* 1977). Glycolipids, especially ganglioside GT_1, inhibit the lectin-induced mitogenesis of thymocytes. It is also known that lymphocytes from tumor patients are partially blocked in their mitogenic reactivity by plasma membrane-associated sialoglycoconjugates. This type of mechanism could also render useless the immune surveillance of tumor patients (KLOPPEL *et al.* 1979, BENNETT and SCHMID 1980, GÖRÖG *et al.* 1980, O'KENNEDY *et al.* 1980). Many studies have been reported on experimental tumor therapy by the injection of desialylated tumor cells. Various research groups have shown that the immunogenic properties of tumor cells are increased by treatment with neuraminidase. Thus, tumor cells desialylated *in vitro* show a decreased transplantability when reinjected, i.e. decreased tumor growth is observed. Other authors have been unable to observe an increased immunogenicity of desialylated tumor cells (see section VI).

Finally it is now proposed that the increased immune reaction after neuraminidase treatment is not only directed against the desialylated tumor cells, but also against unspecific, i.e. primarily Neu5Ac-free, antigens. Whether the increased sialic acid content on the surface of malignant or transformed cells will eventually give a better insight into the pathomechanism of malignant transformation and allow effective therapeutic measures is not yet known (SIMMONS and RIOS 1975, CANTRELL *et al.* 1976, SANSING and KOLLMORGEN 1976, BEKESI *et al.* 1977, BRAZIL and McLAUGHLIN 1978, PIMM *et al.* 1978, BRANDT *et al.* 1981).

Even in the absence of unequivocal proof, it has been repeatedly stated that lectins of plants and invertebrates have a defensive role. They would then correspond functionally to the immunoglobulins of higher organisms, but their structures are absolutely different. Lectins and immunoglobulins show many similarities in their binding characteristics (KÖTTGEN 1977, KÖTTGEN *et al.* 1979). For these reasons two lectins will be discussed here, which show a preferential binding affinity for neuraminic acid residues. Limulin, a neuraminic acid-binding lectin from the haemolymph of the horseshoe crab (Limulus polyphemus) is a glycoprotein with a molecular weight of about 400,000 D. It consists of 18 noncovalently bound subunits. In the presence of Ca^{2+}, it shows a preferred affinity for Neu5Gc, whereas Neu5Ac is bound only weakly or not at all. The free carboxyl group of neuraminic acid and a free hydroxyl on C-4 are required for binding, whereas the side chain of Neu5Gc is not necessary for binding (KAPLAN *et al.* 1977, MAGET-DANA *et al.* 1981).

In addition to binding N-acetylglucosamine, wheat germ agglutinin (WGA) also binds neuraminic acid. This lectin preferentially binds Neu5Ac residues and does not require Ca^{2+}. The binding affinity is strongly pH-dependent (optimum at pH 7.5 ± 0.5). The acetamido group of Neu5Ac participates in the binding, but not the carboxyl group. Bound Neu5Ac can also be competitively displaced by N-acetylglucosamine. Succinylation of the lectin totally abolishes the interaction between WGA and Neu5Ac, whereas the capacity for binding N-acetylglucosamine remains unaffected. By using native and succinylated WGA, the Neu5Ac content of surface glycoconjugates can therefore be quantified. Further ligand specificities of WGA are discussed elsewhere (BHAVANANDAN and KATLIC 1979, ERNI *et al.* 1980, MAGET-DANA *et al.* 1981). Apart from their different sugar specifities, lectins can also be differentiated according to their mitogenic or non-mitogenic capacity. Like limulin, WGA is non-mitogenic. WGA even has antimitogenic activity, since by binding to cell surfaces it is able to prevent previously initiated mitogenesis. This is of biological significance, because the mitogen-stimulated cell secretes soluble inhibitors, which inhibit DNA synthesis by attachment to plasma membrane receptors. It is these receptors, which are important in the regulation of mitogenesis, that are probably occupied by WGA (GREENE *et al.* 1976, GORDON *et al.*1980, GREENE and WALDMANN 1980, KURISU *et al.* 1980).

Maintenance of pregnancy, i.e. maternal tolerance to foreign tissue of the fetus, calls for crucial immunological adaptations, which also involve the metabolism of sialic acid. Endocrinologically controlled changes in sialic acid content of the cervical mucus of the uterus during the menstrual cycle are of primary importance. Under the influence of oestrogens, the mucus has a thin, watery consistency. This increased uptake of water by mucin is due to a marked increase in terminal Neu5Ac residues, which gives the individual glycoprotein molecules an attenuate, rod-shaped structure. After ovulation, i.e. under the influence of progesterone, fucose-rich glycoproteins are produced, which form a very viscous, tightly meshed network, cross-linked by disulfide bridges. Therefore, up to the time of ovulation, migration of spermatozoa into the uterus is facilitated by the linear shape of the sialo-glycoproteins, whereas migration through the cervical canal is prevented by the convoluted arrangement of glycoprotein molecules in the progesterone phase (HATCHER and SCHWARZMANN 1977, ELSTEIN 1978, GIBBONS 1978). The uterine endometrium is rich in neuraminidase, which removes the sialic acid residues that are present in high concentration on the acrosome of spermatozoa (SRIVASTAVA and FAROOQUI 1980). This possibly facilitates the establishment of contact with the egg. According to other authors, primary contact and adhesion may occur through the agency of a sialyltransferase associated with the spermatozoa (DURR *et al.* 1977). Implantation and invasion by the trophoblast in the decidua of the uterus is not biochemically fully understood. This process is, however, of great interest, because the mechanism of tumor invasion is probably similar. Glycoproteins of the trophoblast are much more extensively sialylated than those of the fetus, but unlike malignant, transformed cells, the trophoblast does not possess increased levels of fucoproteins (BROWNELL 1977, GROTH and KADEN 1977, WADA *et al.* 1977, LIU *et al.* 1978, PADMANABHAN *et al.* 1978, WHYTE and LOKE 1978, ROSATI and DE SANTIS 1980). Despite good evidence, it is still not certain that the immune barrier between mother and child is achieved solely by sialic acid-rich glycoproteins. Of at least equal importance is the fact that the maternal organism synthesizes immunosuppressive factors in the early phase of pregnancy (BEER and BILLINGHAM 1978, NOONAN *et al.* 1979).

The same is true for the developing fetus. One example is the synthesis of Neu5Ac-rich α_1-fetoprotein, which induces the development of T-suppressor cells (see above). Since plasma membrane-bound glycoconjugates also become increasingly sialylated, it is also possible that the receptor-mediated induction of natural killer cells is decreased or blocked by increased sialylation and consequent masking of GM_1.

Furthermore, the role of sialic acid and/or neuraminidase in infectious diseases should be mentioned. Many bacteria and viruses possess neuraminidase and there is evidence that these neuraminidases and the corresponding sialo-glycoconjugates of the host cell are factors in the primary adhesion between host organism and bacterial cell or virus particle (CARTWRIGHT and BROWN 1977, ROGENTINE *et al.* 1977, BERGMAN *et al.* 1978, FINNE 1978, PAULSON *et al.* 1979, KNOP *et al.* 1978, LEPRAT and MICHEL-BRIAND 1980). In addition to promoting adhesion, the neuraminidase probably also facilitates the function of bacterial proteases, because desialylation greatly increases the susceptibility of glycoproteins to the action of proteolytic enzymes. Correspondingly, a relationship between the pathogenicity and neuraminidase activity is frequently observed. The action of bacterial neuraminidase can also induce immune mechanisms. The hepatic binding protein-induced activation of cytotoxic desialo-lymphocytes was referred to earlier. A further important example is the haemolytic-uraemic syndrome, which is caused chiefly by pneumococcal neuraminidase. This condition is characterized by a haemolytic anemia, thrombocytopenia and kidney failure. The neuraminidase unmasks the so-called T-antigen on the erythrocytes, thrombocytes and endothelial cells; the circulating blood contains IgM antibodies to T-antigens, even in healthy adults. In conjunction with complement, this antigen-antibody reaction causes heavy cell lysis in the above cell systems (FISCHER and POSCHMANN 1976, SEGER *et al.* 1980). Infections by neuraminidase-producing bacteria during pregnancy often leads to abortion. This reaction is, however, probably not caused by destruction of the earlier-mentioned immune barrier, but rather by a lesion in the decidua of the uterus. Many bacteria possess on their surface receptor proteins that behave like lectins; these establish contact with appropriate glycoconjugates of the host organism, and thereby initiate the infection process. These lectins show specificity chiefly for galactose and mannose residues (HART 1980, WEIR 1980). An exception is found in some mycoplasms, which act as partially pathogenic surface parasites, adhering to the epithelial structures of the respiratory and urogenital tract. *Mycoplasma pneumoniae* and *M. gallisepticum* bind with high specificity to the Neu5Ac residues of glycoproteins. *M. pneumoniae* probably possesses additional, possibly hydrophobic, binding sites. Since neither sialyltransferase nor neuraminidase activity can be demonstrated on *mycoplasma*, the Neu5Ac-specific receptors are probably lectin-like proteins (BANAI *et al.* 1978, GLASGOW and HILL 1980, HART 1980). It should be mentioned here that *cholera* and *tetanus* toxins have lectin-like activity, and bind to GM_1 (RICHARDS *et al.* 1979), as will be discussed elsewhere. The effect of virus infection on protein glycosylation in the host organism is discussed elsewhere.

VI. Immunotherapy with Neuraminidase-Treated Cells

In the past decades numerous attempts have been made to find new ways and to establish new techniques which are suitable to stimulate the immune system of animals and human beings suffering from malignant and metastatic tumors. These studies were based on the observation that acute bacterial infection can reduce tumor proliferation and might even induce complete tumor regression (NAUTS *et al.* 1963, NAUTS 1978). Especially immunotherapy with BCG (Bacillus Calmette-Guérin) seemed to be promising in cases of lymphoblastic leukemia or malignant melanoma, since some patients responded quite favourable to treatment (MATHÉ *et al.* 1969, MORTON *et al.* 1970, GUTTERMAN *et al.* 1973). Besides there was evidence that intratumor injection of BCG can lead to an increase of antitumor antibodies (MORTON *et al.* 1970). In the meantime, however, more extensive studies have revealed that medical treatment with BCG or *Clostridium parvum* can both cause tumor regression and enhancement (SPARKS *et al.* 1973, SPARKS and BREEDING 1974, MATHÉ 1978, MORALES 1980, VOGLER 1980). At the moment no valid criteria are available by which it is possible to predict the effect of BCG with certainty, since the underlying mechanisms, e.g. the stimulation of the reticuloendothelial system and the release of mediator molecules (OLD 1981) are only partly understood. A different approach using desialylated tumor cells or lymphocytes as immunopotentiators has generated widespread interest, because preliminary results were quite encouraging (SEDLACEK and SEILER 1978). The first indications that neuraminidase treatment increases the immunogenicity of tumor cells can be dated back to the sixties. Incubation of tumor cells with neuraminidase caused a substantial decrease in oncogenicity (SANDFORD 1967), increased the life-span of the recipient mice (BAGSHAWE and CURRIES 1968) when compared to controls or rendered the cells more immunogenic, thus protecting the animals from a subsequent challenge with untreated cells (LINDEMANN and KLEIN 1967). Most of these studies have been done with neuraminidase from *Vibrio cholerae* (VCN) or more seldom from *Clostridium perfringens* (CPN), because these enzymes can liberate sialic acids from a wide spectrum of different substrates. VCN and partly CPN are able to hydrolyse 2-3, 2-6, and 2-8 ketosidic linkages especially between galactose or N-acetylglucosamine and Neu5Ac (DRZENIEK 1967, 1972).

The principle procedure for preparing desialylated cells is briefly outlined in Fig. 1. The tumor cells are incubated in the presence of neuraminidase, then, if necessary, exposed to cytostatic agents or X-rays to prevent further proliferation, and finally the cells are reinjected into the animal (or human being). VCN does not only split the ketosidic linkages, but simultaneously enzyme molecules are bound to the plasma membrane of the malignant (or benign) cell (SEDLACEK and SEILER 1974, LÜBEN *et al.* 1976) with the consequence of increasing or decreasing the immunogenicity. There are indications that VCN is attached to terminal galactose residues of surface glycoconjugates (SCHNEIDER *et al.* 1979) or is bound by the ganglioside GM_1 (CESTARO *et al.* 1978). The binding to GM_1 could lead to a transient inactivation of neuraminidase, because at least soluble VCN forms a complex with GM_1 which leads to a marked reduction of enzyme activity. Complete activity, however, is restored in the presence of albumin. Consequently, the number of units of neuraminidase applied and the time of incubation choosen have a decisive influence on the effectiveness of neuraminidase treatment.

Active VCN can split off Neu5Ac from neighbouring cells (PETITOU *et al.* 1977) or alternatively, could stimulate T lymphocytes (HAN 1972) and macrophages (KNOP *et al.* 1978). Furthermore, desialylated malignant cells can be "recognized" by macrophages, since cultured leukemia L 1220 cells pre-treated with VCN become susceptible to the synergistic effect of peritoneal macrophages (SETHI and BRANDIS 1973). Though the described cytotoxic effect is non-phagocytic there are valid indications, that macrophages can bind desialylated cells via a galactose-specific lectin (SCHLEPPER-SCHÄFER *et al.* 1980, NAGAMURA and KOLB 1980, KÜSTER and SCHAUER 1981). It is noteworthy that the cytotoxic effect is not restricted to macrophages, but has also been described for lymphocytes. Treatment of unseparated peripheral blood lymphocytes with neuraminidase caused a 20 to 30-fold increase of their attachment to carcinoma cells paralleled by a substantial elevation of cytotoxic activity. This enhancement seems to depend on the interaction of neuraminidase-treated T lymphocytes, target carcinoma cells and natural killer cells (GALILI and SCHLESINGER 1978). If the activity of these killer cells, however, is impaired, the growth rate of transplantable tumors increases. An antiserum with antinatural killer cell activity has been raised and the antibody responsible for this inactivation was shown to be directed against membrane-bound asialo-GM_1 (KASAI *et al.* 1981). Intravenous injection of minute amounts of anti-asialo-GM_1 blocked nearly completely the killer activity, while a concomitant enhanced proliferation of the lymphoma cells was observed. Hence, an extremely low titer of antibodies directed against asialo-GM_1 could switch off an important regulatory mechanism (KASAI *et al.* 1981).

These few examples clearly show that different reasons could be responsible for the observed immune stimulating effect of VCN-treated (tumor) cells and thus might explain the contradictory results described in literature (SIMMONS *et al.* 1971 a, SIMMONS *et al.* 1971 b, BEKESI *et al.* 1971, WILSON *et al.* 1974, ALBRIGHT *et al.* 1975, SEDLACEK *et al.* 1975, BEKESI *et al.* 1976, ALLEY and SNODGRASS 1977, BRAZIL and MCLAUGHLIN 1978, SPENCE *et al.* 1978, PIMM *et al.* 1978, PINCUS *et al.* 1981).

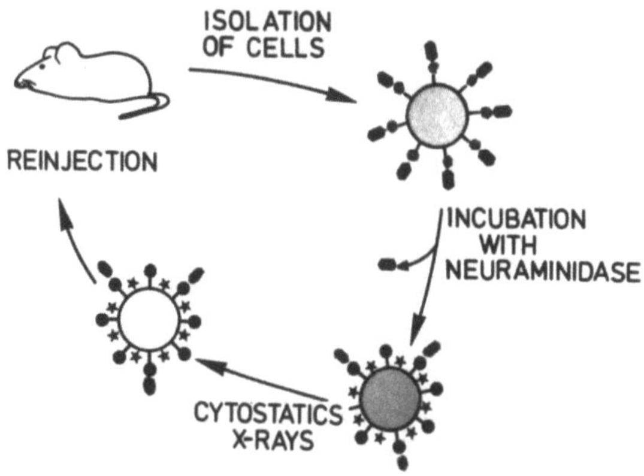

Fig. 1. Basic procedure of immunotherapy with neuraminidase-treated cells. Most of the Neu5Ac residues are split off, while part of the neuraminidase molecules (black asterisks) is still bound to the cell surface. Further proliferation of the malignant cell is, if necessary, suppressed by cytostatic agents or radiation and finally the "asialo"-cells are re-injected into the animal (or human being).

Besides, another detail of the experimental system renders it difficult or even impossible to compare and evaluate the different findings and observations properly, namely the number of desialylated tumor cells inoculated. Data summarized in Table 3 demonstrate unequivocally that only a narrow range of VCN-treated cells is suitable for therapy, while improper doses are either ineffective or increase the risk of metastases (WILSON *et al.* 1974, SEDLACEK and SEILER 1978). Similarly, apart from other factors, the efficacy of intratumor injection of VCN depends on the amount applied so that the therapeutic effect can be positive, negative or is negligible (SPARKS and BREEDING 1974, BINDER *et al.* 1975, ALLEY and SNODGRASS 1977).

These findings have rendered it difficult for many years to make use of neuraminidase-treatment for human immunotherapy. But recent results indicate that especially in the field of myelocytic or lymphocytic leukemia, therapy with VCN-treated autologous and allogeneic myeloblasts (HOLLAND and BEKESI 1976, BEKESI and HOLLAND 1977, BEKESI and HOLLAND 1978) is able to lengthen the duration of remission. 1–1.5 × 10^8 VCN-treated blast cells were injected intradermally in vicinity to node- bearing areas. The immunization was given after the patients had been subjected to chemotherapy at a time when leucocyte recovery was well in progress. Patients treated according to this scheme stayed 3 to 4-times longer in remission compared to a group that only received chemotherapy.

Table 3. *Effect of cell number on therapeutic response*
Fibrosarcoma (mice) (WILSON *et al.* 1974)

3 × 10^4 cells:	complete protection against pulmonary metastases
1 × 10^4 cells 5 × 10^4 cells	increase of incidence of metastases

Mammary tumors (dog) (SEDLACEK and SEILER 1978)

2 × 10^7 cells:	tumor regression, no metastases
2 × 10^6 cells:	transient effect
1 × 10^8 cells:	enhanced tumor growth

Since the patient's own immunological status probably has a decisive influence on the effectiveness of immunotherapy, different protocols have been designed (a) to check the most effective dose of VCN-treated cells and (b) to test the immunocompetence of the individual. A simple, but promising way is to inject a different number of cells in approximately 50 different intradermal sites (HOLLAND and BEKESI 1976, BEKESI and HOLLAND 1977). Similarly, but with regard to the adjuvant effect of neuraminidase itself, a procedure called "chessboard vaccination" has been proposed (SEILER and SEDLACEK 1978). Various amounts of VCN are combined with different numbers of cells and these mixtures are injected immediately thereafter into separate intradermal sites, i.e. a preincubation of the cells with neuraminidase and a subsequent washing are omitted. Based on successful studies with tumor-bearing dogs (SEDLACEK *et al.* 1979) it is conceivable that this procedure could be a possible way for human immunotherapy. Simultaneously, chessboard vaccination could provide information about the immunological state of the individual (ROSENFELD *et al.* 1978).

In summary, tumor immunotherapy with neuraminidase-treated cells could improve medical treatment of cancer, if the following presuppositions are fulfilled: (a) Low tumor burden or the tumor load has to be reduced by surgery, chemotherapy or irradiation, (b) a still "immune-competent" organism and (c) optimized doses of tumor cells and VCN, because—as WILSON (1974) has put it "the greatest danger for clinical application of immunotherapy is the risk of stimulating tumor growth in a patient with only minimal residual disease".

VII. Gangliosides

Gangliosides are a group of sphingoglycolipids characterized by the presence of ceramide, Neu5Ac or Neu5Gc and different other sugars, e.g. glucose, galactose and N-acetylgalactosamine (Fig. 2) (KLENK 1935, 1942, KUHN and WIEGANDT

1963). At least 40 different ganglioside structures have been established during the last 20 years (HÄKKINEN and KULONEN 1963, KUHN and WIEGANDT 1964, HAKOMORI 1966, TETTAMANTI *et al.* 1964, SVENNERHOLM 1963, PENICK and McCLUER 1965, ANDO and YU 1979) and quite recently even fucose-containing

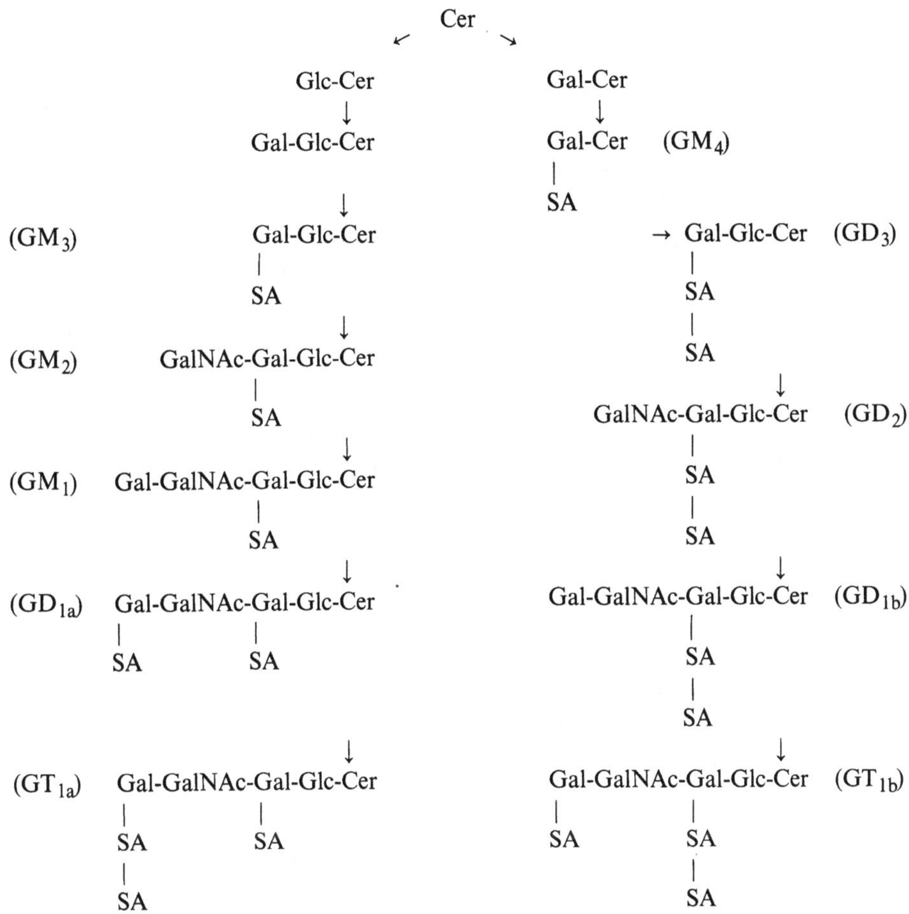

Fig. 2. Biosynthetic pathway of some major human gangliosides. *SA*, Sialic acid.

gangliosides have been identified (SUZUKI *et al.* 1975, GHIDONI *et al.* 1976, FREDMAN *et al.* 1981) (see also chapter B). These glycosphingolipids are concentrated in the plasma membrane of nerve cells, especially in the region of nerve endings and dendrites (HANNSSON *et al.* 1977, LEDEEN 1978) and there is good evidence that they are involved in the release of neurotransmitters (BRUNNGRABER 1979, SVENNERHOLM 1980 a). Different findings indicate that brain gangliosides are mainly synthesized in the Golgi apparatus and perhaps in the smooth endoplasmic reticulum (NG and DAIN 1977), though there is still some controversy about this point (ROSENBERG 1979) (see also chapter I). After completion of biosynthesis, gangliosides are transported along the axon and the dendrite tree to the nerve endings (SVENNERHOLM 1980 b), so that relatively high concentrations are found in the synaptic plasma membrane (BRECKENRIDGE *et al.* 1972, AVROVA *et al.* 1973). Gangliosides have been reported to modulate ion fluxes (DAVRAINVILLE and GAYET 1965) and to be involved in the excitability of cerebral tissue (BALAKRISHNAN and MCILWAIN 1961), but they are not prerequisite for sodium and potassium ion transport. They possess a strong and a weaker binding site for Ca^{2+}; Ca^{2+} bound to the weaker site can be replaced by Mg^{2+}, K^+ or acetylcholine (RAHMANN *et al.* 1978). There are valid indications that the complex formation with Ca^{2+}, due to the presence of Neu5Ac, has a substantial influence on synaptic transmission (RAHMANN *et al.* 1976) and hence can alter cellular activity. Both the glycerol side chain and the negatively charged carboxylate ion are thought to be responsible for Neu5Ac-Ca^{2+} interaction (BEHR and LEHN 1972, JAQUES *et al.* 1977). If hydroxyl groups of the side chain are blocked by ligands the coordination with Ca^{2+} is substantially reduced. Colominic acid, for example, a polymer of Neu5Ac, in which the monomers are linked by α(2-8)-ketosidic bonds, does not show the typical Ca^{2+}-induced shifts in the [^{13}C]NMR spectra (JAQUES *et al.* 1977).

It has been discussed (SVENNERHOLM 1980 a) that the action potential leads to a release of Ca^{2+} from gangliosides at the synaptic junctions thereby causing "conformational" changes of the presynaptical membrane. The liberated Ca^{2+} is transported in specific channels into the presynapse and this intra-presynaptical increase of the ion-concentration will induce the contraction of microfilaments and the release of transmitter substances.

In the case of neurological diseases, e.g. Alzheimer's disease, senile dementia or schizophrenia, a general decrease of gangliosides (and cerebrosides) has been described to occur in the grey matter of frontal, temporal and hippocampal lobes of human brain (CHERAYIL 1969, BASS 1981).

If the schizophrenic state is treated successfully, the Neu5Ac content will reach normal values again (CAMPBELL *et al.* 1967) and lithium, which improves certain schizophrenic disorders and manic-depressive psychosis, induces an increase of Neu5Ac in synaptosomal gangliosides (and glycopeptides) (EDELFORS 1981). Administration of Neu5Ac to rats is associated with an increase in the content of cerebral and cerebellar ganglioside-bound Neu5Ac and the studies have shown that treated rats learned the maze quicker than untreated controls (MORGAN and WINICK 1980). Besides, different findings indicate that thermal adaption of vertebrates to lower temperatures leads to a poly-sialylation of neuronal membranes, which is due to a reduced turnover of multisialogangliosides (RÖSNER *et al.* 1979, RAHMANN and HILBIG 1981, RAHMANN 1981). These findings indicate that gangliosides play an important role in normal brain function, behaviour, and cortical degeneration is apparently accompanied by a decrease of these compounds.

Gangliosides can stimulate pig ileum and rabbit duodenum (BOGOCH *et al.* 1962, KIRSCHNER and VOGT 1961), while desialylation leads to a loss of this biological activity. Similarly, desialylated gangliosides block the respiration of brain mitochondria (MKHEYAN and SOTSKII 1970), but it is worth mentioning that part of sialic acid is resistant to neuraminidase treatment (SCHAUER *et al.* 1980).

Gangliosides are not uniformly distributed within the human body. Especially high concentrations are found in the cerebral cortex of human brain (SVENNERHOLM 1980b), while extraneural organs, plasma (DACREMONT 1972) and cerebrospinal fluid (GINNS and FRENCH 1980) contain much less. In spleen and placenta only 5 to 10% of the brain concentration has been determined and the content of the mucosa of the small intestine is even far beyond the value (0.2%) (SVENNERHOLM 1980b). It is, however, necessary to emphasize that gangliosides are ubiquitous in mammalian tissue; they are normal constituents of plasma membranes and are also present in subcellular organelles (KEENAN *et al.* 1972, 1974). Hence it is intelligible that alterations of cellular functions which affect the cell surface could influence both the content and the pattern of gangliosides. Plasma membranes from the Morris hepatoma 5123tc exhibit an 8-fold increase of total gangliosides compared to normal liver (DNISTRIAN *et al.* 1975, 1979). This elevation is mainly due to an accumulation of monosialogangliosides and disialogangliosides (DNISTRIAN *et al.* 1975) and it appears that there exists a correlation between the level of disialogangliosides and the degree of malignancy (SIDDIQUI and HAKOMORI 1970). Similar results have been described for some other hepatoma lines (CHEEMA *et al.* 1970). Viral or chemical transformation of cells often leads to a simplification of the ganglioside pattern (HAKOMORI and MURAKAMI 1968, BRADY and MORA 1970, YOGEESWARAN *et al.* 1972) with a concomitant increase of the ganglioside GM_3 (YOGEESWARAN *et al.* 1972, DIRINGER *et al.* 1972, KEENAN and DOAK 1973). In general there is a decrease of gangliosides with an oligosaccharide chain longer than sialyllactose (BRADY *et al.* 1973). These differences are mostly due to deficiencies in certain enzymes necessary for ganglioside biosynthesis (CUMAR *et al.* 1970, BRADY *et al.* 1973). In the case of chemically induced mammary carcinomas the formation of disialogangliosides was depressed with a resultant accumulation of GM_1, because the respective sialyltransferase was reduced by 90% (KEENAN and MORRÉ 1973). It has been suggested that ganglioside deletions are late events in the tumorigenic transformation, possibly due to a cascade of events which precedes and continues beyond malignancy (MORRÉ *et al.* 1979). It is still questionable whether there exists a general relationship between alterations of the ganglioside pattern and the degree of malignancy, though some data indicate that an increased ganglioside content may be the sign of a high metastatic potential (SIDDIQUI *et al.* 1978). Conversely, similar human adenocarcinomas can either show a normal or a modified ganglioside pattern (SIDDIQUI *et al.* 1978) and comparative studies on human brain tumors revealed a rather inverse relationship between the ganglioside concentration and the tendency to form metastases (KOSTIC and BUCHHEIT 1970). Moreover, even acute inflammatory processes of the liver, induced by D-galactosamine, can suppress the synthesis of GM_1 and GD_1 by 95% (RUPPRECHT *et al.* 1976).

In the plasma membranes of rat liver the higher gangliosides GT_{1b}, GD_{1a}, and GD_{1b} are degraded with a half-life of about 48 hrs, while GM_3-bound sialic acid did not show any degradation at all (in the time-interval of 12 to 72 hrs after injection of the radioactive labeled precursor N-acetylmannosamine) (HAFERMAAS *et al.* 1982). Similar results have been obtained from studies on the turnover of plasma membrane glycoproteins (KREISEL *et al.* 1980, TAUBER *et al.* 1982). This new characteristic indicates that the regulation of the biological functions of glycoconjugates is most likely associated with the different turnover of terminal carbohydrates and core components.

Evidence is accumulating that gangliosides play an important role as receptors for different toxins and hormones (Table 4) and especially VAN HEYNINGEN and coworkers and SVENNERHOLM and coworkers contributed a great deal to broaden our understanding of toxin-ganglioside interaction. 1971 VAN HEYNINGEN *et al.* could demonstrate that crude mixtures of brain gangliosides block the action of cholera toxin and more detailed studies finally revealed that GM_1 is responsible for this inhibitory effect (HOLMGREN *et al.* 1973, KING and VAN HEYNINGEN 1973, CUATRECASAS 1973). Cholera toxin exhibits a high degree of specificity and affinity for GM_1, because a 50% inhibition of the toxin is already achieved when the molar ratio of cholera toxin to GM_1 is 1 : 1. Moreover, there appears to exist a direct relationship between the GM_1 content of the plasma membrane and the number of toxin molecules attached to the cell (HOLMGREN *et al.* 1975). An unsubstituted carboxyl group of the Neu5Ac moiety and a terminal galactose residue are necessary for optimal binding (SATTLER *et al.* 1977, MOSS *et al.* 1977). Unlike cholera toxin, tetanus toxin is bound by quite different gangliosides with a preference for GT_{1b}, GQ_{1b}, and GD_{1b} (VAN HEYNINGEN and MALLANBY 1971, HOLMGREN *et al.* 1980), but even GM_1 is a suitable receptor. Studies of HOLMGREN *et al.* (1980) indicate that a disialosyl group and a terminal galactose residue are prerequisites for a strong interaction between toxin and ganglioside.

Table 4. *Gangliosides as receptors or as part of the receptor complex*

Ligand	Major affinity to ganglioside
Cholera toxin	GM_1
Tetanus toxin	GT_{1b}, GD_{1b}, GQ_{1b}
Botulinus toxin	GT_1
Escherichia coli toxin	GM_1
Dopamine	GM_1
Thyrotropin (TSH)	GD_{1b}, GT_1
Serotonin	GD_3
Luteotropin (LH)	GT_1, GD_{1b}
Human chorionic gonadotropin (hCG)	GT_1

Membrane gangliosides play an essential role in the binding and action of hormones and they have even been implicated in mediating the antiviral activity of interferon (VENGRIS *et al.* 1976). The present data indicate that gangliosides can be part of the receptor complex. The thyrotropin receptor, for example, consists of a glycoprotein and a ganglioside (KOHN 1977) and it seems that the interaction between both components is necessary to discriminate between structurally related hormones (KOHN *et al.* 1980). The glycoprotein component apparently has the function of a high affinity recognition site, while gangliosides can induce conformational changes of the ligand (ALOJ *et al.* 1979). Binding is mediated by

Neu5Ac, because when solubilized thyrotropin receptor is exposed to Sepharose-coupled neuraminidase, binding activity is reduced by 45%, and it has been shown that fragments of the receptor contain 10% Neu5Ac (TATE *et al.* 1975). It is worth mentioning, however, that the addition of concanavalin A leads to a similar loss of binding activity. Therefore it is reasonable to assume that besides Neu5Ac other sugars participate in receptor ligand-interaction.

VIII. Mucins of the Gastrointestinal Tract

Nearly two decades ago FAILLARD and PRIBILLA (1962, 1964) could demonstrate that sialic acid protects the intrinsic factor from proteolytic degradation. Moreover, release of Neu5Ac led to a loss of affinity to vitamin B_{12}.

Sialic acids also have a central importance in the gastrointestinal tract and in the bronchial system. As for mucins, they form a continuous lining on the inside surface of these hollow organs. There, by virtue of their viscous and elastic properties, they function as protective agents and lubricants. In addition, as integral components of the plasma membrane of the bronchial, gastric and intestinal epithelia, they are involved in transport, secretion and other metabolic processes. For a long time it was thought that the mucin properties of these glycoproteins were due to their pronounced ability to bind water, which was a function of their neuraminic acid content (for review, see CLAMP 1978). On the contrary, we now know that at least the rheological properties are not altered by desialylation (GUSLANDI 1981). It is more probable that the mucin properties are determined by both the protein and the total carbohydrate component as a functional unit. On the other hand, another early known characteristic of certain mucins, the inhibition of virus haemagglutination, is certainly determined by the neuraminic acid component (GOTTSCHALK 1960a, 1960b, HERB *et al.* 1979). The salivary mucins, like other mucins of the gastrointestinal and bronchial system, consist of a protein core with varying quantities of attached carbohydrate side chains (on average one carbohydrate residue per four amino acids). The carbohydrate may constitute up to 75% of the total molecule. Sialic acid and N-acetylgalactosamine have been found in all salivary mucins investigated so far, although not every carbohydrate chain necessarily carries a terminal sialic acid. Other common sugar residues are N-acetylglucosamine, galactose and fucose, while mannose occurs much less frequently. There are typical species-specific differences in the neuraminic acid and fucose contents. In contrast to the predominantly high proportion of sialic acid in other mucins, rabbit salivary mucins possess a higher proportion of fucose (HERB *et al.* 1979). These mucins also possess 4-O-sulfated N-acetylglucosamine residues, which are seldom encountered in other mucin types. Possibly these sulfate residues compensate for the absence of sialic acid in the formation of the protective layer of negative charges. Whereas neuraminic acid is found exclusively as its N-acetyl derivative in complex glycoproteins, the mucin-type glycoproteins contain a high proportion of Neu5Gc. According to work by BUSCHER *et al.* 1977, free Neu5Ac is hydroxylated to N-glycolylneuraminic acid by a soluble cytosolic monooxygenase. Protein-bound Neu5Ac, on the other hand, is hydroxylated by a Golgi membrane-bound enzyme. At least 40% of Neu5Ac is hydroxylated before transfer to the Golgi vesicle membrane, i.e. before attachment to the glycoprotein (BUSCHER *et al.* 1977). The biological significance of the existence of two different processes for the derivatization of neuraminic acid is not known. Recently, disialosyl groups [NeuAcylα(2-8)NeuAcyl-], which are known in glycoconjugates of the central nervous system, were demonstrated in the glycoproteins from the salivary gland of the pig (SLOMIANY *et al.* 1978). In addition to the above-mentioned species specificity, the mucins also exhibit tissue specificity and specific inter-individual differences (PIGMAN 1977).

To summarize, the mucins function mechanically as protective and lubricating agents, and in addition they are involved in immunological activity, Ca^{2+}-binding, agglutination of bacteria and inhibition of virus-induced haemagglutination. The importance of the neuraminic acid residues in these functions is not known in every case. In the inhibition of haemagglutination, the neuraminic acid residues of the mucin probably act by the neuraminidase activity of the virus particle. It is also suggested that sialic acid residues function in biological regulation through their ability to bind Ca^{2+}. Regulation of the carbohydrate composition of mucins by exogenous and endogenous stimuli is incompletely understood. There are indications that the ratio of neuraminic acid to fucose and hexosamines is altered by sympathetic or parasympathetic stimulation of the salivary glands (PIGMAN 1977). However, this alteration is possibly due to the stimulation of different cell systems, rather than a change in the glycosylation of a single mucin. The same applies to the appearance of differently glycosylated (or sialylated) mucins following the injection of pilocarpine or isoproterenol. The mucins of the stomach do not differ fundamentally in their structure from the salivary mucins described above. Gastric mucins that have so far been characterized from different animal species show a similar amino acid and carbohydrate composition. Micro-heterogeneity of glycoproteins, which is frequently observed, is due to varations in the contents of sialic acid and/or sulfate. In this respect, it is difficult to ascertain the actual state of the mucins *in vivo*, because sialic acid and fucose residues are easily released from these glycoproteins by gastric acids. The basic structure consists of four peptide chains linked covalently by S-S bridges (LABAT-ROBERT and DECAENS 1979, HOROWITZ 1977, GALLAGHER and CORFIELD 1978). It is still not known how neuraminic acid residues contribute to the acid-protective properties of the gel formed by the gastric mucus. Some authors report that the sulfate groups of the glycoproteins inhibit pepsin and thereby protect the gastric mucosa from proteolytic degradation (MARTIN *et al.* 1969), but this mechanism is disputed (ALLEN 1978). Similarly, the protective effect of certain prostaglandins against ulceration is not conclusively explained. Work by different authors confirms, however, that the intragastral instillation of prostaglandin E_2 or its analogues promotes a marked increase in the secretion of Neu5Ac-rich mucins (BOLTON *et al.* 1976, DOMSCHKE *et al.* 1978, JOHANNSON and KOLLBERG 1979). On the other hand, ulcerative drugs (e.g. phenylbutazone, indometacin, spironolactone) cause changes in the pattern of mucin glycosylation that do not involve neuraminic acid (KÖTTGEN *et al.* 1979, AZUUMI *et al.* 1980).

Intrinsic factor should also be mentioned. This specific glycoprotein of the gastric juice binds vitamin B_{12} and is inactivated by desialylation. The molecule can be reactivated by resialylation or replacement of the negative charge by various cation exchangers (HOROWITZ 1977). Gamma-glutamyl transferase (GGT) is a variably sialylated glycoprotein of the brush border plasma membrane from jejunum and ileum. Whereas the GGT of the cryptic zone is strongly sialylated, the villus zone contains a sialic acid-deficient variant. Fetal liver also synthesizes Neu5Ac-rich enzyme variants, whereas the enzyme from adult liver shows reduced levels of sialylation. This agrees with the proposal that the cells of the cryptic zone of the intestinal brush border migrate into the villus zone as they mature (KÖTTGEN *et al.* 1976, 1978). Functionally, however, there is no difference between the two enzyme variants.

In cystic fibrosis, an hereditary disease with malfunction of the exocrine pancreas and the bronchial system, it has been reported that the glycoproteins have an altered composition with respect to the relative proportions of fucose and Neu5Ac (ALHADEFF and CIMINO 1978, BEN-YOSEPH *et al.* 1981). In addition to serum glycoproteins, the glycoconjugates of duodenal and bronchial secretions also show an increased content of sialic acid and a decreased content of fucose. At the same time, an increase in the activity of galactosyl transferase and a decrease in sialyl transferase are observed. The extent to which this is responsible for the increased glycoprotein secretion with subsequent obstruction of the glandular ducts is not known. The importance of neuraminic acid metabolism in malignant transformation in the bronchial and intestinal tracts, and the appearance in the mucosa of O-acylated variants of neuraminic acid in cases of tumor of the large intestine, are treated elsewhere (chapter I).

IX. Hormones

Only few data are available on the hormonal influence on both biosynthesis and function of sialic acid-containing glycoconjugates. The involvement of Neu5Ac in hormone action or stability could be shown for erythropoietin, human chorionic gonadotropin and follicle-stimulating hormone (GOTTSCHALK 1960), whose biological functions are destroyed after removal of sialic acid. Recognition and uptake of serotonin by platelets is decreased after desialylation. Also the binding of the tissue hormone histamine by plasmapexin is mediated by sialic acid. The binding of thyrotropin to gangliosides (MULLIN *et al.* 1976) and the assumption that gangliosides act as hormone receptor in thyreoidea (MELDOLESI *et al.* 1976) is still controverse (PACUSZKA *et al.* 1978). For further details see section VII. Recent publications show that oestrogens lead to an increase of protein-bound sialic acid in serum, either during pregnancy (GOOD *et al.* 1971, 1974, GROTH and KADEN 1977, RAJAN *et al.* 1980) or during prolonged application of contraceptiva (KLINGER *et al.* 1981) followed by a decline at parturition or omission of the drug. Also in dogs the administration of protein-bound glucocorticoids causes an increased content of sialic acid in serum. Adrenalectomy of mice is accompanied by a reduced sialic acid content in blood and in liver, whereas administration of corticosteroids and ACTH leads to an increase in the content of sialic acid in liver and myocardium.

X. Concluding Remarks and Outlook

Sialic acid is apparently involved in many different benign or malignant diseases. Part of the findings have been summarized and discussed in the preceding sections. Although data on sialic acids are growing, our knowledge on the biological significance of these compounds is still insufficient. The following main functions, however, can be deduced from the available publications:

— responsibility for the negative charge of many cells and soluble glycoconjugates, thus preventing aggregation, agglutination or precipitation
— protection of blood cells and (serum) glycoconjugates from premature degradation
— masking of antigenic sites
— recognition sites for information transfer
— participation in the signal transmission in nerve cells.

In contrast to the well-documented results on the role of sialic acid for the survival and degradation of serum glycoproteins or blood cells, the role of sialic acid in the regulation of adhesiveness is only partly understood. The main reason for this unsatisfactory situation is the lack of data about purified, defined plasma membrane proteins. Promising attempts in this direction have been made for instance by GLENNEY *et al.* (1979). The partial characterization of five classes of membrane glycoproteins from Novikoff hepatoma cells show structural differences in the non-reducing termini of the oligosaccharide chain. Besides, the finding of markedly different turnover rates of the protein and carbohydrate portion of glycoproteins isolated from the plasma membrane of rat liver, could turn out to be of primary importance (KREISEL *et al.* 1980, TAUBER *et al.* 1982). This new feature could determine both the function and the life-time of at least some glycoproteins. Moreover, the topographical distribution and orientation within the membrane should give further information of the function of the entire glycoprotein. Defined membrane areas, so-called domains (WISHER and EVANS 1975) are composed of different constituents and therefore are endowed with specialized functions (EVANS *et al.* 1980). As shown by an *in vitro* system the adhesion of rat and chicken hepatocytes to gels derivatized with different sugars is dependent on a threshold concentration of the carbohydrate attached to the gel. Below the critical concentration the cells are not bound (SLIFE *et al.* 1980).

A promising approach to elucidate the significance of sialic acids would be (i) to inhibit the transfer of sialic acids, (ii) to modify the endogeneous acceptors (BÜCHSEL *et al.* 1980) so that sialic acids cannot be added to the nascent glycoconjugate, or alternatively, (iii) to block the synthesis of sialic acid by sugar analogues (GRÜNHOLZ *et al.* 1981). Since the activity of various neuraminidases can be inhibited by 2-deoxy-2,3-dehydro-N-acetylneuraminic acid (MEINDL and TUPPY 1969, VEH and SCHAUER 1978) a suitable tool is also available to inhibit the first step of sialoglycoprotein degradation.

Bibliography

ACKERMANN, G. A., 1972: Z. Zellforsch. **134**, 153—166.

AKAGAWA, K. S., MOMOI, T., NAGAI, Y., TOKUNAGA, T., 1981: FEBS Lett. **130**, 80—84.

ALBRIGHT, L., MADIGAN, J. C., GASTON, M. R., HOUCHENS, D. P., 1975: Cancer Res. **35**, 658—665.

ALHADEFF, J. A., CIMINO, G., 1978: Clin. Genet. **13**, 207—212.

ALLEN, A., 1978: Brit. Med. J. **34**, 28—33.

ALLEY, C. D., SNODGRASS, M. J., 1977: Cancer Res. **37**, 95—101.

ALOJ, S. M., LEE, G., CONSIGLIO, E., FORMISANO, S., MINTON, A. P., KOHN, L. D., 1979: J. Biol. Chem. **254**, 9030—9039.

AMINOFF, D., VOR DER BRUEGGE, W. F., BELL, W. C., SARPOLIS, K., WILLIAMS, R., 1977: Proc. Natl. Acad. Sci. U.S.A. **74**, 1521—1524.

ANDO, S., YU, R. K., 1979: J. Biol. Chem. **254**, 12224—12229.

ANSTEE, D. J., 1981: Semin. Hematol. **18**, 13—31.

APFFEL, C. A., PETERS, J. H., 1970: J. Theoret. Biol. **26**, 47—59.

ASHWELL, G., MORELL, A. G., 1974: Enzymol. **41**, 99—128.

ASSMANN, G., BREWER, H. B., jr., 1974: Proc. Natl. Acad. Sci. U.S.A. **71**, 989—993.

ATKINSON, P. H., HAKIMI, J., 1980: In: The Biochemistry of Glycoproteins and Glycosaminoglycans (LENNARZ, W. J., ed.), pp. 191—239. New York-London: Plenum Press.

AVROVA, N. F., CHENYKAEVA, E. Y., OBUKHOVA, E. L., 1973: J. Neurochem. **20**, 997.

AZUUMI, Y., OHARA, S., ISHIHARA, K., OKABE, H., HOTTA, K., 1980: Gut **21**, 533—536.

BAGSHAWE, K. D., CURRIE, G. A., 1968: Nature **218**, 1254—1255.

BALAKRISHAN, S., McILWAIN, H., 1961: Biochem. J. **81**, 72—78.

BALDUINI, C. L., RICEVUTI, G., SOSSO, M., ASCARI, E., BALDUINI, C., 1978: Hoppe-Seyler's Z. Physiol. Chem. **359**, 1573—1577.

— — — — BROVELLI, A., BALDUINI, C., 1977: Acta Haematol. **57**, 178—187.

BANAI, M., KAHANE, I., RAZIN, S., BREDT, W., 1978: Infect. Immun. **21**, 365—372.

BASS, N. H., 1981: In: Gangliosides in Neurological and Neuromuscular Function, Development, and Repair, pp. 29—43. New York: Raven Press.

BAUER, CH., BÜCHSEL, R., MORRIS, H. P., REUTTER, W., 1980: Cancer Res. **40**, 2026—2032.

— LUKASCHEK, R., REUTTER, W., 1974: Biochem. J. **142**, 221—230.

— VISCHER, P., GRÜNHOLZ, H.-J., REUTTER, W., 1977: Cancer Res. **27**, 1513—1518.

BAUVOIS, B., CARTRON, J.-P., NURDEN, A., CAEN, J., 1981: Vox Sang. **40**, 71—78.

BAYER, P. M., GANZINGER, U., BURGER, E. C., GERGELY, TH., GABL, F., MOSER, K., 1977: Wien. klin. Wschr. **89**, 16—17.

BEEK, VAN, W. P., SMETS, L. A., EMMELOT, P., 1973: Cancer Res. **33**, 2913—2922.

BEER, A. E., BILLINGHAM, R. E., 1978: Federation Proc. **37**, 2374—2378.

BEHR, J.-P., LEHN, J. M., 1972: FEBS Lett. **22**, 178—180.

BEKESI, J. G., HOLLAND, J. F., 1977: Recent Results Cancer Res. **62**, 78—89.

— — 1978: In: Immunotherapy of Malignant Diseases, pp. 375—388. Stuttgart-New York: Schattauer.

— ROBOZ, J. P., HOLLAND, J. F., 1976: Isr. J. Med. Sci. **12**, 288—303.

— HOLLAND, J. F., ROBOZ, J. P., 1977: Med. Clin. N. Amer. **61**, 1083—1100.

— ST. ARNEAULT, G., HOLLAND, J. F., 1971: Cancer Res. **31**, 2130—2132.

BENNETT, M., SCHMID, K., 1980: Proc. Natl. Acad. Sci. U.S.A. **77**, 6109—6113.

BEN-YOSEPH, Y., DEFRANCO, C. L., NADLER, H. L., 1981: Pediatr. Res. **15**, 839—842.

BERGMAN, L. W., HARRIS, E., KUEHL, M., 1981: J. Biol. Chem. **256**, 701—706.

BERGMAN, R., MORENO-LÓPEZ, J., MÖLLERBERG, L., MOREIN, B., 1978: Res. Vet. Sci. **25**, 193—199.

BERNACKI, R. J., KIM, U., 1977: Science **195**, 577—579.

BEUKERS, H., DEIERKAUF, F. A., BLOM, C. P., DEIERKAUF, M., SCHEFFERS, C. C., RIEMERSMA, J. C., 1980: Chem.-Biol. Interact. **33**, 91—100.

BEZKOROVAINY, A., 1965: Biochim. Biophys. Acta **101**, 336—342.

BHAVANANDAN, V. P., KATLIC, A. W., 1979: J. Biol. Chem. **254**, 4000—4008.

BINDER, P., PERROT, L., BEAUDRY, Y., BOTTEX, C., FONTANGES, R., 1975: C. R. Acad. Sci., (D) Paris **281**, 1545—1547.

BLAJCHMAN, M. A., SENYI, A. F., HIRSH, J., GENTON, E., 1981: J. Clin. Invest. **68**, 1289—1294.

BLUMENTHAL, R., KLAUSNER, R. D., WEINSTEIN, J. N., 1980: Nature **288**, 333—338.
BOGOCH, S., PAASONEN, M. K., TRENDELENBERG, U., 1962: Br. J. Pharmacol. **18**, 325—330.
BÖHMER, F. D., KIRSCHKE, H., BOHLEY, P., SCHÖN, R., 1979: Acta Biol. Med. Ger. **38**, 1521—1526.
BOLTON, J. P., PALMER, D., COHEN, M. M., 1976: Surgical Forum **27**, 402—403.
BOSMANN, H. B., 1972: Biochim. Biophys. Acta **279**, 456—474.
— 1974: Vox Sang. **26**, 497—512.
— 1972: Biochem. Biophys. Res. Commun. **49**, 1256—1262.
— HALL, T. C., 1974: Proc. Natl. Acad. Sci. U.S.A. **71**, 1833—1837.
— HILF, R., 1974: FEBS Lett. **44**, 313—316.
— SPATARO, A. C., MYERS, M. W., BERNACKI, R. J., HILLAMN, M. J., CAPUTI, S. E., 1975: Res. Commun. Chem. Pathol. Pharmacol. **12**, 499—512.
BOWLES, D. J., HANKE, D. E., 1976: FEBS Lett. **66**, 16—20.
— — 1977: FEBS Lett. **82**, 34—38.
— ROTMAN, A., 1978: FEBS Lett. **90**, 283—285.
BRADLEY, W. P., BLASCO, A. P., WEISS, J. F., ALEXANDER, J. C., SILVERMAN, N. A., CHRETIEN, P. B., 1977: Cancer **40**, 2264—2272.
BRADY, R. O., FISHMAN, P. H., MORA, P. T., 1973: Fed. Proc. **32**, 102—108.
— MORA, P. T., 1970: Biochim. Biophys. Acta **218**, 308—319.
BRAMWELL, M. E., HARRIS, H., 1978: Proc. Royal Soc. (London) **B 201**, 87—106.
BRANDT, A. E., JAMESON, A. K., PINCUS, J. H., 1981: Cancer Res. **41**, 3077—3081.
BRAZIL, J., McLAUGHLIN, H., 1978: Europ. J. Cancer **14**, 757—760.
BRECKENRIDGE, W., GOMBOS, G., MORGAN, I., 1972: Biochem. Biophys. Res. Commun. **266**, 695—700.
BRIDGES, K., HARFORD, J., ASHWELL, G., KLAUSNER, R. D., 1982: Proc. Natl. Acad. Sci. U.S.A. **79**, 350—354.
BRIGGS, D. W., FISHER, J. W., GEORGE, W. J., 1974: Amer. J. Physiol. **227**, 1385—1388.
BROWNELL, A. G., 1977: J. Supramolec. Struct. **7**, 223—234.
BRUNNGRABER, E. G., 1979: In: Neurochemistry of Aminosugars, pp. 90—127. Springfield, Ill.: Ch. C Thomas.
BÜCHSEL, R., HASSELS-VISCHER, B., TAUBER, R., REUTTER, W., 1980: Eur. J. Biochem. **111**, 445—453.
BUNTING, R. W., PEERSCHKE, E. I., ZUCKER, M., 1978: Blood **52**, 643—653.
BURGER, M. M., FINNE, J., MATTER, A., HASKOVEC, C., TAO, T. W., 1979: In: Tumor Markers; Impact and Prospects (BOELSMA, E., RÜMKE, P. H., eds.), Applied Methods in Oncology, Vol. II, pp. 213—223. Amsterdam: Elsevier.
BUSCHER, H.-P., CASALS-STENZEL, J., SCHAUER, R., METRES-VENTURA, P., 1977: Eur. J. Biochem. **77**, 297—310.
BUTTERS, T. D., DEVALIA, V., APLIN, J. D., HUGHES, R. C., 1980: J. Cell Sci. **44**, 33—58.
CAEN, J. P., NURDEN, A. T., JEANNEAU, C., MICHEL, H., TOMBELEM, G., LEVY-TOLEDANO, S., SULTAN, Y., VALENSI, F., BERNARD, J., 1976: J. Lab. Clin. Med. **87**, 586—596.
CAHAN, L. D., PAULSON, J. C., 1980: Virology **103**, 505—509.
CAMPBELL, R., BOGOCH, S., SCOLARO, M. J., BELVAL, P. C., 1967: Amer. J. Psychiat. **8**, 952—962.
CANTRELL, J. L., KILLION, J. J., KOLLMORGEN, G. M., 1976: Cancer Res. **36**, 3051—3057.
CARRAWAY, K. L., 1975: Biochim. Biophys. Acta **415**, 379—410.
CARTRON, J. P., NURDEN, A. T., 1979: Nature **282**, 621—623.
CARTWRIGHT, B., BROWN, F., 1977: J. gen. Virol. **35**, 197—199.
CARUBELLI, R., GRIFFIN, M. J., 1967: Science **157**, 693—694.
CESTARO, B., CERVATO, G., TETTAMANTI, G., 1978: Bulletin of Molecular Biology and Medicine **3**, 159—169.

CHATTERJEE, S. K., BHATTACHARYA, M., BARLOW, J. J., 1978: Biochem. Biophys. Res. Commun. **80**, 826—832.
— 1979: Eur. J. Cancer **15**, 1350—1356.
CHEEMA, P., YOGEESWARAN, G., MORRIS, H. P., MURRAY, R. K., 1970: FEBS Lett. **11**, 181—184.
CHEN, W.-T., SINGER, S. J., 1980: Proc. Natl. Acad. Sci. U.S.A. **77**, 7318—7322.
CHERAYIL, G. D., 1969: J. Neurochem. **16**, 913—920.
CHOI, S.-J., SIMONE, J. V., JOURNEY, L. J., 1972: Br. J. Haematol. **22**, 93—101.
CLAMP, J. R. (ed.), 1978: Mucus (Brit. Med. Bull., Vol. 34).
CLEMETSON, K. J., NAIM, H. Y., LÜSCHER, E. F., 1981: Proc. Natl. Acad. Sci. U.S.A. **78**, 2712—2716.
COHEN, I., POTTER, E. V., GLASER, T., ENTWHISTLE, R., DAVIS, L., CHEDIAK, J., ANDERSON, B., 1981: J. Lab. Clin. Med. **97**, 97—134.
COHEN, N. S., EKHOLM, J. E., LUTHRA, M. G., HANAHAN, D. J., 1976. Biochim. Biophys. Acta **419**, 229—242.
COOK, G. M. W., HEARD, D. H., SEAMAN, G. V. F., 1961: Nature **191**, 44—47.
COSTELLO, M., FIEDEL, B. A., GEWURZ, H., 1979: Nature **281**, 677—678.
CUATRECASAS, P., 1973: Biochemistry **12**, 3547—3558.
CUMAR, F. A., BRADY, R. O., KOLODNY, E. H., MCFARLAND, V. W., MORA, P. T., 1970: Proc. Natl. Acad. Sci. U.S.A. **67**, 757—764.
CURTIS, A. S. G., 1967: The Cell Surface: Its Molecular Role in Morphogenesis. New York: Academic Press.
DACREMONT, G., 1972: Clin. Chim. Acta **37**, 449—454.
DANON, D., MARIKOVSKY, Y., 1961: Compt. Rend. Acad. Sci. **253**, 1271—1273.
DAO, T. L., IP, C., PATEL, J., 1980: J. Natl. Cancer Inst. **65**, 529—534.
DAVRAINVILLE, J. L., GAYET, J., 1965: J. Neurochem. **12**, 771—782.
DEBANNE, M. T., REGOECZI, E., 1981: J. Biol. Chem. **256**, 11266—11272.
DEN, H., SCHULZ, A. M., BASU, M., ROSEMAN, S., 1971: J. Biol. Chem. **246**, 2721—2723.
DESAI, P. R., SPRINGER, G. F., 1980: J. Immunogenet. **7**, 149—155.
DIRINGER, H., STROBEL, G., KOCH, M. A., 1972: Hoppe-Seyler's Z. Physiol. Chem. **353**, 1769—1774.
DNISTRIAN, A. M., SKIPSI, V. P., BARLAY, M., ESSNER, E. S., STOCK, C. C., 1975: Biochem. Biophys. Res. Commun. **64**, 367—375.
— — — STOCK, C. C., 1979: J. Natl. Cancer Inst. **62**, 367—370.
DOBROSSY, L., PAVELIC, Z. P., BERNACKI, R. J., 1981: Cancer Res. **41**, 2262—2266.
DOMSCHKE, W., DOMSCHKE, S., HORNIG, D., DEMLING, L., 1978: Acta Hepato-Gastroent. **25**, 292—294.
DRZENIEK, R., 1972a: Biochem. Biophys. Res. Commun. **26**, 631—638.
— 1972b: Topics Microbiol. Immunol. **59**, 35—74.
DUCDODON, M., QUASH, G. A., 1981: Immunology **42**, 401—408.
DUROCHER, J. R., PAYNE, R. C., CONRAD, M. E., 1975: Blood **45**, 11—19.
DURR, R., SHUR, B., ROTH, S., 1977: Nature **265**, 547—548.
DYATLOVITSKAYA, E. V., ZABLOTSKAYA, A. E., AZIZOV, Y. M., BERGELSON, L. D., 1980: Eur. J. Biochem. **110**, 475—483.
EDDY, B. E., ROWE, W. P., HATLEY, J. W., STEWART, S. E., HUEBNER, R. J., 1958: Virology **6**, 290—291.
EDELFORS, S., 1981: Acta Pharmacol. et Toxicol. **48**, 61—64.
ELSTEIN, M., 1978: Brit. Med. Bull. **34**, 83—86.
ERNI, B., DE BOECK, H., LOONTIENS, F. G., SHARON, N., 1980: FEBS Lett. **120**, 149—154.
EVANS, H., FLINT, N., VISCHER, P., 1980: Biochem. J. **192**, 903—910.

EYLAR, E. H., MADOFF, M. A., BRODY, O. V., ONCLEY, J. L., 1962: J. Biol. Chem. **237**, 1992—2000.
FAILLARD, H., PRIBILLA, W., POSTH, H. E., 1962: Hoppe-Seyler's Z. Physiol. Chem. **327**, 100—108.
— — 1964: Klin. Wschr. **42**, 686—693.
FILIPOVIC, I., SCHWARZMANN, G., MRAZ, W., WIEGANDT, H., BUDDECKE, E., 1979: Eur. J. Biochem. **93**, 51—55.
FINNÉ, E., 1978: Endokrinologie **72**, 363—364.
FISCHER, K., POSCHMANN, A., 1976: Dtsch. med. Wschr. **101**, 1731—1733.
FREDMAN, P., MANSSON, J.-E., SVENNERHOLM, L., SAMUELSON, B., PASCHER, I., PIMLOTT, W., KARLSSON, K.-A., KLINGHARDT, G. W., 1981: Eur. J. Biochem. **116**, 553—564.
FRIED, H., CAHEN, L. D., PAULSON, J. C., 1981: Virology **109**, 188—192.
FURTHMAYR, H., 1978: Nature **271**, 519—520.
GAHMBERG, C. G., HAKOMORI, S., 1975: J. Biol. Chem. **250**, 2438—2455.
— ITAYA, K., HAKAMORI, S., 1976: Methods Membr. Biol. **7**, 179—210.
GALILI, U., SCHLESINGER, M., 1978: Cancer Immunol. Immunother. **4**, 33—39.
GALLAGHER, J. T., CORFIELD, A. P., 1978: TIBS **3**, 38—41.
GANZINGER, U., 1977: Wien. klin. Wschr. **89**, 594—597.
GARDNER, E., jr., WRIGHT, C.-S., WILLIAMS, B. Z., 1961: J. Lab. Clin. Med. **58**, 743—750.
GARTNER, T. K., GERRARD, J. M., WHITE, J. G., WILLIAMS, D. C., 1981: Blood **58**, 153—157.
— PHILLIPS, D. R., WILLIAMS, D. C., 1980: FEBS Lett. **113**, 196—200.
— WILLIAMS, D. C., MINION, F. C., PHILLIPS, D. R., 1978: Science **200**, 1281—1283.
— — PHILLIPS, D. R., 1977: Biochem. Biophys. Res. Commun. **79**, 592—599.
GASTEL, VAN, C., DEWIT, C. D., 1970: Vox Sang. **19**, 105—112.
GATI, W. P., STRAUB, P. W., 1978: J. Biol. Chem. **253**, 1315—1321.
GATTEGNO, L., BLADIER, D., CORNILLOT, P., 1975: Hoppe-Seyler's Z. Physiol. Chem. **356**, 391—397.
— PERRET, G., FABIA, F., CORNILLOT, P., 1981: Mech. Ageing Develop. **16**, 205—219.
GHIDONI, R., SONNINO, S., TETTAMANTI, G., WIEGANDT, H., ZAMBOTTI, V., 1976: J. Neurochem. **27**, 511—515.
GIBBONS, R. A., 1978: Brit. Med. Bull. **34**, 34—36.
GINNS, E., FRENCH, J., 1980: J. Neurochem. **35**, 977—982.
GLASGOW, L. R., HILL, R. L., 1980: Infect. Immun. **30**, 353—361.
GLENNEY, J. R., jr., ALLISON, J. P., HIXON, D. C., WALBORG, E. F., jr., 1979: J. Biol. Chem. **254**, 9247—9253.
GLICK, M. C., BUCK, C. A., 1973: Biochemistry **12**, 85—90.
GLODE, M. P., LEWIN, E. B., SUTTON, A., LE, C. T., GOTSCHLICH, E. C., ROBBINS, J. B., 1979: J. Infect. Dis. **139**, 52—59.
GOLDSTEIN, J. L., ANDERSON, R. G. W., BROWN, M. S., 1979: Nature **279**, 679—685.
GOOD, W., 1974: J. Obstet. Gynaec. **81**, 878—887.
— HANCOCK, K. W., MACDONALD, H. N., 1971: J. Obstet. Gynaec. **78**, 628—631.
GORDON, L. K., HAMILL, B., PARKER, C. W., 1980: J. Immunol. **125**, 814—819.
GÖRÖG, P., KOVÁCS, I. B., BORN, G. V. R., 1980: Br. J. exp. Path. **61**, 490—496.
GOTTSCHALK, A., 1960a: Nature **186**, 949—951.
— 1960b: The Chemistry and Biology of Sialic Acids and Related Substances. Cambridge: University Press.
GRALNICK, H. R., COLLER, B. S., SULTAN, Y., 1976: Science **192**, 56—59.
— GIVELBER, H., ABRAMS, E., 1978: N. Engl. J. Med. **299**, 221—226.
GREENBERG, J. P., PACKHAM, M. A., GUCCIONE, M. A., RAND, M. L., REIMERS, H.-J., MUSTARD, J. F., 1979: Blood **53**, 916—926.
GREENE, W. C., PARKER, C. M., PARKER, C. W., 1976: J. Biol. Chem. **251**, 4017—4025.

Greene, W. C., Waldmann, T. A., 1980: I. Immunol. **124**, 2979—2987.
Greenwalt, T. J., Steane, E. A., 1973: Br. J. Haemotol. **25**, 207—**000**.
Grimes, W. J., 1970: Biochemistry **9**, 5083—5092.
— 1973: Biochemistry **12**, 990—996.
Groth, J., Kaden, J., 1977: Acta Biol. Med. Germ. **36**, 1495—1498.
Grünholz, H.-J., 1978: Thesis, Freiburg i. Br., Federal Republic of Germany.
— Harms, E., Opetz, M., Reutter, W., 1981: Carbohydr. Res. **96**, 259—270.
Guslandi, M., 1981: Clin. Chin. Acta **117**, 3—5.
Gutterman, J. U., McBride, C., Freireich, E. J., Mavligit, G., Frei, E., Hersh, E. M., 1973: Lancet **1**, 1208—1212.
Hafermaas, R., Köttgen, E., Bauer, Ch., Reutter, W., 1982: Proceedings of the FEBS Meeting, Athens.
Hakomori, S. I., 1966: J. Lipid Res. **7**, 789—792.
— Murakami, W. T., 1968: Proc. Natl. Acad. Sci. U.S.A. **59**, 254—261.
Häkkinen, H.-M., Kulonen, E., 1963: Nature **198**, 995.
Halbhuber, K.-J., Helmke, U., Geyer, G., 1972: Folia Haematol. **97**, 196—203.
Han, T., 1972: Transplantation **14**, 515—517.
Hannsson, H. A., Holgren, J., Svennerholm, L., 1977: Proc. Natl. Acad. Sci. U.S.A. **74**, 3782—3786.
Harford, J., Ashwell, G., 1981: Proc. Natl. Acad. Sci. U.S.A. **78**, 1557—1561.
Harms, E., Kreisel, W., Morris, H. P., Reutter, W., 1973: Eur. J. Biochem. **32**, 254—262.
Hart, D. A., 1980: Am. J. Clin. Nutr. **33**, 2416—2425.
Hartley, J. W., Rowe, W. P., Chanock, R. M., Andrews, B. E., 1959: J. Exp. Med. **110**, 81—89.
Harvey, H. A., Lipton, A., White, D., Davidson, E., 1981: Cancer **47**, 324—327.
Hatcher, V. B., Schwarzmann, G. O. H., 1977: Fertility and Sterility **28**, 628—632.
Hayes, M. L., Castellino, F. J., 1979: J. Biol. Chem. **254**, 8768—8771.
Herb, A., Wu, A. M., Moschera, J., 1979: Molec. Cell. Biochem. **23**, 27—43.
Herberman, R. B., 1977: Biochim. Biophys. Acta **473**, 93—119.
Henderson, M., Kessel, D., 1977: Cancer **39**, 1129—1134.
Heyningen, van, W. E., Carpenter, C. C. J., Pierce, N. F., Greenaugh, W. B., 1971: J. Infect. Dis. **124**, 415—418.
— Mellanby, J., 1971: In: Microbiol. Toxins, Vol. 2A: Bacterial Protein Toxins, pp. 69—108. New York: Academic Press.
Higashi, H., Naiki, M., Matuo, S., Okouchi, K., 1977: Biochem. Biophys. Res. Commun. **79**, 388—395.
Hirsch, R. L., Griffin, D. E., Winkelstein, J. A., 1981: J. Immunol. **127**, 1740—1743.
Hoessli, D., Bron, C., Pink, R. L., 1980: Nature **283**, 576—578.
Hogan-Ryan, A., Fennelly, J. J., Jones, M., Cantwell, B., Duffy, M., 1980: Br. J. Cancer **41**, 587—592.
— — 1981: Eur. J. Cancer Clin. Oncol. **17**, 843—844.
Holland, J. F., Bekesi, J. G., 1976: Symposium in Immunotherapy in Malignant Disease, Medical Clinics of North America **60**, 539—549.
Holmgren, J., Elwing, H., Fredman, P., Strannegård, O., Svennerholm, L., 1980: In: Structure and Function of Gangliosides (Svennerholm, L., Mandel, P., Dreyfus, H., Urban, P.-F., eds.), pp. 453—470. New York: Plenum Press.
— Lönnroth, I., Mannsson, J. E., Svennerholm, L., 1975: Proc. Natl. Acad. Sci. U.S.A. **72**, 2520—2524.
— — Svennerholm, L., 1973: Infect, Immun. **8**, 208—214.
— Svennerholm, L., Elwing, H., Fredman, P., Strannegård, D., 1980: Proc. Natl. Acad. Sci. U.S.A. **77**, 1947—1950.

HOROWITZ, M. I., 1977: In: The Glycoconjugates, Vol. I, pp. 189—213. New York: Academic Press.
HOSSNER, K. L., BILLIAR, R. B., 1979: Biochim. Biophys. Acta **585**, 543—553.
HUDGIN, R. L., MURRAY, R. K., PINTERIC, L., MORRIS, H. P., SCHACHTER, H., 1971: Canadian J. Biochem. **49**, 61—70.
— PRICER, W. E., ASHWELL, G., STOCKERT, R. J., MORELL, A., 1974: J. Biol. Chem. **249**, 5536—5543.
HUGHES, R. C., PENA, S. D. J., CLARK, J., DOURMASHKIN, R. R., 1979: Exptl. Cell Res. **221**, 307—314.
— — 1982: In: Carbohydrate Metabolism and Its Disorders (RANDLE, P. J., STEINER, D. F., WHELAN, W. J., eds.), Vol. 3, in press.
— — VISCHER, P., 1979b: Cell Adhesion and Motility (CURTIS, A., PITTS, J. D., eds.), pp. 329—356. Cambridge: University Press.
HYMES, A. J., MULLINAX, G. L., MULLINAX, F., 1979: J. Biol. Chem. **254**, 3148—3151.
HYNES, R. O., 1976: Biochim. Biophys. Acta **458**, 73—107.
IP, C., DAO, L. T., 1977: Cancer Res. **37**, 577—579.
— — 1978: Cancer Res. **38**, 723—728.
JANCIK, J. M., SCHAUER, R., ANDRES, K. H., VON DÜRING, M., 1978: Cell Tissue Res. **186**, 209—226.
JAQUES, L. W., BROWN, E. B., BARRETT, J. M., BREY, W. S., WELLNER, W., 1977: J. Biol. Chem. **252**, 4533—4538.
JEANLOZ, R. W., CODINGTON, J. F., 1976: In: Biological Roles of Sialic Acid (ROSENBERG, A., SCHENGRUND, C.-L., eds.), pp. 201—238. New York: Plenum Press.
JOHANSSON, C., KOLLBERG, B., 1979: Eur. J. Clin. Invest. **9**, 229—232.
KAHANE, I., POLLIACK, A., RACHMILEWITZ, E. A., BAYER, E. A., SKUTELSKY, E., 1978: Nature **271**, 674—675.
KAHANE, J., BEN-CHETRIT, E., SHIFTER, A., RACHMILEWITZ, E. A., 1980: Biochim. Biophys. Res. Commun. **596**, 10—17.
KANWAR, J. S., FARQUHAR, M. G., 1980: Lab. Invest. **42**, 375—384.
KAO, K.-J., PIZZO, S. V., MCKEE, P. A., 1980: J. Biol. Chem. **255**, 10134—10139.
KAPLAN, R., LI, S. S.-L., KEHOE, J. M., 1977: Biochem. **16**, 4297—4303.
KASAI, M., IWAMORI, M., NAGAI, Y., OKUMURA, K., TADA, T., 1980: Eur. J. Immunol. **10**, 175—180.
— YONEDA, T., HABU, S., MARUYAMA, Y., OKUMURA, K., TOKUNAGA, T., 1981: Nature **291**, 334—335.
KAUFMANN, S., SCHAUER, R., HAHN, H., 1981: Immunobiol. **160**, 184—195.
KAWASAKI, T., ASHWELL, G., 1976: J. Biol. Chem. **251**, 1296—1302.
KAY, M. M. B., 1975: Proc. Natl. Acad. Sci. U.S.A. **72**, 3521—3525.
KAZATCHKINE, M. D., FEARON, D. T., AUSTEN, K. F., 1979: J. Immunol. **122**, 75—81.
KEENAN, T. W., DOAK, R. L., 1973: FEBS Lett. **37**, 124—128.
— HUANG, C. M., MORRÉ, D. J., 1972: Biochem. Biophys. Res. Commun. **47**, 1277.
— MORRÉ, D. J., 1973: Science **182**, 935—937.
— — BASU, S., 1974: J. Biol. Chem. **249**, 310—315.
KEMP, R. B., 1970: J. Cell Sci. **6**, 751—766.
KEPPLER, D., LESCH, R., REUTTER, W., DECKER, K., 1968: Exp. Mol. Pathol. **9**, 279—290.
KESSEL, D., ALLEN, J., 1975: Cancer Res. **35**, 670—672.
KIKUCHI, O., KIKUCHI, H., TSUIKI, S., 1971: Biochim. Biophys. Acta **252**, 357—368.
KIM, Y. S., ISAACS, R., PEDOMO, J. M., 1974: Proc. Natl. Acad. Sci. U.S.A. **71**, 4869—4873.
KING, C. A., VAN HEYNINGEN, W. E., 1973: J. Infect. Dis. **127**, 639—647.
KIRSCHNER, H., VOGT, W., 1961: Biochem. Pharmacol. **8**, 224—234.

KISS, P., SCHARBERT, F., KATTERMANN, R., 1981: Hoppe-Seyler's Z. Physiol. Chem. **362**, 897—902.

— KATTERMANN, R., 1979: In: Glycoconjugates (SCHAUER, R., BOER, P., BUDDECKE, E., KRAMER, M. F., VLIEGENTHART, J. F. G., WIEGANDT, H., eds.), pp. 316—317. Stuttgart: G. Thieme.

KLAUSNER, R. D., BRIDGES, K., TSUNOO, H., BLUMENTHAL, R., WEINSTEIN, J. N., ASHWELL, G., 1980: Proc. Natl. Acad. Sci. U.S.A. **77**, 5087—5091.

KLENK, E., 1935: Hoppe-Seyler's Z. Physiol. Chem. **235**, 24—36.

— 1942: Hoppe-Seyler's Z. Physiol. Chem. **273**, 76—86.

KLINGEMANN, H.-G., EGBRING, R., HAVEMANN, K., 1980: Klin. Wochenschr. **58**, 533—535.

KLINGER, G., KUNSTMANN, F. W., STELZNER, H., 1981: Zbl. Gynäkol. **103**, 36—40.

KLOPPEL, T. M., MORRÉ, D. J., JACOBSEN, L. B., 1979: J. Supramolec. Struct. **11**, 485—492.

KNOP, J., AX, W., SEDLACEK, H. H., SEILER, F. R., 1978: Immunol. **34**, 555—563.

KOHN, L. D., CONSIGLIO, E., DEWOLF, M. J., GROLLMAN, E. F., LEDLEY, F. D., LEE, G., MORRIS, N. P., 1980: In: Structure and Functions of Gangliosides, pp. 487—503. New York: Plenum Press.

— 1977: In: Horizons in Biochemistry and Biophysis (QUAGLIARIELL, E., ed.), Vol. 3, pp. 123—163. Reading, Mass.: Addison-Wesley.

KOLB, H., SCHLEPPER-SCHÄFER, J., REUTTER, W., 1979: unpublished.

— KOLB-BACHOFEN, V., 1978: Biophys. Res. Commun. **85**, 678—683.

— — SCHLEPPER-SCHÄFER, J., 1979: Biol. Cell **36**, 301—308.

— VOGT, D., HERBERTZ, L., CORFIELD, A. P., SCHAUER, R., SCHLEPPER-SCHÄFER, J., 1980: Hoppe-Seyler's Z. Physiol. Chem. **361**, 1749—1750.

— FRIEDRICH, E., SÜSS, R., 1981: Hoppe-Seyler's Z. Physiol. Chem. **362**, 1609—1614.

— SCHMIDT, C., KOLB-BACHOFEN, V., KOLB, H.-A., 1978: Exptl. Cell Res. **113**, 319—325.

KOLODNY, G. M., 1972: Exptl. Cell. Res. **70**, 196—202.

KOLSET, S. O., TOLLESHAUG, H., BERG, T., 1979: Exp. Cell Res. **122**, 159—167.

KONDO, Y., SATO, K., UEYAMA, Y., OHSAWA, N., 1981: Cancer Res. **41**, 2912—2916.

KOSTIC, D., BUCHHEIT, 1970: Life Sci. **9**, Part II, 589—596.

KOTELIANSKY, V. E., LEYTIN, V. L., SVIRIDOV, D. D., REPIN, V. S., SMIRNOV, V. N., 1981: FEBS Lett. **123**, 59—62.

KOVÁCS, I. B., GÖRÖG, P., 1979: Thrombos. Haemostas. **42**, 1187—1192.

KÖTTGEN, E., 1977: Klin. Wschr. **55**, 359—373.

— BAUER, CH., GEROK, W., W., 1979: Analyt. Biochem. **96**, 391—394.

— 1980: Internist **21**, 231—235.

— FABRICIUS, H.-A., STAHN, R., GEROK, W., 1981: Klin. Wschr. **59**, 669—674.

— BAUER, CH., REUTTER, W., GEROK, W., 1979: Klin. Wschr. **57**, 199—214.

— REUTTER, W., GEROK, W., 1976: Biochem. Biophys. Res. Comm. **72**, 61—66.

— — — 1978: Eur. J. Biochem. **82**, 279—284.

KRAEMER, P. M., 1966: J. Cell Physiol. **68**, 85—90.

KREISEL, W., VOLK, B., BÜCHSEL, R., REUTTER, W., 1980: Proc. Natl. Acad. Sci. U.S.A. **78**, 1828—1831.

KRISHNARAJ, R., SAAT, Y. A., KEMP, R. G., 1980: Cancer Res. **40**, 2808—2813.

KUHN, R., WIEGANDT, H., 1963: Chem. Ber. **96**, 866—880.

— — 1964: Z. Naturforsch. **19b**, 256—257.

KUNICKI, T. J., JOHNSON, M. M., ASTER, R. H., 1978: J. Clin. Invest. **62**, 716—719.

KUPPERS, R. C., HENNEY, C. S., 1979: J. Immunol. **122**, 1834—1840.

KURISU, M., YAMAZAKI, M., MIZUNO, D., 1980: Cancer Res. **40**, 3798—3803.

KÜSTER, J. M., SCHAUER, R., 1981: Hoppe-Seyler's Z. Physiol. Chem. **362**, 1507—1514.

LABAT-ROBERT, J., DECAENS, C., 1979: Path. Biol. **27**, 241—247.

LANDAW, S. A., TANFORDE, T., SCHOOLEY, J. C., 1973: Clin. Res. **21**, 266.

LAWLER, J. W., SLAYTER, H. S., COLIGAN, J. E., 1978: J. Biol. Chem. **253**, 8609—8616.
LEBLOND-LAROUCHE, L., MORAIS, R., NIGAM, V. N., KARASAKI, 1975: Arch. Biochem. Biophys. **167**, 1—12.
LEDEEN, R. W., 1978: J. Supramol. Struct. **8**, 1—17.
LEPRAT, R., MICHEL-BRIAND, Y., 1980: Ann. Microbiol. **131**, 209—222.
LESTER, E. P., MILLER, J. B., BARON, J. M., YACHNIN, S., 1978: Immunol. **34**, 189—198.
— — YACHNIN, S., 1976: Proc. Natl. Acad. Sci. U.S.A. **73**, 4645—4648.
LICHTHARDT, D., DAHR, W., ROELCKE, D., 1981: Hoppe-Seyler's Z. Physiol. Chem. **362**, 219.
LIJNEN, H. R., VAN HOEF, B., COLLEN, D., 1981: Eur. J. Biochem. **120**, 149—154.
LINDENMANN, J., KLEIN, P. A., 1967: Recent Results in Cancer Research **9**, 66—84.
LIU, C. K., SCHMIED, R., GREENSPAN, E. M., WAXMANN, S., 1978: Biochim. Biophys. Acta **522**, 375—384.
LIU, T.-Y., GOTTSCHLICH, E. C., EGAN, W., ROBBINS, J. B., 1977: J. Infect. Dis. Suppl. **136**, S71 – S77.
LLOYD, C. W., 1975: Biol. Rev. **50**, 325—350.
LÜBEN, G., SEDLACEK, H. H., SEILER, F. R., 1976: Behring Inst. Mitt. **59**, 30—37.
LUI, S. W. L., MUN, H., 1980: Arch. Virol. **63**, 31—41.
LUNER, S. J., SZKLAREK, D., KNOX, R. J., SEAMAN, G. V. F., JOSEFOWICZ, J. Y., WARE, B. R., 1977: Nature **269**, 719—721.
LÜSCHER, E. F., 1978: Bibliotheca haemat. **45**, 28—33.
LUTZ, H. U., FEHR, J., 1979: J. Biol. Chem. **25**, 11177—11180.
MAGET-DANA, R., VEH, R. W., SANDER, M., ROCHE, A.-C., SCHAUER, R., MONSIGNY, M., 1981: Eur. J. Biochem. **114**, 11—16.
MARINEZ, J., PALASCAK, J. E., KWASNIAK, D., 1978: J. Clin. Invest. **61**, 535—538.
MARSHALL, J. S., GREEN, A. M., PENSKY, J., WILLIAMS, S., ZINN, A., CARLSON, D. M., 1974: J. Clin. Invest. **54**, 555—562.
— WILLIAMS, S. T., 1978: Biochim. Biophys. Acta **543**, 41—52.
MARTIN, F., MATHIAN, R., BERARD, A., LAMBERT, R., 1969: Digestion **2**, 103—112.
MATHÉ, G., AMIEL, J. L., SCHWARZBERG, L., SCHNEIDER, M., CHAN, A., SCHLIEMBERGER, J. R., HAYAT, M., DE VASSAL, F., 1969: Lancet **1**, 697—699.
— 1978: Cancer Immunol. Immunother. **5**, 149—152.
MATSUUCHI, L., SHARON, J., MORRISON, S. L., 1981: J. Immunol. **127**, 2188—2190.
MCGREGOR, J. L., CLEMETSON, K. J., JAMES, E., CAPITANO, A., GREENLAND, T., LÜSCHER, F., DECHAVANNE, M., 1981: Eur. J. Biochem. **116**, 379—388.
MEAGER, A., UNGKITCHANNKIT, A., HUGHES, R. C., 1976: Biochem. J. **154**, 113—124.
MEEZAN, E., WU, H. C., BLACK, P. H., ROBBINS, P. W., 1969: Biochemistry **8**, 2518—2524.
MEINDL, P., TUPPY, H., 1969: Hoppe-Seyler's Z. Physiol. Chem. **350**, 1088—1092.
MINTA, J. O., 1981: J. Immunol. **126**, 245—249.
MIZRAHI, A., O'MALLEY, A., CARTER, W. A., TAKATSUKI, A., TAMURA, G., SULKOWSKI, E., 1978: J. Biol. Chem. **253**, 7612—7615.
MIZUOCHI, T., YAMASHITA, K., FUJIKAWA, K., KISIEL, W., KOBATA, A., 1979: J. Biol. Chem. **254**, 6419—6425.
— — TITANI, K., KOBATA, A., 1980: J. Biol. Chem. **255**, 3526—3531.
MKHEYAN, E. E., SOTSKII, O. P., 1970: Vop. Biokhim. Mozga **6**, 219.
MONACO, F., ROBBINS, J., 1973: J. Biol. Chem. **248**, 2328—2336.
MORALES, A., 1980: Cancer Immunol. Immunother. **9**, 69—72.
MORELL, A. G., IRVINE, R. A., STERNLIEB, I., SCHEINBERG, I. H., ASHWELL, G., 1968: J. Biol. Chem. **243**, 155—159.
MORGAN, B. L., WINICK, M., 1980: J. Nutr. **110**, 416—424.
MORRÉ, D. J., KLOPPEL, T. M., WALTER, V., 1979: In: Glycoconjugates (SCHAUER, R., BOER, P., BUDDECKE, E., KRAMER, M. F., VLIEGENTHART, J. F. G., WIEGANDT, H., eds.), pp. 617 – 618. Stuttgart: G. Thieme.

MORTON, D. L., EILBER, F. R., MALGREM, R. A., WOOD, W. C., 1970: Surgery **68**, 158—164.
MOSHER, D. F., VAHERI, A., CHOATE, J. J., GAHMBERG, C. G., 1979: Blood **53**, 437—445.
MOSS, J., MANGANIELLO, V. C., FISHMAN, P. H., 1977: Biochemistry **16**, 1876—1881.
MÜLLER, E., FRANCO, M. W., SCHAUER, R., 1981: Hoppe-Seyler's Z. Physiol. Chem. **362**, 1615—1620.
MULLIN, B. R., FISHMAN, P. H., LEE, G., ALOJ, S. M., LEDLEY, F. D., WINAND, R. J., KOHN, C. D., BRADY, R. O., 1976: Proc. Natl. Acad. Sci. U.S.A. **73**, 842—846.
NAGAMURA, Y., KOLB, H., 1980: FEBS Lett. **115**, 59—62.
NAUTS, H. C., FOWLER, G. A., BOGATKO, F. H., 1953: Acta Med. Scand. **145**, Suppl. 276, 103.
— 1978: Dev. Biol. Stand. **38**, 487—494.
NEUFELD, E. F., ASHWELL, G., 1980: In: The Biochemistry of Glycoproteins and Proteoglycans (LENNARZ, W. J., ed.), pp. 241—266. New York: Plenum Press.
— — 1980: In: The Biochemistry of Glycoproteins and Proteoglycans (LENNARZ, W. J., ed.), pp. 241—266. New York: Plenum Press.
NG, S., DAIN, J., 1977: J. Neurochem. **29**, 1085.
NICOLAI, VON, H., MÜLLER, H. E., ZILLIKEN, F., 1980: FEBS Lett. **117**, 107—110.
NOONAN, F. P., HALLIDAY, W. J., MORTON, H., CLUNIE, G. J. A., 1979: Nature **278**, 649—651.
NORDT, F. J., FRANCO, M., CORFIELD, A., SCHAUER, R., RUHENSTROTH-BAUER, G., 1981: Blut **42**, 95—98.
NOVOGRODSKY, A., ASHWELL, G., 1977: Proc. Natl. Acad. Sci. U.S.A. **74**, 676.
— SUTHANTHIRAN, M., SALTZ, B., NEWMAN, D., RUBIN, A. L., STENZEL, K. H., 1980: J. Exp. Med. **151**, 755—760.
NURDEN, A. T., CAEN, J. P., 1976: Thromb. Memost. **35**, 139—150.
— — 1978: Br. J. Haemat. **38**, 155—160.
— DUPUIS, D., KUNICKI, T. J., CAEN, J. P., 1981: J. Clin. Invest. **67**, 1431—1440.
NYDEGGER, U. E., FEARON, D. T., AUSTEN, K. F., 1978: Proc. Natl. Acad. Sci. U.S.A. **75**, 6078—6082.
OHTA, N., PARDEE, A. B., MCAUSLAN, B. R., BURGER, M. M., 1968: Biochim. Biophys. Acta **158**, 98—102.
O'KENNEDY, R., SMITH, H., THORNES, R. D., CORRIGAN, A., 1980: Europ. J. Cancer **16**, 1163—1169.
OKUMURA, T., HASITZ, M., JAMIESON, G. A., 1978: J. Biol. Chem. **253**, 3435—3443.
OLD, L. J., 1981: Cancer Res. **41**, 361—375.
ORDINAS, A., MARAGALL, S., CASTILLO, R., NURDEN, A. T., 19**00**: Throm. Res. **13**, 297—302.
ØRSKOV, F., ØRSKOV, L., SUTTON, A., SCHNEERSON, R., LIN, W., EGAN, W., HOFF, G. E., ROBBINS, J. B., 1979: J. Exp. Med. **149**, 669—685.
PACKHAM, M. A., GUCCIONE, M. A., KINLOUGH-RATHBONE, R. L., MUSTARD, J. F., 1980: Blood **56**, 876—880.
PACUSZKA, T., MOSS, J., FISHMAN, P. H., 1978: J. Biol. Chem. **253**, 5103—5108.
PADMANABHAN, M., HEDGE, U. C., RAO, S. S., 1978: Indian J. Med. Res. **67**, 234—238.
PAINTER, R. G., WHITE, A., 1976: Proc. Natl. Acad. Sci. U.S.A. **73**, 837—841.
PALASCAK, J. E., MARTINEZ, J., 1977: J. Clin. Invest. **60**, 89—95.
PARISH, C. R., O'NEILL, H. C., HIGGINS, T. J., 1981: Immunol. Today **2**, 98—101.
PATSCHEKE, H., 1981: Klin. Wschr. **59**, 451—457.
PATSCHEKE, H., BROSSMER, R., WÖRNER, P., 1977: Biochem. Biophys. Res. Commun. **75**, 200—205.
PAULSON, J. C., SADLER, J. E., HILL, R. L., 1979: J. Biol. Chem. **254**, 2120—2124.
PEARLSTEIN, E., SALK, P. L., YOGEESWARAN, G., KARPATKIN, S., 1980: Proc. Natl. Acad. Sci. U.S.A. **77**, 4336—4339.
PENA, S. D., HUGHES, R. C., 1978: Nature (London) **276**, 80—83.
PENA, S. D. J., MILLS, G., HUGHES, R. C., ALPIN, J. D., 1980: Biochem. J. **189**, 337—347.

PENICK, R. J., McCLUER, R. H., 1965: Biochim. Biophys. Acta **106**, 435—438.
PERONA, G., CORTESI, S., XODO, P., SCANDELLARI, C., GHIOTTO, G., DE SANDRE, G., 1964: Acta Isotopica **4**, 287—295.
PETERS, K., RICHARDS, F. M., 1977: Ann. Rev. Biochem. **46**, 523—551.
PETITOU, M., ROSENFELD, C., SINAY, P., 1977: Cancer Immunol. Immunother. **2**, 135—137.
PHILLIPS, D. R., AGIN, P. P., 1977: J. Clin. Invest. **60**, 535—545.
PIGMAN, W., 1977: The Glycoconjugates, Vol. I, pp. 189—213. New York: Academic Press.
PIMM, M. V., COOK, A. J., BALDWIN, R. W., 1978: Eur. J. Cancer **14**, 869—878.
PINCUS, J. H., JAMESON, A. K., BRANDT, A. E., 1981: Cancer Res. **41**, 3082—3086.
PLETCHER, C. H., RESNICK, R. M., WEI, G. J., BLOOMFIELD, V. A., NELSESTUEN, G. L., 1980: J. Biol. Chem. **255**, 7433—7438.
PODOLSAK, B., BRUNSWIG, D., 1976: Klin. Wschr. **54**, 613—617.
POUYSSÉGUR, J., YAMADA, K. M., 1978: Cell **13**, 139—150.
PRICER, W. E., ASHWELL, G., 1976: J. Biol. Chem. **251**, 7539—7544.
PRICER, W. E., jr., ASHWELL, G., 1971: J. Biol. Chem. **246**, 4825—4833.
RAHMANN, H., 1981: Zool. Jb. Physiol. **85**, 209—248.
— HILBIG, R., 1981: J. therm. Biol. **6**, 315—319.
— RÖSNER, H., BREER, H., 1976: J. Theor. Biol. **57**, 231—237.
— — PROBST, W., 1978: Krankenhausarzt **51**, 503—505.
RAJAN, R., IAYARAMAN, S., RAO, S. S., 1980: J. med. Primatol. **9**, 361—367.
RAZ, A., MCLELLAN, W. L., HART, J. R., BUCANA, C. D., HOYER, L. C., SELA, B.-A., DRAGSTEN, P., FIEDLER, J., 1980: Cancer Res. **40**, 1645—1651.
REGOECZI, E., RAYLOR, P., HATTON, M. W., WONG, K.-L., KOJ, A., 1978: Biochem. J. **174**, 171—178.
REUTTER, W., BAUER, CH., 1978: In: Morris Hepatomas: Mechanisms of Regulation (MORRIS, H. P., CRISS, W. E., eds.), pp. 405—437. New York: Plenum Press.
— KREISEL, W., LESCH, R., 1970: Hoppe-Seyler's Z. Physiol. Chem. **351**, 1320.
— LESCH, R., KEPPLER, D., DECKER, K., 1968: Naturwissensch. **55**, 497—498.
— TAUBER, R., VISCHER, P., GRÜNHOLZ, H.-J., HARMS, E., BAUER, CH., 1978: In: Protein Turnover and Lysosome Function, pp. 779—790. New York: Academic Press.
RICHARDS, R. L., MOSS, J., ALVING, C. R., FISHMAN, P. H., BRADY, R. O., 1979: Proc. Natl. Acad. Sci. U.S.A. **76**, 1673—1679.
RIGGS, M. G., INGRAM, V. M., 1977: Biochem. Biophys. Res. Commun. **74**, 191—198.
ROELCKE, D., PRUZANSKI, W., EBERT, W., RÖMER, W., FISCHER, E., LENHARD, V., RAUTERBERG, E., 1980: Blood **55**, 677—681.
ROGENTINE, G. N., DOHERTY, C. M., PINCUS, S. H., 1977: J. Immunol. **119**, 1652—1654.
RONQUIST, G., RIMSTEN, A., WESTMAN, M., CERVÉN, E., 1980: Acta Chir. Scand. **146**, 247—252.
ROSATI, F., DE SANTIS, R., 1980: Nature **283**, 762—764.
ROSENBERG, A., 1979: In: Complex Carbohydrates of Nervous Tissue (MARGOLIS, R. U., MARGOLIS, R. K., eds.), pp. 25—43. New York: Plenum Press.
— SCHENGRUND, C.-L., 1976: Biological Roles of Sialic Acid. New York-London: Plenum Press.
ROSENFELD, C., PICO, J. L., PETITOU, M., SHARIFF, A., CHOQUET, C., DRIANCOURT, C., TUY, F., 1978: In: Immunotherapy of Malignant Diseases, pp. 279—288. Stuttgart-New York: Schattauer.
RÖSNER, H., BREER, H., HILBIG, R., RAHMANN, H., 1979: J. Therm. Biol. **4**, 69—73.
RUPPRECHT, E., HANS, CH., LEONARD, G., DECKER, K., 1976: Biochim. Biophys. Acta **450**, 45—56.
SACHS, L., FOGEL, M., WINCOUR, E., 1959: Nature (London) **183**, 663—664.
SADLER, J. E., PAULSON, J. C., HILL, R. L., 1979: J. Biol. Chem. **254**, 2112—2119.

SAITO, M., SATOH, H., UKITA, 1974: Biochim. Biophys. Acta **362**, 549—557.
SAKARIASSEN, K. S., BOLHUIS, P. A., SIXMA, J. J., 1979: Nature **279**, 636—638.
SANDFORD, B. H., 1967: Transplantation **5**, 1273—1279.
SANSING, W. A., KOLLMORGEN, G. M., 1976: J. Natl. Cancer Inst. **56**, 1113—1118.
SATTLER, J., SCHWARZMANN, G., STAERK, J., ZIEGLER, W., WIEGANDT, H., 1977: Hoppe-Seyler's Z. Physiol. Chem. **358**, 159—163.
SAWAMURA, T., KAWASATO, S., SHIOZAKI, Y., SAMESHIMA, Y., NAKADA, H., TASHIRO, Y., 1981: Gastroenterology **81**, 527—533.
SCHAUER, R., 1982: Adv. Carbohydr. Chem. Biochem. **40**, 131—234.
— VEH, R. W., SANDER, M., CORFIELD, A. P., WIEGANDT, H., 1980: Adv. Exp. Med. Biol. **125**, 283—294.
SCHENGRUND, C.-L., JENSEN, D. S., ROSENBERG, A., 1972: J. Biol. Chem. **247**, 2742—2746.
SCHLEPPER-SCHÄFER, J., KOLB-BACHOFEN, V., KOLB, H., 1980: Biochem. J. **186**, 827—831.
SCHNEIDER, D. R., SEDLACEK, H. H., SEILER, F. R., 1979: In: Glycoconjugates, pp. 354—355. Stuttgart: G. Thieme.
SCHWARTING, G. A., GAJEWSKI, A., 1981: J. Immunol. **126**, 2403—2407.
SCHWARTZ, A. L., FRIDOVICH, S. E., KNOWLES, B. B., LODISH, H. F., 1981a: J. Biol. Chem. **256**, 8878—8881.
— MARSHAK-ROTHSTEIN, A., RUP, D., LODISH, H. F., 1981b: Proc. Natl. Acad. Sci. U.S.A. **78**, 3348—3352.
— RUP, D., LODISH, H. F., 1980: J. Biol. Chem. **255**, 9033—9036.
SCHWICK, H. G., HEIDE, K., HAUPT, H., 1977: In: The Glycoconjugates, Vol. I, pp. 261—321. New York: Academic Press.
SCIALLA, S. J., SPECKART, S. F., HAUT, M. J., KIMBALL, D. B., 1979: Cancer Res. **39**, 2031—2035.
SEDLACEK, H. H., MEESMANN, H., SEILER, F. R., 1975: Int. J. Cancer **15**, 409—416.
— SEILER, F. R., 1974: Behring Inst. Mitt. **55**, 254—257.
— — 1978: Cancer Immunol. Immunother. **5**, 153—163.
— WEISE, M., LEMMER, A., SEILER, F. R., 1979: Cancer Immunol. Immunother. **6**, 47—58.
SEGER, R., JOLLER, P., BAERLOCHER, K., HITZIG, W. H., 1980: Schweiz. med. Wschr. **110**, 1454—1456.
SEILER, F. R., SEDLACEK, H. H., 1978: In: Immunotherapy of Malignant Diseases, pp. 479—488. Stuttgart-New York: Schattauer.
SELA, B.-A., 1981: Eur. J. Immunol. **11**, 347—349.
SEMENZATO, G., SARASIN, P., AMADORI, G., GASPAROTTO, G., 1977: Experientia **33**, 1520—1521.
SETHI, K. K., BRANDIS, H., 1973: Europ. J. Cancer **9**, 809—817.
SHATILL, S. J., BENNETT, J. S., 1981: Ann. Int. Med. **94**, 108—118.
SHIMIZU, S., FUNAKOSHI, I., 1970: Biochim. Acta **203**, 167—169.
SHIU, R. P., POUYSSÉGUR, J., PASTAN, J., 1977: Proc. Natl. Acad. Sci. U.S.A. **74**, 3840—3844.
SIDDIQUI, B., HAKOMORI, S., 1970: Cancer Res. **30**, 2930—2936.
— WHITEHEAD, J. S., KIM, Y. S., 1978: J. Biol. Chem. **253**, 2168—2175.
SIDEBOTTOM, E., 1980: In vitro **16**, 77—86.
SILVER, H. K. B., KAMRIN, K. A., ARCHIBALD, E. L., SALINAS, F. A., 1979: Cancer Res. **39**, 5036—5042.
SIMMONS, R. L., RIOS, A., LUNDGREN, G., 1971a: Surgery **70**, 38—46.
— — RAY, P. K., 1971b: Nature New Biol. **231**, 179—181.
— — 1975: Transplant. Proceed. **VII**, 247—251.
SLIFE, C. W., KUHLENSCHMIDT, M. S., ROSEMAN, S., 1980: In: Communications of Liver Cell (POPPER, H., BIANCHI, L., GUDAT, F., REUTTER, W., eds.), pp. 403—415. London: MTP Press.

Slomiany, B. L., Slomiany, A., Herp, A., 1978: Eur. J. Biochem. **90**, 255—260.
Smith, D. F., Ginsburg, V., 1980: J. Biol. Chem. **255**, 55—59.
Sobue, G., Kosaka, A., 1980: Hepato-Gastroenterol. **27**, 200—203.
Sodetz, J. M., Paulson, J. C., Pizzo, S. V., McKee, P. A., 1978: J. Biol. Chem. **253**, 7202—7206.
— Pizzo, S. V., McKee, P. A., 1977: J. Biol. Chem. **252**, 5538—5546.
Solum, N. O., Hagen, I., Gjemdal, T., 1977: Thromb. Hemost. **38**, 914—923.
— — 1978: Bibliotheca haemat. **45**, 22—27.
Sparks, F. C., Breeding, J. H., 1974: Cancer Res. **34**, 3262—3269.
— Silverstein, M. J., Hunt, J. S., Haskell, C. M., Pilch, Y. H., Morton, D. L., 1973: New Engl. J. Med. **289**, 827—830.
Spence, R. J., Simon, R. M., Baker, A. R., 1978: J. Natl. Cancer Inst. **60**, 451—459.
Spiegel, S., Wilchek, M., 1981: J. Immunol. **127**, 572—575.
Srivastava, P. N., Farooqui, A. A., 1980: Biol. Reprod. **22**, 858—863.
Steer, C. J., Ashwell, G., 1980: J. Biol. Chem. **255**, 3008—3013.
— Clarenburg, R., 1979: J. Biol. Chem. **254**, 4457—4461.
Stewart, S. E., Eddy, B. E., Haas, V. M., Borgese, N. G., 1957: Ann. N.Y. Acad. Sci. **68**, 419—429.
Stewart, W. B., Petenyi, C. W., Rose, H. M., 1955: Blood **10**, 228—234.
Suzuki, A., Ishizuka, I., Yamakawa, T., 1975: J. Biochem. (Tokyo) **78**, 947—954.
Svennerholm, L., 1963: J. Neurochem. **10**, 613—623.
— 1980a: In: Structure and Function of Gangliosides (Svennerholm, L., Mandel, P., Dreyfus, H., Urban, P.-F., eds.), pp. 533—544. New York: Plenum Press.
— 1980b: In: Cholera and Related Diarrheas, pp. 80—87. Basel: Karger.
Tanabe, T., Pricer, W. E., Ashwell, G., 1979: J. Biol. Chem. **254**, 1038—1043.
Tarutani, O., Schulman, S., 1971: Biochim. Biophys. Acta **236**, 384—390.
Tate, R. L., Holmes, J. M., Kohn, L. D., Winand, R. J., 1975: J. Biol. Chem. **250**, 6527—6533.
Tatumi, K., Suzuki, Y., Sinohara, H., 1979: Biochim. Biophys. Acta **583**, 504—511.
Tauber, R., Park, Ch.-S., Hofmann, W., Reutter, W., 1982: Eur. J. Cell Biol. **27**, 31.
Tettamanti, G., Bertona, L., Zambotti, V., 1964: Biochim. Biophys. Acta **84**, 756—758.
Tishkoff, G. H., 1966: Blood **28**, 229—240.
Tolleshaug, H., Berg, T., 1979: Biochem. Pharmacol. **28**, 2919—2922.
— — Holte, K., 1980: Eur. J. Cell Biol. **23**, 104—109.
— 1981: Int. J. Biochem. **13**, 45—51.
— Chindemi, P. A., Regoeczi, E., 1981: J. Biol. Chem. **256**, 6526—6528.
— Berg, T., Nilsson, M., Norum, K. R., 1977: Biochim. Biophys. Acta **499**, 73—84.
Tonelli, Q., Meints, R. H., 1977: Science **195**, 897—898.
Vaheri, A., Mosher, D. F., 1978: Biochim. Biophys. Acta **516**, 1—25.
— Ruoslahti, E., Nordling, S., 1972: Nature New Biol. **238**, 211—213.
— Vartio, T., Stenman, S., Saksela, O., Hedman, K., Alitalc, K., 1980: Proteinases and Tumor Invasion (Sträuli, P., *et al.*, eds.), pp. 49—57. New York: Raven Press.
Vasiliev, J. M., Gelfand, J. M., 1977: Int. Rev. Cytol. **50**, 159—274.
Veh, R. W., Schauer, R., 1978: In: Enzymes of Lipid Metabolism (Gatt, S., Freysz, L., Mandel, P., eds.), pp. 447—462. New York: Plenum Press.
Vengris, V. E., Reynolds, F. H., jr., Hollenberg, M. D., Pitha, P. M., 1976: Virology **72**, 486—493.
Verbert, A., Cacan, R., Debeire, P., Montreuil, J., 1977: FEBS Lett. **74**, 234—2238.
Vilarem, M. J., Jouanneau, J., Bourillon, R., 1981: Biochem. Biophys. Res. Commun. **98**, 7—14.
Vischer, P., Hughes, R. C., 1979: In: Glycoconjugates, pp. 283—284. Stuttgart: G. Thieme.

VISCHER, P., REUTTER, W., 1978: Eur. J. Biochem. **84**, 363—368.
VISSER, A., EMMELOT, P., 1973: J. Membrane Biol. **14**, 73—84.
VOGLER, W. R., 1980: Cancer Immunol. Immunother. **9**, 15—21.
VOLK, B., KREISEL, W., GEROK, W., REUTTER, W., 1982: FEBS Letters, submitted.
WADA, H. G., GÓRNICKI, S. Z., SUSSMAN, H. H., 1977: J. Supramolec. Struct. **6**, 473—484.
WALL, D. A., WILSON, G., HUBBARD, A. L., 1980: Cell **21**, 79—93.
WALTER, H., KORB, E. F., TAMBLYN, C. H., SEAMAN, G. V. F., 1980: Biochem. Biophys. Res. Commun. **97**, 107—113.
WARREN, L., FUHRER, J. P., BUCK, C. A., 1972: Proc. Natl. Acad. Sci. U.S.A. **69**, 1838—1842.
WATKINS, E., jr., 1974: Behring Inst. Mitt. **55**, 355—370.
WEIGEL, P. H., 1980: J. Biol. Chem. **255**, 6111—6120.
— OKA, J. A., 1981: J. Biol. Chem. **256**, 2615—2617.
— SCHNAAR, R. L., KUHLENSCHMIDT, M. S., SCHMELL, E., LEE, R. T., ROSEMAN, S., 1979: J. Biol. Chem. **254**, 10830—10838.
WEIR, D. M., 1980: Immunology today, August 1980, 45—51.
WEISS, L., 1973: J. Natl. Cancer Inst. **50**, 3—19.
— 1967: The Cell Periphery, Metastasis and Other Contact Phenomena. Amsterdam: North-Holland.
WHYTE, A., LOKE, Y. W., 1978: J. Exp. Med. **148**, 1087—1092.
WILSON, R. E., SONIS, S. T., GODRICK, E. A., 1974: Behring Inst. Mitt. **55**, 334—342.
WINKELHAKE, J. L., KUNICKI, T. J., ELCOMBE, B. M., ASTER, R. H., 1980: J. Biol. Chem. **255**, 2822—2828.
WINZLER, R. J., 1955: Methods Biochem. Anal. **2**, 279—311.
WISHER, M. H., EVANS, W. H., 1975: Biochem. J. **146**, 375—388.
WOODRUFF, J. J., GESNER, B. M., 1969: J. Exptl. Med. **129**, 551—567.
— 1974: Cell Immunol. **13**, 378—384.
— WOODRUFF, J. F., 1976: J. Immunol. **117**, 852—858.
WU, H. C., MEEZAN, E., BLACK, P. H., ROBBINS, P. W., 1969: Biochemistry **8**, 2509—2517.
WU, K. K., KU, C. S. L., 1978: Thromb. Res. **13**, 183—192.
YAARI, A., 1969: Blood **33**, 159—163.
YACHNIN, S., GARDNER, P. H., 1961: Br. J. Haematol. **7**, 464—490.
YAMADA, K. M., OLDEN, K., 1978: Nature (London) **275**, 179—184.
YASUE, S., HANDA, S., MIYAGAWA, S., INOUE, J., HASEGAWA, A., YAMAKAWA, T., 1978: J. Biochem. **83**, 1101—1107.
YOGEESWARAN, G., STEIN, B. S., SEBASTIAN, H., 1978: Cancer Res. **38**, 1336—1344.
— SHEININ, R., WHERETT, J. R., MURRAY, R. K., 1972: J. Biol. Chem. **247**, 5146—5158.
YOSHIDA, A., YAMAGUCHI, Y. F., DAVÉ, V., 1979: Blood **54**, 344—350.
ZIMMERMAN, E. F., VOORTING-HAWKIN, M., MICHAEL, J. G., 1977: Nature **265**, 354—356.

K. Sialidoses

MICHAEL CANTZ

Department of Pathochemistry, University of Heidelberg, Heidelberg, Federal Republic of Germany

With 1 Figure

Contents

I. Introduction

The term sialidosis has been introduced recently to denote a group of patients with an inherited defect in the catabolism of sialic acid-containing oligosaccharides and glycoproteins. Sialidosis, therefore, belongs to the category of the so-called oligosaccharidoses (MAROTEAUX and HUMBEL 1976), metabolic diseases which are characterized by an excessive accumulation of glycoprotein-derived oligosaccharides.

In a patient classified as mucolipidosis I, CANTZ *et al.* (1977) and SPRANGER *et al.* (1977) demonstrated a severe deficiency of an "acid" neuraminidase (sialidase; N-acetyl neuraminic acid hydrolase, E.C. 3.2.1.18) in his cultured fibroblasts. The patient had a neurodegenerative disorder with myoclonus, skeletal changes like in Hurler disease, and cherry-red spots in the maculae of his eyes. In addition to the neuraminidase defect, the fibroblasts of the patient accumulated abnormal amounts of sialic acid-containing compounds. The patient excreted excessive quantities of sialyloligosaccharides in the urine (MICHALSKI *et al.* 1977). Fibroblasts from the parents of another such patient had activities of neuraminidase which were intermediate between patients and controls (CANTZ and

MESSER 1979). This gene dosage effect indicated that the neuraminidase deficiency was primary. A similar neuraminidase deficiency was demonstrated in an infant with clinical symptoms suggestive of mucolipidosis I (KELLY and GRAETZ 1977) and in patients with "nephrosialidosis", a combination of the symptoms of mucolipidosis I with renal disease (MAROTEAUX *et al.* 1978 a). In these patients, there was likewise a massive excretion of sialyloligosaccharides in the urine.

Besides this group of dysmorphic patients there is another type of sialidosis with normal physical appearance, absent or mild mental retardation, and manifestation in adolescence. DURAND *et al.* (1977), O'BRIEN (1977), THOMAS *et al.* (1978) and RAPIN *et al.* (1978) showed that patients, who presented mainly with ophthalmological problems, or with the so-called cherry-red spot—myoclonus syndrome, had a profoundly diminished neuraminidase activity and excreted abnormal amounts of sialyloligosaccharides in their urine.

In addition to these two types of sialidosis, there are patients with a combined defect of neuraminidase and β-galactosidase (WENGER *et al.* 1978, ANDRIA *et al.* 1978, OKADA *et al.* 1979), possibly caused by a common defect in the biosynthetic processing of the two enzymes (HOOGEVEEN *et al.* 1980). A neuraminidase deficiency has also been observed in patients with mucolipidosis II (I-cell disease) and mucolipidosis III (STRECKER *et al.* 1976, THOMAS *et al.* 1976). In these disorders, however, the neuraminidase deficiency is but one of many lysosomal hydrolase deficiencies, presumably due to a defect in the proper compartmentalization of these enzymes (NEUFELD 1974).

In this article, emphasis will be given to the description of the biochemical basis of the sialidoses, whereas the clinical features will be treated relatively briefly. It will be seen that the biochemical findings not only are of help in explaining the pathogenesis of the sialidoses but in addition provide insight into the metabolism of sialylated compounds, and into the genetics and specificity of mammalian neuraminidases. For a more detailed discussion of clinical phenotypes the reader is referred to the review of LOWDEN and O'BRIEN (1979).

II. Clinical Manifestations and Classification

As mentioned in the preceding chapter, sialidosis seems to manifest itself in two principal phenotypes: a severe form resembling a mucopolysaccharidosis, and a milder form with mainly neurological symptoms.

In 1968, BÉRARD and coworkers and SPRANGER *et al.* described patients who exhibited a clinical picture resembling Hurler disease with coarse facial features, skeletal deformities, short stature, mental retardation, cherry-red macular spots, and neurological problems, yet who excreted normal amounts of mucopolysaccharides in the urine. Histological and ultrastructural studies revealed that neuronal, visceral and mesenchymal tissues accumulated within lysosomes abnormal amounts of compounds thought to consist of acid mucopolysaccharides and glycolipids (BÉRARD *et al.* 1968, SPRANGER *et al.* 1968, FREITAG *et al.* 1971). These patients were therefore classified under the name "lipomucopolysaccharidosis", but were later reclassified, together with morphologically similar disorders, as mucolipidosis I by SPRANGER and WIEDEMANN (1970). A further patient with the clinical symptoms of mucolipidosis I was identified as a neuraminidase deficiency disease and thus recognized as a sialidosis (CANTZ *et al.* 1977, SPRANGER *et al.* 1977). Of the two original patients of SPRANGER *et al.* (1968), patient no. 1 (R.H.) had died at the age of 21 years before the diagnosis sialidosis was made. However, his cultured fibroblasts were retrieved from the freezer and, after reculturing, were found to be deficient in neuraminidase activity, whereas fibroblasts of his parents showed intermediate neuraminidase activities (CANTZ and MESSER 1979). Patient no. 2 (H.D.) of SPRANGER *et al.* (1968), on the other

Table 1. *Major clinical findings in the sialidoses*

	Sialidosis		Combined β-galactosidase/ neuraminidase deficiency
	mild form[1]	severe form[1]	
Age of onset	8-29 years	0-2 years	2-10 years
Hurler phenotype	—	++	++
Skeletal dysplasia	—	++	++
Mental retardation	±	++	±
Renal disease	—	±	—
Hepatosplenomegaly	—	++	
Cherry-red macular spots	++	++	++
Fine corneal opacities	±	+	+
Myoclonus	+	+	+
Seizures	±	+	±
Course	slowly progressive	death between 2-21 years of age	progressive

[1] The mild and severe forms correspond to types 1 and 2 (infantile onset), respectively, of the classification of LOWDEN and O'BRIEN (1979).

++ present; + mostly present; ± occasionally present; — absent.

hand, turned out to have a deficiency of α-mannosidase rather than of neuraminidase, i.e. he had a mannosidosis (GEHLER *et al.* 1975). Interestingly, this patient did not exhibit a cherry-red macular spot, pointing to the importance of this symptom as a diagnostic hallmark of sialidosis. Patients with the clinical features of mucolipidosis I and primary neuraminidase deficiency were also described by MAROTEAUX *et al.* (1978b) and by WINTER *et al.* (1980). KELLY and GRAETZ (1977) described an eight months old infant who presented symptoms which may develop into the clinical picture of mucolipidosis I and who had a complete deficiency of neuraminidase activity in his cultured fibroblasts. MAROTEAUX *et al.* (1978a) published two patients who, in addition to features of mucolipidosis I, had a renal disease with renal insufficiency and were therefore classified under the name of "nephrosialidosis". A nephrotic syndrome in conjunction with congenital ascites, pericardial effusion, organomegaly, skeletal changes and delayed development was also observed in an infant boy who had a neuraminidase deficiency in his cultured fibroblasts, and who died at 22 months of age (AYLSWORTH *et al.* 1980); although this child's fibroblasts showed a β-galactosidase activity of 60% of normal, this reduction seems insufficient to warrant a classification as a combined neuraminidase/β-galactosidase deficiency. The main clinical symptoms of all these patients with the severe form of sialidosis are summarized in Table 1.

Prior to the elucidation of the enzymatic defect, some patients with the mild form of sialidosis had been classified as cherry-red spot—myoclonus syndrome. DURAND *et al.* (1977) described 22- and 13-year-old siblings who exhibited mainly ophthalmological problems including a progressive reduction of visual acuity, colour blindness, bilateral cherry-red macular spots, punctate opacities of the lens, and minimal neurological symptoms. There was a drastic reduction of neuraminidase activity in their leukocytes and cultured fibroblasts. O'BRIEN (1977), THOMAS *et al.* (1978) and RAPIN *et al.* (1978) published patients who presented at 8–29 years of age with decreasing visual acuity, myoclonus, cherry-red spots, occasionally punctate opacities of the lens, but with normal growth and skeletal system, normal facies, and no mental retardation. About half of the patients had seizures of the grand mal type followed by loss of consciousness. In cultured fibroblasts derived from these patients there was a marked decrease of neuraminidase activity. This deficiency, however, seems not as complete as in the severe form of sialidosis (O'BRIEN and WARNER 1980), offering a possible explanation for the milder course of this disease type. The main clinical features of the mild form of sialidosis are summarized in Table 1.

Apart from the mild and severe forms of primary neuraminidase deficiency there is a group of patients with a clinical picture resembling sialidosis, yet who exhibit a deficiency of β-galactosidase activity in a variety of tissues (GOLDBERG *et al.* 1971, ORII *et al.* 1972, YAMAMOTO *et al.* 1974, LOONEN *et al.* 1974). Such patients, sometimes classified as the Goldberg syndrome, were later found to exhibit a deficiency of neuraminidase in addition to that of β-galactosidase (WENGER *et al.* 1978, ANDRIA *et al.* 1978, OKADA *et al.* 1979, THOMAS *et al.* 1979, MAIRE and NIVELON-CHEVALLIER 1981). Although it was quite clear that the β-galactosidase deficiency was secondary, since obligate heterozygotes had a normal (instead of intermediate) β-galactosidase activity, it is at present doubtful whether neuraminidase is the gene product primarily affected. Thus, HOOGEVEEN *et al.* (1980) found genetic complementation in somatic cell hybrids derived from fibroblast cultures of combined β-galactosidase/neuraminidase deficiency and "classical" sialidosis and, in additional experiments discussed below, provided evidence that the combined deficiency may be due to defective posttranslational modification of these enzymes. Therefore, these patients will not be classified as a sialidosis proper, in contrast to the review of LOWDEN and O'BRIEN (1979), but will tentatively be labelled "combined β-galactosidase/neuraminidase deficiency" (HOOGEVEEN *et al.* 1980) until a more appropriate name is available. Such classification seems also justified on clinical grounds, since patients with the combined deficiency exhibit a phenotype intermediate between the mild and severe forms of sialidosis (Table 1).

In fact, all of the patients classified by LOWDEN and O'BRIEN (1979) as sialidosis type 2, juvenile onset form, exhibited a β-galactosidase deficiency when it was looked for.

III. The Metabolic Defect

1. Storage Material

Patients with sialidosis store in their tissues and excrete in the urine excessive amounts of sialic acid-containing compounds. As shown in Table 2, there is a 3- to 14-fold elevation in total sialic acid content in patients with the severe form of sialidosis as compared to healthy individuals. More than 80% of this sialic acid is in the bound form. An increase in "bound" sialic acid was also observed by other investigators in fibroblasts, leukocytes, and urine in patients with both the severe

Table 2. *Sialic acid content of tissues*
Total sialic acid content of tissue extracts was determined as described (CANTZ *et al.* 1977) and is expressed as nM Neu5Ac/mg protein

	Sialidosis, severe form	Normal controls		
		mean value	range	(n)
Cultured fibroblasts	16.4[1]	3.3	1.7-5.6	(7)
Leukocytes	9.2[1]	3.0	2.1-3.4	(5)
Liver	32.9[2]	3.7		(1)
Urine	3.39[1,3]	0.25[3]		(1)

[1] Patient D.F.; [2] patient R.H.; [3] μM Neu5Ac/mg creatinine.

and mild forms of sialidosis (KELLY and GRAETZ 1977, DURAND *et al.* 1977) as well as in patients with the combined β-galactosidase/neuraminidase deficiency (WENGER *et al.* 1978, OKADA *et al.* 1979, THOMAS *et al.* 1979, MIYATAKE *et al.* 1979).

So far, there is little information on the chemical nature of the sialyl compounds stored in sialidosis tissues. In brain tissue of patients with the severe form, BÉRARD *et al.* (1968) found a normal ganglioside content and pattern, whereas PALLMANN and colleagues (1980) found somewhat increased amounts of certain gangliosides, particularly of the gangliosides GM_1, GD_3 and GT_{1b}. Other biochemical investigations on tissue concentrations of defined sialyl compounds have not been reported.

There is, however, a remarkable knowledge of the detailed chemical structures of the sialic acid-containing oligosaccharides excreted in the patient's urine. Thus, STRECKER *et al.* (1977), MICHALSKI *et al.* (1977) and STRECKER (1980) characterized a total of 20 different sialyloligosaccharides found in the urine of both severe and mild cases of sialidosis. When compared to the normal, the excretion of these compounds was increased about 80- to 800-fold. The complete structures of the 10 major sialyloligosaccharides excreted by a patient with the severe form are shown

```
I     α-Neu5Ac-(2 → 3)-β-Gal-(1 → 4)-β-GlcNAc-(1 → 2)-α-Man-(1 → 3)-β-Man-(1 → 4)-GlcNAc

II    α-Neu5Ac-(2 → 6)-β-Gal-(1 → 4)-β-GlcNAc-(1 → 2)-α-Man-(1 → 3)-β-Man-(1 → 4)-GlcNAc

III   α-Neu5Ac-(2 → 6)-β-Gal-(1 → 4)-β-GlcNAc-(1 → 2)-α-Man-(1 → 3)
                                                                  \
                                                                   β-Man-(1 → 4)-GlcNAc
                                                                  /
                                                      α-Man-(1 → 6)

IV    α-Neu5Ac-(2 → 3)-β-Gal-(1 → 4)-β-GlcNAc-(1 → 2)-α-Man-(1 → 3)
                                                                  \
                                                                   β-Man-(1 → 4)-GlcNAc
                                                                  /
                        β-Gal-(1 → 4)-β-GlcNAc-(1 → 2)-α-Man-(1 → 6)

V     α-Neu5Ac-(2 → 6)-β-Gal-(1 → 4)-β-GlcNAc-(1 → 2)-α-Man-(1 → 3)
                                                                  \
                                                                   β-Man-(1 → 4)-GlcNAc
                                                                  /
                        β-Gal-(1 → 4)-β-GlcNAc-(1 → 2)-α-Man-(1 → 6)

VI    α-Neu5Ac-(2 → 6)-β-Gal-(1 → 4)-β-GlcNAc-(1 → 2)
                                                     \
                                                      α-Man-(1 → 3)-β-Man-(1 → 4)-GlcNAc
                                                     /
      α-Neu5Ac-(2 → 3)-β-Gal-(1 → 4)-β-GlcNAc-(1 → 4)

VII   α-Neu5Ac-(2 → 3)-β-Gal-(1 → 4)-β-GlcNAc-(1 → 2)-α-Man-(1 → 3)
                                                                  \
                                                                   β-Man-(1 → 4)-GlcNAc
                                                                  /
      α-Neu5Ac-(2 → 3)-β-Gal-(1 → 4)-β-GlcNAc-(1 → 2)-α-Man-(1 → 6)

VIII  α-Neu5Ac-(2 → 3)-β-Gal-(1 → 4)-β-GlcNAc-(1 → 2)-α-Man-(1 → 3)
                                                                  \
                                                                   β-Man-(1 → 4)-GlcNAc
                                                                  /
      α-Neu5Ac-(2 → 6)-β-Gal-(1 → 4)-β-GlcNAc-(1 → 2)-α-Man-(1 → 6)

IX    α-Neu5Ac-(2 → 6)-β-Gal-(1 → 4)-β-GlcNAc-(1 → 2)-α-Man-(1 → 3)
                                                                  \
                                                                   β-Man-(1 → 4)-GlcNAc
                                                                  /
      α-Neu5Ac-(2 → 6)-β-Gal-(1 → 4)-β-GlcNAc-(1 → 2)-α-Man-(1 → 6)

X     α-Neu5Ac-(2 → 3)-β-Gal-(1 → 4)-β-GlcNAc-(1 → 2 or 4)                    1
                                                          \
                                                           α-Man-(1 → 3 or 6)
                                                          /                  \
      α-Neu5Ac-(2 → 6)-β-Gal-(1 → 4)-β-GlcNAc-(1 → 4 or 2)                    β-Man-(1 → 4)-GlcNAc
                                                                             /
      α-Neu5Ac-(2 → 6)-β-Gal-(1 → 4)-β-GlcNAc-(1 → 2)-α-Man-(1 → 6 or 3)
```

Fig. 1. Structures of the 10 major sialyloligosaccharides excreted in the urine of a patient with the severe form of sialidosis. From MICHALSKI *et al.* (1977).

in Fig. 1. They closely resemble the glycan portions of the complex type found in many glycoproteins. Invariably, the N-acetylneuraminic acid residues are located at a non-reducing terminus, linked either α(2-3) or α(2-6) to galactose. The reducing sugar residue is always N-acetylglucosamine, suggesting that each saccharide results from an endo-β-acetylglucosaminidase acting on the chitobiosyl linkage of the parent sialoglycopeptides. It is rather likely that sialyloligosaccharides of similar structures are a major part of the storage substances observed within the lysosomes of the various tissues of sialidosis patients.

2. The Neuraminidase Defect

The deficiency of a neuraminidase as the basis of sialidosis was first suggested by Cantz *et al.* (1977) and by Spranger *et al.* (1977) from their observations of a nearly complete lack of this enzyme activity in conjunction with excessive storage of sialoglycoconjugates in fibroblasts cultured from the skin of a patient with mucolipidosis I. Independently, similar observations were published by Kelly and Graetz (1977) and Maroteaux *et al.* (1978 b). Durand and colleagues (1977) and O'Brien (1977) found a neuraminidase deficiency and an increased urinary excretion of sialyloligosaccharides in adolescents and young adults who exhibited mainly ophthalmological symptoms, or the cherry-red spot—myoclonus syndrome. The substrates used to demonstrate the neuraminidase deficiency included oligosaccharides like sialyllactose and sialyl-hexasaccharides, the glycoprotein fetuin, and the synthetic compound 3′-methoxyphenyl-N-acetyl-neuraminic acid.

In mammalian tissues, sialic acids are constituents of glycoproteins and glycolipids (gangliosides). They usually occupy a terminal non-reducing position on the carbohydrate side chains and are linked to the penultimate sugar via α(2-3), α(2-6) and sometimes α(2-8) glycosidic bonds, the latter to another sialic acid residue in the chain (see chapters B, G of this book). The catabolism of these sialoglycoconjugates is thought to occur in lysosomes through the action of neuraminidases and other hydrolases. Relatively little was known of the substrate specificity of the neuraminidases. Thus, it was an open question if there are different neuraminidases for glycoprotein and ganglioside substrates, or if the different sialosyl linkages are cleaved by enzymes of distinct specificity. The investigation of the specificity of the neuraminidase deficiency in the sialidoses was therefore of interest for both, an understanding of the pathogenesis of the diseases as well as the specificity and genetics of human neuraminidases.

Strecker *et al.* (1977) and Durand *et al.* (1977) noted that among the urinary sialyloligosaccharides excreted by patients with both severe and mild forms of sialidosis, about 80% of the sialosyl linkages were of the α(2-6) type, the remainder being of the α(2-3) type. This, and the finding in leukocytes and fibroblasts of a neuraminidase deficiency towards sialyloligosaccharides with α(2-6) sialosyl linkage, but normal activity towards an α(2-3) sialyloligosaccharide, led these authors to postulate the existence of two different neuraminidases, of which only the α(2-6) neuraminidase would be deficient in sialidosis. The α(2-3) neuraminidase, on the other hand, was claimed to be the primarily deficient enzyme in mucolipidosis II (Strecker and Michalski 1978). Such an explanation must be regarded with skepticism, however, because other authors reported different findings. O'BRIEN (1977) found a markedly deficient neuraminidase activity towards both α(2-3) and α(2-6) sialosyl linkages of the oligosaccharide substrates in patients with the mild form of sialidosis, and FRISCH and NEUFELD (1979) and CANTZ and MESSER (1979) observed a nearly complete lack of activity towards both kinds of linkages of the sialyllactitol or sialyl-hexasaccharide substrates in patients with the severe and mild forms of sialidosis. It therefore appears more likely that both α(2-3) and α(2-6) oligosaccharide neuraminidase activities are the properties of a single enzyme. In some patients, there may be higher residual activity towards the α(2-3) isomer (LOWDEN and O'BRIEN 1979), perhaps explaining the findings mentioned above. That the activities towards both kinds of linkages are the properties of one single enzyme was also suggested by the observation of identical thermal inactivation curves for each activity (CANTZ *et al.* 1981). The same kind of oligosaccharide neuraminidase deficiency was observed in patients with the combined β-galactosidase/neuraminidase deficiency (WENGER *et al.* 1978, OKADA *et al.* 1979). Using the synthetic substrate 4-methylumbelliferyl-α-D-N-acetylneuraminic acid, O'BRIEN and WARNER (1980) found about ten times higher residual neuraminidase activities in cultured fibroblasts from patients with mild as compared to the severe form, suggesting a correlation between clinical phenotype and the severity of enzymatic deficiency. In two Italian patients with mild sialidosis, the same authors observed a K_m for the residual neuraminidase activity of one sixth of normal, whereas in other mild cases they found a normal value.

In contrast to the findings with the oligosaccharide or synthetic substrates, cultured fibroblasts from patients with sialidosis showed essentially normal neuraminidase activities towards ganglioside substrates. The gangliosides used in these studies included GM_3 and GD_{1a} with terminal α(2-3) sialosyl linkage, and GD_3 with terminal α(2-8) sialosyl linkage (CANTZ and MESSER 1979, CANTZ 1980, CAIMI *et al.* 1979, WENGER *et al.* 1978, OKADA *et al.* 1979). These results provide genetic evidence for the existence of separate neuraminidases with specificity for either oligosaccharides and glycoproteins, or for gangliosides, respectively. They are in line with recent observations of SANDER *et al.* (1979) of separate glycoprotein- and glycolipid-specific neuraminidase activities in horse liver (see chapter I).

3. *Pathogenesis*

In both the severe and mild forms of sialidosis there is a deficiency of a single neuraminidase specific for α(2-3) and α(2-6) sialosyl linkages of oligosaccharides and glycoproteins. By analogy with other lysosomal storage diseases, it may be assumed that the deficient neuraminidase is of lysosomal localization. Indeed, preliminary evidence suggests that the oligosaccharide neuraminidase activity of cultured fibroblasts shows a subcellular distribution pattern which closely resembles that of the lysosomal marker enzyme β-N-acetylhexosaminidase (CANTZ *et al.* 1981). Since this neuraminidase has a role in cleaving neuraminic acid residues from sialylated oligosaccharides and glycoproteins, its deficiency leads to the lysosomal accumulation and eventual excretion in the urine of oligosaccharides like those shown in Fig. 1. In the absence of neuraminidase, these oligosaccharides cannot be degraded to their constituent monosaccharides because all of the other glycosidases are exo-enzymes, cleaving "their" sugar residue only after it has been exposed on the non-reducing end of the saccharide chain.

Evidence for an impaired catabolism of sialylated glycoproteins comes from electrophoretic studies showing increased anodal mobilities of lysosomal and other enzymes known to be glycoproteins in cultured fibroblasts and in leukocytes of patients with sialidosis (SWALLOW *et al.* 1979).

On the other hand, the catabolism of the other class of sialylated compounds, namely the gangliosides, seems not impaired, as there is no evidence of ganglioside storage. Moreover, the neuraminidase activity towards various ganglioside substrates was normal in the patients' fibroblasts.

In the severe and mild forms of sialidosis, the deficient oligosaccharide neuraminidase is very likely to be the primary product of the mutant gene. This assumption is based on the observation of a gene dosage effect, i.e., intermediate neuraminidase activities in obligate heterozygotes (such as parents), and on the finding that no other lysosomal hydrolases seem to be involved. The existence of both severe and mild forms within the same enzyme deficiency disease is not uncommon among the lysosomal storage diseases, one of the best known examples being HURLER and SCHEIE diseases (mucopolysaccharidoses I-H and I-S), which are both due to the deficiency of α-L-iduronidase (McKUSICK *et al.* 1978). A plausible explanation of this phenomenon is allelic mutations leading to enzyme deficiency states with different residual activities and, as a consequence, varying degrees of clinical severity. Experimental support for this assumption comes from studies using somatic cell hybridization techniques. HOOGEVEEN *et al.* (1980) found that the fusions between fibroblasts from patients with severe sialidosis and those with the mild form resulted in no genetic complementation, whereas complementation was observed in hybrids between severe sialidosis and the combined β-galactosidase/neuraminidase deficiency. It was therefore concluded that the severe and mild forms of sialidosis result from mutations within the same gene, and that the combined β-galactosidase/neuraminidase deficiency, on the other hand, is due to the mutation of a different gene. This latter point, however, is in contradiction to the findings of KATO *et al.* (1979), who observed no genetic complementation between sialidosis (mucolipidosis I) and their case of combined β-galactosidase/neuraminidase deficiency.

The pathogenetic mechanism in the combined β-galactosidase/neuraminidase deficiency is at present not understood. Some investigators (WENGER *et al.* 1978, OKADA *et al.* 1979) suggest that the deficiency of β-galactosidase in this disorder is a secondary consequence of the neuraminidase defect, because the β-galactosidase activity was reduced only in some, but not in all tissues, and because obligate heterozygotes showed normal β-galactosidase, but intermediate neuraminidase activities. HOOGEVEEN *et al.* (1980), on the other hand, believe that both β-galactosidase and neuraminidase deficiencies are due to the primary defect of a "factor" involved in the posttranslational processing of the two enzymes. Evidence in favor of this hypothesis comes from their cell hybridization studies described above, and from their experiments showing the presence of a "correction factor" for the neuraminidase deficiency in mixed cultures of sialidosis and combined β-galactosidase/neuraminidase deficiency fibroblasts. This factor was found also in conditioned medium of sialidosis and normal fibroblasts. It seems to be a glycoprotein but not neuraminidase itself. This factor, when added to the culture medium of β-galactosidase/neuraminidase deficient cells, leads to a marked increase in the intracellular neuraminidase activity. It is believed to function in the posttranslational biosynthetic processing of neuraminidase and β-galactosidase. Whatever the precise pathogenetic mechanism, the combined β-galactosidase/neuraminidase deficiency seems sufficiently distinct biochemically and genetically to warrant its classification as a separate entity.

In all of the sialidoses it is at present not understood how the intralysosomal accumulation of sialylated compounds causes the often fatal clinical consequences.

IV. Other Disorders with Neuraminidase Deficiency

Patients with mucolipidosis II (I-cell disease) or mucolipidosis III (pseudo-HURLER polydystrophy) exhibit a deficiency of many lysosomal hydrolases in cultured fibroblasts (and presumably other connective tissue cells), whereas the same enzymes are greatly elevated in extracellular fluids like serum, urine or culture medium (reviewed by MCKUSICK *et al.* 1978). The finding of a neuraminidase deficiency in leukocytes and cultured fibroblasts (STRECKER *et al.* 1976, THOMAS *et al.* 1976, CANTZ *et al.* 1977), and of an excessive urinary excretion of sialyloligosaccharides with a high proportion of α(2-3) linkages (STRECKER *et al.* 1977) prompted STRECKER and MICHALSKI (1978) to speculate that the basic defect in these patients consisted of a deficiency of a neuraminidase specific for α(2-3) sialosyl linkages. This neuraminidase was thought to function in the posttranslational modification of lysosomal enzymes. This seems not so, however, since it was recently shown that the lysosomal enzymes in these disorders are lacking a phosphorylated marker required for their transport to the lysosomal compartment via receptors on plasma- and intracellular membranes (BACH *et al.* 1979a, HASILIK and NEUFELD 1980, HASILIK *et al.* 1981). The defective marker on lysosomal enzymes, in turn, is due to the deficiency of a N-acetylglucosamine-l-phosphate transferase (HASILIK *et al.* 1981, 1982, WAHEED *et al.* 1982). This transferase deficiency is thus the primary biochemical lesion in mucolipidoses II and III, and the neuraminidase deficiency a secondary consequence of this defect, just as the deficiencies of the other hydrolases.

A deficiency of a neuraminidase specific for gangliosides has been found in cultured fibroblasts and amniotic cells from patients with mucolipidosis IV (BACH *et al.* 1979b, 1980). This deficiency is only partial and is thought to involve a lysosomal ganglioside neuraminidase, whereas a non-lysosomal enzyme of similar specificity is partially masking this defect. Cultured fibroblasts accumulate abnormal amounts of gangliosides GM_3 and GD_3 (BACH *et al.* 1975) and, curiously, acid mucopolysaccharides, but no sialyloligosaccharides (BACH *et al.* 1977). Since obligate heterozygotes exhibit intermediate activities, it is likely that the ganglioside neuraminidase deficiency is the primary defect in this disease.

A neuraminidase deficiency has also been observed in a strain of mice (designated SM/J) by POTIER *et al.* (1979a) using the synthetic substrate 4-methylumbelliferyl-α-D-N-acetylneuraminic acid. This deficiency is confined almost exclusively to the liver, other organs showing normal or only slightly diminished activity. In normal mouse liver, a major part of the neuraminidase activity is rather labile, whereas a smaller portion is more stable upon preincubation at 0 °C. The deficiency in SM/J mice seems to involve the labile component. There is no increase in the sialic acid content of the liver. For these reasons, it is uncertain whether this mouse strain can serve as an animal model of human neuraminidase deficiency.

V. Laboratory Diagnosis

When a sialidosis is suspected on the basis of the clinical phenotype, the diagnosis may be ascertained by the demonstration of an excessive urinary excretion of sialyloligosaccharides. This can be done using thin-layer chromatography screening procedures (HOLMES and O'BRIEN 1979, SEWELL 1980), or directly by quantitating the amount of bound sialic acid (OKADA *et al.* 1978). For a final diagnosis, however, it is necessary to demonstrate a deficiency of neuraminidase activity in cultured fibroblasts or in leukocytes. For this purpose, neuraminidase activity can be measured using oligosaccharide substrates such as sialyllactose or sialyl-hexasaccharides, or the synthetic substrate 4-methylumbelliferyl-α-D-N-acetylneuraminic acid. The synthetic substrate enables a very sensitive fluorometric assay and is well suited for diagnostic work (POTIER *et al.* 1979b, O'BRIEN and WARNER 1980, DEN TANDT and LEROY 1980). The sialyloligosaccharide substrates may be of value in cases with residual neuraminidase activity for the investigation of specificity and kinetics, as they represent the natural substrates of the enzyme. Additional determinations of β-galactosidase and, eventually, other lysosomal hydrolase activities are necessary in the differential diagnosis of sialidosis, combined β-galactosidase/neuraminidase deficiency, and the mucolipidoses.

As neuraminidase activity is readily detectable in amniotic cells, a prenatal diagnosis should be possible in a fetus at risk for sialidosis. Indeed, prenatal diagnosis has been performed in a pregnancy at risk for combined β-galactosidase/neuraminidase deficiency, and the diagnosis of an affected fetus was confirmed by assays of neuraminidase and β-galactosidase in tissues from the aborted fetus (KLEIJER *et al.* 1979).

VI. Summary

Sialidoses are genetic diseases due to the deficiency of a lysosomal neuraminidase. The patients accumulate within the lysosomes of their tissues and excrete in the urine excessive quantities of sialyloligosaccharides originating from the catabolism of glycoproteins. The fact that there is no sialoglycolipid (ganglioside) storage, and that the activity of ganglioside-specific neuraminidase(s) is not diminished, suggests that there are different neuraminidases for the catabolism of sialyloligosaccharides (sialoglycoproteins) and of gangliosides. Biochemically, there are two types of sialidosis: sialidosis in the strict sense of the word, characterized by an isolated neuraminidase deficiency; and another type with combined deficiency of neuraminidase and β-galactosidase, thought to be due to a defect in a posttranslational modification step common to both enzymes.

Sialidosis with isolated neuraminidase deficiency occurs in at least two clinical phenotypes: a severe form with HURLER-like appearance, myoclonus, cherry-red spots of the retina, mental retardation and, sometimes, with renal disease, and a mild form with or without mental retardation, with myoclonus and with cherry-red spots. Quite likely, the two forms originate from allelic mutations of the same neuraminidase gene, giving rise to enzyme deficiency states of varying degrees.

The combined β-galactosidase/neuraminidase deficiency, on the other hand, presents a clinical picture which seems intermediate in severity between the mild and severe forms of isolated neuraminidase deficiency. Whereas it is not yet clear whether the neuraminidase deficiency in this disorder is secondary to an unknown processing defect, a defect in such a mechanism has been shown to be responsible for the deficiency of numerous lysosomal hydrolases, including neuraminidase, in patients with mucolipidoses II and III.

Acknowledgements

The help of Mrs. H. Meyer in the prepration of the manuscript is gratefully acknowledged. The work from the author's laboratory was supported by grants from the Deutsche Forschungsgemeinschaft.

Bibliography

ANDRIA, G., DEL GIUDICE, E., REUSER, A. J. J., 1978: Clin. Genet. **14**, 16—23.

AYLSWORTH, A. S., THOMAS, G. H., HOOD, J. L., MALOUF, N., LIBERT, J., 1980: J. Pediatr. **96**, 662—668.

BACH, G., COHEN, M. M., KOHN, G., 1975: Biochem. Biophys. Res. Commun. **61**, 1483—1490.

— ZEIGLER, M., KOHN, G., COHEN, M. M., 1977: Am. J. Hum. Genet. **29**, 610—618.

— BARGAL, R., CANTZ, M., 1979a: Biochem. Biophys. Res. Commun. **91**, 976—981.

— ZEIGLER, M., SCHAAP, T., KOHN, G., 1979b: Biochem. Biophys. Res. Commun. **90**, 1341—1347.

— — KOHN, G., 1980: Clin. Chim. Acta **106**, 121—128.

BÉRARD, M., TOGA, M., BERNARD, R., DUBOIS, D., MARIANI, R., HASSOUN, J., 1968: Path. europ. **3**, 172—183.

CAIMI, L., LOMBARDO, A., PRETI, A., WIESMANN, U., TETTAMANTI, G., 1979: Biochim. Biophys. Acta **571**, 137—146.

CANTZ, M., GEHLER, J., SPRANGER, J., 1977: Biochem. Biophys. Res. Commun. **74**, 732—738.

— MESSER, H., 1979: Eur. J. Biochem. **97**, 113—118.

— 1980: In: Structure and Function of Gangliosides (SVENNERHOLM, L., MANDEL, P., DREYFUS, H., URBAN, P. F., eds.). Advances in Experimental Medicine and Biology, Vol. 125, pp. 415—430. New York-London: Plenum.

— MENDLA, K., BAUMKÖTTER, J., KALAYDJIEVA, L., 1981: In: Sialidases and Sialidoses. Perspectives in Inherited Metabolic Diseases (TETTAMANTI, G., DURAND, P., DI DONATO, S., eds.), Vol. 4, pp. 219—232, Milano: Edi Ermes.

DEN TANDT, W. R., LEROY, J. G., 1980: Hum. Genet. **53**, 383—388.

DURAND, P., GATTI, R., CAVALIERI, S., BORRONE, C., TONDEUR, M., MICHALSKI, J.-C., STRECKER, G., 1977: Helv. paediat. Acta **32**, 391—400.

FREITAG, F., BLÜMCKE, S., SPRANGER, J., 1971: Virchows Arch. Path. **B7**, 189—204.

FRISCH, A., NEUFELD, E. F., 1979: Anal. Biochem. **95**, 222—227.

GEHLER, J., CANTZ, M., O'BRIEN, J. F., TOLKSDORF, M., SPRANGER, J., 1975: In: Birth Defects: Original Article Series (BERGSMA, D., ed.), Vol. XI, no. 6, pp. 269—272. White Plains, N. Y.: The National Foundation.

GOLDBERG, M. F., COTLIER, E., FICHENSCHER, L. G., KENYON, K., ENAT, R., BOROWSKY, S. A., 1971: Arch. Intern. Med. **128**, 387—398.

HASILIK, A., NEUFELD, E. F., 1980: J. Biol. Chem. **255**, 4946—4950.

— WAHEED, A., VON FIGURA, K., 1981: Biochem. Biophys. Res. Commun. **98**, 761—767.

— — CANTZ, M., VON FIGURA, K., 1982: Eur. J. Biochem. **122**, 119—123.

HOLMES, E. W., O'BRIEN, J. S., 1979: Anal. Biochem. **93**, 167—170.

HOOGEVEEN, A. T., VERHEIJEN, F. W., D'AZZO, A., GALJAARD, H., 1980: Nature **285**, 500—502.

KATO, T., OKADA, S., YUTAKA, T., INUI, K., YABUUCHI, H., CHIYO, H., FURUYAMA, J.-I., OKADA, Y., 1979: Biochem. Biophys. Res. Commun. **91**, 114—117.

KELLY, T. E., GRAETZ, G., 1977: Amer. J. Med. Genet. **1**, 31—46.

KLEIJER, W. J., HOOGEVEEN, A., VERHEIJEN, F. W., NIERMEIJER, M. F., GALJAARD, H., O'BRIEN, J. S., WARNER, T. G., 1979: Clin. Genet. **16**, 60—61.

LOONEN, M. C. B., LUGT, L. V. B., FRANKE, C. L., 1974: Lancet **II**, 785.

LOWDEN, J. A., O'BRIEN, J. S., 1979: Am. J. Hum. Genet. **31**, 1—18.

MAIRE, I., NIVELON-CHEVALLIER, A., 1981: J. Inh. Metab. Dis. **4**, 221—223.

MAROTEAUX, P., HUMBEL, R., 1976: Arch. Franç. Péd. **23**, 641.

— — STRECKER, G., MICHALSKI, J.-C., MANDE, R., 1978a: Arch. Franç. Péd. **35**, 819—829.

— POISSONNIER, M., TONDEUR, M., STRECKER, G., LEMONNIER, M., 1978b: Arch. Franç. Péd. **35**, 280—291.

McKUSICK, V. A., NEUFELD, E. F., KELLY, T. E., 1978: In: The Metabolic Basis of Inherited Disease (STANBURY, J. B., WYNGAARDEN, J. B., FREDRICKSON, D. S., eds.), pp. 1282—1307. New York: McGraw-Hill.

MICHALSKI, J.-C., STRECKER, G., FOURNET, B., CANTZ, M., SPRANGER, J., 1977: FEBS Lett. **79**, 101—104.

MIYATAKE, T., YAMADA, T., SUZUKI, M., PALLMANN, B., SANDHOFF, K., ARIGA, T., ATSUMI, T., 1979: FEBS Lett. **97**, 257—259.

NEUFELD, E. F., 1974: In: Progress in Medical Genetics (STEINBERG, A. G., BEARN, A. G., eds.), Vol. X, pp. 81—101. New York-London: Grune and Stratton.

O'BRIEN, J. S., 1977: Biochem. Biophys. Res. Commun. **79**, 1136—1141.

— WARNER, T. G., 1980: Clin. Genet. **17**, 35—38.

OKADA, S., KATO, T., MIURA, S., YABUUCHI, H., NISHIGAKI, M., KOBATA, A., CHIYO, H., FURUYAMA, J.-I., 1978: Clin. Chim. Acta **86**, 159—167.

— YUTAKA, T., KATO, T., IKEHARA, C., YABUUCHI, H., OKAWA, M., INUI, M., CHIYO, H., 1979: Eur. J. Pediatr. **130**, 239—249.

ORII, T., MINAMI, R., SUKEGAWA, K., SATO, S., TSUGAWA, S., HORINO, K., MIURA, R., NAKAO, T., 1972: Tohoku J. Exp. Med. **107**, 303—315.

PALLMANN, B., SANDHOFF, K., BERRA, B., MIYATAKE, T., 1980: In: Structure and Function of Gangliosides (SVENNERHOLM, L., MANDEL, P., DREYFUS, H., URBAN, P. F., eds.), Advances in Experimental Medicine and Biology, Vol. 125, pp. 401—414. New York-London: Plenum Press.

POTIER, M., LU SHUN YAN, D., WOMACK, J. E., 1979a: FEBS Lett. **108**, 345—348.

— MAMELI, L., BÉLISLE, M., DALLAIRE, L., MELANÇON, S. B., 1979b: Anal. Biochem. **94**, 287—296.

RAPIN, I., GOLDFISCHER, S., KATZMAN, R., ENGEL, J., O'BRIEN, J. S., 1978: Ann. Neurol. **3**, 234—242.

SANDER, M., VEH, R. W., SCHAUER, R., 1979: In: Glycoconjugates (SCHAUER, R., BOER, P., BUDDECKE, E., KRAMER, M. F., VLIEGENTHART, J. F. G., WIEGANDT, H., eds.), pp. 358—359. Stuttgart: G. Thieme.

SEWELL, A. C., 1980: Eur. J. Pediat. **134**, 183—194.

SPRANGER, J., WIEDEMANN, H.-R., TOLKSDORF, M., GRAUCOB, E., CAESAR, R., 1968: Zschr. Kinderheilk. **103**, 285—306.

— — 1970: Humangenetik **9**, 113—139.

— GEHLER, J., CANTZ, M., 1977: Amer. J. Med. Genet. **1**, 21—29.

STRECKER, G., MICHALSKI, J.-C., MONTREUIL, J., FARRIAUX, J.-P., 1976: Biomed. **25**, 238—239.

— PEERS, M. C., MICHALSKI, J.-C., HONDI-ASSAH, T., FOURNET, B., SPIK, G., MONTREUIL, J., FARRIAUX, J.-P., MAROTEAUX, P., DURAND, P., 1977: Eur. J. Biochem. **75**, 391—403.

— MICHALSKI, J.-C., 1978: FEBS Lett. **85**, 20—24.

— 1980: In: Structure and Function of Gangliosides (SVENNERHOLM, L., MANDEL, P., DREYFUS, H., URBAN, P. F., eds.), Advances in Experimental Medicine and Biology, Vol. 125, pp. 371—384. New York-London: Plenum Press.

SWALLOW, D. M., EVANS, L., STEWART, G., THOMAS, P. K., ABRAMS, J. D., 1979: Ann. Hum. Genet. (London) **43**, 27—35.

THOMAS, G., TILLER, G. E., REYNOLDS, L. W., MILLER, C. S., BACE, J. W., 1976: Biochem. Biophys. Res. Commun. **71**, 188—195.

THOMAS, G. H., TIPTON, R. E., CH'IEN, T., REYNOLDS, L. W., MILLER, C. S., 1978: Clin. Genet. **13**, 369—379.

— GOLDBERG, M. F., MILLER, C. S., REYNOLDS, L. W., 1979: Clin. Genet. **16**, 323—330.

WAHEED, A., HASILIK, A., CANTZ, M., VON FIGURA, K., 1982: Hoppe-Seyler's Z. Physiol. Chem. **363**, 169—178.

WENGER, D. A., TARBY, T. J., WHARTON, C., 1978: Biochem. Biophys. Res. Commun. **82**, 589—595.

WINTER, R. M., SWALLOW, D. M., BARAITSER, M., PURKISS, P., 1980: Clin. Genet. **18**, 203—210.

YAMAMOTO, A., ADACHI, S., KAWAMURA, S., TAKAHASHI, M., KITANI, T., OHTORI, T., SHINJI, Y., NISHIKAWA, M., 1974: Arch. Intern. Med. **134**, 627—634.

Subject Index

Cell Biology Monographs

Volume 1: **The Lytic Compartment of Plant Cells**
By **Ph. Matile**
1975. 59 figures and 40 plates. XIII, 183 pages.
ISBN 3-211-81296-2

Volume 2: **The Golgi Apparatus**
By **W. G. Whaley**
1975. 97 figures. XII, 190 pages.
ISBN 3-211-81315-2

Volume 3: **Lysosomes: A Survey**
By **E. Holtzman**
1976. 56 figures. XI, 298 pages.
ISBN 3-211-81316-0

Volume 4: **Mitochondria: Structure, Biogenesis and Transducing Functions**
By **H. Tedeschi**
1976. 18 figures. X, 164 pages.
ISBN 3-211-81317-9

Volume 5: **Microbodies/Peroxisomen pflanzlicher Zellen**
Morphologie, Biochemie, Funktion und Entwicklung eines Zellorganells
Von **B. Gerhardt**
1978. 59 Abbildungen. IX, 283 Seiten.
ISBN 3-211-81436-1

Volume 6: **Patterns of Chloroplast Reproduction**
A Developmental Approach to Protoplasmic Plant Anatomy
By **Th. Butterfass**
1979. 28 figures. XIV, 205 pages.
ISBN 3-211-81541-4

Cell Biology Monographs

Volume 7: **Peroxisomes and Related Particles in Animal Tissues**
By **P. Böck, R. Kramar, M. Pavelka**
1980. 60 figures. XIII, 239 pages.
ISBN 3-211-81582-1

Volume 8: **Cytomorphogenesis in Plants**
Edited by **O. Kiermayer**
1981. 202 figures. X, 439 pages.
ISBN 3-211-81613-5

Volume 9: **The Protozoan Nucleus**
Morphology and Evolution
Revised from the 1978 Russian Edition
Translated by N. Bobrov and M. Verkhovtseva
By **I. B. Raikov**
1982. 1 portrait and 116 figures. XV, 474 pages.
ISBN 3-211-81678-X

Springer-Verlag Wien New York

MIX
Papier aus verantwortungsvollen Quellen
Paper from responsible sources
FSC® C105338

If you have any concerns about our products,
you can contact us on
ProductSafety@springernature.com

In case Publisher is established outside the EU,
the EU authorized representative is:
Springer Nature Customer Service Center GmbH
Europaplatz 3, 69115 Heidelberg, Germany

Printed by Libri Plureos GmbH
in Hamburg, Germany